Rapid Detection, Characterization, and Enumeration of

Foodborne Pathogens

Rapid Detection, Characterization, and Enumeration of **Foodborne** Pathogens

Edited by

J. Hoorfar
National Food Institute,
Technical University of Denmark
Søborg, Denmark

ASM PRESS
Washington, DC

American Society for Microbiology
1752 N Street, N.W.
Washington, DC 20036-2904

Library of Congress Cataloging-in-Publication Data

Rapid detection, characterization, and enumeration of foodborne pathogens / edited by J. Hoorfar.

p. ; cm.

Includes bibliographical references and index.

ISBN 978-1-55581-542-4

1. Food—Microbiology. 2. Food poisoning. 3. Food contamination. I. Hoorfar, J. (Jeffrey) II. American Society for Microbiology.
[DNLM: 1. Food Analysis—methods. 2. Food Microbiology. 3. Foodborne Diseases—prevention & control. 4. Microbiological Techniques—trends. 5. Time Factors. QW 85]

QR115.R357 2011
664.001'579—dc22

2010046836

Printed in the United States of America

1 0 9 8 7 6 5 4 3 2

Address editorial correspondence to: ASM Press, 1752 N St., N.W., Washington, DC 20036-2904, U.S.A.

Send orders to: ASM Press, P.O. Box 605, Herndon, VA 20172, U.S.A.
Phone: 800-546-2416; 703-661-1593
Fax: 703-661-1501
Email: Books@asmusa.org
Online: estore.asm.org

CONTENTS

CONTRIBUTORS

H. J. M. Aarts
Laboratory for Zoonoses and Environmental Microbiology (LZO), National Institute for Public Health (RIVM), Centre for Infectious Disease Control, P.O. Box 1, 3720 BA, Bilthoven, The Netherlands

G. C. Barker
Food Research Institute, Norwich Research Park, Colney, Norwich, NR4 7UA, United Kingdom

L. Beutin
National Reference Laboratory for Escherichia coli, Federal Institute for Risk Assessment (BfR), Diedersdorfer Weg 1, D-12277 Berlin, Germany

A. K. Bhunia
Department of Food Sciences, Purdue University, 745 Agriculture Mall Drive, West Lafayette, IN 47907

S. Bidawid
Microbiology Research Division, Health Canada, Food Directorate, Bureau of Microbial Hazards, Sir F.G. Banting Research Centre, Tunney's Pasture, Ottawa, Ontario, Canada K1A 0K9

B. Bisha
Department of Animal Sciences, Center for Meat Safety and Quality, Colorado State University, Fort Collins, CO 80523

A. Bosch
Virus Entèrics, Department of Microbiology, University of Barcelona, Avda. Diagonal 645, 08028 Barcelona, Spain

S. Cahill
Nutrition and Consumer Protection Division, Food and Agriculture Organization of the United Nations, Viale delle Terme di Caracalla, 00183 Rome, Italy

C. Carroll
School of Natural Sciences, National University of Ireland, Galway, University Road, Galway City, Ireland

M. Carter
Silliker Inc., 160 Armory Drive, South Holland, IL 6047

B. B. Christensen
National Food Institute, Technical University of Denmark, Mørkhøj Bygade 19, DK-2860 Søborg, Denmark

L. S. Christensen
Danish National Food Institute, Division of Microbiology and Risk Assessment, Technical University of Denmark, Mørkhøj Bygade 28, Building H, 2860 Søborg, Denmark

R. Clarke
Nutrition and Consumer Protection Division, Food and Agriculture Organization of the United Nations, Viale delle Terme di Caracalla, 00183 Rome, Italy

J. S. Dickson
Department of Animal Science, Inter-Department Program in Microbiology, 215F Meat Laboratory, Iowa State University, Ames, IA 50011

B. R. Dixon
Health Canada, Microbiology Research Division, Bureau of Microbial Hazards, Food Directorate, Ottawa, ON, K1A 0K9, Canada

D. H. D'Souza
Department of Food Science and Technology, University of Tennessee-Knoxville, 2605 River Drive, Room 102 FSPB, Knoxville, TN 37996-4591

J. P. Dubey
U.S. Department of Agriculture, Agricultural Research Service, Animal and Natural Resources Institute, Animal Parasitic Diseases Laboratory, Beltsville, MD 20705

M. Ehling-Schulz
Food Microbiology Unit, Clinic of Ruminants, Department for Farm Animals and Veterinary Public Health, Veterinaerplatz 1, A-1210 Vienna, Austria

P. Fach
AFSSA (French Food Safety Agency), Laboratory for Study and Research on Food Quality and Processes (LERQAP), 23 Av. du Général De Gaulle, Fr-94706 Maisons-Alfort, France

S. Fanning
WHO-Collaborating Centre for Research, Reference & Training on *Cronobacter*, Centre for Food Safety, UCD Veterinary Sciences Centre, University College Dublin, Belfield, Dublin 4, Ireland

R. Fayer
U.S. Department of Agriculture, Agricultural Research Service, Animal and Natural Resources Institute, Environmental Microbial and Food Safety Laboratory, Beltsville, MD 20705

A. Fazil
Public Health Agency of Canada, 160 Research Lane, Unit 206, Guelph, Ontario, N1G 5B2, Canada

P. C. H. Feng
U.S. Food & Drug Administration, HFS-711, 5100 Paint Branch Parkway, College Park, MD 20740

P. I. Fields
Enteric Diseases Laboratory Branch, Division of Foodborne, Waterborne, and Environmental Diseases, Centers for Disease Control and Prevention, 1600 Clifton Road, Mailstop C03, Atlanta, GA 30333

C. Fitzgerald
Enteric Diseases Laboratory Branch, Division of Foodborne, Waterborne, and Environmental Diseases, Centers for Disease Control and Prevention, 1600 Clifton Road, Mailstop C03, Atlanta, GA 30333

P. Fratamico
U.S. Department of Agriculture, Agricultural Research Service, Eastern Regional Research Center, Wyndmoor, PA 19038

M. Fredriksson-Ahomaa
Department of Food Hygiene and Environmental Health, Faculty of Veterinary Medicine, P.O. Box 66, FI-00014 University of Helsinki, Finland

I. Gardner
Department of Medicine and Epidemiology, School of Veterinary Medicine, University of California, Davis, CA 95616

L. D. Goodridge
Department of Animal Sciences, Center for Meat Safety and Quality, Colorado State University, Fort Collins, CO 80523

K. Grant
Health Protection Agency, Centre for Infections, London, NW9 5HT, United Kingdom

P. E. Granum
Department of Food Safety and Infection Biology, Norwegian School of Veterinary Science, N-0033 Oslo, Norway

M. Greiner
Federal Institute for Risk Assessment, Scientific Services, Epidemiology, Biostatistics and Mathematical Modelling, D-14195 Berlin, Germany

M. Griffiths
Canadian Research Institute for Food Safety, University of Guelph, Guelph, Ontario, Canada, N1G 2W1

P. Häggblom
Department of Chemstry, Environment and Animal Feed, National Veterinary Institute, SE-75189 Uppsala, Sweden

J. Hedman
Department of Biology, Swedish National Laboratory of Forensic Science (SKL), SE-581 94 Linköping, Sweden

I. Hein
Institute of Milk Hygiene, Milk Technology and Food Safety, Department for Farm Animals and Veterinary Public Health, Veterinaerplatz 1, A-1210 Vienna, Austria

D. E. Hill
U.S. Department of Agriculture, Agricultural Research Service, Animal and Natural Resources Institute, Environmental Microbial and Food Safety Laboratory, Beltsville, MD 20705

A. D. Hitchins
U.S. Food and Drug Administration, Center for Food Safety and Applied Nutrition, 5100 Paint Branch Parkway, College Park, MD 20740-3835

J. Hoorfar
National Food Institute, Technical University of Denmark, Mørkhøj Bygade 19, DK-2860 Søborg, Denmark

C. Iversen
Centres for Food Safety & Foodborne Zoonomics, UCD Veterinary Sciences Centre, University College Dublin, Belfield, Dublin 4, Ireland

L. Jaykus
Department of Food, Bioprocessing and Nutrition Sciences, North Carolina State University, Raleigh, NC 27695-7624

H. Joosten
Nestlé Research Center, Vers-chez-les-Blanc, 1000 Lausanne 26, Switzerland

K. N. Jordan
Teagasc, Moorepark Food Research Centre, Moorepark, Fermoy, Co. Cork, Ireland

M. H. Josefsen
National Food Institute, Technical University of Denmark, Mørkhøj Bygade 19, DK-2860 Søborg, Denmark

K. E. Kniel
Department of Animal and Food Sciences, University of Delaware, Newark, DE 19716

H. Korkeala
Department of Food Hygiene and Environmental Health, Faculty of Veterinary Medicine, FI-00014 University of Helsinki, Finland

J. L. Kornacki
Kornacki Microbiology Solutions, Inc., P.O. Box 163, McFarland, WI 53558

R. G. Labbé
Department of Food Science, University of Massachusetts, Amherst, MA 01003

F. S. Le Guyader
Laboratoire de Microbiologie-LNR, Departement EMP, IFREMER, 1 rue de l'Ile d'Yeu, BP 21105, 44311 Nantes cedex 03, France

D. Lees
European Community Reference Laboratory, CEFAS Laboratory, Barrack Road, The Nothe, Weymouth, Dorset, DT4 8UB, United Kingdom

C. Löfström
National Food Institute, Technical University of Denmark, Mørkhøj Bygade 19, DK-2860 Søborg, Denmark

B. Malorny
Federal Institute for Risk Assessment, National Reference Laboratory for Salmonella, Diedersdorfer Weg 1, D-12277 Berlin, Germany

V. Þ. Marteinsson
Matis—Icelandic Food Research and Biothech R&D, Vínlandsleið 12, 113 Reykjavík, Iceland

J. Marugg
Nestlé Research Center, Vers-chez-les-Blanc, 1000 Lausanne 26, Switzerland

J. McQuiston
Enteric Diseases Laboratory Branch, Division of Foodborne, Waterborne, and Environmental Diseases, Centers for Disease Control and Prevention, 1600 Clifton Road, Mailstop C03, Atlanta, GA 30333

U. Messelhäusser
Bavarian Health and Food Safety Authority, Veterinärstr. 2, D-85764 Oberschleißheim, Germany

S. Morabito
Food-Borne Zoonoses Unit, Department of Food Safety and Veterinary Public Health, Istituto Superiore di Sanità, 00161 Rome, Italy

J. Nakayama
Department of Bioscience and Biotechnology, Faculty of Agriculture, Graduate School Kyushu University, 6-10-1 Hakozaki Higashi-ku, Fukuoka, Japan 812-8581

T. Nesbakken
Norwegian School of Veterinary Science, Dept. of Food Safety and Infection Biology, P.O. Box 8146 Dep., 0033 Oslo, Norway

S. O'Brien
Centres for Food Safety & Foodborne Zoonomics, UCD Veterinary Sciences Centre, University College Dublin, Belfield, Dublin 4, Ireland

R. O'Kennedy
School of Biotechnology, Dublin City University, Dublin 9, Ireland

E. Olsson Engvall
National Veterinary Institute, SE-751 89 Uppsala, Sweden

F. Pagotto
Bureau of Microbial Hazards Health Products and Food Branch, P/L 2204E Room E412, Health Canada, 251 Sir Frederick Banting Driveway, Ottawa, Ontario K1A 0K9, Canada

S. D. Pillai
Food Safety and Environmental Microbiology Program, Texas A&M University, Room 418B Kleberg Center, MS 2472, College Station, TX 77843-2472

S. Qvist
NordVal, c/o National Veterinary Institute, P.O. Box 750, Sentrum N-0106, Oslo, Norway

P. Rådström
Department of Applied Microbiology, Lund University, Veterinärplatz 1, SE-221 00 Lund, Sweden

C. H. Rambo
Food Safety and Environmental Microbiology Program, Texas A&M University, Room 418B Kleberg Center, MS 2472, College Station, TX 77843-2472

A. Rasooly
Division of Biology, Office of Science and Engineering, FDA, Silver Spring, MD 20993, and Cancer Diagnosis Program, National Cancer Institute, 6130 Executive Blvd., Rockville, MD 20852

E. Reynisson
Matis—Icelandic Food Research and Biothech R&D, Vínlandsleið 12, 113 Reykjavík, Iceland

T. Ross
Tasmanian Institute of Agricultural Research, School of Agricultural Science, University of Tasmania, Private Bag 54, Hobart 7001, Tasmania, Australia

P. Rossmanith
Christian Doppler Laboratory for Molecular Food Analytics, University of Veterinary Medicine, Veterinärplatz 1, A-1210 Vienna, Austria

K. Rudi
Norwegian University of Life Sciences, P.O. Box 5003, No-1432 Ås, and Hedmark University College, 2318 Hamar, Norway

N. Sakamoto
Department of Bioscience and Biotechnology, Faculty of Agriculture, Graduate School Kyushu University, 6-10-1 Hakozaki Higashi-ku, Fukuoka, Japan, 812-8581

M. Sanaa
Agence Française de Sécurité Sanitaire des Aliments, Directorate of Risk Assessment, 27-31, avenue du Général Leclerc, BP 19-94701 Maisons-Alfort Cédex, France

M. Santín
U.S. Department of Agriculture, Agricultural Research Service, Animal and Natural Resources Institute, Environmental Microbial and Food Safety Laboratory, Beltsville, MD 20705

M. Skurnik
Haartman Institute, Department of Bacteriology and Immunology, P.O. Box 21, FI-00014 University of Helsinki, Finland

R. Stephan
Institute for Food Safety and Hygiene, University of Zurich, Winterthurerstrasse 272, CH-8057 Zurich, Switzerland

B. Stessl
Institute of Milk Hygiene, Milk Technology and Food Safety, Department for Farm Animals and Veterinary Public Health, Veterinaerplatz 1, A-1210 Vienna, Austria

S. Thisted Lambertz
National Food Administration, P.O. Box 622, SE-75126 Uppsala, Sweden

H. Vigre
Department of Microbiology and Risk Assessment, National Food Institute, Technical University of Denmark, DK-2860 Søborg, Denmark

M. Wagner
Institute of Milk Hygiene, Milk Technology and Food Science, Department of Veterinary Public Health and Food Science, University of Veterinary Medicine, Veterinärplatz 1, A-1210 Vienna, Austria

P. Whyte
Centres for Food Safety & Foodborne Zoonomics, UCD Veterinary Sciences Centre, University College Dublin, Belfield, Dublin 4, Ireland

C. Wolf-Hall
Department of Veterinary and Microbiological Sciences, Great Plains Institute of Food Safety, 1523 Centennial Blvd., P.O. Box 6050-Dept. 7690, North Dakota State University, Fargo, ND 58108

D. L. F. Wong
Food Safety, Zoonoses and Foodborne Diseases (FOS), Sustainable Development and Healthy Environments (SDE), World Health Organization, 20 Avenue Appia, CH-1211 Geneva 27, Switzerland

H. Zhao
Great Plains Institute of Food Safety, 1523 Centennial Blvd., P.O. Box 6050-Dept. 7690, North Dakota State University, Fargo, ND 58108

C. Zweifel
Institute for Food Safety and Hygiene, University of Zurich, Winterthurerstrasse 272, CH-8057 Zurich, Switzerland

M. H. Zwietering
Laboratory of Food Microbiology, Agrotechnology & Food Sciences Group, Wageningen Agricultural University, Postbus 8129, 6700EV, Wageningen, The Netherlands

PREFACE

The current increase in public attention to food safety, especially foodborne microbes, has increased research into new, rapid methods of detection of foodborne pathogens. Faster and better detection and characterization of pathogens are the cornerstones of the fight against foodborne pathogens.

The food industry and control authorities are putting substantial efforts into testing in order to ensure and document safer foods, prevent product recalls, and limit economic losses. But how are the current rapid methods validated, what is the cost-benefit of using rapid methods, and which rapid method is the appropriate one to choose? These are some of the questions that this book attempts to answer.

A group of 85 experts from 50 research centers provides readers with the latest developments in this field. One of the book's unique aspects is that it focuses on commodities rather than pathogens. Thus, it is organized according to food production lines rather than types of pathogens. For each production line, rapid methods are described for a number of important target pathogens.

Another unique aspect of the book is the presentation of open-formula, noncommercial protocols. There are many commercial kits available, and new ones are frequently introduced while others are withdrawn. Some kits are better validated than others, and some perform better on certain types of samples. It is not the intention of this book to promote specific commercial kits, but rather to discuss the scientific basis for new methods. For this reason, and to avoid bias toward certain products, we have done our best to limit mention of commercial kits.

The book is organized into seven sections and contains a total of 28 chapters. Section I is a state-of-the-art review of the latest laboratory technologies that can accelerate test results. Section II, entitled "Critical Considerations before Setting Up Rapid Methods," serves as an introduction to the field of rapid methods and provides an overview of the critical issues.

Section III deals with the sample types, testing considerations, and main foodborne pathogens in the meat production chain. In Section IV, the latest rapid methods concerned with the dairy production chain are reviewed, along with practical implications for sampling schemes and pathogens of concern to the public health. Section V provides the reader with the latest developments in the testing of fresh produce, water, and seafood, which is considered as an emerging public health issue due to the increasing international trade of fresh produce.

In the sixth section, the latest testing issues in the food service and catering industry are discussed and guidelines for rapid testing are given. In the last section, "Conclusions," the final chapter of this book looks to the future of rapid methods by explaining research needs and discussing emerging areas.

I hope those readers who work in the food industry and end-use laboratories will find these protocols useful and implement them for actual testing. For those scientists developing new methods and adopting new protocols, the discussions especially in Section II will be helpful with the less elucidated aspects, such as statistics, sampling plans, validation, and so on. For those students who are involved in test development both at the undergraduate and postgraduate levels, this book will help them to understand important aspects of laboratory work. Finally, by consulting this book, quality control managers will become more familiar with the principles of testing when they are deciding to use a new test or detecting a different pathogen. This book is written in the hope that it makes the food your company produces safer.

J. Hoorfar

ACKNOWLEDGMENTS

In today's demanding work culture, it is not easy to find time for extracurricular work such as book contributions; in particular, writing updated reviews can be quite time-consuming. That is why I express my heartfelt gratitude to the large panel of my colleagues who contributed to the chapters. Their enthusiasm to share expertise and knowledge has made this book possible.

Coordinating the European project BIOTRACER project (FOOD-2006-CT-036272), one of the world's largest food microbiology research projects, provided me with the opportunity to collaborate closely with many excellent scientists from different disciplines of microbial food safety. One of the outcomes of this collaboration is this book. The European project BIOTRACER supported a large number of scientists and much of the work that is described in different chapters of this book.

The advice and dedication of senior editor Eleanor Riemer and the editorial staff at ASM Press have contributed to the quality of this book.

I thank my dedicated administrative team in Copenhagen, Iben Bang-Berthelsen, Solveig Lind Bouquin, and Jeffrey Skiby, who supported me directly and indirectly through editorial support or by relieving me from some of the day-to-day work in order to focus on this book project.

Last but not least, I thank my family for putting up with me sitting many nights in front of a computer to do this work.

J. Hoorfar

THE PUBLIC HEALTH, INDUSTRIAL, AND GLOBAL SIGNIFICANCE OF RAPID MICROBIOLOGICAL FOOD TESTING

J. Hoorfar, S. Cahill, R. Clarke, G. C. Barker, A. Fazil, D. L. F. Wong, and P. C. H. Feng

1

SIGNIFICANCE OF MICROBIOLOGICAL FOOD TESTING

Historically, microbiological testing of foods has been an integral part of food production and food safety regulations. It was most often applied for end product analysis, which we now recognize as an ineffective means of assessment, due to logistical complexities in sampling and the heterogeneous distribution of contamination. The implementation of the hazard analysis and critical control points (HACCP) system has shifted the burden from testing to process control as a means of ensuring food safety. However, microbiological testing remains a critical tool in process control monitoring, quality control, and surveillance and in providing inputs to risk assessment. Furthermore, with the increasing emphasis on "farm-to-fork" approaches to ensure the safety of foods from the production of raw materials, to processing, distribution, and eventually to the consumer table, environmental sampling and analysis have also become routine in food production as well as in outbreak tracing and tracking. As food safety management further develops, microbiological testing will continue to play an important role in assessing whether Food Safety Objectives (maximum levels of hazards at the point of consumption) are achieved (Cole and Tompkin, 2005). However, traditional microbiological culture-based methods are limited, particularly in their ability to provide timely data to meet current food safety objectives. This chapter covers the basic arguments for why we are increasingly turning to rapid methods to meet the microbiological testing needs in all facets of food production, including regulatory compliance and food-related outbreaks. Rapid methods are here defined as alternative microbiological testing methods that are able to provide reliable test results in a shorter time than those obtained by culture cultivation.

J. Hoorfar, National Food Institute, Technical University of Denmark (DTU), Mørkhøj Bygade 28, DK-2860 Søborg, Denmark. *S. Cahill and R. Clarke,* Nutrition and Consumer Protection Division, Food and Agriculture Organization of the United Nations, Viale delle Terme di Caracalla, 00183 Rome, Italy. *G. C. Barker,* Food Research Institute, Norwich Research Park, Colney, Norwich, NR4 7UA, United Kingdom. *A. Fazil,* Public Health Agency of Canada, 160 Research Lane, Unit 206, Guelph, Ontario, N1G 5B2, Canada. *D. L. F. Wong,* Food Safety, Zoonoses and Foodborne Diseases, Sustainable Development and Healthy Environments, World Health Organization, 20 Avenue Appia, CH-1211 Geneva 27, Switzerland. *P. C. H. Feng,* U.S. Food & Drug Administration, HFS-711, 5100 Paint Branch Parkway, College Park, MD 20740.

Rapid Detection, Identification, and Quantification of Foodborne Pathogens, ed. by J. Hoorfar

Reasons for Testing

Microbiological analysis of foods is often performed with two primary purposes: (i) to establish the absence of pathogens or toxins to ensure the safety of foods and (ii) to enumerate total or indicator microbial load to monitor the effectiveness of processing and to verify product quality control and shelf life stability.

In addition, there are a number of secondary purposes for testing: (i) current food safety emphasis on "farm to fork" (Scheule and Sneed, 2001) has initiated increases in field testing, (ii) more cost-effective and increasing numbers of tests are needed for large-scale data collection for risk assessment and risk management, and (iii) surveillance and monitoring at the primary production levels require increased screening methods.

These basic food testing programs have long been performed with traditional microbiological methods, which due to the complexities in food testing have evolved into laborious and time- and material-intensive procedures. For instance, and with a few exceptions, pathogen testing is usually done as a "presence/absence analysis" to conform to the regulatory limits of "absence in foods" established by most countries in the world. Nevertheless, analysis for pathogens in foods is challenging due to the complexities of food matrices; the heterogeneous distribution of low levels of pathogens, if present; the high levels of normal flora bacteria, especially in raw foods; the presence of ingredients that can interfere with assays; and the bacterial stress or injury that may occur from food processing procedures. As a result, food samples often had to be processed through a series of culture enrichments to compensate for these logistical problems prior to analysis. Although effective, culture enrichment is time-consuming and often extends the analysis to several days.

In contrast, an analysis for total or indicator bacteria in foods is done as enumeration assays, so culture enrichment is usually not used. However, recent reports show that it is possible both to enrich a sample and to enumerate by pathogen-specific real-time PCR the target pathogens, if the enrichment time is short (Josefsen et al., 2004; Krämer et al., 2010).

In general, the levels of bacteria in foods vary greatly, ranging from low or absent in some processed foods to exceeding log 7 CFU/g or CFU/ml in some raw foods. As a result, a food homogenate has to be subjected to media- and labor-intensive, multifold, serial dilutions followed by lengthy incubation periods, in order to obtain an accurate count. Thus, regardless of whether testing for pathogens or enumerating for indicator organisms in foods, conventional microbiological methods are inadequate to provide quick assessments on the microbiological safety of foods and inadequate to determine whether the product meets established microbial specifications (Feng, 2001).

Traditional microbiological methods are not fast enough to keep up with the pace of today's food production and distribution networks. Distribution networks are able to get food from the farm to virtually any table around the globe in a remarkably short time. Faster detection of foods contaminated with specific pathogen strains can mean faster action in preventing tainted food from being delivered and consumed and thereby minimizing the risk of an outbreak of foodborne illness.

Rapid detection is especially critical in situations where products have already been set in motion. For instance, once the food product has been shipped, the producer may not know whether it is still in transit on a truck or on a store shelf or even in a consumer's pantry. Thus, when identifying and tracing the contaminated product, time is of the essence. However, this has become complex, because some foods, such as black pepper, peanut butter, or hydrolyzed vegetable protein, can be used as base ingredients in other food products. *Salmonella* contamination of these base ingredients has led to thousands of product recalls as well as many cases of salmonellosis (Centers for Disease Control and Prevention, 2009).

New Platforms, New Challenges

The advent of biotechnology introduced many new technologies that led to the emergence of rapid diagnostic methods in the 1980s and forever altered food testing practices (Lübeck and Hoorfar, 2002). These assays, often referred to as "rapid methods," comprise many different technologies including the use of specialized enzyme substrates, ATP, antibodies, and DNA, with equally diverse formats ranging from simple differential/selective plating media to the use of sophisticated instruments, as in automated enzyme-linked immunosorbent assay (ELISA) and real-time PCR.

Some of these assays take only minutes to perform, while others merely reduce the time of analysis in comparison with traditional methods; thus, how "rapid" these assays are is subject to interpretation. However, and without a doubt, rapid methods are faster and often simpler and require less hands-on manipulation than culture-based (traditional) microbiological methods. While convenient, the application of rapid methods to food testing is not without complications, and there are many logistical issues to consider, including validation and regulatory implications (Feng, 2007). Furthermore, there are no assays without limitations; hence, the user needs to be aware of the advantages and disadvantages of these rapid methods in order to maximize the benefits of their application.

Screening versus Confirmation

One of the primary uses of rapid methods is for fast screening of large numbers of samples, whereby negative results are accepted and the products can be released for sale. This has been, until now, the real strength of rapid methods.

This practice is most useful and beneficial to manufacturers, but from a food safety standpoint, the occurrence of false-negative results is of great concern, as these will not be identified until the product has been implicated in infections or outbreaks. Moreover, in screening applications, positive results are only regarded as "presumptive" and need to be confirmed, often by traditional microbiological methods. In regulatory or outbreak settings, confirmation goes even further, to the point where an isolate is obtained, identified, and subtyped.

Although tedious, confirmation is a critical component of food testing, as unconfirmed false positives not only lead to needless destruction of product but can also mislead epidemiological outbreak investigations. The need for confirmation may not pose great inconveniences because the majority of the samples are expected to be negative. Even so, for large manufacturers, the need to confirm even 1% of "presumptive" positives may add to significant time, labor, and cost expenditures. Hence, false-positive and -negative rates associated with an assay can have a severe economic and food safety impact.

Rapid methods use many different technologies and assay formats, and expectedly, their detection efficiencies can be food dependent. For example, a food may contain ingredients that inhibit PCR but have no effect on antigen-antibody reactions, and the converse may occur in another food. Furthermore, these methods have different detection sensitivities that can range from 1 to 6 log organisms; since the level of contamination in food varies, it is not surprising that an assay may be effective in some situations but not in others. It is critical, therefore, that assays are comparatively evaluated or validated to ensure their effectiveness in the intended testing situation prior to implementation (Feng, 2001; Malorny et al., 2003).

Better Detection Sensitivity and Zero Tolerance

Aside from speed, rapid methods also can have better detection sensitivities than traditional culture-based methods. For example, the previous standard method used to detect staphylococcal enterotoxins was the Microslide gel diffusion assay that had a sensitivity of 10 μg.

This is poor compared with the 10-pg sensitivity that can be attained with some rapid assays such as ELISA (Feng, 2007).

However, as detection methods improve, sensitivity also increases, and this creates some interesting logistical challenges (product recall and/or export disputes) to the food industry and the regulatory agencies. Most countries have adopted the regulatory position of "zero tolerance" or "absence" for most pathogens in foods, but the determination of "absence" is method dependent, and the differences in sensitivities have given rise to situations where foods previously analyzed by traditional methods and found to conform to the requirement of "absence" no longer meet specifications when more sensitive methods are used. In other words, as assays become even more sensitive, so does the stringency of "absence," which poses interesting challenges to the industry, who may have to modify their processing and quality control procedures in order to comply with increased test sensitivity (Feng, 2001). It also inevitably leads to discussions on which concentration levels in food still pose a threat to public health.

Logistical Issues of Rapid Methods

It is not uncommon to hear academic scientists and test kit manufacturers lamenting that the "industry" or regulatory agencies are unwilling to adopt rapid molecular pathogen detection technologies. In order to understand who would require rigorous microbial pathogen monitoring and who would be forced or required to institute these monitoring programs, it is necessary to understand the inherent "market forces" at play as to who would be willing to invest in education and technology adoption and why. For example, the value of a commodity increases significantly as it moves through the supply chain from the farm to the retail level. Therefore, the return on investment of a particular intervention strategy or program is quite small for growers at the farm level compared with the return on investment at the retail stage. Thus, the increased value of the commodity at the retail level provides the necessary impetus for ensuring that the commodity meets the food safety expectations of the customer. The necessary incentive therefore rests mainly with the distributors and the retail industry to invest in education and adoption of new technologies to protect their brand name and customer base and to avoid recalls and associated economic ramifications.

However, the adoption and implementation of rapid methods by industries are not easy, and there are many other complex issues to consider in the worldwide implementation of rapid methods for food testing. These include factors such as local economic situations and assay costs, distribution and availability of assays worldwide, shelf life stability and ruggedness of the assays and reagents to time and temperature abuses that may occur during shipping, and import regulation and tariffs, just to name a few (Feng and Heredia, 2009).

One factor that seems to be a major hindrance to the widespread use of rapid methods is the lack of reference material (Madej et al., 2010) and harmonized standard methods and validation protocols (Malorny et al., 2003), which not only affects usage but also severely impacts international trade. These and other practical and logistical issues on the application of rapid methods in food testing will be examined in this chapter.

GLOBAL ASPECTS

Rapid Methods and the Global Health Perspective

Global health is the health of populations in a global context and transcends the perspectives and concerns of individual nations (Brown et al., 2006). The term reflects the realization that the health of populations around the world is interconnected. Though travel and trade are often still regarded in terms of monetary values, it is increasingly recognized that these are also main drivers of modern global health. Humans, animals, raw materials, and food products are being transported faster and in greater volumes than ever before. As such, the national health issues of one country, e.g., food safety and zoonoses, can have global implications.

Despite efforts to understand the mechanisms that determine the occurrence and emergence of zoonotic pathogens in the food chain, it is still not possible to predict zoonotic events with any great certainty. Intervention and control strategies are mainly targeted towards known threats and hazardous practices. Because zoonotic pathogens move rapidly through human, animal, and food vectors, the timely detection and identification of these pathogens are of great importance to prevent its spread locally, nationally, and internationally.

Rapid testing results are critical to minimize the number of infected humans who are travelling or to prevent contaminated batches of food from further distribution. In addition, emerging zoonotic threats, such as severe acute respiratory syndrome and H1N1, or bioterrorism attacks on the food supply all serve to illustrate the need for rapid diagnostic methods in relation to safeguarding global health. Hence, rapid detection to identify causative agents at an early stage coupled with fast communication of food contamination can minimize the impact of an "attack," reduce further distribution and spread, and prevent an international crisis. To protect human health globally from cross-border events related to international travel and trade, the International Health Regulations (IHRs) were developed (Anonymous, 2005). The IHRs aim to prevent, protect against, control, and respond to the international spread of disease while avoiding unnecessary interference with international traffic and trade. These regulations, which entered into force on 15 June 2007, require countries to report disease outbreaks and public health events that may be of international concern to the World Health Organization. In order to meet the requirements of the IHR (Anonymous, 2005), many countries will have to strengthen disease prevention programs, increase surveillance reporting, and become more vigilant in their disease control and outbreak response systems.

Significance of Rapid Methods for International Trade and Regulation

The ability to trade food between countries is subject to a strong legal framework that is consistent with the rules outlined in the World Trade Organization's *Agreement on the Application of Sanitary and Phytosanitary Measures* (World Trade Organization, 1995) aimed at protecting human, animal, and plant life or health.

Benchmark standards are developed at the international level by the Codex Alimentarius Commission, the International Plant Protection Convention, and the World Organization for Animal Health. However, each exporting and importing country is responsible for putting in place systems and measures that aim at ensuring that the food traded nationally or internationally is safe for consumers. Given that the value of exported agriculture commodities was approximately USD 876 billion in 2007, about 70% of which was food (Food and Agriculture Organization, 2009), this is an enormous task.

Even though modern approaches to food control place increased emphasis on process control, testing remains an important component of any system that aims to produce safe food. The food sector is in fact the largest market for microbiological tests. Testing is also a valuable tool for import control as part of programs to check and assess compliance with import requirements. For example, in the year from March 2009 to March 2010 the European Rapid Alert System for Food and Feed issued 325 notifications related to *Salmonella* alone in food and feed based on testing within the European Union's import control program (RASFF, 2010). Monthly import refusal reports from the U.S. Food and Drug Administration reflect a similar importance of testing in their import control program (U.S. Food and Drug Administration, 2010).

Another important factor is the increase in international trade of perishable products such as fresh fruit and vegetables (Food and Agriculture Organization, 2009). With conventional test methods, irrespective of the outcome, the lengthy delay in getting a result can mean that by the time a shipment is released, it is no longer marketable. The ability to test and subsequently transport food quickly is a critical element of international trade today; hence, the availability of rapid test methods offers a major advantage.

However, there is also the potential that as we begin to use rapid methods, which are often more sensitive than conventional methods, pathogens and other contaminants will be detected at lower levels, leading to the rejection of larger amounts of imported food. The application of rapid methods needs to be linked to policy considerations and the potential need to address the implications of detection of, and the risk associated with, very low levels of pathogens/contaminants. This adjustment has already occurred in the area of testing for veterinary drug residues, where the detection of very low levels of residues by new methods led to a major trade disruption before the risk management and policy aspects were addressed to reflect the change in detection levels (Food and Agriculture Organization, 2004).

The availability of a rapid method alone does not mean that regulators are immediately going to take it on board as a standard method. The performance of the method compared with conventional culture-based methods is an important consideration for regulators, in relation not only to their own methods but also to those used by regulators in export markets. There are numerous bodies that can certify the performance of such methods or provide guidance on their validation compared with currently accepted methods, which can facilitate equivalence; but this can be time-consuming. Other issues that also need to be considered include the cost and availability, reliability and ease of use, and the applicability of the method to the food items to be tested. In addition, the users should be cognizant of the possibility that a method validated by one country for the analysis of a particular product may not be applicable to the same product in another country. Different countries can have different production and processing practices that may result in foods containing different ingredients, pathogens, or levels and types of normal flora bacteria. Thus, the acceptance and use of such methods by official authorities are, usually for valid reasons, slow and tend to lag behind the use of such methods by the private sector.

Rapid Methods and Developing Countries

The contribution of developing countries to international food trade is increasing. In 2007, almost 40% of agricultural commodities in international trade came from developing countries, 87% of which were food items (Food and Agriculture Organization, 2009). For many of these countries, increased access to international markets is an important element of their development strategies. This can be neither achieved nor sustained without addressing food safety, and thus, there is a strong interest and demand for testing of foods from these countries, particularly given the challenges posed to food safety management by highly fragmented production systems, contaminated water supplies, unreliable electricity supplies, weak storage facilities, and poor road infrastructure. For example, in the African region, some of the food safety challenges include weak foodborne disease surveillance, inability of small- and medium-scale producers to provide safe food, outdated food regulations, weak law enforcement, inadequate capacity for safe food storage, and inadequate cooperation among stakeholders (Anonymous, 2007).

Over the last decade, many developing countries have made significant improvements to their national systems of food control. However, in many cases, considerable work is still required for these systems to afford the level of public health protection from foodborne hazards achieved in many industrialized countries. Lack of reliable and efficient laboratory services is a major constraint facing both regulators and food chain operators in these countries. High capital costs for investment in testing facilities, high running costs due to inadequately trained human resources, and distant technical services for the repair and maintenance of equipment and for provision of supplies and consumables contribute to the reluctance of authorities to invest in the development of laboratory infrastructure. Even when such infrastructure exists within a region or country, getting samples to a laboratory in good condition may be problematic. Capacity building, which also includes improved analytical skills for monitoring raw material going into food production and control of foods on the market, is critical. Here, cost-effective, noncomplex rapid methods, such as simple, disposable assays that do not require instrumentation can play a key role in improving public health and food safety in developing countries.

Furthermore, the use of quick and easy-to-use tests, particularly for screening purposes, can increase the number of samples to be tested, thus improving the overall statistical reliability of testing programs. The potential to reliably and accurately test product on site ("in the field"), be it a border inspection post, market, or primary producer, can provide many benefits.

If rapid field tests are to offer value, then they need to be robust and reliable under the storage conditions of test kits and use in developing countries, taking into consideration the often high ambient temperatures and lack of refrigeration. In addition, they must be affordable and readily accessible. Many field tests (diagnostic kits) are manufactured by a few selected countries but have limited distribution abroad. The customs and importation requirements of various countries can also affect the accessibility of diagnostic kits (Feng and Heredia, 2009).

For developing countries that are still building their food control systems and laboratory infrastructure, the availability of rapid and automated methods, validated for use under their conditions, can influence the design of their laboratory networks, their training programs for food inspectors and laboratory personnel, and the guidance provided to food producers in relation to their hygiene monitoring. However, as only one component of a food control system, consideration on how to deal with the rapid availability of greater volumes of results from possibly more sensitive testing procedures will be an important aspect of the application of rapid methods.

RAPID METHODS AS PART OF RISK ASSESSMENTS AND RISK MANAGEMENT STRATEGIES

The primary objective of risk assessment and the process of modeling contamination in the food chain is improved decision making and risk management actions, such as identifying intervention points to reduce risk effectively. This is typically achieved by developing a model of a food production system that can span the entire "farm-to-fork" continuum.

Models use mathematical equations to describe the system and rely on data to inform it. The quality of the data is a key factor in ensuring that the model reflects reality and consequently that decision-making advice can go beyond vague risk management guidelines to more specific interventions and their expected risk reduction. Typically, however, the information and data that have been available for modeling the food chain have been insufficient, uncertain, and too often qualitative in nature (Food and Agriculture Organization/World Health Organization, 2002, 2009; Duffy et al., 2006).

Increased Data and Information

Rapid methods offer the promise of improving the risk assessment and modeling process by increasing the amount of data and information available for use in developing risk assessments as well as reducing the uncertainty associated with the data that are used.

The process of developing a risk assessment model can improve our understanding of the food system under investigation by closely examining how different parts of the system interact with each other and by evaluating what data exist or what information we have about the system. A key part of developing a model is to use pathogen contamination data at the start of the process and simulate in the model how this changes as the food product moves through the food production system. Thus, despite the fact that a model has included the use of expert opinion, data from surrogates, and data from other countries or other systems as proxies (Food and Agriculture Organization/World Health Organization, 2003, 2008; Boone et al., 2009; Ross and Sumner, 2002), without good, reliable quantitative data on pathogen presence and levels, the model is left to describe contamination with a high degree of uncertainty.

The absence of quantitative data for use in risk assessment models has been a concern in those models developed to date. The use of rapid methods, by reducing the time and cost associated with collecting quantitative data, offers the potential to dramatically increase both the quantity and quality of the quantitative data available for risk modeling activities.

The potential of rapid methods to be more sensitive can skew risk assessments. While the quantification of a pathogen in a food product is an important part of the risk modeling process, an equally important part of the process is the characterization of the prevalence of contamination. A risk model needs to determine the probability of exposure as well as the amount of exposure to a pathogen. Historically, there has been recognition that the probability of exposure that was typically used was a function of the method, and risk assessments have taken these data at face value. However, a problem that has often been ignored is diagnostic method sensitivity, which often has resulted in underestimations (Lindqvist et al., 2002; Oscar, 2004).

The problem is exacerbated when the product is assumed to be uncontaminated in the risk model and when it subsequently experiences time and temperature abuse that allows for growth or, even more significantly, is mixed with uncontaminated product that subsequently allows for growth. In the former situation, a product that was assumed to have no risk can in fact, expose an individual to a dangerous infectious dose, and in the latter case, multiple people that would have been estimated to have no risk could be at significant risk of illness.

Outbreak Data

In order to translate exposure to pathogens to a health risk outcome, risk assessments require dose-response models that predict the probability of infection or illness upon exposure to a dose of pathogen. While human feeding trial data from the 1960s do exist, it is unlikely now, due to ethical considerations, that any new dose-response data will be collected by using human feeding trials. As a result, risk assessments are faced with the intractable problem of either a complete lack of information on the probability of infection for some pathogens or a large uncertainty associated with other pathogens and no way of addressing this information gap. One source of information that can help overcome this problem is to make use of outbreak data. However, the ability to obtain samples of food and process them rapidly and with sufficient sensitivity to estimate a dose that was likely to have caused illness during an outbreak has been historically difficult to achieve. The development and use of rapid methods can clearly contribute to the last two issues (rapid processing and sensitivity) and should increase the likelihood that outbreak data can contribute to the overall understanding of dose-response and improve the ability of risk models to estimate the probability of illness.

Decreasing Uncertainty

The concentration of a pathogen in a food product is a fundamental part of the risk-modeling process; however, the sensitivity of traditional microbiological methods has often meant that in many studies, many samples are known to be positive but are below the minimum detection limit. This left censoring of the data often has to be corrected for by using statistical techniques including (i) assuming that all data below the detection limit are actually equal to the detection limit, or (ii) assuming that all data below the detection limit are only slightly greater than 0, or (iii) more elegant statistical techniques (Dechesne and Soyeux, 2007). Regardless of the technique used to correct for this left censoring, the end result is that the risk estimates generated can be over- or underestimates, depending upon organism characteristics, such as virulence and growth rates, as well as consumer and food characteristics. With lower detection limits and greater sensitivity, rapid methods have the ability to quantify pathogens at lower levels, allowing risk assessments to better characterize the true statistical distribution of pathogens and, as a result, better quantify and estimate the risk.

Oftentimes, outbreaks of a foodborne illness are a result of a chain of breakdowns in the system. The low-level contamination that might have been previously ignored can quickly turn out to be the decisive factor. Finally, the very nature of traditional, culture-based methods means that the strains of a pathogen that are isolated and counted are in fact, those that grew the fastest. This does not necessarily mean that they are the only ones that pose the greatest public health threat. The use of rapid methods allows for a more accurate characterization of the pathogens that are present in the food system, which translates into a more reliable risk model and risk estimate.

Risk-Based Decision-Making

In traditional food safety management systems, decision making is an activity that translates evidence, in the form of results from surveillance or from hazard control monitors, into actions such as targeted communications, resource allocations, or direct interventions. The actions are chosen by a decision maker or risk manager to achieve simple, predetermined goals in relation to food safety. For instance, a zero tolerance for microbial pathogens is a common, simple objective for the decision process involving ready-to-eat foods. However, the introduction of rapid methods, which provide more sensitive results more quickly, has increased the information supply to the food safety decision maker. This added complexity has prompted a reassessment of the existing decision-making process.

Decision theory recognizes two kinds of decision-making processes: normative decision making, which is usually associated with a rational framework and is based on quantitative expressions; and descriptive decision making, which predominantly engages the experience and heuristics of a specialized decision maker. Although food safety decision makers, depending on the particular scenario, will engage in both processes, the introduction of rapid methods increasingly emphasizes normative decision making. In turn, this trend leads to increased roles for probabilistic modeling and for quantitative risk assessment as supports for food safety management activity (Anonymous, 2009).

Generally, the information relating to food safety management decisions includes uncertainty, i.e., information or data are often incomplete or imprecise. Uncertainty clouds decision making and often leads to compromise and, in relation to safety, to precautionary conservatism. In many respects, rapid methods, with relatively high throughput, can be used to reduce the traditional information uncertainties and, hence, to strengthen decision making. However, rapid methods also, and often, introduce additional uncertainty associated with the interpretation of results, where the decision maker has to include validation of the rapid method into the decision-making process. A screening process designed to eliminate large numbers of negative samples, rapidly and economically, from further monitoring illustrates the need for additional considerations in decision making. In this case, the possibility of false negatives is crucial in order to understand safety decisions. The probability of false negatives, established from controlled trials, can most easily be included in probabilistic modeling and hence into normative decision-making processes.

An additional piece of information that reduces the uncertainty associated with the interpretation of results from rapid methods is validation, which is required in many situations that involve regulatory decisions. Food safety regulations are still, predominantly, based on traditional (culture-based) methods, so rapid-method-based decisions may not have an immediate value. The lack of recognized standards precludes the introduction of novel decision making.

Integrated Information

Rapid methods for microbial testing of foods often introduce timescales and volumes for information generation that are incompatible with descriptive decision making. Information supplies from rapid methods routinely engage automated storage and retrieval systems that have structured access and multiuse functions. In this way, rapid methods promote the integration of multiple information sources in decision making and, in combination with the normative approach, contribute to cooperation and consistency of decision makers. As an example, an in-line monitor for contamination might support real-time decision making in relation to intervention and also, in combination with other food chain information sources, might contribute to a database that supports targeted tracking or tracing investigations on a longer timescale.

Increased sensitivity is usually the most visible feature of rapid (often molecular) methods for food analysis. Increased measurement sensitivity drives changes in the criteria that can be used for food safety management decisions, particularly zero tolerance conditions. In the face of increased sensitivity, criteria that relate to the presence or absence of food contamination must be modified to accommodate inevitable beliefs concerning small but potentially harmless levels of contamination. This extension of the decision space requires integration of uncertain beliefs about human dose-response behavior and probabilistic information about the limit of detection. In this case, decisions based on simple threshold criteria are replaced by more complex decisions based on the probabilities of a range of outcomes, and these decisions are most suited to normative processes. The increased sensitivity of rapid methods for enumeration of microbes in foods often provides a more sophisticated set of options, including distinct action levels, for consideration by the decision maker.

Rapid Methods and Food Safety Management

Each of these decision-making situations, reflecting the impact of rapid methods on food safety management decisions, highlights complex information integration and a process based on probabilistic modeling. Other disciplines, particularly finance, embrace complex decision making by combining probabilistic models with a utility formalism (Kjaerulff and Madsen, 2008). Utility reflects the multifaceted criteria, such as cost and benefit, that form the basis for decision making under uncertainty, and the increasing application of rapid methods for gathering multivariate information about food safety prompts an improved search for appropriate utilities.

As well as modifying established food safety management decision making, some rapid methods, particularly subtyping technologies, generate completely new decision-making opportunities. Type matching, based on rapid molecular methods, can provide evidence concerning the likely source associated with particular contamination observations (Barker et al., 2009). In this case, a probabilistic understanding of the discriminatory power of the typing can be combined with prior beliefs about possible sources and their strengths to support decisions about source tracing. Increasingly, tracing decisions (source level inferences) contribute to resource allocations for in-line preventative controls or contribute to considerations of liability in relation to foodborne hazards.

Rapid methods are a major driver in developing decision processes that relate to food safety management. The balance between the uncertainties associated with increased information and with challenging interpretations and the effect on decision making is central to the use of rapid methods and, increasingly, is addressed by the use of probabilistic models for foodborne hazard scenarios.

The increased sensitivity of technologies employed to measure contamination levels in food can be used to support a more flexible approach to management decisions as opposed to providing an overpowering sense of quantitative precision. Additionally, several new decision structures, such as biotracing, are supported by accessible and economic rapid detection and typing techniques. Rapid methods for microbial testing and the associated information and data supplies are driving a progression away from isolated, uncoordinated decision making towards automated decisions that are integrated into broader operational food safety management.

CONCLUDING REMARKS

The traditional argument for developing and using rapid methods has been faster detection of pathogens, but it is becoming increasingly evident that rapid methods can contribute to other aspects of food safety, public health, and global trade. Another voice of support for the use of rapid methods has come from the recent concerns about bioterrorism attacks, which require much faster response and highly innovative technologies. This has moved rapid methods further up on the agenda of regulators, health policy makers, and research organizations. The following chapters in the book will explore in more detail the various aspects of rapid methods, in general, and important pathogen-bearing commodities, in particular.

REFERENCES

Anonymous. 2005. *International Health Regulations 2005.* World Health Organization, Geneva, Switzerland.

Anonymous. 2007. *Food Safety and Health: a Strategy for the WHO African Region.* World Health Organization, Geneva, Switzerland.

Anonymous. 2009. *National Research Council of the National Academies. Science and Decisions.* The National Academies Press, Washington, DC.

Barker, G. C., N. Gomez, and J. Smid. 2009. An introduction to biotracing in food chain systems. *Trends Food Sci. Technol.* **20:**220–226.

Boone, I., Y. Van der Stede, K. Bollaerts, W. Messens, D. Vose, G. Daube, M. Aerts, and K. Mintiens. 2009. Expert judgement in a risk assessment model for Salmonella spp. in pork: the performance of different weighting schemes. *Prev. Vet. Med.* **92:**224–234.

Brown, T., M. Cueto, and E. Fee. 2006. The World Health Organization and the transition from "International" to "Global" public health. *Am. J. Public Health* **96:**62–72.

Centers for Disease Control and Prevention. 2009. Multistate outbreak of *Salmonella* infections associated with peanut butter and peanut butter-containing products—United States, 2008–2009. *MMWR Morb. Mortal. Wkly. Rep.* **58:**1–6.

Cole, M. B., and R. B. Tompkin. 2005. Microbiological performance objectives and criteria, p. 673–693. *In* J. Sofos (ed.), *Improving the Safety of Fresh Meat.* Woodhead Publishing Ltd., Cambridge, England.

Dechesne, M., and E. Soyeux. 2007. Assessment of source water pathogen contamination. *J. Water Health* **5**(Suppl. 1)**:**39–50.

Duffy, G., E. Cummins, P. Nally, S. O'Brien, and F. Butler. 2006. A review of quantitative microbial risk assessment in the management of *Escherichia coli* O157:H7 on beef. *Meat Sci.* **74:**76–88.

Feng, P. 2001. Development and impact of rapid methods for detection of foodborne pathogens, p. 775–796. *In* M. P. Doyle, L. R. Beuchat, and T. J. Montville (ed.), *Food Microbiology: Fundamentals and Frontiers,* 2nd ed. ASM Press, Washington, DC.

Feng, P. 2007. Rapid methods for the detection of foodborne pathogens: current and next-generation technologies, p. 911–934. *In* M. P. Doyle and L. R. Beuchat (ed.), *Food Microbiology: Fundamentals and Frontiers,* 3rd ed. ASM Press, Washington, DC.

Feng, P., and N. Heredia. 2009. Rapid methods for food borne bacterial enumeration and pathogen detection, p. 547–560. *In* N. Heredia, I. Wesley, and J. Garcia (ed.), *Microbiologically Safe Foods.* Wiley and Sons, Hoboken, NJ.

Food and Agriculture Organization. 2004. *Report of an FAO/WHO Technical Workshop on Residues of Veterinary Drugs without ADI/MRL, 24 to 26 August 2004, Bangkok, Thailand.* Food and Agriculture Organization, Rome, Italy.

Food and Agriculture Organization. 2009. *FAO Statistical Yearbook 2009.* http://www.fao.org/economic/ess/publications-studies/statistical-yearbook/fao-statistical-yearbook-2009/en/. Food and Agriculture Organization, Rome, Italy. Accessed 23 April 2010.

Food and Agriculture Organization/World Health Organization. 2002. *Risk Assessments of* Salmonella *in Eggs and Broiler Chickens: Technical Report. Microbiological Risk Assessment Series No. 2.* Food and Agriculture Organization, Rome, Italy.

Food and Agriculture Organization/World Health Organization. 2003. *Hazard Characterization for Pathogens in Food and Water: Guidelines. Microbiological Risk Assessment Series No. 3.* Food and Agriculture Organization, Rome, Italy.

Food and Agriculture Organization/World Health Organization. 2008. *Exposure Assessment of Microbiological Hazards in Food: Guidelines. Microbiological Risk Assessment Series No.* 7. Food and Agriculture Organization, Rome, Italy.

Food and Agriculture Organization/World Health Organization. 2009. *Risk Assessment on* Campylobacter *spp. in Broiler Chickens: Technical Report. Microbiological Risk Assessment Series No. 12.* Food and Agriculture Organization, Rome, Italy.

Josefsen, M. H., N. R. Jacobsen, and J. Hoorfar. 2004. Enrichment followed by quantitative PCR for rapid detection and as a tool for quantitative risk assessment of foodborne thermotolerant campylobacters. *Appl. Environ. Microbiol.* **70:**3588–3592.

Kjaerulff, U. B., and A. L. Madsen. 2008. *Bayesian Networks and Influence Diagrams.* Springer, New York, NY.

Krämer, N., C. Löfström, H. Vigre, J. Hoorfar, C. Bunge, and B. Malorny. 18 September 2010. A novel strategy to obtain quantitative data for modelling: combined enrichment and real-time PCR for enumeration of salmonellae from pig carcasses. *Int. J. Food Microbiol.*, Epub ahead of print. doi: 10.1016/j.ijfoodmicro.2010.08.026.

Lindqvist, R., S. Sylven, and I. Vagsholm. 2002. Quantitative microbial risk assessment exemplified by *Staphylococcus aureus* in unripened cheese made from raw milk. *Int. J. Food Microbiol.* **78:**155–170.

Lübeck, P. S., and J. Hoorfar. 2002. PCR technology for detection of food borne pathogens, p. 65–84. *In* K. Sachse and J. Frey (ed.), *Methods in Molecular Biology*, vol. 216. *PCR Detection of Microbial Pathogens: Methods and Protocols.* Humana Press Inc., Totowa, NJ.

Madej, R., J. Davis, M. J. Holden, S. Kwang, E. Labourier, and G. J. Schneider. 2010. International standards and reference materials for quantitative molecular infectious disease testing. *J. Mol. Diagn.* **12:**133–143.

Malorny, B., P. Tassios, P. Rådström, N. Cook, M. Wagner, and J. Hoorfar. 2003. Standardization of diagnostic PCR for the detection of foodborne pathogens. *Int. J. Food Microbiol.* **83:**39–48.

Oscar, T. P. 2004. A quantitative risk assessment model for *Salmonella* and whole chickens. *Int. J. Food Microbiol.* **93:**231–247.

RASFF. 2010. RASFF portal. https://webgate.ec.europa.eu/rasff-window/portal/. Accessed 23 April 2010.

Ross, T., and J. Sumner. 2002. A simple, spreadsheet-based, food safety risk assessment tool. *Int. J. Food Microbiol.* **77:**39–53.

Scheule, B., and J. Sneed. 2001. From farm to fork: critical control points for food safety. *J. Nutr. Recipe Menu Dev.* **3**(2):3–23.

U.S. Food and Drug Administration. 2010. Import refusal reports for OASIS. http://www.accessdata.fda.gov/scripts/ImportRefusals/ir_index.cfm. Accessed 23 April 2010.

World Trade Organization. 1995. The results of the Uruguay round of multilateral trade negotiations: the legal texts, p. 69–84. *Agreement on the Application of Sanitary and Phytosanitary Measures.* World Trade Organization, Geneva, Switzerland.

ACCELERATED GENERIC TECHNIQUES

I

STRENGTHS AND SHORTCOMINGS OF ADVANCED DETECTION TECHNOLOGIES

L. D. Goodridge, P. Fratamico, L. S. Christensen, M. Griffith, J. Hoorfar, M. Carter, A. K. Bhunia, and R. O'Kennedy

2

Due to the low infectious dose of many pathogens, the presence of even low concentrations of pathogenic bacteria in food is likely to cause disease (Ryan, 2004; Teunis et al., 2004). Additionally, the interference caused by components of food, environmental, or fecal samples leads to increased background signals and false-positive results, thereby reducing the diagnostic specificity of the methodology (Selby, 1999; de Boer and Heuvelink, 2000). To overcome the sensitivity and specificity issues, current confirmatory foodborne pathogen detection methods generally require an initial, time-consuming growth step in culture media, followed by isolation on solid media, biochemical identification, and molecular or serological conformation.

L. D. Goodridge, Department of Animal Sciences, Center for Meat Safety and Quality, Colorado State University, Fort Collins, CO 80523. *P. Fratamico,* U.S. Department of Agriculture, Agricultural Research Service, Wyndmoor, PA 19038-0000. *L. S. Christensen and J. Hoorfar,* Danish National Food Institute, Division of Microbiology and Risk Assessment, Technical University of Denmark, Mørkhøj Bygade 28, 2860 Søborg, Denmark. *M. Griffiths,* Canadian Research Institute for Food Safety, University of Guelph, Guelph, Ontario, Canada, N1G 2W1. *M. Carter,* Silliker Inc., 160 Armory Drive South Holland, IL 6047. *A. K. Bhunia,* Department of Food Sciences, Purdue University, West Lafayette, IN 47907. *R. O'Kennedy,* School of Biotechnology, Dublin City University, Dublin 9, Ireland.

Recent advances in microbial detection technologies have aimed to effect faster and more sensitive identification of the target pathogen. In addition, such tests can be applicable to more cost-effective testing in a field or food processing environment. Regardless of the many advances in foodborne pathogen testing, culture detection of the target organism still remains the gold standard by which all other detection methods are assessed. Culture methods, which include the enrichment of bacterial concentrations in liquid and on solid media, have been modernized to include selective and differential agents including fluorogenic and chromogenic substrates within solid media, allowing the simultaneous identification of several bacterial species following incubation.

Newer methods, often referred to as rapid methods, continue to gain popularity in response to the never-ending challenge of decreasing the time to detection. While some of these assays have been automated to reduce hands-on manipulations, all current assays used to detect specific pathogens in foods still require an enrichment step prior to analysis. Also, many advances have been made in diagnostic methodology in recent years, but the ideal detection method, which should be capable of rapidly detecting and confirming the presence of pathogens directly from complex food samples with no false-positive or false-negative results, remains an elusive challenge.

MODERNIZED CULTURE METHODS

Culture methods still dominate applications in which detection and enumeration of the target foodborne microorganism are required. In such applications, incorporation of fluorogenic or chromogenic substrates into the media has facilitated enumeration and detection directly on the agar plate. These media are characterized as containing enzyme substrates linked to a chromogen (color reaction), a fluorogen (light reaction), or a combination of both. The introduction of many of these differential media has led to improved accuracy and more rapid detection of target organisms, often reducing the need for isolation of pure cultures and confirmatory tests.

Generally, four groups of fluorogenic and chromogenic compounds have been used in differential media, including the fluorogenic dyes, whereby an increase of fluorescence results from adsorbance or intercalation of fluorescent dye onto or into the DNA or protein of bacterial cells; pH-based fluorescent or chromogenic indicators, whereby microbiological biochemical activity results in a change in intensity of fluorescence or absorbance of a pH indicator; redox potential indicators, which become colorless and subsequently fluorescent upon reduction; and fluorogenic and chromogenic enzyme substrates, which are metabolized by the target microorganism to release the chromogen/fluorogen, which results in a color change in the medium and/or fluorescence under long-wave UV light (Manafi, 1996). The substance generated during the enzymatic reaction either is adsorbed directly onto the microbial cells, thereby specifically labeling the microorganisms, or diffuses into the culture medium, giving a pronounced color or fluorescence. Other substances confer resuscitation advantages to aid in the recovery of sublethally damaged cells, and inclusion of such selective agents has facilitated the development of media that are both selective and differential.

While all of the major food pathogens can be detected and enumerated by using selective and differential media in a wide variety of food samples, the methodology for detection of each pathogen can vary due to the sample matrix, the presence of large amounts of background microflora, differential requirements of the target organism for oxygen, and growth temperature. Regardless, most methods typically include a preenrichment step of 24 h or more. This allows time for any sublethally damaged pathogens to resuscitate and achieve a high enough concentration such that sufficient numbers of the pathogen are transferred onto the selective and/or differential medium when a subsample is plated out.

RAPID DETECTION TECHNOLOGIES

Rapid-detection-based technologies can reduce the time and labor involved in screening food products for the presence of pathogens. Many of the rapid tests can be completed within 24 h, with high throughput, thereby reducing the labor involved in the testing process. These assays can be broadly grouped into three categories including immunologically based methods, nucleic acid-based assays, and biosensors.

Immunological Methods

The specific binding of antibodies to their particular antigen, as well as the speed and simplicity of this interaction, has driven the production of a variety of antibody assays and detection formats based on immunochemistry. These tests originated in the 1970s and 1980s, when major advances in research (such as the ability to manufacture monoclonal antibodies) were transferred rapidly to applied areas, including the diagnostic field (Feng, 1996).

There are several broad formats of antibody assays (Feng, 1997), including latex agglutination, in which antibody-coated, colored latex beads or colloidal gold particles are used for quick identification of foodborne pathogens following culture enrichment (D'Aoust et al., 1991; Feng, 1997). Reverse passive latex agglutination is related to latex agglutination and is used mostly in testing for the presence of toxins in food extracts or for toxin production by pure bacterial cultures (Feng, 1997). Immunoprecipitation-based assays function on the principle of agar diffusion; an enrichment sample is introduced to the testing device and migrates through a gel matrix with the antibody, and if the specific antigen is present, a visible line of precipitation is formed (Moberg et al., 1988).

The enzyme-linked immunosorbent assay (ELISA) is the most prevalent antibody assay format used for pathogen detection in foods (Candish, 1991). In the ELISA, which is usually designed in a sandwich format, an antibody is bound to a solid matrix such as a microplate well and facilitates capture of the antigen from enrichment cultures. A second (typically monoclonal) antibody conjugated to an enzyme is bound to the target, followed by the addition of a chromogenic, fluorescent, or chemiluminescent enzyme substrate, which is used for visualization of the bound targets (Feng, 1997).

A modification of the ELISA is the lateral flow assay (LFA), a technology made popular by the plethora of home pregnancy tests that are commercially available. LFAs, also referred to as capillary attraction tests, are usually single-use, disposable cartridge tests in the form of either immunochromatographic line assays or enzyme immunoassays that generate detectable colored end products which give a presence/absence answer. Like the ELISA, the LFA is performed in a sandwich format, but the second antibody is conjugated to colored latex beads or to colloidal gold instead of an enzyme. A small volume (typically 100 μl) of enrichment sample is introduced to the sample port and is wicked across a series of chambers by capillary action to obtain results (Lim et al., 2005). A positive result is obtained from the capture of the labeled antigen-antibody complex with a second immobilized antibody and the formation of a line or pattern in the appropriate result window (Murray et al., 2003). Also incorporated are control reagents, which result in the formation of a control line in a separate window, indicating that the test worked correctly. LFAs are extremely simple, require no washing or manipulation, and are completed within 10 min following culture enrichment. However, these assays have a poor detection sensitivity compared to nucleic acid-based methods (see below).

Antibodies coupled to magnetic beads are used in immunomagnetic separation (IMS) technology to capture pathogens from preenrichment media (Bilir Ormanci et al., 2008; Olsvik et al., 1994; Uyttendaele et al., 2000). IMS was developed as an alternative or complement to selective enrichment, and instead of antibiotics or harsh reagents that can cause stress or injury to the target cells, an antibody is used to concentrate the target, which can then be plated onto selective and differential media or detected using a rapid immunological or molecular assay. While there are many different immunological formats that facilitate rapid detection of foodborne pathogens, the majority of tests used in the food chain are based upon the lateral flow, ELISA, and IMS concentration formats.

LFAs

LFAs have continued to gain popularity for their ability to detect pathogens in food. Shim et al. (2008) combined IMS technology with LFA technology and created a test that is capable of rapidly detecting *Listeria monocytogenes* in a cost-effective and potentially on-site method. An LFA test strip that was able to detect all 11 *L. monocytogenes* serotypes tested was produced. When used without IMS, it took 24 h to detect 10^2 CFU/10 g of meat originally inoculated onto meat samples. The IMS system was shown to be specific for all *Listeria* spp. tested. The combination of IMS and the LFA reduced the test time to 14 h. The developed test is rapid compared to other immunoassays and commercially available kits, as results can be read the same day.

There are several commercially available LFA kits for detection of *Escherichia coli* O157:H7 and *Salmonella* spp. (Tokarskyy and Marshall, 2008). One of the issues associated with LFAs is the potential for a high rate of false positives, due to interference from the sample matrix. As such, these assays often have to be optimized for detection of bacteria from a specific food matrix. Seo and coworkers (2003) developed an LFA system to rapidly detect *Salmonella enterica* serovar Enteritidis and other group D salmonellae from eggs. In this work, a series of studies were conducted to optimize the test procedure for raw eggs by evaluating different sample preparations against a panel of group D *Salmonella* organisms. The minimum concentration of serovar Enteritidis to generate a positive reaction was $\sim 10^7$ cells/ml in pure culture, and no cross-reactivity was detected with two other *Salmonella* serovars. A potential issue was recognized due to the viscosity of the sample matrix.

In order for an LFA to function properly, the sample has to be effectively wicked down the membrane within the LFA so that the organism present in the sample can interact with the detection antibodies. Whole egg does not effectively transit the panel due to its viscous nature. In an attempt to circumvent this problem, the egg samples were diluted. Serially diluting the egg with phosphate-buffered saline reduced the viscosity of egg contents and allowed more effective wicking of sample. However, it was also observed that the intensity of the positive band appeared weaker as the dilution was increased, resulting in a decrease in sensitivity to approximately 10^8 cells/ml. To address this issue, the authors developed a novel sample preparation method, in which a mixture of fatty acids was prepared by combining 2 parts of oleic acid and 1 part of caprylic acid (vol/vol). One milliliter of the inoculated egg contents and 0.5 ml of the fatty acid mixture were combined in a microcentrifuge tube for 1 min to allow the acid to react and form an ester with reactive OH groups. The reaction formed water as a byproduct and also broke up the structure of the egg contents as it released the bacteria into the aqueous phase. After 2 min of incubation at room temperature, the tube was centrifuged for 5 min at $10{,}000 \times g$. Centrifugation formed an oily layer with yellow carotenes and oil solubles on the top, a soapy hard layer of esters in the middle, and a clear aqueous layer with soluble bacterial antigen and a pellet of bacterial debris on the bottom. The aqueous layer containing soluble antigen was removed with a 1-ml syringe, and 100 μl of the solution was then applied to the LFA sample port. The detection limit of the LFA was increased approximately 10-fold when the extraction method was used. The results of this study indicated that the LFA could be successfully used to detect serovar Enteritidis in eggs in conjunction with various enrichment methods and also highlighted the fact that sample manipulation and assay optimization are essential to ensure optimal specificity and sensitivity when rapid assays are used to test food samples for the presence of pathogenic microorganisms.

While there are only a few diagnostic methods that have been designed to allow for point of detection testing, the LFAs represent perhaps the easiest format with which testing can be accomplished outside the laboratory. Nevertheless, in the food microbiology sector, the main applications of LFAs are likely to remain within the laboratory, since testing for foodborne bacterial pathogens generally involves at least one enrichment step before the assay strip is used to confirm the presence or absence of the pathogen.

In addition to being developed as a rapid assay in their own right, LFAs may also be useful for rapid confirmation of the identity of pure bacterial isolates from cultural microbiological testing. Although LFAs are easier to perform and more rapid than classic instrument-based immunoassays, they are typically not as sensitive and have the potential for higher rates of false positives.

ELISAs

ELISAs are widely used to detect in foods the presence of bacterial enterotoxins that are likely to lead to foodborne intoxications. ELISAs have been developed and used to detect botulinum toxin and staphylococcal enterotoxins (SETs), although their applicability for detection of botulinum toxin is limited because they are not as sensitive as the mouse bioassay (Swaminathan and Feng, 1994). In contrast, the SETs are amenable to detection by immunoassays because the concentration of enterotoxin required to produce illness in sensitive individuals is easily detectable by ELISA (Notermans and Wernars, 1991). Still, the food matrix and extraction procedure used to isolate the toxin prior to detection can affect the sensitivity of the ELISA, as can the presence of certain bacterial proteins such as protein A (Notermans and Wernars, 1991).

Wieneke (1991) evaluated the performances of three ELISA kits for their ability to detect SETs in foods implicated in outbreaks of food intoxications and reported that none of the assays were clearly superior to the others.

ELISAs are the most widely used type of immunoassay for detection of enterohemorrhagic *E. coli* (EHEC) and Stx (de Boer and Heuvelink, 2000). In order to detect all EHEC serotypes and not just *E. coli* O157:H7, ELISAs that screen for Stx are commonly utilized. Several commercially available Shiga toxin-producing *E. coli* ELISA kits have been developed and have been reported to compare well in sensitivity and specificity with other EHEC detection methods such as PCR and verocytotoxin cell culture assay (Mackenzie et al., 1998; Beutin et al., 1999; Gavin et al., 2004; Willford et al., 2009).

Although all of the ELISA experiments reported require overnight enrichment to generate positive results, it has been demonstrated that enrichments as short as 7 h could be utilized to produce positive results in EHEC-contaminated foods (Czajka and Batt, 1996).

In contrast to what is used for detection of Shiga toxin-producing *E. coli*, the most commonly used ELISAs for the detection of *L. monocytogenes* are based on the detection of whole cells. In this scenario, *L. monocytogenes* cells are enriched from the food sample, heat killed, and then detected by an ELISA. One disadvantage to the use of whole-cell ELISAs for detection of *L. monocytogenes* is the fact that many of the cell surface antigens are genus specific rather than *L. monocytogenes* specific (Feldstine et al., 1997a, 1997b). This is problematic, because *L. monocytogenes* is the only *Listeria* species pathogenic to humans (Gray et al., 1966), with the rare exception of other *Listeria* spp. that have caused or have the potential to cause listeriosis (Johnson et al., 2004). In an attempt to develop *L. monocytogenes*-specific ELISAs, some researchers designed tests to detect the flagella of the bacteria rather than the whole cells (Kim et al., 2005). However, these assays require sample manipulation prior to testing, including washing steps and extraction of the flagella, and the specificity of these tests is still only to the genus level (Kim et al., 2005).

Other ELISAs using the O and H antigens have been developed (Palumbo et al., 2003). Whole-cell protein extracts (Bourry et al., 1997) and detection of the listeriolysin toxin have also been employed as detection targets in *L. monocytogenes* ELISAs (Matar et al., 1992; Paoli et al., 2004).

Other researchers have evaluated foods for the presence of *L. monocytogenes* with the use of ELISAs (Mattingly et al., 1988; Curiale et al., 1994), and a number of commercially available ELISA methods for detection of *L. monocytogenes* in food and environmental samples have been validated by one or more recognized formal validation studies (Dunbar et al., 2003; Sewell et al., 2003). However, because the antibodies used in these tests may cross-react with other *Listeria* spp., the ELISAs are only considered presumptive for *L. monocytogenes*. To confirm the presence of *L. monocytogenes* in a sample, enrichment cultures that are positive by these methods should be streaked onto selective media and suspect colonies should be biochemically identified as *L. monocytogenes* according to the FDA method (Hitchins, 2003) (see chapter 18 in this book).

Bohaychuk and colleagues (2005) compared ELISAs, PCRs, and LFAs with conventional culture methodology for the detection of *Salmonella, Campylobacter, Listeria,* and *E. coli* O157:H7 inoculated into raw and processed meat and poultry products. Recommended protocols were modified so that the same enrichment broths used in the culture methods were also used in the ELISAs, PCRs, and LFAs. The percent agreement between the rapid technologies and culture methods ranged from 80 to 100% depending on the pathogen detected and the method used. ELISA, PCR, and LFA methods all performed well, with no statistical difference, compared with the culture method for the detection of *E. coli* O157:H7. ELISA performed better for the detection of *Salmonella*, with sensitivity and specificity rates of 100%. PCR performed better for the detection of *Campylobacter jejuni*, with 100% agreement with the culture method. Although the LFAs were statistically different from the cultural methods for *Salmonella* and *Listeria* because of false-positive results, the tests did not produce any false negatives, indicating that this method would be suitable for screening meat and poultry products for these pathogens.

Concentration by IMS

IMS utilizes antibody-coated paramagnetic nanobeads that are impregnated with an iron core. This structure allows the beads to become magnetized when placed in the presence of a magnet but to lose their magnetism when the magnet is removed, facilitating their free movement through a liquid or semisolid food sample. In IMS, the paramagnetic antibody-coated beads bind to the target antigen (bacteria, toxins, or parasites) and are specifically separated out of solution by placing the tube containing the food sample and the IMS beads into a magnet. The beads are drawn to the magnet, allowing concentration of the target antigen and also removal of the background, nontarget antigens (de Boer and Heuvelink, 2000; Feng, 2006). The concentrated sample is then screened by an appropriate detection method to provide confirmatory results. Therefore, IMS represents a concentration method and not a detection method for pathogens.

IMS has been shown to be a sensitive method for isolation of *E. coli* O157 from artificially mixed bacterial cultures, fresh produce, meat samples, and bovine feces (Himathongkham et al., 2007; Tsai et al., 2000; Grif et al., 1998; Gooding and Choudary, 1997; Johnson et al., 1995). It has been demonstrated that the use of IMS coupled with sorbitol-MacConkey agar media resulted in 100 times greater sensitivity of detection from bovine fecal and minced meat samples than the use of sorbitol-MacConkey agar plating alone (Chapman et al., 1994; Wright et al., 1994).

The production of IMS beads possessing antibodies specific against major non-O157 EHEC serogroups has improved the detection of these EHEC serotypes (de Boer and Heuvelink, 2000). For example, the use of IMS followed by LA was found to be 2.5 times more sensitive for detection of EHEC serogroup O26 in bovine feces than a comparative PCR/DNA probe method. Additionally, samples from naturally infected cattle produced positive results for EHEC serogroups O26, O103, and O145 via the IMS-LA method, while PCR/DNA probes could only identify EHEC serogroups O26 and O103 in the same samples (Jenkins et al., 2003). These studies highlight the ability of IMS to improve the detection of non-O157 EHEC strains in a complex matrix.

IMS protocols have also been developed for other foodborne pathogens. Hudson et al. (2001) used immunomagnetic separation to isolate *L. monocytogenes* directly from ham. In this procedure, the food was homogenized with some growth media, the particulate matter was removed, and after a number of washes, particles of bacterial size were pelleted and resuspended in a small volume of buffer. Commercial IMS beads coated with an anti-*Listeria* spp. antibody were added to the buffer solution and incubated to allow binding of the *L. monocytogenes* to the beads. The beads were trapped on a magnet and washed, and DNA was extracted for amplification of *L. monocytogenes*-specific genes by the PCR. The IMS procedures reduced the detection time to approximately 1 day but were limited in terms of sensitivity, since the recovery of cells on the beads was only about 20% of those initially added (Hudson et al., 2001). IMS on average allowed detection of 1 to 2 CFU/g food sample, but the results were somewhat variable in terms of sensitivity, having a detection limit ranging from 0.1 CFU/g to greater than 5.7 CFU/g (Hudson et al., 2001).

IMS separation was compared to conventional cultivation methods to determine the specificity and sensitivity of detection of *L. monocytogenes* from turkey meat samples (Bilir Ormanci et al., 2008). A total of 180 turkey samples were tested. The conventional methods detected *L. monocytogenes* in 15 (8.3%) of the samples, and the IMS method detected *L. monocytogenes* in 22 (12.2%) of the samples. The combination of the conventional and IMS methods revealed that 23 (12.7%) of the samples were contaminated with *L. monocytogenes*, which was confirmed by PCR. The IMS method was found to be more sensitive than the conventional culture methods, and the authors noted that it was more rapid as well, by reducing the detection period by approximately 1 day (Bilir Ormanci et al., 2008).

The use of IMS to separate *Salmonella* from foods has been less convincing. Skjerve and Olsvik (1991) investigated the use of IMS to isolate *Salmonella* from foods and reported that *Salmonella* could be separated from different inoculated foods including soft cheese (Brie), pasteurized whole milk, whole-milk powder, yogurt, chicken liver, vacuum-packed sliced ham, and a raw vegetable blend with a sensitivity of 10 to 20 *Salmonella* cells/g of the original sample. IMS of inoculated *Salmonella* cells from milk, milk powder, ham, soft cheese, and vegetable blend was successful. However, the IMS method appeared less useful for some food items. For example, in the yogurt and chicken liver samples, a substantial loss of beads was observed during initial separation and washing. In addition, the remaining IMS beads did not form a distinct pellet after magnetic concentration, and consequently, recovery after IMS was poor. This observation has also been found by others. Rybolt and coworkers (2004) compared four methods (IMS, tetrathionate broth, Rappaport-Vassiliadis R10 broth, and a secondary enrichment procedure) for their ability to isolate *Salmonella* in four different spiked sample matrices including Butterfield's solution, poultry litter, broiler crops, and carcass rinses. All four methods were able to detect *Salmonella* in Butterfield's solution, poultry litter, carcass rinse, and broiler crop matrices. For the broiler litter and Butterfield's solution, there were no ($P > 0.05$) differences among the four isolation protocols. However, in the carcass rinse and crop samples, there were no differences among the isolation of *Salmonella* when using Rappaport-Vassiliadis R10 broth, the secondary enrichment procedure, or tetrathionate broth, but all three were ($P \leq 0.05$) more successful at recovering *Salmonella* than the IMS method.

Bacteriophages

Antibody-based detection methods are considered to be the gold standard of affinity-based testing methods for detection of foodborne pathogens. However, the use of antibodies in diagnostics can be problematic, due in part to their physical structure and the methods used to synthesize the antibodies (Tombelli et al., 2007). For example, the generation of antibodies against nonimmunogenic molecules is often difficult, and antibodies are typically identified under physiological conditions that limit the extension to which the antibodies can be functionalized (Tombelli et al., 2007). Furthermore, batch-to-batch variation in binding activity (avidity) is typically observed with antibodies. Finally, antibodies are large proteins that can undergo irreversible denaturation when exposed to high temperatures (Tombelli et al., 2007).

Due to the issues observed with using antibodies to concentrate and detect foodborne pathogens, other affinity capture methods have been proposed. One such alternative takes advantage of the affinity that a bacteriophage has for its host cell. More specifically, components of the bacteriophage tail, which facilitates attachment to the bacterial cell, have been utilized in a new class of affinity-based bacterial concentration and detection assays. A bacterial capture and concentration method based on this principle has been developed for several foodborne pathogens. The method employs biotinylated phage proteins that attach to the target cells. The bacterial cell-biotinylated phage protein complex is mixed with streptavidin-coated paramagnetic beads, and the cells bind to the beads via the biotinylated phage proteins. The cells can then be removed from the solution by using magnetic separation. The complexes are washed in order to remove possible food residues and can be directly plated onto an appropriate selective and differential agar, or alternatively, the target bacteria can be detected by a rapid immunological or molecular method.

Hammack (2008) recently evaluated the bacteriophage-based capture method for its ability to capture *Listeria*. In an inclusivity study, all 60 *Listeria* spp. strains tested were positive with the assay. In the exclusivity study, all 42 non-*Listeria* strains grown in brain heart infusion broth were negative with the assay. In internal and independent laboratory studies, the alternative method compared favorably to the International Organization for Standardization 11290-1:1996 method (Anonymous, 1996) for the detection and enumeration of *L. monocytogenes* cells. Salami, smoked salmon, and Camembert were naturally and artificially contaminated with different *Listeria* species. No significant difference was detected between the methods. The overall sensitivity rate was 97%, and the specificity was 100% (Hammack, 2008).

The bacteriophage binding technology has also been exploited to produce ELISA-like assays in which binding proteins from bacteriophage tails are used for the targeted capture and detection of bacteria from a sample (Byrne, 2008).

Nucleic Acid-Based Methods Used in the Food Chain

Over the past 2 decades, major advances have been made in the development of nucleic acid-based methods for detection of pathogens in foods. These methods offer several advantages over traditional microbiological methods, including increased speed, sensitivity, and specificity. Both DNA and RNA can be analyzed by molecular methods to assess the presence of specific pathogens in a sample and to determine viability. Furthermore, nucleic acid-based methods can be used to quantify pathogens in food and environmental samples. Methods based on the PCR, nucleic acid hybridization, and microarray technology have greatly expedited the process of pathogen detection and enumeration.

PCR

PCR-based assays are now accepted methods for rapidly confirming the presence or absence of specific pathogens in foods and other types of samples. The significant feature of the PCR is that a unique template sequence of the target organism(s) can be repeatedly replicated in the presence of DNA from nontarget microorganisms found in food samples. Commonly, a target gene sequence unique to the species of interest is selected, and it is frequently a virulence gene. Information on PCR technology and use of the PCR for pathogen detection is available from a number of sources (Sambrook and Russell, 2001; Sachse and Frey, 2003). Following the PCR, the amplified product, the length of which is defined by the location of the primer binding sites, can be detected in a number of ways.

Until recently, the most commonly used procedure for product detection was agarose gel electrophoresis with ethidium bromide staining, which allows visualization of products due to the ability of the stain to intercalate into the DNA. Accurate size determination of the amplicons is accomplished by comparison to a set of size markers that are applied to adjacent wells of the same gel. Other nucleic acid stains such as SYBR Green have also been used in gel electrophoresis to visualize the amplicon. ELISA formats have also been developed for detection of amplicons. For example, PCR amplification can be performed with a digoxigenin-labeled primer and a biotinylated probe. The biotin-digoxigenin hybrids can then be quantified through binding to streptavidin-coated microtiter plates followed by detection using a peroxidase-labeled antidigoxigenin antibody (Gutierrez et al., 1998). Detection of an *L. monocytogenes*-specific *hlyA* gene product was achieved using an antibody directed to RNA-DNA hybrids formed upon hybridization with a probe bound to a microtiter plate (Blais, 1994). PCR products are labeled with biotin and ruthenium(II) trisbipyridal [Ru(bpy)3(2+)] and then captured by streptavidin-coated magnetic beads prior to electrochemiluminescent detection. In recent years, conventional PCR has advanced from end point detection of products to detection while the reaction is occurring. With "real-time" PCR, amplification and detection of product occur during the PCR (see below).

Multiplex PCR assays enable simultaneous amplification of multiple targets of interest in a single reaction using two or more primer pairs, resulting in savings in time, labor, and cost. However, compared to PCR assays in which one sequence is amplified, multiplex assays require more extensive optimization of annealing conditions and reagent concentrations for maximal amplification efficiency and to avoid the formation of spurious amplification products and uneven amplification of target sequences. Primers with nearly identical annealing temperatures and which do not display homology either internally or to one another should be used.

Multiplex PCR assays have been used to detect and/or identify one organism by amplification of more than one gene, or multiple organisms can be detected simultaneously by targeting unique sequences from each organism (Bhaduri and Cottrell, 1997; Fratamico et al., 1995; Fratamico and Strobaugh, 1998).

There are a number of advantages of PCR-based methods for detection of foodborne pathogens over traditional culture methods and immunoassays. With the choice of appropriate primers targeting unique genetic markers, PCR assays can be designed to be very specific for the target organism. This contrasts with cross-reactions of the antisera that can occur with immunoassays. Due to the relatively high sensitivity of PCR assays, preenrichment and enrichment steps could potentially be eliminated completely or shortened considerably, so that test results are available within the same working day (Josefsen et al., 2007). Instead of plating and performing biochemical and serological tests, the PCR may be performed directly on food samples after a short enrichment step (Koch et al., 1993). Viruses and parasites are more difficult to recover from foods, and enrichment culturing in liquid medium, as is performed to increase levels of bacteria, is not possible with foodborne viruses and parasites. Nucleic acid-based amplification systems are a promising alternative to cumbersome, time-consuming, and less sensitive traditional methods used for detection and identification of viruses and parasites. The PCR is also a functional tool for detection and identification of bacteria that cannot be identified by culture techniques because they are nonculturable or cannot be distinguished from other bacteria by culture techniques (Deng et al., 1996; Relman et al., 1992; Tsen et al., 1998). Furthermore, PCR-based methods are amenable to automation, leading potentially to more cost-effective, high-throughput testing of foods for the presence of undesirable microorganisms.

The usefulness of the PCR for detection of microorganisms in food or other complex matrices is limited by the presence of substances such as bilirubin, bile salts, hemoglobin degradation products, polyphenolic compounds, proteinases, complex polysaccharides, and fat, which inhibit the DNA polymerase, bind magnesium, or denature the DNA (Monteiro et al., 1997; Rossen et al., 1992). Sensitivity is dramatically decreased if the PCR is performed on crude samples containing inhibitors; therefore, sample preparation steps and/or DNA extraction are often required prior to performing the PCR.

Various sample preparation methods have been used, including IMS, centrifugation, or filtration, and a number of DNA extraction kits and reagents are available. Appropriate controls should be performed in each assay to assess if negative results could be due to the presence of PCR inhibitors in the sample. The PCR can detect DNA from nonviable as well as from live target cells. With enrichment culturing, however, only live cells multiply, thus diluting out the presence of dead cells. Nevertheless, the PCR may not be suitable for testing pasteurized, cooked, or irradiated samples, which may contain considerable numbers of dead microorganisms (Arnal et al., 1998; Josephson et al., 1993). Detection of only viable cells can be accomplished by detecting bacterial mRNA with reverse transcription PCR (RT-PCR) (see below). However, the instability of mRNA and reproducibility are problems encountered with the use of RT-PCR on food samples. Another technique used to discriminate DNA from live and dead cells is referred to as ethidium monoazide-PCR. The ethidium monoazide can penetrate cells with compromised cell membranes and covalently binds to the DNA, preventing its amplification by the PCR (He and Chen, 2010).

Real-time PCR

Real-time PCR systems rely upon detection and quantification of signal generated from a fluorescent reporter, which increases in direct proportion to the amount of PCR product. The cycle threshold (C_t) is defined as the first cycle in which there is a significant increase in fluorescence above a specified threshold, and plotting fluorescence against cycle number yields a curve that represents the accumulation of PCR product over the duration of the PCR. A standard curve can be generated from C_t values for a series of reactions using dilutions containing known quantities of target DNA. The amount of target in unknown samples can be derived by measuring the C_t and using the standard curve to determine the starting copy number.

A number of fluorescence systems have been employed for real-time PCR, and among the various chemistries available, intercalating dyes, such as SYBR Green I or SYTO 9 are the most economical and convenient to use. SYBR Green I is a thermostable intercalating cyanine dye that binds to double-stranded DNA and emits a fluorescent signal that increases proportionately with the amount of PCR product. However, one drawback is that the dye is nonspecific; thus, any nonspecific product that is obtained will result in an incorrect increase in the fluorescent signal. To overcome this drawback, an additional time-temperature program after the PCR is used to generate a melting curve to determine the quality and accuracy of the real-time PCR data. Melt curves can distinguish between specific and nonspecific products based on the melting temperature of the amplicons, since each amplification product melts at different temperatures based on the length and GC content.

TaqMan probes are short oligonucleotides that contain a 5′ reporter dye and a 3′ quenching dye. When intact, the fluorescence of the reporter dye is suppressed by the quencher. During the PCR, the TaqMan probe hybridizes to a complementary sequence on the template between the two primer binding sites. When *Taq* polymerase encounters the bound probe during extension from the primer, it cleaves the probe due to its 5′-to-3′ exonuclease activity, separating the reporter dye from the quencher. The fluorescence increases in proportion to the quantity of amplicons generated. TaqMan-based real-time PCR assays have been used for detection of foodborne pathogens (Cheung et al., 2004; Krause et al., 2006; Sharma, 2002; Vishnubhatla et al., 2000; Ward and Bej, 2006).

Minor groove binder probes, which have a minor groove binder ligand at the 5′ end and a fluorophore at the 3′ end, provide higher specificity and sensitivity than do TaqMan probes (Afonina et al., 2002; Reynisson et al., 2006; Yao et al., 2006). Other real-time PCR systems employ molecular beacons, which are stem-and-loop shaped probes consisting of a PCR product-specific portion (loop, 20 to 24 nucleotides) and complementary stem sequences (4 to 6 nucleotides) that hold the probes in a hairpin configuration at low temperatures. At high temperatures, the PCR product and probe are single stranded, and as the temperature is lowered the molecular beacon hybridizes to the PCR product, resulting in a conformational change that spatially separates the quencher from the reporter dye, and a light signal from the reporter dye can be detected (Chen et al., 2000; Liming and Bhagwat, 2004; Vet et al., 1999).

The fluorescence resonance energy transfer system employs two probes, one labeled with a donor fluorochrome (fluorescein) at the 3′ end and the other labeled with an acceptor dye at the 5′ end. The probes are designed to hybridize to the template during the annealing phase of the PCR in a head-to-tail arrangement. When both are hybridized so that they are in close proximity to one another, the energy emitted from the donor excites the acceptor fluorophore, which then emits red fluorescent light at a different wavelength, which can be detected (Palladino et al., 2003; Reischl et al., 2003). Other fluorescence systems employ Scorpion probes (Stroup et al., 2006), LUX (acronym for "light upon extension") primers (Gilmour et al., 2006), and Amplifluors (Rodríguez-Lázaro et al., 2004a).

In real-time multiplex PCR assays, probes that bind to their respective PCR products are labeled with different fluorophores that have unique emission spectra. Josefsen et al. (2004) compared real-time PCRs using two different instruments for detection of thermotolerant *Campylobacter* species *C. jejuni, C. coli*, and *C. lari* in naturally contaminated chicken rinse samples. There was good agreement between PCR assays performed with both instruments and with an International Standard Organization-based culture method. A multiplex real-time RT-PCR assay was used for the simultaneous detection of noroviruses of genogroups I and II, human astroviruses, and enteroviruses by use of the LightCycler and the SYBR Green dye (Beuret, 2004). The real-time PCR was 10 times more sensitive than conventional endpoint PCR. Fratamico and DebRoy (2010) employed multiplex PCR assays using TaqMan probes and primers targeting the stx_1, stx_2, and wzy_{O157} genes in combination with probes and primers targeting either the $fliC_{h7}$ or the *eae* genes to detect *E. coli* O157:H7 in apple cider, raw milk, lettuce, and ground beef. The pathogen was detected in samples inoculated with 2 or 20 CFU/g or 25 ml after both 8 h and 20 h of enrichment.

Quantitative PCR

Quantitative PCR is based on the exponential increase in the initial DNA amount with the number of PCR cycles performed rather than on the end point signal. A standard curve is set up by using serial dilutions of known numbers of target copies, which is then used to determine an unknown amount of DNA in a sample. Proper assay controls are important to the reproducibility of quantitative PCRs. Quantitative PCR methods to determine the level of pathogens in foods, carcass swabs/rinses, and environmental samples have been described (Arnal et al., 1998; Fernández et al., 2006; Guy et al., 2006; Park et al., 2009; Wolffs et al., 2006). Takahashi et al. (2005) developed a quantitative real-time PCR-based assay targeting the *toxR* gene of *Vibrio vulnificus* and used it to detect and quantify the pathogen in seawater and oyster samples. A standard curve was constructed from samples containing known concentrations of *V. vulnificus*, and this curve was then used as a reference standard for extrapolating quantitative information from seawater and oyster samples. The assay could detect as few as 10 cells/ml of seawater and oyster homogenate.

Methods for quantitative PCR include competitive PCR, which involves the addition of various amounts of a DNA sequence bearing the same primer annealing sites as the template. The degree to which the added DNA competes will depend on its initial concentration relative to that of the test sequence. The amount of initial target sequence can be obtained from the point on the curve where target and standard values are equal (Li and

Drake, 2001). The endogenous standard assay utilizes an endogenous sequence, which is expressed at a relatively constant level in all of the samples. The level of amplification of the target sequence is then compared to that of the standard. However, problems with this technique are that (i) it is a comparative technique, and the actual copy number cannot easily be determined; and (ii) the endogenous standard may be expressed at levels far higher than the target sequence, making dilutions of the sample necessary. The synthetic internal standard method involves the addition of synthetic standards designed to carry very slight sequence variations, which make them readily distinguishable from the sequences of interest. Usually, a unique restriction site is created in the synthetic standard, such that after amplification and restriction enzyme digestion, only the standard will yield two small digestion products.

RT-PCR

RT-PCR is used for RNA detection and quantification. Since RNA cannot serve as a template for the PCR, RT is combined with the PCR to make RNA into a cDNA molecule suitable for PCR amplification. The RT-PCR can differentiate between viable and nonviable bacterial cells if mRNA serves as the target of amplification. However, bacterial mRNA has a very short half-life and is rapidly degraded with processes that render the cells nonviable. Klein and Juneja (1997) used RT-PCR for detection of viable *L. monocytogenes* based on the *iap* gene. A 2-h enrichment was necessary to increase the sensitivity of the RT-PCR, and their results showed that nonviable cells would have been detected following short enrichment periods if PCR had been used. Sheridan et al. (1998) studied the relationship between viability and detection of specific *E. coli* mRNAs. They found that the type of cell inactivation treatment and subsequent holding conditions influenced the ability to detect mRNA targets; however, the presence of 16S rRNA was detected in samples containing dead cells, and it persisted for long periods.

RT-PCR was used to detect viable *S. enterica* serovar Typhimurium in pork by using a SYBR Green I real-time RT-PCR assay (Techathuvanan et al., 2010). RT-PCR assays are used for detection of RNA viruses, such as hepatitis A virus. As an example, real-time RT-PCR was used to identify oysters contaminated with noroviruses as the cause of an international outbreak and to quantify the level of viruses in the oysters (Le Guyader et al., 2006).

Novel or Alternative Nucleic Acid Amplification Methods

Various alternative nucleic acid target amplification methods have been developed. The ligase chain reaction (LCR) involves the formation of new target DNA molecules by a thermostable DNA ligase, two adjacent forward primers, and their complements. Typically, there is a gap of 1 to 3 bases between adjacent primers, and these act as templates for ligation by the DNA ligase, which is intolerant to base mismatches. The LCR is cyclic, with ligated product serving as template for the next reaction; product is formed without DNA replication. LCR is useful for discriminating between DNA sequences differing in a single base pair and thus is useful for single nucleotide polymorphism detection.

Wiedmann et al. (1992) used a PCR-coupled LCR assay to distinguish *L. monocytogenes* from other *Listeria* species by targeting a single base pair difference in the *Listeria* 16S rRNA gene. The LCR is also used in clinical diagnoses, and a kit for detection of *Mycobacterium tuberculosis* is commercially available (León Muiños et al., 2004).

The isothermal strand displacement amplification (SDA) technique can result in about 10^8-fold amplification of the target after 2 h at ca. 40°C. A restriction enzyme is used to nick a hemiphosphorothioated recognition site, and a DNA polymerase initiates synthesis of a new DNA strand at the nick while displacing the existing strand. The modified target sequences are amplified exponentially by repeated nicking, strand displacement, and priming of displaced strands. Ge et al. (2002)

combined strand displacement amplification of the *hlyA* gene of EHEC including O157:H7 and other serotypes with detection of product by fluorescence polarization, enabling detection of the pathogen within 35 min of reaction initiation with a detection limit of 4.3 bacteria.

With the nucleic acid sequence-based amplification (NASBA) technique, there is isothermal (37 to 42°C) amplification of an RNA template. The reaction consists of continuous cycles of reverse transcriptase-mediated synthesis of cDNA from an RNA target sequence, followed by in vitro transcription by RNA polymerase from the double-stranded cDNA template. The coordination of three enzymes are required: reverse transcriptase to synthesize the cDNA, RNase H to digest the RNA strand of RNA:DNA heteroduplexes, and T7 RNA polymerase to synthesize as many as 100 copies of RNA used as substrate in the next cycle. NASBA targeting mRNA of *L. monocytogenes* inoculated into dairy and egg products successfully detected the organism at an initial inoculum level of <10 CFU/g (Blais et al., 1997).

NASBA has been employed for detection of *Campylobacter* spp., *L. monocytogenes, Salmonella* spp., *Cryptosporidium parvum*, and foodborne viruses by using 16S rRNA and mRNA as target molecules (Cook, 2003). The technique was used to detect viable *S. enterica* in fresh meats, poultry, and other foods by targeting mRNA transcribed from the *Salmonella dnaK* gene by using a NASBA assay with electrochemiluminescence detection of product (D'Souza and Jaykus, 2003; Simpkins et al., 2000). Real-time molecular beacon-based NASBA assays were used to detect *C. jejuni* and *C. coli* in chicken (Churruca et al., 2007) and *Mycobacterium avium* subsp. *paratuberculosis* in water and milk (Rodríguez-Lázaro et al., 2004b). A multiplex NASBA method for simultaneous detection of hepatitis A and genogroup I and II noroviruses in ready-to-eat foods has been described (Jean et al., 2004).

Rolling-circle amplification (RCA) and ramification amplification (RAM) employ processes similar to viral replication. RCA uses restriction digestion and exonuclease digestion to produce single-stranded DNA targets, which can then hybridize to specific oligonucleotides called padlock probes. The padlock probes circularize following hybridization to target DNA or RNA sequences and are then ligated to provide a template for RCA employing a DNA polymerase with strand displacement activity. The padlock probe contains a specific target sequence that is detected using a fluorescent complementary probe. RCA was used to detect and enumerate microorganisms (Maruyama et al., 2005; Wang et al., 2005). If a second strand displacement primer is encoded in the circular probe, the process grows exponentially, and the technique is referred to as RAM. RAM results in a large branching or ramified DNA complex. Using a circular probe specific for the Shiga toxin 2 gene, Li et al. (2005a) were able to detect as few as 10 *E. coli* O157:H7 cells per reaction.

The Q-beta replicase amplification method was named after the enzyme that replicates the RNA genome of bacteriophage $Q\beta$ and is used for amplification of RNA probe molecules. A section of the $Q\beta$ genome is replaced with an RNA probe specific for the target sequence, such that the unique folded structure of the genome is not altered. Probe molecules, which specifically anneal to the target sequence, are enzymatically replicated and then detected (Wolcott, 1992). The level of amplification can approach 10^9-fold in a 30-min incubation time; however, nonspecific amplification is a limitation of $Q\beta$ replicase assays.

Notomi et al. (2000) described a method referred to as loop-mediated isothermal amplification (LAMP) for DNA amplification using DNA polymerase with strand displacement activity and four to six primers to amplify the target gene under isothermal conditions with high specificity, efficiency, and rapidity. In the first step, DNA synthesis from the two sets of inner primers and outer primers results in the formation of a dumbbell-like DNA structure. Subsequently, one inner primer hybridizes to the loop on the product and initiates strand displacement synthesis, yielding the original stem-loop structure and another with a longer stem. The amplification products are stem-loop DNA structures with several inverted repeats of the target and cauliflower-like structures with multiple loops. The cycling reaction, performed under a single temperature of 60 to 65°C, results in accumulation of 10^9 copies in less than 1 h. Amplification products can be detected by turbidity because a by-product of the reactions, magnesium pyrophosphate, is insoluble, and turbidity correlates with the amount of DNA synthesized.

Alternatively, DNA intercalating dyes can be used to measure fluorescence generated with the LAMP reaction. Ohtsuka et al. (2005) used a LAMP assay to detect *Salmonella* in naturally contaminated liquid egg samples. *Salmonella* was detected in a higher number of samples via the LAMP assay than via a conventional PCR assay. A LAMP assay targeting *stx* genes was more sensitive for detection of Shiga toxin-producing *E. coli* O157 and O26 in ground beef and alfalfa sprouts than plating following IMS (Hara-Kudo et al., 2008).

Isothermal and chimeric primer-initiated amplification of nucleic acids (ICAN) is an isothermal amplification method that employs chimera primers composed of RNA at the 3′ end and DNA at the 5′ end, RNase H, and DNA polymerase. First, the primers hybridize to the target gene, and nucleotide replication is initiated by the DNA polymerase that has SDA. An intermediate product is formed after a template exchange reaction. RNase H then cuts out RNA portions derived from the chimeric primers. This is followed by an elongation reaction beginning from the site at which the RNA portions were cut out and is accompanied by strand displacement and template switching. This chain reaction is repeated, amplifying genes of interest with high specificity and efficiency. Isogai et al. (2005) described the detection of *Salmonella* spp. from chicken carcasses, egg yolk, and cattle fecal samples targeting the *invA* gene by the PCR and by the ICAN method combined with detection by ELISA. The number of bacteria detectable by the ICAN technique was less than 50 CFU/ml of chicken rinse solution. Twelve of 14 chicken rinse samples were positive by the ICAN technique; however, only 7 of the 14 samples were positive by the standard PCR assay.

DNA Hybridization and DNA Microarrays

Nucleic acid hybridization techniques employing labeled gene probes for the detection and identification of foodborne pathogens have been described (Dooley, 1994; Olsen et al., 1995). Hybridization probes are labeled nucleotide sequences ranging in length from 15 to 20 to several thousand base pairs that hybridize to complementary target sequences. The resulting nucleic acid hybrids can then be detected in a number of ways, including isotopic, chemiluminescent, chromogenic, or fluorescent techniques. The sensitivity of gene probe hybridization assays is in the range of 10^4 to10^6 target gene copies. Several gene probe assay kits for detection of *Listeria* spp., *E. coli, Salmonella, Staphylococcus, Campylobacter,* and *Yersinia enterocolitica* are commercially available. Lehtola et al. (2006) employed peptide nucleic acid probes in a fluorescence in situ hybridization method to detect *Mycobacterium avium* in spiked water samples and biofilms containing the pathogen.

As a nucleic acid-based detection technology, DNA microarrays offer the potential for parallel detection of many pathogens simultaneously or many genetic sequences of a single pathogen simultaneously, providing the capability of obtaining detailed genomic information on the pathogen. DNA arrays employed for pathogen detection consist of large numbers of probes (oligonucleotides or PCR products) immobilized on specially treated glass slides. DNA samples for analysis are chemically labeled with fluorescent dyes or are labeled during PCR amplification and are then hybridized with their complementary sequences on the chip. Following hybridization and washing steps, the arrays are examined using a high-resolution scanner. Microarrays can be used to detect labeled PCR products by hybridization to an array composed of pathogen-specific probes.

Sergeev et al. (2004) used degenerate primers to amplify variable regions of *Staphylococcus aureus* enterotoxin genes and hybridized chemically labeled single-stranded DNA (derived from PCRs specific for enterotoxin genes) to microarrays onto which probes specific for each of the different enterotoxin genes were bound. Call et al. (2001) were able to detect 55 CFU of *E. coli* O157:H7 per milliliter of chicken rinse without an enrichment step by using immunomagnetic capture followed by PCR amplification and hybridization of the products onto a microarray containing oligonucleotide probes (25- to 30-mer) complementary for four virulence genes of *E. coli* O157:H7. The microarray assay was 32-fold more sensitive than gel electrophoresis for PCR product detection.

Chizhikov et al. (2001) detected microbial virulence factors (*eaeA, slt-I, slt-II, fliC, rfbE,* and *ipaH*) of multiple pathogens by hybridizing Cy5-labeled fluorescent PCR products to the gene-specific oligonucleotides spotted on microarrays. A similar approach was used by Keramas et al. (2004) to detect *Campylobacter* spp. directly from fecal cloacal swabs targeting the 16S rRNA, the 16S-23S rRNA intergenic region, and specific *Campylobacter* genes. Other investigators have also reported on the use of DNA microarrays combined with the PCR to detect pathogens, including *Yersinia enterocolitica* and *Enterobacter sakazakii* (now *Cronobacter sakazakii*), in food and for identification of pathogens to the species level (Liu et al., 2006; Myers et al., 2006; Volokov et al., 2002, 2003).

A DNA microarray was developed to detect enterotoxigenic *E. coli* (ETEC) enterotoxin genes and the 19 most common ETEC serogroups (Wang et al., 2010). Genomic DNA was labeled using a random PCR strategy and was hybridized to probes complementary to sequences in ETEC O antigen *wzx* genes and to ETEC toxin genes spotted on glass slides.

Cost, need for skilled personnel, limited sample throughput, and reproducibility issues are obstacles in the widespread use of DNA microarrays for pathogen detection, identification, and typing; however, alternative microarray formats, such as bead arrays/suspension microarray technology may circumvent the cost and high-throughput limitations. Specific capture moieties are covalently coupled to the surface of microspheres that are embedded with different ratios of red and infrared dyes to yield unique spectral addresses.

Within an automated flow cytometer, lasers excite the internal dyes that identify each microsphere particle and also any reporter dye captured during the assay. This technology allows rapid multiplexing of up to 100 unique assays within a single sample. A multiplexed bead-based suspension assay was used for molecular determination (but not detection) of common *Salmonella* serogroups (Fitzgerald et al., 2007). Bead arrays were also used to screen environmental samples for biothreat agents and to detect respiratory viruses (Washington et al., 2010; Wilson et al., 2005).

Lab-on-a-Chip PCR

Nanotechnology, as it applies to diagnostics, is a rapidly developing new analysis technology that is driving the development of nanochip, nanoarray, and lab-on-a-chip systems. Procedures involving on-chip PCR amplification and detection of products have been described (Kricka, 1998, 2001; Nickisch-Rosenegk et al., 2005). The PCR described by Nickisch-Rosenegk and coworkers (2005) involved PCR amplification with primers immobilized on a chip, and the products were visualized through the use of intercalating dyes or by confocal microarray scanning or fluorescence microscopy using Cy5-dye fluorescence of the modified free primer. Moreover, microchip devices are being developed that integrate cell lysis, multiplex PCR amplification, electrophoretic separation of PCR products, and product detection (Cheng et al., 1996; Waters et al., 1998; Woolley et al.,

1996). An integrated plastic microfluidic device combined DNA amplification and injection of amplicons through a gel valve, followed by on-chip electrophoretic separation and detection of labeled PCR products using laser-induced fluorescence (Koh et al., 2003). The system was used to detect *E. coli* O157 and *S. enterica* serovar Typhimurium, and the limit of detection was six copies of target DNA.

Liu et al. (2004) described a completely self-contained integrated biochip device that was used to perform sample preparation, followed by PCR and product detection. *E. coli* was inoculated into whole-blood samples, and after IMS and cell lysis, target sequences were detected following PCR and DNA-microarray detection. Westin et al. (2000) developed a microchip array with primers anchored to the surface, which were used to perform multiplex SDA-based amplification and detection in situ on the chip. Cremonesi et al. (2009) employed a DNA chip based on a ligation detection reaction-universal array assay to detect and differentiate a number of pathogens directly from milk. A review by Liu et al. (2009) describes advances in integrated microfluidic pathogen detection platforms and microfluidic processing steps. Microminiaturization of nucleic acid-based techniques will have a significant impact on diagnostic and food testing, potentially enabling the assays to be performed by relatively unskilled operators in nonlaboratory settings.

BIOSENSORS

Various strategies have been used to detect multiple pathogens simultaneously on assay platforms such as those using nucleic acid- or protein-based microarray systems (Gehring et al., 2006; Kim et al., 2009; Suo et al., 2010). These systems require prior knowledge of the agents so that a specific set of probes (genetic or antibody) can be used on the array. The array-based assays are robust because multiple probes for a given target pathogen can be immobilized on the array platform. Still, these methods have several drawbacks, including that they may fail to differentiate pathogens from nonpathogens or active toxins from inactive toxins, they may give false-positive results due to nonspecific binding of probes (Byrne et al., 2009; Moreno et al., 2001), they do not provide any information on the biological activity of the toxins or the biothreat select agents, and finally, they are unable to detect or identify unknown or genetically altered agents (Rasco and Bledsoe, 2005). To overcome such problems, alternative label or probe-free detection methods can be used. Functional biosensing systems such as cell-based sensors (Banerjee and Bhunia, 2009; Ziegler, 2000) and spectral technologies such as flow cytometry (Kim et al., 2009), surface enhanced Raman scattering (Kudelski, 2008), Fourier transform infrared spectroscopy (Berthomieu and Hienerwadel, 2009), and light-scattering technology (Banada et al., 2009) are considered promising alternatives (Bhunia, 2008; Bhunia et al., 2010).

Cell-Based Sensors

In cell-based sensors, bacterial or mammalian cells serve as transducers, and often these cells are engineered to exhibit pathogen/toxin-specific responses. When these cells are exposed to microbes, toxins, or hazards, they will emit agent-specific signals in the form of luminescence or color, which can be sensitively detected using a spectrometer. The specificity of the sensors depends on the construction of engineered prokaryotic or eukaryotic cells. In this section, the focus will be on the development and use of mammalian cell-based biosensors (CBBs) for detection of pathogens and toxins.

In the event of a foodborne outbreak resulting from natural or intentional contamination, prompt acquisition of functional properties of infective agents would help institute proper response to prevent further illnesses or fatalities. Thus, a rapid and sensitive method capable of detecting a broad range of pathogens or toxins would be very attractive. Generally, this can be achieved by the use of a functional biosensing element such as cultured mammalian cells that are sensitive to pathogens or toxins at minute quantities (Banerjee and Bhunia, 2009; Stenger et al., 2001). A mammalian CBB monitors the aberrations in normal physiological activities of mammalian cells upon exposure to toxicant(s); this method is considered an emerging and promising approach to determine the functionality or biological activity of an analyte (Banerjee and Bhunia, 2009; Stenger et al., 2001; Rider et al., 2003; Tencza and Sipe, 2004; Ziegler, 2000).

Major advantages of CBBs include the ability to distinguish viable from nonviable microbes, pathogenic from nonpathogenic species of the same genera, and active from inactive toxins. A living cell system can provide some of the key physiologically relevant information about a pathogen/toxin such as those that cause membrane damage, alter signal transduction pathways, promote receptor-ligand interactions, activate secondary messenger pathways, and promote changes in gene expression in response to a foreign substance (Banerjee and Bhunia, 2009; O'Shaughnessy and Pancrazio, 2007; Stenger et al., 2001).

Recent developments in the development of CBBs include the use of a B-lymphocyte cell line, Ped-2E9, in a three-dimensional cell culturing configuration in a 96-well plate for detection of membrane active toxins such as listeriolysin O (*L. monocytogenes*), enterotoxin (*Bacillus* spp.), α-hemolysin (*Staphylococcus aureus*), phospholipase C (*Clostridium perfringens*), and cytolysin from the sea anemone *Stoichactis helianthus* (Banerjee, 2008; Banerjee et al., 2008; Bhunia et al., 1994; Gray et al., 2005; Kim et al., 2009; Shroyer and Bhunia, 2003) at nanogram quantities.

A novel genetically engineered B-cell-based biosensor expressing pathogen-specific antibodies (receptors) on the surface was also reported for detection of *E. coli* O157:H7, *Bacillus anthracis, Yersinia pestis*, and vaccinia virus (Rider et al., 2003). Using membrane engineering technology, Kintzios et al. (2001) inserted pathogen-specific bioreceptors into mammalian cell membranes by electroporation and measured the changes in membrane potential by using a bioelectric recognition assay as specific interaction of viruses with the CBB occurred. Likewise, mast cells were also engineered to express chimeric protein on the cell membrane to monitor specific interaction with *L. monocytogenes* and *E. coli*, which emitted fluorescence signals (Curtis et al., 2008).

Light Scattering

Spectral signature-based analytical methods allow pathogen detection and identification primarily based on biophysical properties of microorganisms. Lasers have been used widely to generate spectral signatures of biological materials including microorganisms for detection and identification purposes. Depending on the methods and application, spectral methods may use antibody or nucleic acid probes to provide specificity (Banada and Bhunia, 2008) and fluorophores to emit signals (Bhunia, 2008). Examples of such applications include flow cytometry (Kim et al., 2009), fiber optic sensors (Bosch et al., 2007; Leung et al., 2007), surface enhanced Raman scattering (Kudelski, 2008), and Fourier transform infrared spectroscopy (Berthomieu and Hienerwadel, 2009).

In a novel application, a laser was used to generate scatter signature of bacterial colonies growing on a petri dish in real time (Banada et al., 2007, 2009). The light scattering technology designated BARDOT (an acronym for "bacterial rapid detection using optical light scattering technology") generates forward scattering patterns that are highly sensitive to minute genetic differences in bacteria and to resulting biophysical differences in their colonies (Bae et al., 2007; Banada et al., 2009). BARDOT identifies unknown bacteria based on the built-in scatter image library. Also, one can easily build image libraries based on specific applications. BARDOT can detect multiple pathogens in a single run, since it does not require any probes or labeling reagents.

In addition to detection of microorganisms, another use of light scattering is for microbiological quantification. For example, Jones and colleagues (1998) studied the quantification of yeast cells by use of light scattering. The authors investigated the relationship between cell counts and light scattering across a wide range of concentrations of yeast cells in suspension. A supervised learning model was used to predict cell counts by light scattering.

Regardless of the type, biosensor applications are gaining popularity and are regarded as the future of pathogen detection technologies. Traditional pathogen detection technologies, while sensitive enough, are presently too slow, costly, or complicated to fulfill all the needs of the food industry (Lazcka et al., 2007). While more research is needed to overcome issues such as the complexity of the systems, their cost, and the need for sample concentration and enrichment, their potential for label-free, rapid, simple, and inexpensive detection of pathogens makes them an exciting and important new technology (Acharya et al., 2006).

BIOCHIPS

Biochips generally consist of glass or plastic surfaces to which capture molecules such as antibodies, antibody fragments, DNA, RNA, aptamers, or peptide nucleic acids are attached. The binding of captured analytes can then be detected optically (e.g., by using surface plasmon resonance, fluorescent labels, or enzymes that generate an optically active product), electrochemically, or through mass changes. The advantages of the use of biochips include small sample volume requirements, capacity to measure multiple analytes simultaneously, incorporation of standard/controls with in situ correction for matrix-associated effects, ability to link to microfluidics-based formats for sample preparation, handling, and detection, very high sensitivity, and the availability of multiple formats and potential applications.

In any assay system, key features are specificity, sensitivity, and robustness in complex matrices. To achieve optimal performance, the incorporation of a specific recognition or binding ligand is often a prerequisite. This recognition unit may play a number of roles including facilitating capture and concentration of the analyte and the eventual measurement process. Antibodies have played a major role as recognition ligands due to their specificity, and a range of antibody types, i.e., polyclonal, monoclonal, and more recently, recombinant, are commonly used (Conroy et al., 2009; Leonard et al., 2007). The last antibody type offers potential advantages including the availability of a range of derivative forms such as Fab, scFv, and scAb and bifunctional, bispecific, and fused forms with either fluorescent proteins or enzymes attached that can act as labels. The capacity for broadening or improving the specificity and the ability to incorporate a range of tags to assist isolation (e.g., His), immobilization (e.g., His and biotin), and orientation (His, biotin, and hemagglutinin) provide significant advantages over conventional antibody preparations. In addition, the incorporation of labels (e.g., enzymes or fluorescent proteins) and the ability to greatly enhance specificity (through high-throughput screening and highly selective genetic engineering using chain shuffling, site-directed mutagenesis, and error-prone PCR), linked to improved methods for modeling antigen-antibody interactions, have provided powerful tools to customize antibody derivatives for the particular application, and very significant increases in sensitivity and specificity have been achieved. Recent advances in new production methods, including use of ribosomal display and rapid selection, have reduced dependence on phage display and some associated issues with patenting. All these factors should lead to a very significant growth in the use of recombinant antibodies and other genetically improved binders. Other developments include increased use of a variety of hosts for production of antibody and antibody-derived fragments including chickens, sharks, and cameloids (camels, alpaca, and llama). Improvement in in vivo (e.g., in plants) and in vitro (e.g., cell-free) systems for large-scale production should also involve significant cost reductions.

However, there are a range of other binding ligands that are increasingly used, including aptamers and peptide nucleic acids. There are huge numbers of aptamers now available offering high sensitivity, specificity, stability, and the capacity to work successfully in complex materials (Daniels et al., 2004). Peptide nucleic acids also have significant potential, offering good specificity and stability and wide applications. Other binders that will increasingly be used include lectins, molecularly imprinted polymers, cell surface receptors, immobilized membrane-based binding complexes, immobilized ion channels, cells, enzyme complexes, and tissue extracts (O'Kennedy et al., 2010).

A key issue that must be addressed is the capture and concentration of analytes that are present in relatively low levels often in very large matrix volumes. Increasingly, capture ligands bound to plastic or other surfaces, columns, beads, or magnetic particles are used. These provide the capacity to bind multiple analytes on the surface simultaneously, allowing assay multiplexing. Use of beads coated with antibodies or other binders also greatly increases the possibility of capturing and enumerating low-level analytes. An advantage of the use of metal particles with attached binders is that they may be concentrated and collected by simple magnetic capture.

In relation to detection, the choice of label may well change from more traditional fluorescent or enzymatic labels to quantum dots and a range of other nanoparticles. A new focus is on fluorescent proteins, allowing a major increase in the use of bar coding for multiple component analysis (Conroy et al., 2009; O'Kennedy et al., 2010; Hearty et al., 2010).

There is also a marked increase in the use of hyphenated systems involving the exploitation of the advantages of both nucleic acids and binding ligands such as antibodies. Examples include immuno-PCR, whereby an antibody is labeled with a nucleic acid of a particular sequence. This sequence can be detected and amplified, leading to dramatic improvements in sensitivity. This approach also offers the potential to use multiple antibodies with different nucleic acid label sequences that can be independently amplified and determined, thus providing bar code-based analysis (Hearty et al., 2010). Other markers to be exploited will include a range of RNA-derived entities including transfer-messenger RNA and small interfering RNA (Dimov et al., 2008).

All of these systems will be employed in conventional formats including ELISA, fluorophore-linked immunosorbent assay, LFAs, and multiplexed bead-based assays. Increased use of high-sensitivity electrochemical detection systems, impedance, optical systems such as surface plasmon resonance, and carbon nanotubes will be evident (Dimov et al., 2008; Barton et al., 2009; Stapleton et al., 2009; Lynam et al., 2009). Biochips, CD-based formats, arrays and other miniaturized platforms will see significant future developments (Jokers et al., 2010). These systems will incorporate a range of isolation techniques, assays, and detection strategies, and the need for reagents will be minimized, thus allowing point-of-use analysis, lower costs, and the capacity for wireless and continuing monitoring of processes (Jokers et al., 2010).

DETECTION OF PATHOGENS IN AIR

Thus far, this review has focused on methods to isolate and detect pathogens in food samples. The presence of pathogens in air and the transmission of infections in air is an intriguing phenomenon, which, although subject to a never-ending debate, incidentally plays prominent epidemiological roles in husbandry and transmission of zoonotic microorganisms from the primary sources of infection, i.e., animals. Examples of pathogens of domestic animals for which an airborne route of transmission is well documented are foot-and-mouth disease virus

(Gloster et al., 2005; Christensen et al., 2005) and Aujeszky's disease virus (Christensen et al., 1990), and airborne transmission is known to play a role—or suggested to play a role—for a number of zoonotic agents such as influenza A virus (Hammond et al., 1989; Tellier, 2006) potentially from pigs and poultry; severe acute respiratory syndrome (Li et al., 2005b); *Coxiella burnetii*, in particular from cattle (Acha and Szyfres, 2003; European Centre for Disease Prevention and Control, 2010); and Hendra virus (Playford et al., 2010).

The nature of aerosolization can be either as liquid droplets present in the exhaust of individuals with respiratory infections or as solid particles originating from the skin of animals or debris (i.e., in buildings due to the activity of animals). Detection in air might be a promising approach in situations where microorganisms entering an airborne state at the source of a subsequent spread often will be present in large quantities and with a uniform dispersal.

Air-Based Detection Technologies

The detection of microorganisms in air traditionally has been accomplished by sampling of airborne particles with subsequent analysis of the samples by a vast variety of detection methods. Principles of air sampling include solid and liquid impaction (May and Harper, 1957; Andersen, 1958; Errington and Powell, 1969), filter-based samplers (Fields et al., 1974), and electrostatic absorption (Gerone et al., 1966). As these methods of monitoring traditionally included separate steps of sampling, sample preparation, and detection analysis, such methods are both time-consuming and laborious.

Biosensor technologies suitable for pathogen detection that allow the integration of sampling and detection analysis are currently being developed and are expected to be soon made available. Some devices are expected to be suitable for noninvasive detection in the exhaust of infected individuals, while other devices will be suitable for permanent installation in buildings such as stables for automatic semicontinuous monitoring for adventitious microorganisms in the building. In the case of appearance of an adventitious pathogen a signal will be generated and transmitted to an appropriate receiver. Installation of an integrated detection device in ventilation systems in modern stables is a limited expense compared to the benefits of an early warning against infections such as the former OIE (World Organization for Animal Health) List A infections, zoonotic microorganism infections, and nonexotic infections, for which appropriate intervention can reduce disease spread, production losses, or disease in humans.

Biosensors integrating sampling and detection analyses might include a wide variety of combinations of unity operation solutions among which only some are suitable for detection in air. The principles of detection analysis could include immunoaffinity capture as well as nucleic acid amplification technologies, and principles of signal transduction include surface plasmon resonance (Phillips and Cheng, 2007), quartz crystal microbalance (Cooper and Singleton, 2007), and cantilevers (Waggoner and Craighead, 2007), as well as electrochemical-DNA (Lai et al., 2006). The former three principles can be used in combination with immunoaffinity capture (Skottrup et al., 2008) as well as nucleic acid amplification, while electrochemical-DNA signal transduction can be used only in combination with nucleic acid amplification. Suitable for detection in air are combinations that allow a large-volume throughput with the adsorption of the droplet nuclei at the focus of interaction between analytes and sensing elements such as in the chamber of a microchip.

One integrated principle of sampling and detection analysis named "integrated lab-on-a-chip" (ILOC) technology recently has been presented, which integrates within a single microchip the unity operations of electrostatic sampling of aerosols, electrolysis of microorganisms, and amplification of pathogen-specific genomic targets by PCR (Jensen et al., 2005). A signal is generated in the chip and can then be transmitted to a computer on location, a cell phone, or a satellite.

Feasibility of Monitoring in Air

The ILOC technology was recently used for sampling and detection of *Campylobacter* in broiler stables (Olsen et al., 2009). Due to its microaerophilic nature, *Campylobacter* is not considered an airborne pathogen. Hence, the application in broiler houses demonstrates the possibility of real-time monitoring of airborne infection transmission; the mere presence in air of debris of microorganisms may serve as an indicator of microorganisms colonizing herds and, hence, might be an interesting target of sampling (Olsen et al., 2009). Due to the frequent sectioning in modern farming of synchronized populations of animals, high concentrations of airborne microorganisms may build up quickly. These both stress the need for stringent monitoring in primary production at relevant time points and the advantage of this approach to exclude the presence at the source of adventitious pathogens in the farm-to-fork chain.

Another potential benefit of monitoring in air is that in ventilated rooms the size fraction of particles of $<10\ \mu M$ with a capacity to remain airborne for extended periods of time might quickly be widely distributed in the space of the room. This means that the introduction of an infectious organism in one section of a building might quickly be reflected by its presence in air at any spot in the building. This probably is the explanation why monitoring for the presence of *Campylobacter* in the air of broiler houses using the ILOC technology allowed the detection at the same time or before detection in fecal samples collected conventionally by wearing socks on the outside of the shoes (Olsen et al., 2009).

CONCLUSIONS: APPLICATIONS AND LIMITATIONS OF NEW DETECTION TECHNOLOGIES

Almost all rapid methods are designed to detect a single target, which makes them ideal for use in quality control programs to quickly screen large numbers of food samples for the presence of a particular pathogen or toxin. A positive result by a rapid method, however, is only regarded as presumptive and must be confirmed by standard methods. Still, the emphasis in food microbiology is shifting from methods that detect a single target to methods that detect the presence of multiple pathogens simultaneously. Such methods may also require that different classes of microorganisms (bacteria, viruses, and parasites) be detected at the same time. It is likely that such requirements will pose challenges to the current rapid methods, due to issues associated with the fact that some microorganisms (bacteria) can be enriched, while others (viruses and parasites) are difficult, if not impossible, to culture. As such, it is likely that molecular amplification assays including PCR, LAMP, and NASBA will dominate future diagnostic development, when multiplex detection is a requirement. Most rapid methods can be completed within a few minutes to a few hours, so they are more rapid than traditional methods. However, such methods still lack sufficient diagnostic sensitivity and specificity for direct testing, and foods still need to be culture enriched prior to analysis, ultimately increasing the total detection time. Although enrichment is a limitation in terms of assay speed, it is necessary for several reasons beyond increasing bacterial numbers, including diluting the effects of inhibitors, allowing the differentiation of viable from nonviable cells, and allowing for resuscitation of injured cells.

Numerous studies have shown that rapid assays perform better in some foods than in others, mainly due to differences in the bacterial microflora and in the sample matrix. As such, all rapid assays should be carefully validated in the respective food prior to initiating testing.

Also, the types of information obtained when using the various rapid methods must be realized. For example, DNA-based assays detect genetic targets (such as virulence genes), which only indicates that bacteria with those gene sequences are present. But they do not indicate whether the gene is actually expressed, or whether the bacterium is viable. Also, the need for isolation of the target bacteria in situations including food recalls means that cultural methods need to be employed in addition to DNA-based methods during outbreak investigations. Also, molecular methods cannot detect proteins such as enterotoxins, which can cause foodborne intoxications.

Finally, the cost and ease of use of rapid assays remain challenging obstacles to surmount. Rapid assays need to be cost-effective to facilitate their employment within the food chain. Similarly, although such methods have been developed to facilitate their ease of use, the need for sample manipulation and enrichment protocols prior to detection often complicates the use of rapid assays in food testing. This means that the majority of rapid assays are still largely confined to the laboratory, where they are performed by skilled technicians.

Notwithstanding the limitations of rapid assays, the development of this class of detection methods continues to increase. In the future, assays will need to be developed that focus on overlooked pathogens including viruses and parasites, for which the current rapid methods are limited to molecular assays. Biosensors, at the center of the marriage between biology and electronics, should be designed with the food matrix in mind and should be able to test larger, rather than smaller, volumes of sample. Also, research must commence to find a truly viable alternative to bacterial enrichment. Such an invention will revolutionize the food diagnostic field, leading to real-time analysis of food within the food chain.

REFERENCES

Acha, P. N., and B. Szyfres. 2003. *Zoonoses and Communicable Diseases Common to Men and Animals*, 3rd ed., vol. II. *Chlamydioses, Rickettsioses and Viroses*, p. 16–27. Pan American Health Organization, Washington, DC.

Acharya, G., C. Chun-Li, and C. Savran. 2006. An optical biosensor for rapid and label-free detection of [bacterial] cells. *J. Am. Chem. Soc.* **128:** 3862–3863.

Afonina, I. A., M. W. Reed, E. Lusby, L. G. Shiskina, and Y. S. Belousov. 2002. Minor groove binder conjugated DNA probes for quantitative DNA detection by hybridization-triggered fluorescence. *BioTechniques* **32:**940–944.

Andersen, A. A. 1958. New sampler for the collection, sizing, and enumeration of viable airborne particles. *J. Bacteriol.* **76:**471–484.

Anonymous. 1996. *Microbiology of Food and Animal Feeding Stuffs—Horizontal Method for the Detection and Enumeration of* Listeria monocytogenes*—Part 1. Detection Method (ISO 11290-1:1996)*. International Organization for Standardization, Geneva, Switzerland.

Arnal, C., J. M. Crance, G. Gantzer, L. Schwartzbrod, R. Deloince, and S. Billaudel. 1998. Persistence of infectious hepatitis A virus and its genome in artificial seawater. *Zentralbl. Hyg. Umweltmed.* **201:**279–284.

Bae, E., P. P. Banada, K. Huff, A. K. Bhunia, J. P. Robinson, and E. D. Hirleman. 2007. Bio-physical modeling of forward scattering from bacterial colonies using scalar diffraction theory. *Appl. Optics* **46:**3639–3648.

Banada, P. P., and A. K. Bhunia. 2008. Antibodies and immunoassays for detection of bacterial pathogens, p. 567–602. *In* M. Zourob, S. Elwary, and A. Turner (ed.), *Principles of Bacterial Detection: Biosensors, Recognition Receptors and Microsystems*. Cambridge University, Manchester, England.

Banada, P. P., S. L. Guo, B. Bayraktar, E. Bae, B. Rajwa, J. P. Robinson, E. D. Hirleman, and A. K. Bhunia. 2007. Optical forward-scattering for detection of *Listeria monocytogenes* and other *Listeria* species. *Biosens. Bioelectron.* **22:**1664–1671.

Banada, P. P., K. Huff, E. Bae, B. Rajwa, A. Aroonnual, B. Bayraktar, A. Adil, J. P. Robinson, E. D. Hirleman, and A. K. Bhunia. 2009. Label-free detection of multiple bacterial pathogens using light-scattering sensor. *Biosens. Bioelectron.* **24:**1685–1692.

Banerjee, P. 2008. Mammalian cell-based biosensor for rapid screening of pathogenic bacteria and toxins. Ph.D. thesis. Purdue University, West Lafayette, IN.

Banerjee, P., and A. K. Bhunia. 2009. Mammalian cell-based biosensors for pathogens and toxins. *Trends Biotechnol.* **27:**179–188.

Banerjee, P., D. Lenz, J. P. Robinson, J. L. Rickus, and A. K. Bhunia. 2008. A novel and simple cell-based detection system with a collagen-encapsulated B-lymphocyte cell line as a biosensor for rapid detection of pathogens and toxins. *Lab. Investig.* **88:**196–206.

Barton, A. C., S. D. Collyer, F. Davis, G. Z. Garifallou, G. Tsekenis, E. Tully, R.

O'Kennedy, T. D. Gibson, P. A. Millner, and S. P. J. Higson. 2009. Labeless AC impedimetric antibody-based sensors with pgml(−1) sensitivities for point-of-care biomedical applications. *Biosens. Bioelectron.* **24:**1090–1095.

Berthomieu, C., and R. Hienerwadel. 2009. Fourier transform infrared (FTIR) spectroscopy. *Photosynth. Res.* **101:**157–170.

Beuret, C. 2004. Simultaneous detection of enteric viruses by multiplex real-time RT-PCR. *J. Virol. Methods* **115:**1–8.

Beutin, L., K. Gleier, and S. Zimmermann. 1999. Comparative evaluation of different diagnostic methods for detection of Shiga-toxin producing strains of *Escherichia coli* STEC in human clinical stool specimens, p. 89–93. *In* G. Duffy, P. Garvey, J. Coia, Y. Wasteson, and D. A. McDowell (ed.), *Verocytotoxigenic* E. coli *in Europe. Concerted Action CT98-3935.* 1. Methods for verocytotoxigenic E. coli. Teagasc, The National Food Centre, Dublin, Ireland.

Bhaduri, S., and B. Cottrell. 1997. Direct detection and isolation of plasmid-bearing virulent serotypes of *Yersinia enterocolitica* from various foods. *Appl. Environ. Microbiol.* **63:**4952–4955.

Bhunia, A. K. 2008. Biosensors and bio-based methods for the separation and detection of foodborne pathogens, p. 1–44. *In* S. Taylor (ed.), *Advances in Food and Nutrition Research*, vol. 54. Elsevier, Inc., New York, NY.

Bhunia, A. K., V. Nanduri, E. Bae, and E. D. Hirleman. 2010. Biosensors: foodborne pathogen detection, p. 23–39. *In* M. C. Flickinger (ed.), *Encyclopedia of Industrial Biotechnology: Bioprocess, Bioseparation, and Cell Technology*. John Wiley & Sons, Inc., Hoboken, NJ.

Bhunia, A. K., P. J. Steele, D. G. Westbrook, L. A. Bly, T. P. Maloney, and M. G. Johnson. 1994. A six-hour in vitro virulence assay for *Listeria monocytogenes* using myeloma and hybridoma cells from murine and human sources. *Microb. Pathog.* **16:**99–110.

Bilir Ormanci, F. S., I. Erol, N. D. Ayaz, O. Iseri, and D. Sariguzel. 2008. Immunomagnetic separation and PCR detection of *Listeria monocytogenes* in turkey deli meat and antibiotic resistance of the isolates. *Br. Poult. Sci.* **49:**560–565.

Blais, B. W. 1994. Transcriptional enhancement of the *Listeria monocytogenes* PCR and simple immunoenzymatic assay of the product using anti-RNA: DNA antibodies. *Appl. Environ. Microbiol.* **60:**348–352.

Blais, B. W., G. Turner, R. Sooknanan, and L. T. Malek. 1997. A nucleic acid sequence-based amplification system for detection of *Listeria monocytogenes hlyA* sequences. *Appl. Environ. Microbiol.* **63:**310–313.

Bohaychuk, V. M., G. E. Gensler, R. K. King, J. T. Wu, and L. M. McMullen. 2005. Evaluation of detection methods for screening meat and poultry products for the presence of foodborne pathogens. *J. Food Prot.* **68:**2637–2647.

Bosch, M., A. Sánchez, F. Rojas, and C. Ojeda. 2007. Recent development in optical fiber biosensors. *Sensors* **7:**797–859.

Bourry, A., T. Cochard, and B. Poutrel. 1997. Serological diagnosis of bovine, caprine, and ovine mastitis caused by *Listeria monocytogenes* by using an enzyme-linked immunosorbent assay. *J. Clin. Microbiol.* **35:**1606–1608.

Byrne, B., E. Stack, N. Gilmartin, and R. O. Kennedy. 2009. Antibody-based sensors: principles, problems and potential for detection of pathogens and associated toxins. *Sensors* **9:**4407–4445.

Byrne, J. 2008. New phage offers rapid *E. coli* detection in meat, says BioMerieux. *Food Quality News.* (Online.) http://www.foodqualitynews.com / Product-Categories / Testing / New-phage-offers - rapid - E. - coli - detection - in - meat - says - BioMerieux.

Call, D. R., F. J. Brockman, and D. P. Chandler. 2001. Detecting and genotyping *Escherichia coli* O157:H7 using multiplexed PCR and nucleic acid microarrays. *Int. J. Food Microbiol.* **67:**71–80.

Candish, A. A. G. 1991. Immunological methods in food microbiology. *Food Microbiol.* **8:**1–14.

Chapman, P. A., D. J. Wright, and C. A. Siddons. 1994. A comparison of immunomagnetic separation and direct culture for the isolation of verocytotoxin-producing *Escherichia coli* from bovine faeces. *J. Med. Microbiol.* **40:**424–427.

Chen, W., G. Martinez, and A. Mulchandani. 2000. Molecular beacons: a real-time polymerase chain reaction assay for detecting *Salmonella. Anal. Biochem.* **280:**166–172.

Cheng, J., M. A. Shoffner, G. E. Hvichia, L. J. Kricka, and P. Wilding. 1996. Chip PCR. II. Investigation of different PCR amplification systems in microfabricated silicon-glass chips. *Nucleic Acids Res.* **24:**380–385.

Cheung, P. W., C. W. Chan, W. Wong, T. L. Cheung, and K. M. Kam. 2004. Evaluation of two real-time polymerase chain reaction pathogen detection kits for *Salmonella* spp. in food. *Lett. Appl. Microbiol.* **39:**509–515.

Chizhikov, V., A. Rasooly, K. Chumakov, and D. D. Levy. 2001. Microarray analysis of microbial virulence factors. *Appl. Environ. Microbiol.* **67:** 3258–3263.

Christensen, L. S., J. Mousing, S. Mortensen, K. J. Soerensen, S. B. Strandbygaard, C. A. Henriksen, and J. B. Andersen. 1990. Evidence of long distance airborne transmission of Aujeszky's disease (pseudorabies) virus. *Vet. Rec.* **127:**471–474.

Christensen, L. S., P. Normann, S. Thykier-Nielsen, J. H. Sørensen, K. de Stricker, and S. Rosenørn. 2005. Analysis of the epidemiological dynamics during the 1982–1983 epidemic of foot-and-mouth disease (FMD) in Denmark based on molecular high-resolution strain identification. *J. Gen. Virol.* **86:**2577–2584.

Churruca, E., C. Girbau, I. Martínez, E. Mateo, R. Alonso, and A. Fernández-Astorga. 2007. Detection of *Campylobacter jejuni* and *Campylobacter coli* in chicken meat samples by real-time nucleic acid sequence-based amplification with molecular beacons. *Int. J. Food Microbiol.* **117:**85–90.

Conroy, P. J., S. Hearty, P. Leonard, and R. O'Kennedy. 2009. Antibody production, design and use for biosensor-based applications. *Semin. Cell Dev. Biol.* **20:**10–26.

Cook, N. 2003. The use of NASBA for the detection of microbial pathogens in food and environmental samples. *J. Microbiol. Methods* **53:**165–174.

Cooper, M. A., and V. T. Singleton. 2007. A survey of the 2001 to 2005 quarts crystal microbalance biosensor literature: applications of acoustic physics to the analysis of biomolecular interactions. *J. Mol. Recognit.* **20:**154–184.

Cremonesi, P., G. Pisoni, M. Severgnini, C. Consolandi, P. Moroni, M. Raschetti, and B. Castiglioni. 2009. Pathogen detection in milk samples by ligation detection reaction-mediated universal array method. *J. Dairy Sci.* **92:**3027–3039.

Curiale, M. S., W. Lepper, and B. Robison. 1994. Enzyme-linked immunoassay for detection of *Listeria monocytogenes* in dairy products, seafoods, and meats: collaborative study. *J. AOAC Int.* **77:** 1472–1489.

Curtis, T., R. M. Z. G. Naal, C. Batt, J. Tabb, and D. Holowka. 2008. Development of a mast cell-based biosensor. *Biosens. Bioelectron.* **23:**1024–1031.

Czajka, J., and C. Batt. 1996. A solid phase fluorescent capillary immunoassay for the detection of *Escherichia coli* O157:H7 in ground beef and apple cider. *J. Appl. Bacteriol.* **81:**601–607.

Daniels, D. A., H. Chen, B. J. Hicke, K. M. Swiderek, and L. Gold. 2004. A tenascin-C aptamer identified by tumor cell SELEX: systematic evolution of ligands by exponential enrichment. *Proc. Natl. Acad. Sci. USA* **100:**15416–15421.

D'Aoust, J. Y., A. M. Sewell, and P. Greco. 1991. Commercial latex agglutination kits for the detection of foodborne *Salmonella*. *J. Food Prot.* **54:** 725–730.

de Boer, E., and A. E. Heuvelink. 2000. Methods for the detection and isolation of Shiga-toxin producing *Escherichia coli*. *J. Appl. Microbiol. Symp. Suppl.* **88:**133S–143S.

Deng, M. Y., D. O. Cliver, S. P. Day, and P. M. Fratamico. 1996. Enterotoxigenic *Escherichia coli* detected in foods by PCR and an enzyme-linked oligonucleotide probe. *Int. J. Food Microbiol.* **30:** 217–229.

Dimov, I. K., J. L. Garcia-Cordero, J. O'Grady, C. R. Poulsen, C. Viguier, L. Kent, P. Daly, B. Lincoln, M. Maher, R. O'Kennedy, T. J. Smith, A. J. Ricco, and L. P. Lee. 2008. Integrated microfluidic tmRNA purification and real-time NASBA device for molecular diagnostics. *Lab Chip* **8:**2071–2078.

Dooley, J. S. G. 1994. Nucleic acid probes for the food industry. *Biotechnol. Adv.* **12:**669–677.

D'Souza, D. H., and L. A. Jaykus. 2003. Nucleic acid sequence based amplification for the rapid and sensitive detection of *Salmonella enterica* from foods. *J. Appl. Microbiol.* **95:**1343–1350.

Dunbar, S. A., C. A. Vander Zee, K. G. Oliver, K. L. Karem, and J. W. Jacobson. 2003. Quantitative, multiplexed detection of bacterial pathogens: DNA and protein applications of the Luminex LabMAP(TM) system. *J. Microbiol. Methods* **53:**245–252.

Errington, F. P., and E. O. Powell. 1969. A cyclone sampler for aerosol sampling in the field. *J. Hyg. Camb.* **67:**387–399.

European Centre for Disease Prevention and Control. 2010. *Risk Assessment on Q Fever*. European Centre for Disease Prevention and Control, Stockholm, Sweden. doi:10.2900/28860.

Feldstine, P. T., A. H. Lienau, R. L. Forgey, and R. D. Calhoon. 1997a. Assurance polyclonal enzyme immunoassay for detection of *Listeria monocytogenes* and related *Listeria* species in selected foods: collaborative study. *J. AOAC Int.* **80:**775–790.

Feldstine, P. T., A. H. Lienau, R. L. Forgey, and R. D. Calhoon. 1997b. Visual immunoprecipitate assay (VIP) for *Listeria monocytogenes* and related *Listeria* species detection in selected foods: collaborative study. *J. AOAC Int.* **80:**791–805.

Feng, P. 1996. Emergence of rapid methods for identifying microbial pathogens in foods. *J. Assoc. Off. Anal. Chem. Int.* **79:**809–812.

Feng, P. 1997. Impact of molecular biology on the detection of foodborne pathogens. *Mol. Biotechnol.* **7:**267–278.

Feng, P. C. H. 2006. Overview of rapid methods for the detection of foodborne pathogens and toxins, p. 14–27. *In* V. K. Juneja, J. P. Cherry & M. H. Tunick (ed.), *Advances in Microbial Food Safety*. Oxford University Press, Washington, DC.

Fernández, M., B. del Río, D. M. Linares, M. C. Martín, and M. A. Alvarez. 2006. Real-time polymerase chain reaction for quantitative detection of histamine-producing bacteria: use in cheese production. *J. Dairy Sci.* **89:**3763–3769.

Fields, N. D., G. S. Oxborrow, J. R. Puleo, and C. M. Herring. 1974. Evaluation of membrane filter field monitors for microbiological air sampling. *Appl. Microbiol.* **27:**517–520.

Fitzgerald, C., M. Collins, S. van Duyne, M. Mikoleit, T. Brown, and P. Fields. 2007. Multiplex, bead-based suspension array for molecular determination of common *Salmonella* serogroups. *J. Clin. Microbiol.* **45:**3323–3334.

Fratamico, P. M., and C. DebRoy. 2010. Detection of *Escherichia coli* O157:H7 in food using real-time multiplex PCR assays targeting the stx_1, stx_2, wzy_{O157}, and the $fliC_{h7}$ or *eae* genes. *Food Anal. Methods.* doi:10.1007/s12161-010-9143-7.

Fratamico, P. M., S. K. Sackitey, M. Wiedmann, and M. Y. Deng. 1995. Detection of *Escherichia coli* O157:H7 by multiplex PCR. *J. Clin. Microbiol.* **33:**2188–2191.

Fratamico, P. M., and T. P. Strobaugh. 1998. Simultaneous detection of *Salmonella* spp. and *Escherichia coli* O157:H7 by multiplex PCR. *J. Ind. Microbiol. Biotechnol.* **21:**92–98.

Gavin, P. J., L. R. Peterson, A. C. Pasquariello, J. Blackburn, M. G. Hamming, K. J. Kuo, and R. B. Thomson. 2004. Evaluation of performance and potential clinical impact of ProSpecT Shiga toxin *Escherichia coli* microplate assay for detection of Shiga toxin-producing *E. coli* in stool samples. *J. Clin. Microbiol.* **42:**1652–1656.

Ge, B., C. Larkin, S. Ahn, M. Jolley, M. Nasir, J. Meng, and R. H. Hall. 2002. Identification of *Escherichia coli* O157:H7 and other enterohemorrhagic serotypes by EHEC-hlyA targeting, strand displacement amplification, and fluorescence polarization. *Mol. Cell. Probes* **16:**85–92.

Gehring, A. G., D. M. Albin, A. K. Bhunia, S. A. Reed, S. I. Tu, and J. Uknalis. 2006. Antibody microarray detection of *Escherichia coli* O157:H7: quantification, assay limitations, and capture efficiency. *Anal. Chem.* **78:**6601–6607.

Gerone, P. J., R. B. Couch, G. V. Keefer, R. G. Douglas, E. B. Derrenbacher, and V. Knight. 1966. Assessment of experimental and natural viral aerosols. *Bacteriol. Rev.* **30:**576–588.

Gilmour, M. W., D. M. Dracz, A. K. Andrysiak, C. G. Clark, S. Tyson, A. Severini, and L. K. Ng. 2006. Use of the *espZ* gene encoded in the locus of enterocyte effacement for molecular typing of Shiga toxin-producing *Escherichia coli*. *J. Clin. Microbiol.* **44:**449–458.

Gloster, J., A. Freshwater, R. F. Selelrs, and S. Alexandersen. 2005. Reassessing the likelihood of airborne spread of foot-and-mouth disease at the start of the 1967–1968 UK foot-and-mouth disease epidemic. *Epidemiol. Infect.* **133:**767–783.

Gooding, C. M., and P. V. Choudary. 1997. Rapid and sensitive immunomagnetic separation-polymerase chain reaction method for the detection of *Escherichia coli* O157:H7 in raw milk and ice-cream. *J. Dairy Res.* **64:**87–93.

Gray, K. M., P. P. Banada, E. O'Neal, and A. K. Bhunia. 2005. Rapid Ped-2E9 cell-based cytotoxicity analysis and genotyping of *Bacillus* species. *J. Clin. Microbiol.* **43:**5865–5872.

Gray, M. L., and A. H. Killinger. 1966. *Listeria monocytogenes* and listeric infections. *Microbiol. Mol. Biol. Rev.* **30:**309–382.

Grif, K., M. P. Dierich, and F. Allerberger. 1998. Dynabeads plus 3M Petrifilm HEC versus Vitek Immunodiagnostic Assay System for detection of *E. coli* O157 in minced meat. *Lett. Appl. Microbiol.* **26:** 199–204.

Gutierrez, R., T. Garcia, I. Gonzalez, B. Sanz, P. E. Hernandez, and R. Martin. 1998. Quantitative detection of meat spoilage bacteria by using the polymerase chain reaction (PCR) and an enzyme linked immunosorbent assay (ELISA). *Lett. Appl. Microbiol.* **26:**372–376.

Guy, R. A., A. Kapoor, J. Holicka, D. Shepherd, and P. A. Horgen. 2006. A rapid molecular-based assay for direct quantification of viable bacteria in slaughterhouses. *J. Food Prot.* **69:** 1265–1272.

Hammack, T. S. 2008. Committee on Microbiology and Extraneous Methods. *J. AOAC Int.* **91:** 23b–31b.

Hammond, G. W., R. L. Raddatz, and D. E. Gelskey. 1989. Impact of atmospheric dispersion and transport of viral aerosols on the epidemiology of influenza. *Rev. Infect. Dis.* **11:**494–497.

Hara-Kudo, Y., N. Konishi, K. Ohtsuka, R. Hirramatsu, H. Tanaka, H. Konum, and K. Takatori. 2008. Detection of verotoxigenic *Escherichia coli* O157 and O26 in food by plating methods and LAMP method: a collaborative study. *Int. J. Food Microbiol.* **122:**156–161.

He, Y., and C. Y. Chen. 2010. Quantitative analysis of viable, stressed and dead cells of *Campylobacter jejuni* strain 81-176. *Food Microbiol.* **27:**439–446.

Hearty, S., P. Leonard, and R. O'Kennedy. 2010. Nanomedicine: barcodes check out prostate cancer. *Nat. Nanotechnol.* **5:**9–10.

Himathongkham, S., M. L. Dodd, J. K. Yee, D. K. Lau, R. G. Bryant, A. S. Badoiu, H. K. Lau, L. S. Guthertz, L. Crawford-Miksza, and M. A. Soliman. 2007. Recirculating immunomagnetic separation and optimal enrichment conditions for enhanced detection and recovery of low levels of *Escherichia coli* O157:H7 from fresh leafy produce and surface water. *J. Food Prot.* **70:**2717–2724.

Hitchins, A. D. 2003. *Listeria monocytogenes,* p. 10.01–10.13. *In* U.S. Food and Drug Administration (ed.), *Bacteriological Analytical Manual,* 8th ed. AOAC International, Washington, DC.

Hudson, J. A., R. J. Lake, M. G. Savill, P. Scholes, and R. E. McCormick. 2001. Rapid detection of *Listeria monocytogenes* in ham samples using immunomagnetic separation followed by polymerase chain reaction. *J. Appl. Microbiol.* **90:** 614–621.

Isogai, E., C. Makungu, J. Yabe, P. Sinkala, A. Nambota, H. Isogai, H. Fukushi, M. Silungwe, C. Mubita, M. Syakalima, B. M. Hang'ombe, S. Kozaki, and J. Yasuda. 2005. Detection of *Salmonella invA* by isothermal and chimeric primer-initiated amplification of nucleic acids (ICAN) in Zambia. *Comp. Immunol. Microbiol. Infect. Dis.* **28:**363–370.

Jean, J., D. H. D'Souza, and L. A. Jaykus. 2004. Multiplex nucleic acid sequence-based amplification for simultaneous detection of several enteric viruses in model ready-to-eat foods. *Appl. Environ. Microbiol.* **70:**6603–6610.

Jenkins, C., M. C. Pearce, A. W. Smith, H. I. Knight, D. J. Shaw, T. Cheasty, G. Foster, G. J. Gunn, G. Dougan, H. R. Smith, and G. Frankel. 2003. Detection of *Escherichia coli* serogroups O26, O103, O111, and O145 from bovine faeces using immunomagnetic separation and PCR/DNA probe techniques. *Lett. Appl. Microbiol.* **37:**207–212.

Jensen, G. B., L. Thomsen, and O. R. Veltman. September 2005. Method, chip, device and integrated system for detection of biological materials. International patent WO2005083426.

Johnson, J., K. Jinneman, G. Stelma, B. G. Smith, D. Lye, J. Messer, J. Ulaszek, L. Evsen, S. Gendel, R. W. Bennett, B. Swaminathan, J. Pruckler, A. Steigerwalt, S. Kathariou, S. Yildirim, D. Volokhov, A. Rasooly, V. Chizhikov, M. Wiedmann, E. Fortes, R. E. Duvall, and A. D. Hitchins. 2004. Natural atypical *Listeria innocua* strains with *Listeria monocytogenes* pathogenicity island 1 genes. *Appl. Environ. Microbiol.* **70:**4256–4266.

Johnson, R. P., R. J. Durham, S. T. Johnson, L. A. MacDonald, S. R. Jeffrey, and B. T. Butman. 1995. Detection of *Escherichia coli* O157: H7 in meat by an enzyme-linked immunosorbent assay, EHEC-Tek. *Appl. Environ. Microbiol.* **61:** 386–388.

Jokers, J. V., J. W. Jacobson, B. D. Bhagwandin, P. N. Floriano, N. Christodoulides, and J. T. McDevitt. 2010. Programmable nano-bio-chip sensors: analytical meets clinical. *Anal. Chem.* **82:** 1571–1579.

Jones, A., D. Young, J. Taylor, D. B. Kell, and J. J. Rowland. 1998. Quantification of microbial productivity via multi-angle light scattering and supervised learning. *Biotechnol. Bioeng.* **59:**131–143.

Josefsen, M. H., N. R. Jacobsen, and J. Hoorfar. 2004. Enrichment followed by quantitative PCR both for rapid detection and as a tool for quantitative risk assessment of foodborne thermotolerant campylobacters. *Appl. Environ. Microbiol.* **70:**3588–3592.

Josefsen, M. H., M. Krause, F. Hansen, and J. Hoorfar. 2007. Optimization of a 12-hour TaqMan PCR-based method for detection of *Salmonella* in meat. *Appl. Environ. Microbiol.* **73:**3040–3048.

Josephson, K. L., C. P. Gerba, and I. L. Pepper. 1993. Polymerase chain reaction detection of nonviable bacterial pathogens. *Appl. Environ. Microbiol.* **59:**3513–3515.

Keramas, G., D. D. Bang, M. Lund, M. Madsen, H. Bunkenborg, P. Telleman, and C. B. V. Christensen. 2004. Use of culture, PCR analysis, and DNA microarrays for detection of *Campylobacter jejuni* and *Campylobacter coli* from chicken feces. *J. Clin. Microbiol.* **42:**3985–3991.

Kim, J. S., G. P. Anderson, J. S. Erickson, J. P. Golden, M. Nasir, and F. S. Ligler. 2009. Multiplexed detection of bacteria and toxins using a microflow cytometer. *Anal. Chem.* **81:**5426–5432.

Kim, S. H., M. K. Park, J. Y. Kim, P. D. Chuong, Y. S. Lee, B. S. Yoon, K. K. Hwang, and K. L. Yoon. 2005. Development of a sandwich ELISA for the detection of *Listeria* spp. using specific flagella antibodies. *J. Vet. Sci.* **6:**41–46.

Kintzios, S., E. Pistola, P. Panagiotopoulos, M. Bomsel, N. Alexandropoulos, F. Bem, G. Ekonomou, J. Biselis, and R. Levin. 2001. Bioelectric recognition assay (BERA). *Biosens. Bioelectron.* **16:**325–336.

Klein, P. G., and V. K. Juneja. 1997. Sensitive detection of viable *Listeria monocytogenes* by reverse transcription-PCR. *Appl. Environ. Microbiol.* **63:** 4441–4448.

Koch, W. H., W. L. Payne, B. A. Wentz, and T. A. Cebula. 1993. Rapid polymerase chain reaction method for detection of *Vibrio cholerae* in foods. *Appl. Environ. Microbiol.* **59:**556–560.

Koh, C. G., W. Tan, M. Q. Zhao, A. J. Ricco, and Z. H. Fan. 2003. Integrating polymerase chain reaction, valving, and electrophoresis in a plastic device for bacterial detection. *Anal. Chem.* **75:**4591–4598.

Krause, M., M. H. Josefsen, M. Lund, N. R. Jacobsen, L. Brorsen, M. Moos, A. Stockmarr, and J. Hoorfar. 2006. Comparative, collaborative, and on-site validation of a TaqMan PCR method as a tool for certified production of fresh,

Campylobacter-free chickens. *Appl. Environ. Microbiol.* **72:**5463–5468.

Kricka, L. J. 1998. Miniaturization of analytical system. *Clin. Chem.* **44:**2008–2014.

Kricka, L. J. 2001. Microchips, microarrays, biochips and nanochips: personal laboratories for the 21st century. *Clin. Chim. Acta* **307:**219–223.

Kudelski, A. 2008. Analytical applications of Raman spectroscopy. *Talanta* **76:**1–8.

Lai, R. Y., E. T. Lagally, S. H. Lee, H. T. Soh, K. W. Plaxco, and A. J. Heeger. 2006. Rapid, sequence-specific detection of unpurified PCR amplicons via a reusable, electrochemical sensor. *Proc. Natl. Acad. Sci. USA* **103:**4017–4021.

Lazcka, O., F. J. D. Campo, and F. X. Muñoz. 2007. Pathogen detection: a perspective of traditional methods and biosensors. *Biosens. Bioelectron.* **22:**1205–1217.

Le Guyader, F. S., F. Bon, D. DeMedici, S. Parnaudeau, A. Bertone, S. Crudeli, A. Doyle, M. Zidane, E. Suffredini, E. Kohli, F. Maddalo, M. Monini, A. Gallay, M. Pommepuy, P. Pothier, and F. M. Ruggeri. 2006. Detection of multiple noroviruses associated with an international gastroenteritis outbreak linked to oyster consumption. *J. Clin. Microbiol.* **44:**3878–3882.

Lehtola, M. J., E. Torvinen, I. T. Miettinen, and C. W. Keevil. 2006. Fluorescence in situ hybridization using peptide nucleic acid probes for rapid detection of *Mycobacterium avium* subsp. *avium* and *Mycobacterium avium* subsp. *paratuberculosis* in potable-water biofilms. *Appl. Environ. Microbiol.* **72:**848–853.

Leonard, P., P. Säfsten, S. Hearty, B. McDonnell, W. I. Finlay, and R. O'Kennedy. 2007. High throughput ranking of recombinant avian scFv antibody fragments from crude lysates using the Biacore A100. *J. Immunol. Methods* **323:**172–179.

León Muiños, E., M. L. Pérez Del Molino, F. L. Lado, M. Pazo Nuñez, and F. Pardo. 2004. Use of ligase chain reaction for the rapid diagnosis of lymph node tuberculosis. *Scand. J. Infect. Dis.* **36:**724–726.

Leung, A., P. M. Shankar, and R. Mutharasan. 2007. A review of fiber-optic biosensors. *Sens. Actuat. B* **125:**688–703.

Li, F., C. Zhao, W. Zhang, S. Cui, J. Meng, J. Wu, and D. Y. Zhang. 2005a. Use of ramification amplification assay for detection of *Escherichia coli* O157:H7 and other *E. coli* Shiga toxin-producing strains. *J. Clin. Microbiol.* **43:**6086–6090.

Li, W., and M. A. Drake. 2001. Development of a quantitative competitive PCR assay for detection and quantification of *Escherichia coli* O157:H7 cells. *Appl. Environ. Microbiol.* **67:**3291–3294.

Li, Y., X. Huang, I. T. Yu, T. W. Wong, and H. Qian. 2005b. Role of air distribution in SARS transmission during the largest nosocomial outbreak in Hong Kong. *Indoor Air* **15:**83–95.

Lim, D. V., J. M. Simpson, E. A. Kearns, and M. F. Kramer. 2005. Current and developing technologies for monitoring agents of bioterrorism and biowarfare. *Clin. Microbiol. Rev.* **18:**583–607.

Liming, S. H., and A. A. Bhagwat. 2004. Application of a molecular beacon—real-time PCR technology to detect *Salmonella* species contaminating fruits and vegetables. *Int. J. Food Microbiol.* **95:**177–187.

Liu, C., N. C. Cady, and C. A. Batt. 2009. Nucleic acid-based detection of bacterial pathogens using integrated microfluidic platform systems. *Sensors* **9:**3713–3744.

Liu, R. H., J. Yang, R. Lenigk, J. Bonanno, and P. Grodzinski. 2004. Self-contained, fully integrated biochip for sample preparation, polymerase chain reaction amplification, and DNA microarray detection. *Anal. Chem.* **76:**1824–1831.

Liu, Y., Q. Gao, X. Zhang, Y. Hou, J. Yang, and X. Huang. 2006. PCR and oligonucleotide array for detection of *Enterobacter sakazakii* in infant formula. *Mol. Cell. Probes* **20:**11–17.

Lynam, C., N. Gilmartin, A. I. Minett, R. O'Kennedy, and G. Wallace. 2009. Carbon nanotube based transducers for immunoassays. *Carbon* **47:**2337–2343.

Mackenzie, A. M. R., P. Lebel, E. Orrbine, P. Rowe, L. Hyde, F. Chan, W. Johnson, P. N. McLaine, et al. 1998. Sensitivities and specificities of Premier *E. coli* O157 and Premier EHEC enzyme immunoassays for diagnosis of infection with verotoxin (Shiga-like toxin)-producing *Escherichia coli. J. Clin. Microbiol.* **38:**547–551.

Manafi, M. 1996. Fluorogenic and chromogenic enzyme substrates in culture media and identification tests. *Int. J. Food Microbiol.* **31:**45–58.

Maruyama, F., T. Kenzaka, N. Yamaguchi, K. Tani, and M. Nasu. 2005. Visualization and enumeration of bacteria carrying a specific gene sequence by in situ rolling circle amplification. *Appl. Environ. Microbiol.* **71:**7933–7940.

Matar, G. M., W. F. Bibb, L. Helsel, W. Dewitt, and B. Swaminathan. 1992. Immunoaffinity purification, stabilization and comparative characterization of listeriolysin O from *Listeria monocytogenes* serotypes 1/2a and 4b. *Res. Microbiol.* **143:**489–498.

Mattingly, J. A., B. T. Butman, M. C. Plank, R. J. Durham, and B. J. Robison. 1988. Rapid monoclonal antibody-based enzyme-linked immunosorbent assay for detection of *Listeria* in food products. *J. Assoc. Off. Anal. Chem.* **71:**679–681.

May, K. R., and G. J. Harper. 1957. The efficiency of various liquid impinge samplers in bacterial aerosols. *Br. J. Ind. Med.* **14:**287–297.

Moberg, L. J., M. K. Wagner, and L. A. Kellen. 1988. Fluorogenic assay for rapid detection of *Escherichia coli* in chilled and frozen foods: collaborative study. *J. Assoc. Off. Anal. Chem.* **71:**589–602.

Monteiro, L., D. Bonnemaison, A. Vekris, K. G. Petry, J. Bonnet, R. Vidal, J. Cabrita, and F. Mégraud. 1997. Complex polysaccharides as PCR inhibitors in feces: *Helicobacter pylori* model. *J. Clin. Microbiol.* **35:**995–998.

Moreno, Y., M. Hernandez, M. A. Ferrus, J. L. Alonso, S. Botella, R. Montes, and J. Hernandez. 2001. Direct detection of thermotolerant campylobacters in chicken products by PCR and in situ hybridization. *Res. Microbiol.* **152:**577–582.

Murray, P., E. J. Baron, J. H. Jorgensen, M. A. Pfaller, and R. H. Yolken (ed.). 2003. *Manual of Clinical Microbiology,* 8th ed. ASM Press, Washington, DC.

Myers, K. M., J. Gaba, and S. F. Al-Khaldi. 2006. Molecular identification of *Yersinia enterocolitica* isolated from pasteurized whole milk using DNA microarray chip hybridization. *Mol. Cell. Probes* **20:**71–80.

Nickisch-Rosenegk, M., X. Marschan, D. Andresen, A. Abraham, C. Heise, and F. F. Bier. 2005. On-chip PCR amplification of very long templates using immobilized primers on glassy surfaces. *Biosens. Bioelectron.* **20:**1491–1498.

Notermans, S., and K. Wernars. 1991. Immunological methods for detection of foodborne pathogens and their toxins. *Int. J. Food Microbiol.* **12:**91–102.

Notermans, S. H., J. Dufrenne, M. Leimeister-Wachter, E. Domann, and T. Chakraborty. 1991. Phosphatidylinositol-specific phospholipase C activity as a marker to distinguish between pathogenic and nonpathogenic *Listeria* species. *Appl. Environ. Microbiol.* **57:**2666–2670.

Notomi, T., H. Okayama, H. Masubuchi, T. Yonekawa, K. Watanabe, N. Amino, and T. Hase. 2000. Loop-mediated isothermal amplification of DNA. *Nucleic Acids Res.* **28:**E63.

Ohtsuka, K., K. Yanagawa, K. Takatori, and Y. H. Kudo. 2005. Detection of *Salmonella enterica* in naturally contaminated liquid eggs by loop-mediated isothermal amplification, and characterization of *Salmonella* isolate. *Appl. Environ. Microbiol.* **71:**6730–6735.

O'Kennedy, R., S. Townsend, G. G. Donohoe, P. Leonard, S. Hearty, and B. Byrne. 2010. Speedy, small, sensitive and specific—reality or myth for future analytical methods. *Anal. Lett.* **33:**1630–1648.

Olsen, J. E., S. Aabo, W. Hill, S. Notermans, K. Wernar, P. E. Granum, T. Popovic, H. N. Rasmussen, and Ø. Olsvik. 1995. Probes and polymerase chain reaction for detection of food-borne bacterial pathogens. *Int. J. Food Microbiol.* **28:**1–78.

Olsen, K. N., M. Lund, J. Skov, L. S. Christensen, and J. Hoorfar. 2009. Detection of *Campylobacter* bacteria in air samples for continuous real-time monitoring of *Campylobacter* colonization in broiler flocks. *Appl. Environ. Microbiol.* **75:**2074–2078.

Olsvik, O., T. Popovic, E. Skjerve, K. S. Cudjoe, E. Hornes, J. Ugelstad, and M. Uhlen. 1994. Magnetic separation techniques in diagnostic microbiology. *Clin. Microbiol. Rev.* **7:**43–54.

O'Shaughnessy, T. J., and J. J. Pancrazio. 2007. Broadband detection of environmental neurotoxicants. *Anal. Chem.* **79:**8838–8845.

Palladino, S., I. D. Kay, A. M. Costa, E. J. Lambert, and J. P. Flexman. 2003. Real-time PCR for the rapid detection of *vanA* and *vanB* genes. *Diagn. Microbiol. Infect. Dis.* **45:**81–84.

Palumbo, J. D., M. K. Borucki, R. E. Mandrell, and L. Gorski. 2003. Serotyping of *Listeria monocytogenes* by enzyme-linked immunosorbent assay and identification of mixed-serotype cultures by colony immunoblotting. *J. Clin. Microbiol.* **41:**564–571.

Paoli, G. C., C. Y. Chen, and J. D. Brewster. 2004. Single-chain Fv antibody with specificity for *Listeria monocytogenes*. *J. Immunol. Methods* **289:**147–155.

Park, E. J., H. W. Chang, K. H. Kim, Y. D. Nam, S. W. Roh, and J. W. Bae. 2009. Application of quantitative real-time PCR for enumeration of total bacterial, archaeal, and yeast populations in kimchi. *J. Microbiol.* **47:**682–685.

Phillips, K. S., and Q. Cheng. 2007. Development of a "membrane cloaking" method for amperometric enzyme immunoassay and surface plasmon resonance analysis of proteins in serum samples. *Anal. Bioanal. Chem.* **387:**1831–1840.

Playford, E. G., B. McCall, G. Smith, V. Slinko, G. Allen, I. Smith, F. Moore, C. Taylor, Y. H. Kung, and H. Field. 2010. Human Hendra virus encephalitis associated with equine outbreak, Australia, 2008. *Emerg. Infect. Dis.* **16:**219–223.

Rasco, B., and G. E. Bledsoe. 2005. *Bioterrorism and Food Safety.* CRC Press, Boca Raton, FL.

Reischl, U., S. Bretagne, D. Krüger, P. Ernault, and R. J. M. Costa. 2003. Comparison of two DNA targets for the diagnosis of toxoplasmosis by real-time PCR using fluorescence resonance energy transfer hybridization probes. *BMC Infect. Dis.* **3:**7.

Relman, D. A., T. M. Schmidt, R. P. MacDermott, and S. Falkow. 1992. Identification of the uncultured bacillus of Whipple's disease. *N. Engl. J. Med.* **327:**293–301.

Reynisson, E., M. H. Josefsen, M. Krause, and J. Hoorfar. 2006. Novel probes to improve the detection limit of real-time PCR, using *Salmonella* as a model. *J. Microbiol. Methods* **66:**206–216.

Rider, T. H., M. S. Petrovick, F. E. Nargi, J. D. Harper, E. D. Schwoebel, R. H. Mathews, D. J. Blanchard, L. T. Bortolin, A. M. Young, J. Chen, and M. A. Hollis. 2003. A B cell-based sensor for rapid identification of pathogens. *Science* **301:**213–215.

Rodríguez-Lázaro, D., M. Hernández, M. Scortti, T. Esteve, J. A. Vásquez-Boland, and M. Pla. 2004a. Quantitative detection of *Listeria monocytogenes* and *Listeria innocua* by real-time PCR: assessment of *hly, iap,* and *lin02483* targets and AmpliFluor technology. *Appl. Environ. Microbiol.* **70:**1366–1377.

Rodríguez-Lázaro, D., J. Lloyd, A. Herrewegh, J. Ikonomopoulos, M. D'Agostino, M. Pla, and N. Cook. 2004b. A molecular beacon-based real-time NASBA assay for detection of *Mycobacterium avium* subsp. *paratuberculosis* in water and milk. *FEMS Microbiol. Lett.* **237:**119–126.

Rossen, L., P. Nørskov, K. Holmstrøm, and O. F. Rasmussen. 1992. Inhibition of PCR by components of food samples, microbial diagnostic assays and DNA-extraction solutions. *Int. J. Food Microbiol.* **17:**37–45.

Ryan, K. J. 2004. *Enterobacteriaceae. In* K. Ryan & C. Ray (ed.), *Sherris Medical Microbiology, an Introduction to Infectious Diseases,* 4th ed. McGraw-Hill, New York, NY.

Rybolt, R. L., R. W. Wills, J. A. Byrd, T. P. Doler, and R. H. Bailey. 2004. Comparison of four *Salmonella* isolation techniques in four different inoculated matrices. *Poult. Sci.* **83:**1112–1116.

Sachse, K., and J. Frey (ed.). 2003. *PCR Detection of Microbial Pathogens.* Humana Press, Totowa, NJ.

Sambrook, J., and D. W. Russell. 2001. *Molecular Cloning: A Laboratory Manual* (3rd ed.), p. 8.1–8.113. Cold Spring Harbor Laboratory Press, Cold Spring Harbor, NY.

Selby, C. 1999. Interference in immunoassay. *Ann. Clin. Biochem.* **36:**704–721.

Seo, K. H., P. S. Holt, H. D. Stone, and R. K. Gast. 2003. Simple and rapid methods for detecting *Salmonella enteritidis* in raw eggs. *Int. J. Food Microbiol.* **87:**139–144.

Sergeev, N., D. Volokhov, V. Chizhikov, and A. Rasooly. 2004. Simultaneous analysis of multiple staphylococcal enterotoxin genes by an oligonucleotide microarray assay. *J. Clin. Microbiol.* **42:**2134–2143.

Sewell, A. M., D. W. Warburton, A. Boville, E. F. Daley, and K. Mullen. 2003. The development of an efficient and rapid enzyme linked fluorescent assay method for the detection of *Listeria* spp. from foods. *Int. J. Food Microbiol.* **81:**123–129.

Sharma, V. K. 2002. Detection and quantitation of enterohemorrhagic *Escherichia coli* O157, O111, and O26 in beef and bovine feces by real-time polymerase chain reaction. *J. Food Prot.* **65:**1371–1380.

Sheridan, G. E. C., C. I. Masters, J. A. Shallcross, and B. M. Mackey. 1998. Detection of mRNA by reverse transcription-PCR as an indicator of viability in *Escherichia coli* cells. *Appl. Environ. Microbiol.* **64:**1313–1318.

Shim, W. B., J. G. Choi, J. Y. Kim, Z. Y. Yang, K. H. Lee, M. G. Kim, S. D. Ha, K. S. Kim, K. Y. Kim, C. H. Kim, S. A. Eremin, and D. H. Chung. 2008. Enhanced rapidity for qualitative detection of *Listeria monocytogenes* using an enzyme-linked immunosorbent assay and immunochromatography strip test combined with immunomagnetic bead separation. *J. Food Prot.* **71:**781–789.

Shroyer, M. L., and A. K. Bhunia. 2003. Development of a rapid 1-h fluorescence-based cytotoxicity assay for *Listeria* species. *J. Microbiol. Methods* **55**:35–40.

Simpkins, S. A., A. B. Chan, J. Hays, B. Pöpping, and N. Cook. 2000. A RNA transcription-based amplification technique (NASBA) for the detection of viable *Salmonella enterica. Lett. Appl. Microbiol.* **20:**75–79.

Skjerve, E., and O. Olsvik. 1991. Immunomagnetic separation of *Salmonella* from foods. *Int. J. Food Microbiol.* **14:**11–18.

Skottrup, P. D., M. Nicolaisen, and A. F. Justesen. 2008. Toward on-site pathogen detection using antibody-based sensors. *Biosens. Bioelectron.* **24:**339–348.

Stapleton, S., B. Bradshaw, and R. O'Kennedy. 2009. Development of a surface plasmon resonance-based assay for the detection of *Corynebacterium pseudotuberculosis* infection in sheep. *Anal. Chim. Acta* **651:**98–104.

Stenger, D. A., G. W. Gross, E. W. Keefer, K. M. Shaffer, J. D. Andreadis, W. Ma, and J. J. Pancrazio. 2001. Detection of physiologically active compounds using cell-based biosensors. *Trends Biotechnol.* **19:**304–309.

Stroup, S. E., S. Roy, J. Mchele, V. Maro, S. Ntabaguzi, A. Siddique, G. Kang, R. L. Guerrant, B. D. Kirkpatrick, R. Fayer, J. Herbein, H. Ward, R. Haque, and E. R. Houpt. 2006. Real-time PCR detection and speciation of *Cryptosporidium* infection using Scorpion probes. *J. Med. Microbiol.* **55:**1217–1222.

Suo, B., Y. He, G. Paoli, A. Gehring, S.-I. Tu, and X. Shi. 2010. Development of an oligonucleotide-based microarray to detect multiple foodborne pathogens. *Mol. Cell. Probes* **24:**77–86.

Swaminathan, B., and P. Feng. 1994. Rapid detection of food-borne pathogenic bacteria. *Annu. Rev. Microbiol.* **48:**401–426.

Takahashi, H., Y. Hara-Kudo, J. Miyasaka, S. Kumagai, and H. Konuma. 2005. Development of a quantitative real-time polymerase chain reaction targeted to the *toxR* for detection of *Vibrio vulnificus*. *J. Microbiol. Methods* **61:**77–85.

Techathuvanan, C., F. A. Draughon, and D. H. D'Souza. 2010. Real-time reverse transcriptase PCR for the rapid and sensitive detection of *Salmonella* Typhimurium from pork. *J. Food Prot.* **73:** 507–514.

Tellier, R. 2006. Review on aerosol transmission of influenza A virus. *Emerg. Infect. Dis.* **12:**1657–1662.

Tencza, S. B., and M. A. Sipe. 2004. Detection and classification of threat agents via high-content assays of mammalian cells. *J. Appl. Toxicol.* **24:**371–377.

Teunis, P., K. Takumi, and K. Shinagawa. 2004. Dose response for infection by *Escherichia coli* O157: H7 from outbreak data. *Risk Anal.* **24:**401–407.

Tokarskyy, O., and D. L. Marshall. 2008. Immunosensors for rapid detection of *Escherichia coli* O157:H7—perspectives for use in the meat processing industry. *Food Microbiol.* **25:**1–12.

Tombelli, S., M. Minunni, and M. Mascini. 2007. Aptamers-based assays for diagnostics, environmental and food analysis. *Biomol. Eng.* **24:**191–200.

Tsai, W., C. E. Miller, and E. R. Richter. 2000. Determination of the sensitivity of a rapid *Escherichia coli* O157:H7 assay for testing 375-gram composite samples. *Appl. Environ. Microbiol.* **66:**4149–4151.

Tsen, H. Y., L. Z. Jian, and W. R. Chi. 1998. Use of a multiplex PCR system for the simultaneous detection of heat labile toxin I and heat stable toxin II genes of enterotoxigenic *Escherichia coli* in skim milk and porcine stool. *J. Food Prot.* **61:**141–145.

Uyttendaele, M., I. Van Hoorde, and J. Debevere. 2000. The use of immuno-magnetic separation (IMS) as a tool in a sample preparation method for direct detection of *L. monocytogenes* in cheese. *Int. J. Food Microbiol.* **54:**205–212.

Vet, J. A. M., A. R. Majthia, S. A. E. Marras, S. Tyagi, S. Dube, B. J. Poiesz, and F. R. Kramer. 1999. Multiplex detection of four pathogenic retroviruses using molecular beacons. *Proc. Natl. Acad. Sci. USA* **96:**6394–6399.

Vishnubhatla, A., D. Y. C. Fung, R. D. Oberst, M. P. Hays, T. G. Nagaraja, and S. J. A. Flood. 2000. Rapid 5′ nuclease (TaqMan) assay for detection of virulent strains of *Yersinia enterocolitica*. *Appl. Environ. Microbiol.* **66:**4131–4135.

Volokov, D., A., Rasooly, K. Chumakov, and V. Chizhikov. 2002. Identification of *Listeria* species by microarray-based assay. *J. Clin. Microbiol.* **40:**4720–4728.

Volokov, D., V. Chizhikov, K. Chumakov, and A. Rasooly. 2003. Microarray-based identification of thermophilic *Campylobacter jejuni, C. coli, C. lari,* and *C. upsaliensis*. *J. Clin. Microbiol.* **41:**4071–4080.

Waggoner, P. C., and H. G. Craighead. 2007. Micro- and nanomechanical sensors for environmental, chemical and biological detection. *Lab Chip* **7:**1238–1255.

Wang, B., S. J. Potter, Y. Lin, A. L. Cunningham, D. E. Dwyer, Y. Su, X. Ma, Y. Hou, and N. K. Saksena. 2005. Rapid and sensitive detection of severe acute respiratory syndrome coronavirus by rolling circle amplification. *J. Clin. Microbiol.* **43:**2339–2344.

Wang, Q., S. Wang, L. Beutin, B. Cao, L. Feng, and L. Wang. 2010. Development of a DNA microarray for detection and serotyping of enterotoxigenic *Escherichia coli*. *J. Clin. Microbiol.* **48:**2066–2074.

Ward, L. N., and A. K. Bej. 2006. Detection of *Vibrio parahaemolyticus* in shellfish by use of multiplexed real-time PCR with TaqMan fluorescent probes. *Appl. Environ. Microbiol.* **72:**2031–2042.

Washington, C., D. Metzgar, H. Hazbón, L. Binn, A. Lyons, C. Coward, and R. Kuschner. 2010. Multiplexed Luminex xMAP assay for detection and identification of five adenovirus serotypes associated with epidemics of respiratory disease in adults. *J. Clin. Microbiol.* **48:**2217–2222.

Waters, L. C., S. C. Jacobson, N. Kroutchinina, J. Khandurina, R. S. Foote, and J. M. Ramsey. 1998. Multiple sample PCR amplification and electrophoretic analysis on a microchip. *Anal. Chem.* **70:**5172–5176.

Westin, L., X. Xu, C. Miller, L. Wang, C. F. Edman, and M. Nerenberg. 2000. Anchored multiplex amplification on a microelectronic chip array. *Nat. Biotechnol.* **18:**199–204.

Wiedmann, M., J. Czajka, F. Barany, and C. A. Batt. 1992. Discrimination of *Listeria monocytogenes* from other *Listeria* species by ligase chain reaction. *Appl. Environ. Microbiol.* **58:**3443–3447.

Wieneke, A. A. 1991. Comparison of four kits for the detection of staphylococcal enterotoxin in foods from outbreaks of food poisoning. *Int. J. Food Microbiol.* **14:**305–312.

Willford, J., K. Mills, and L. D. Goodridge. 2009. Evaluation of three commercially available ELISA kits for detection of Shiga toxin–producing *Escherichia coli*. *J. Food Prot.* **72:**741–747.

Wilson, W. J., A. M. Erler, S. L. Nasarabadi, E. W. Skowronski, and P. M. Imbro. 2005. A multiplexed PCR-coupled liquid bead array for the simultaneous detection of four biothreat agents. *Mol. Cell. Probes* **19:**137–144.

Wolcott, M. J. 1992. Advances in nucleic acid-based detection methods. *Clin. Microbiol. Rev.* **5:**370–386.

Wolffs, P. F., K. Glencross, R. Thibaudeau, and M. W. Griffiths. 2006. Direct quantitation and detection of salmonellae in biological samples without enrichment, using two-step filtration and real-time PCR. *Appl. Environ. Microbiol.* **72:**3896–3900.

Woolley, A. T., D. Hadley, P. Landre, A. J. deMello, R. A. Mathies, and M. A. Northrup. 1996. Functional integration of PCR amplification and capillary electrophoresis in a microfabricated DNA analysis device. *Anal. Chem.* **68:**4081–4086.

Wright, D. J., P. A. Chapman, and C. A. Siddons. 1994. Immunomagnetic separation as a sensitive method for isolating *Escherichia coli* O157 from food samples. *Epidemiol. Infect.* **113:**31–39.

Yao, Y., C. Nellåker, and H. Karlsson. 2006. Evaluation of minor groove binding probe and Taqman probe PCR assays: influence of mismatches and template complexity on quantification. *Mol. Cell. Probes* **20:**311–316.

Ziegler, C. 2000. Cell-based biosensors. *Fresenius J. Anal. Chem.* **366:**552–559.

CHROMOGENIC AND ACCELERATED CULTURAL METHODS

L. D. Goodridge and B. Bisha

3

Analysis of food samples by cultural and enrichment techniques remains an integral part of the examination of food for the presence or enumeration of foodborne pathogens. Use of such techniques is often required due to the fact that in a practical setting there is a need to amplify the target bacteria to a detectable concentration prior to analysis. It is not unusual for the target microorganism to be present in foods at levels of approximately 1 cell per 100 g (Sperber et al., 2001); however, enrichment methods have afforded the ability to detect as few as one cell per 500 g of food (Silliker and Gabis, 1973). In addition, enrichment techniques aid in the recovery of injured bacteria, which are often present in foods in a stressed condition because some foods may lack the optimal nutrients essential for bacterial growth, the food may present an environment not conducive to growth of the microorganisms, or the bacteria may be damaged during processing of the food (Sperber et al., 2001). In this regard, the enrichment process provides an optimal growth medium and a period of time that can be effectively used by injured/stressed microorganisms to perform repair of cellular damage, allowing for resuscitation of the bacterial cells. Additionally, enrichment processes provide an opportunity for target bacterial cells to proliferate, while the growth of competing background microflora is truncated (Taylor and Silliker, 1961).

EHEC MEDIA

Many differential and selective media have been developed in response to the presence of bacterial pathogens capable of causing illness in humans, by employing unique phenotypic characteristics of these microorganisms to develop detection platforms based on cultural approaches (Table 1). For example, cultural media were originally developed to detect *Escherichia coli* O157:H7 based on the inability of this serotype to ferment sorbitol (March and Ratnam, 1986). The resulting medium, termed sorbitol-MacConkey agar (SMAC), has been modified from MacConkey agar by replacing lactose with sorbitol. Since *E. coli* O157:H7 cannot ferment sorbitol, colonies of this organism are colorless and hence readily differentiated from sorbitol fermenters growing on the same plate. Although very sensitive for detection of *E. coli* O157:H7 (March and Ratnam, 1986; Klein et al., 2004), the identification of non-O157 enterohemorrhagic *E.*

L. D. Goodridge and B. Bisha, Department of Animal Sciences, Center for Meat Safety and Quality, Colorado State University, Fort Collins, CO 80523.

Rapid Detection, Identification, and Quantification of Foodborne Pathogens, ed. by J. Hoorfar

TABLE 1 Examples of representative accelerated cultural methods for pre-enrichment/enrichment, plating, and confirmation of select foodborne pathogens

Pathogen	Preenrichment/enrichment		Plating			References
	Medium	Selective agents	Medium	Selective agents	Differential agents	
EHEC	SEL broth[a]	Acriflavine, cycloheximide, fosfomycin, nalidixic acid	SMAC	Bile salts, crystal violet	Sorbitol, neutral red	March and Ratnam, 1986; Zadik et al., 1993; Jiang et al., 1998; DeBoer and Heuvelink, 2000; Sharma and Carlson, 2000; Funatsu et al., 2002; Klein et al., 2004; Church et al., 2007; Kim and Bhunia, 2008; Jenkins et al., 2008; Kanki et al., 2009
	UPB	Nonselective	CT-SMAC	Cefizime, tellurite, bile salts, crystal violet	Sorbitol, neutral red	
	GNTSB[a]	Nonselective	CHROMagar O157	Potassium tellurite, cefixime, cefsulodin	Proprietary	
			Rainbow agar O157	Potassium tellurite, novobiocin	β-Galactosidase and β-glucuronidase chromogenic substrate	

Listeria spp. and *L. monocytogenes*	SEL broth[a]	Acriflavine, cycloheximide, fosfomycin, nalidixic acid	Lithium chloride phenylethanol moxalactam plating (LPM) agar	Lithium chloride, phenylethanol, moxalactam	Esculin + ferric iron	Notermans et al., 1991; Jiang et al., 1998; Manafi, 2000; Hitchins, 2003; Reissbrod, 2004; El Marrakchi et al., 2005; Gasanov et al., 2005; Greenwood et al., 2005; Becker et al., 2006; Willis et al., 2006; Aragon-Alegro et al., 2008; Kim and Bhunia, 2008; Yu et al., 2010
	UPB[a]	Nonselective	Oxford agar	Lithium chloride, acriflavin, colistin sulfate, cefotetan, cycloheximide, fosfomycin	Esculin + ferric iron	
	SSL broth[a]	Nalidixic acid, lithium chloride, and potassium tellurite	MOX agar	Moxalactam, colistin sulfate, lithium chloride	Esculin + ferric iron	
			PALCAM agar	Lithium chloride, acriflavine, polymyxin, ceftazidime	Esculin + ferric iron, D-mannitol + phenol red	
			ALOA[d]	Lithium chloride, polymyxin B and nalidixic acid, amphotericin	PI-PLC enzyme system, X-glucoside	

(continued next page)

TABLE 1 Examples of representative accelerated cultural methods for pre-enrichment/enrichment, plating, and confirmation of select foodborne pathogens (*Continued*)

Pathogen	Preenrichment/enrichment		Plating			References
	Medium	Selective agents	Medium	Selective agents	Differential agents	
Salmonella spp.	SEB	Acriflavine, cycloheximide, fosfomycin, nalidixic acid	Rambach agar	Sodium desoxycholate	Propylene glycol + X-Gal	Rambach, 1990; Poupart et al., 1991; Pignato et al., 1995; Jiang et al., 1998; Hoorfar and Baggesen, 1998; Cooke et al., 1999; Gaillot et al., 1999; Perry et al., 1999; Miller and Mallinson, 2000; Sharma and Carlson, 2000; Cassar and Cuschieri, 2003; Kim and Bhunia, 2008; Schönenbrücher et al., 2008; Kanki et al., 2009; van Dijk et al., 2009; Yu et al., 2010
	Revive	Nonselective	SM-ID agar		β-Galactosidase substrate + α-glucuronate	
	UPB	Nonselective	CSE agar, ABC medium	Novobiocin	SLPA-octanoate[b]	
	SEL broth[a]	Acriflavine, cycloheximide, fosfomycin, nalidixic acid	MM	Tergitol 4	3,4-Cyclohexenoesculetin-β-D-galactoside and X-Gal	
SSL broth[a]	Nalidixic acid, lithium chloride, and potassium tellurite					
	GNTSB[a]	Nonselective				

Campylobacter spp.	Campy-Cefex agar	Cycloheximide, cefoperazone		Skirrow, 1977; Line, 2001; Murray et al., 2003; Oyarzabal et al., 2005; NACMCF,[c] 2007
	mCampy-Cefex	Cefoperazone, amphotericin B		
	mCCDA	Cefoperazone, amphotericin B		
	CampyFDA agar	Polymyxin B sulfate, trimethoprim, vancomycin, amphotericin, novobiocin, sodium metabisulfate, ferrous sulfate		
	Campy-Line agar	Polymyxin B sulfate, trimethoprim, vancomycin, cycloheximide, cefoperazone	Triphenyltetrazolium chloride	
	Karmali agar	Cefoperazone, vancomycin, cycloheximide		
Vibrio spp.	TCBS	Sodium chloride, bile salts	Bromothymol blue, thymol blue	Nakanishi, 1963; Kobayashi et al., 1963; Daniels et al., 2000a, 2000b; DePaola et al., 2003; Wong, 2003; Su et al., 2005
	BCVM and CHROMagar Vibrio	Proprietary	Proprietary	

[a] Simultaneous enrichment of multiple target pathogens.
[b] SLPA-octanoate, 4-[2-(4-octanoyloxy-3,5-dimethoxyphenyl)-vinyl]-quinolinium-1-(propan-3-yl carboxylic acid) bromide.
[c] NACMCF, National Advisory Committee on Microbiological Criteria for Foods.
[d] ALOA, agar Listeria according to Ottaviani and Agosti.

coli (EHEC) on SMAC is impossible, as these bacteria are incapable of sorbitol fermentation (Jenkins et al., 2008). Since it is estimated that approximately one-third of EHEC disease in the United States is caused by non-O157 EHEC, detection of these potentially life-threatening pathogens remains a critical need (Mead et al., 1999; Jenkins et al., 2003).

In an attempt to address this issue, several selective and differential agars have been developed to recover EHEC from food and fecal samples (Jenkins et al., 2008). The use of antibiotics such as cefixime, tellurite, novobiocin, cefsulodin, and vancomycin, which are selective for EHEC, has found common place in media compositions (de Boer and Heuvelink, 2000; Funatsu et al., 2002; Jenkins et al., 2008). When added to an already selective and differential medium, these antibiotic components remove large proportions of the background microflora to allow for faster isolation and identification of EHEC.

Novicki and coworkers (2000) were able to identify EHEC by utilizing a selective and differential agar supplemented with novobiocin, followed by an enzyme-linked immunosorbent assay (ELISA) as a confirmatory step, and they compared this two-step method to culture on SMAC. On the differential and selective medium, *E. coli* O157:H7 appeared as black colonies, while all other *E. coli* strains including non-O157 EHEC appeared as red, blue, or purple colonies. All presumptive *E. coli* O157:H7 colonies were subsequently screened with an ELISA specific for Shiga toxin (Stx) to make a positive confirmation of EHEC.

Compared to a Vero cell cytotoxic assay, SMAC had sensitivities of 23.5% for the identification of all EHEC serotypes and of 50.0% for the identification of O157:H7 alone. The two-step method was shown to have sensitivities of 76.5 and 100%, respectively. The ELISA alone had a sensitivity of 82.4% in the detection of Stx1 and Stx2. The specificity was 100% in all cases. However, this study highlighted the need to confirm the identity of target pathogens by other methods, even when differential components are incorporated into the media.

According to the U.S. Centers for Disease Control and Prevention, a sample should be plated onto a selective and differential medium such as SMAC (March and Ratnam, 1986), cefixime tellurite-sorbitol MacConkey agar (CT-SMAC), or CHROMagar O157 as a means of isolating Shiga toxin-producing *E. coli* (STEC). Following incubation for 16 to 24 h at 37°C, *E. coli* O157 colonies appear colorless on SMAC or CT-SMAC and are mauve or pink on CHROMagar O157. Both CT-SMAC and CHROMagar O157 are more selective than SMAC, which increases the sensitivity of culture for detection of O157 STEC (Zadik et al., 1993; Church et al., 2007).

For example, Church and colleagues (2007) recently evaluated the performance of CHROMagar O157 versus that of SMAC media for detection of *E. coli* O157 and reported that CHROMagar had a higher sensitivity (96.30%) and negative predictive value (100%) and a better diagnostic efficiency than SMAC. Earlier, Manafi and Kremsmaier (2001) compared several differential agars including Rainbow agar O157 (RB), Biosynth culture medium O157:H7 (BCM O157:H7), and Fluorocult HC (HC) medium with SMAC agar for the isolation of *E. coli* O157:H7 from raw ground beef and unpasteurized milk.

The growth of 466 strains of gram-negative rods isolated from food samples and 46 type culture strains of *E. coli* O157 and non-O157 EHEC was examined on the agar media. The *E. coli* O157:H7 strains were identified by their typical black-gray colonies on RB and blue-black colonies on BCM O157:H7. Examination of the 46 known *E. coli* reference strains showed false-negative results of 3.0% on BCM O157:H7, 8.8% on RB, 5.9% on HC, and 5.9% on SMAC. None of 60 food samples tested yielded *E. coli* O157:H7. Examination of the food samples showed that RB gave the lowest number of false positives.

LISTERIA MEDIA

Various selective and differential media have been developed to effect isolation and detection of *Listeria monocytogenes* from foods. These media take advantage of biochemical characteristics that are specific for *L. monocytogenes* or *Listeria* spp., to allow differentiation from closely related bacteria. For example, lithium chloride phenylethanol moxalactam (LPM) agar, Oxford agar, modified Oxford (MOX) agar, and polymyxin-acriflacine-LiCl-ceftazidime-esculin-mannitol (PALCAM) agar make use of the presence of the enzyme β-D-glucosidase in *Listeria* spp., which acts as an esculinase, and the resulting biochemical reaction results in the formation of black colonies with a black zone surrounding them (Gasanov et al., 2005).

However, the esculinase activity associated with β-D-glucosidase is present in all *Listeria* spp., meaning that *L. monocytogenes* cannot be differentiated from other nonpathogenic species of *Listeria* when grown on the above media (Greenwood et al., 2005; Becker et al., 2006; Willis et al., 2006). In response to this issue, several chromogenic media have been developed that can differentiate between pathogenic and nonpathogenic species of *Listeria* (Greenwood et al., 2005; Becker et al, 2006; Willis et al., 2006; Aragon-Alegro et al., 2008).

Phosphatidylinositol-specific phospholipase C (PI-PLC) is an enzyme that is produced only by the pathogenic *Listeria* spp. *L. monocytogenes* and *Listeria ivanovii* (Notermans et al., 1991). The two species can be differentiated via their sugar fermentation profiles. In one study, 468 strains of *Listeria* spp. were checked for the presence of PI-PLC by overlaying colonies formed on agar plates with L-α-phosphatidylinositol, a substrate for the enzyme (Notermans et al., 1991). PI-PLC-positive isolates exhibited turbid halos around the colonies. None of the 167 nonpathogenic strains of *Listeria* demonstrated PI-PLC activity. Ninety-eight percent (295/301) of *L. monocytogenes* strains tested could be correctly identified on the basis of their ability to produce PI-PLC, and the three most notable serotypes associated with foodborne illness (1/2a, 1/2b, and 4b) were detected 99, 98, and 97% of the time, respectively. While *L. ivanovii* strains were not tested in this study, the authors had previously demonstrated PI-PLC activity in this species, which is pathogenic to animals (Leimeister-Wächter et al., 1991).

L. monocytogenes can be distinguished from *L. ivanovii* by plating onto BCM confirmatory plating medium. On this medium, both species fluoresce but only *L. monocytogenes* will produce acid due to the fermentation of rhamnose. Other non-*Listeria* organisms that produce PI-PLC, including *Bacillus cereus, Bacillus thuringiensis, Staphylococcus aureus*, and yeasts, are eliminated by the presence of selective agents in *L. monocytogenes* plating medium (LMPM) agar (Manafi, 2000).

Several selective and differential agars have been developed for *L. monocytogenes*, based on incorporation of chromogenic substrates within the media. Several media, including BCM, Agar *Listeria* according to Ottaviani and Agosti (ALOA), and Rapid' L. mono agar, have been approved for use by the FDA (Hitchins, 2003). Reissbrodt (2004) evaluated LMPM agar and PALCAM for their abilities to isolate *L. monocytogenes* from meat samples. A total of 1,633 samples of raw sausage batters and fresh fermented spreadable sausage were tested, and LMPM was observed to isolate significantly more *L. monocytogenes* cells than PALCAM (4.8% versus 3.2%). No false positives or false negatives were detected by confirmatory tests (Reissbrodt, 2004).

SALMONELLA MEDIA

As with *E. coli* O157 and *L. monocytogenes*, many selective and differential media have been introduced for the rapid detection of *Salmonella* spp. Traditional cultural isolation from food samples involves a 24-h nonselective preenrichment step followed by selective enrichments, in a variety of liquid media including tetrathionate, Rappaport-Vassiliadis (RV), and selenite cystine broths and incubation for an additional 24 h.

After selective enrichment, the samples are plated onto selective media, such as brilliant green sulfa, bismuth sulfite, xylose lysine deoxycholate (XLD), xylose lysine tergitol 4 (XLT4), modified lysine iron, and Hektoen enteric agars, followed by incubation for a further 24 h. In addition, after incubation, *Salmonella* presumptive positive colonies are subjected to conformational biochemical screening on triple-sugar iron and lysine iron agars, which involves 24 h of incubation (Food and Drug Administration, 2007).

Chromogenic plating media can be utilized for presumptive isolation and detection of *Salmonella* in food (Schönenbrücher et al., 2008). Rambach (1990) described a chromogenic medium for detection of *Salmonella* based on the formation of acid from propylene glycol and hydrolysis of X-Gal (5-bromo-4-chloro-3-indolyl-β-D-galactopyranoside). After incubation of plates at 37°C for 24 h, the formation of acid from propylene glycol causes precipitation of the neutral red in *Salmonella* colonies, yielding a red color. In contrast, coliform colonies are blue (β-D-galactosidase activity) or violet (the formation of acid from propylene glycol and β-D-galactosidase activity) and *Proteus* spp. remain colorless. The inclusion of sodium deoxycholate within the medium inhibits the growth of gram-positive flora.

Salmonella identification (SM-ID) agar is similar to Rambach agar, and *Salmonella* colonies are similarly detected by their distinctive red coloration, while coliforms appear blue, violet, or colorless (Poupart et al., 1991). The medium employs bile salts and brilliant green as selective agents.

One hundred twelve *Salmonella* serovars belonging to subspecies I to VI were examined on Rambach agar and SM-ID medium (Pignato et al., 1995). Colonies exhibiting the typical red coloration were seen with 100 of the 112 serovars assayed on SM-ID, and with 87 of 112 on Rambach agar, suggesting that SM-ID medium is more sensitive than Rambach agar. However, the authors cautioned that the relatively low sensitivity of both media makes them suitable for use only in association with a traditional selective medium.

MUCAP is a confirmation test for *Salmonella* species based on the rapid detection of caprylate esterase, using the fluorogenic substrate 4-methylumbelliferyl-caprylate. In the presence of the esterase, the substrate is cleaved with the release of 4-methylumbelliferone, which produces a strong blue fluorescence when excited by a UV light source. One drop of MUCAP has to be added to each colony tested on Columbia agar and observed under UV light (365 nm) for 1 to 5 min. A strong bluish fluorescence indicates the presence of *Salmonella* spp. (Pontello et al., 1987).

Many chromogenic media are available for detection of *Salmonella*, including *Salmonella* esterase (CSE) agar (Cooke et al., 1999), ABC medium (Perry et al., 1999), CHROMagar *Salmonella* medium (Gaillot et al., 1999), AES *Salmonella* agar plate (ASAP), Oxoid *Salmonella* Chromogen media (OSCM), and Miller-Mallinson agar (MM) (Schönenbrücher et al., 2008). Many comparative studies have been accomplished in which several of the media have been evaluated against each other (Gaillot et al., 1999; Perry et al., 1999; Cassar and Cuschieri, 2003). Many of the comparative studies have used stool samples, which may not be applicable to detection in food samples due to differences in the background microbiota and presence of inhibitory components of the sample matrix.

van Dijk and colleagues (2009) tested CHROMagar *Salmonella* medium, OSCM, and SM ID2 medium against pure cultures of 53 bacteria, including 34 *Salmonella* strains and *Aeromonas hydrophila, Citrobacter freundii, E. coli, Shigella* spp., *Pseudomonas aeruginosa*, and others. Many bacteria produced false-positive results on the different media. On CHROMagar, after 24 h of incubation, *A. hydrophila* and *P. aeruginosa* both produced false-positive results, and after 48 h, other organisms that gave false-positive results were *Acinetobacter baumannii, C. freundii, Shigella dysenteriae*, and *Candida tropicalis*.

On Oxoid chromogenic media, *A. baumannii, A. hydrophila, P. aeruginosa, S. dysenteriae*, and *Candida albicans* were falsely identified as *Salmonella* after 24 h, while *Citrobacter koseri, Morganella morganii, Proteus mirabilis*, and *Shigella boydii* gave false-positive results after 48 h. On SM ID2, organisms that gave false-positive results after 24 h included *A. baumannii* and *P. aeruginosa*, and after 48 h *A. hydrophila, C. albicans*, and *C. tropicalis* also gave false-positive results. The authors also evaluated the chromogenic media for their ability to isolate *Salmonella* from stool samples. Of the 1,339 cultured stool samples that were tested, 32 (2.4%) were positive for *Salmonella* on at least one medium.

Earlier, Schönenbrücher et al. (2008) compared the growth and appearance of 36 bacterial strains (*Salmonella* and 21 other species) on three chromogenic plating media, including ASAP, OSCM, MM, and three traditional agars including brilliant green agar according to Edel and Kampelmacher (BGA), XLD agar, and XLT4 agar. MM agar was the only medium tested on which all *Salmonella* strains exhibited typical colonies. Furthermore, BGA, XLD, and XLT4 all produced false-positive results, and XLD and XLT4 produced false negatives as well. Artificial-inoculation experiments using raw pork ground meat were subsequently conducted. A shortened incubation time of 24 h in RV soya peptone broth yielded a *Salmonella* species recovery of 100% from spiked meat samples. Finally, 286 naturally contaminated raw porcine and bovine minced meat samples and raw poultry meat samples were investigated. Forty-three strains from a total of 39 *Salmonella*-positive samples were found. The authors reported that the chromogenic plating media MM, ASAP, and OSCM gave the best results in contrast to the traditional media BGA, XLD, and XLT4. Other researchers have reported similar results (Miller and Mallinson, 2000).

Hoorfar and Baggesen (1998) evaluated the performance of two 1-day culture methods, *Salmonella* enrichment broth (SEB) and Revive, with the universal preenrichment broth (UPB) on 100 fecal samples from porcine and 100 neck skin samples from poultry. The sensitivity (number of positive cases per method among all the positive cases) of the conventional preenrichment in buffered peptone water (BPW) was found to be 0.77 for swine and 0.66 for poultry samples, while a combination of the BPW method with parallel preenrichment of the same sample in UPB resulted in high sensitivity for swine (0.92) and poultry (0.95) samples. A 2-h preenrichment in the nonselective Revive followed by overnight enrichment in selective broth resulted in a low sensitivity, particularly for the neck skin samples (0.16; $P = 0.001$). The SEB method in the porcine samples resulted in a sensitivity (0.71) comparable to that of the standard method ($P = 0.31$) (Hoorfar and Baggesen, 1998).

CAMPYLOBACTER MEDIA

Selective and differential media have been developed for other pathogens. For example, the U.S. Department of Agriculture Food Safety and Inspection Service recommends the use of Campy-Cefex agar during isolation of *Campylobacter jejuni* and *Campylobacter coli* from poultry rinses and sponge samples (National Advisory Committee on Microbiological Criteria for Foods, 2007). Campy-Cefex agar is based on the modification of brucella agar with antibiotics including cefoperazone, added to inhibit gram-positive and gram-negative organisms, and cycloheximide, added to inhibit the growth of fungi.

Several limitations to the use of this medium are noteworthy. First, because it is a selective but not differential medium, biochemical and/or serological tests should be performed to confirm the identification of any suspect colonies on the plate. Also, cephalothin-sensitive *Campylobacter* spp. such as *Campylobacter fetus* and *Campylobacter upsaliensis* may not be recovered on Campy-Cefex agar because of the addition of cefoperazone (Murray et al., 2003).

Several other selective *Campylobacter* media have been described. For example,

CampyFDA is based on the medium developed by Skirrow (1977), modified with the addition of amphotericin B, sodium pyruvate, sodium metabisulfite, and ferrous sulfate. Campy-Line agar (Line, 2001) is a differential and selective medium that contains triphenyltetrazolium chloride (200 mg/liter). This agar is translucent, and the addition of triphenyltetrazolium chloride imparts a deep red to magenta color to the *Campylobacter* colonies. The contrast of deep red colonies on a translucent background greatly facilitates *Campylobacter* identification and makes it possible to enumerate the bacteria on light boxes or by electronic means (Line, 2001).

Karmali agar is a blood-free, charcoal-based selective medium. Oyarzabal et al. (2005) evaluated several agars for direct enumeration of *Campylobacter* spp. from poultry carcass rinses. In this work, the researchers collected 20 postchill carcass rinses from four poultry processing plants on three successive visits. Samples were plated onto Campy-Cefex agar, modified Campy-Cefex agar (mCampy-Cefex), modified charcoal cefoperazone deoxycholate agar (mCCDA), Campy-Line agar, Karmali agar, and CampyFDA agar.

The combined results from all six plating media showed that 63.3% (152) of the samples were positive for *Campylobacter* spp. Campy-Cefex and its modification, mCampy-Cefex, were the media that produced the best results, although there were no statistical differences compared to CAMPY, mCCDA, or Karmali media. Campy-Line, however, exhibited the lowest counts, most likely due to the large numbers of antimicrobial substances incorporated into this medium.

Line (2001) compared the recovery of campylobacters from carcass rinses on Campy-Cefex, Campy-Line agar, and Campy-Line blood agars and reported that the recovery results were not significantly different between the agars. Although some contaminants could still grow on Campy-Line agar and present red colonies, most of these contaminants were easily distinguished from *Campylobacter* by differences in colony morphology.

VIBRIO MEDIA

Vibrio parahaemolyticus was first isolated in the 1950s following several foodborne outbreaks in Japan (Wong, 2003). This foodborne pathogen is a gram-negative halophile that is ubiquitous in the marine environment and commonly isolated from various seafoods (Wong, 2003). Several oyster-associated outbreaks of *V. parahaemolyticus* (Daniels et al., 2000a, 2000b; Centers for Disease Control and Prevention, 1998, 1999) that occurred in several states during the late 1990s provided the impetus for the development of several cultural and molecular methods, enabling rapid detection and quantification of total and pathogenic *V. parahaemolyticus* in oysters (DePaola et al., 2003). Thiosulfate citrate bile salts sucrose agar (TCBS) is used for the selective isolation of cholera vibrios and *Vibrio parahaemolyticus* from a variety of clinical and nonclinical specimens. Thiosulfate citrate bile sucrose agar (TCBS) is the primary plating medium universally used for the selective isolation of *Vibrio cholerae* and *V. parahaemolyticus* from a variety of clinical and nonclinical specimens, including food. It was developed by Kobayashi et al. (1963), who modified the selective medium of Nakanishi (1963).

Recently, a chromogenic medium, Bio-Chrome *Vibrio* medium (BCVM), was developed to allow differentiation of *V. parahaemolyticus* from other *Vibrio* species based on the formation of unique purple colonies on the medium. Growth of *V. parahaemolyticus* on BCVM can easily be distinguished from blue-green colonies formed by growth of *Vibrio vulnificus, Vibrio cholerae*, and *Vibrio mimicus*.

A study evaluating the selectivity and specificity of BCVM using 179 strains of *Vibrio* spp. found that BCVM was capable of differentiating *V. parahaemolyticus* from other species including *V. vulnificus* and *V. mimicus*. All 148 strains of *V. parahaemolyticus* grew on BCVM, and 145 of them produced purple colonies. The remaining 31 *Vibrio* spp., with one exception, either were unable to grow or produced blue-green or white colonies on BCVM (Su et al., 2005).

Duan and Su (2005) compared BCVM with TCBS for detecting *V. parahaemolyticus* in 296 seawater, sediment, and oyster samples by using a three-tube most-probable-number method and found that the specificities of BCVM and TCBS for *V. parahaemolyticus* detection were 94% and 77%, respectively, while the accuracies of the two media for detecting *V. parahaemolyticus* were 84% (BCVM) and 54% (TCBS). These results indicate that BCVM is more specific and accurate than TCBS in detecting *V. parahaemolyticus*.

ENUMERATION MEDIA

The cultural methods described above are mainly used to quantify the concentrations of foodborne pathogens in food and environmental samples. Enumeration is important in evaluating the degree of pathogenic contamination in foods. There are advantages and disadvantages to the various cultural methods used for enumeration, but almost all are laborious and require at least 24 h before quantitation can occur.

As such, many attempts have been made to improve the efficiency of such methods by making them less labor- and cost-intensive and more convenient to employ. For example, one popular alternative to traditional agar plates is the use of disposable cardboards containing rehydratable media, which eliminates the need for agar plates, constituting savings in storage, incubation, and disposal procedures (Chain and Fung, 1991). The rehydratable medium is embedded onto the cardboard along with a gelling agent, which is soluble in cold water.

Typically, 1 ml of liquid sample is placed on the center of the cardboard, where it mixes with the rehydrated nutrients and the gelling agent. Following solidification and subsequent incubation, bacterial colonies can be directly counted on the cardboard, as in conventional agar plates.

Such products are available for a number of foodborne pathogenic and spoilage microorganisms including yeasts and molds, coliform bacteria, aerobic plate count, *E. coli, Staphylococcus aureus, Enterobacteriaceae,* and *Listeria.* While the majority of classical methods remain time-consuming and labor-intensive, it has nonetheless been suggested that they still be utilized, at least in parallel, with any rapid methodology proposed (Klein et al., 2004).

SIMULTANEOUS ENRICHMENT OF MULTIPLE PATHOGENS

A preenrichment step often facilitates the need for stressed bacteria to resuscitate within a nonselective or moderately selective environment. Since these broths do not include selective agents for any specific target bacteria, the use of preenrichment may allow for simultaneous cultivation of multiple bacteria, thereby increasing the efficacy of detection and decreasing the cost- and labor-intensiveness of cultural techniques.

Several groups have investigated the use of a single broth to simultaneously enrich the concentrations of several bacterial species, followed by downstream detection. For example, Kim and Bhunia (2008) described the development of *Salmonella, Escherichia, Listeria* (SEL) broth, which was designed to allow the simultaneous growth of *Salmonella enterica, E. coli* O157:H7, and *L. monocytogenes.* In this work, the researchers compared SEL broth to individual selective enrichment broths, including RV for *Salmonella enterica,* modified *E. coli* broth with 20 mg of novobiocin/liter (mEC+n) for *E. coli* O157:H7, and Fraser broth for *L. monocytogenes.*

UPB was also tested during the experimental procedures. The results demonstrated that when inoculated at either 10^1 or 10^3 CFU/ml, the growth of each pathogen in SEL was superior to that in their respective individual enrichment broths, except in the case of RV, in which *Salmonella* cells, inoculated at both concentrations, grew equally well. When bacteria from the three species were combined in a cocktail in equal concentrations, the overall growth was proportional to the initial inoculation levels. The growth of *L. monocytogenes* was suppressed when cells of this species were present at lower concentrations than those of the other two species. In addition, SEL was able to resuscitate acid- and cold-stressed cells, and recovery was comparable to that of tryptic soy broth (TSB) containing 6% yeast extract but superior to recovery in the individual selective broths.

Even though the growth of each pathogen in SEL was comparable to that in UPB, fewer numbers of background microflora members were present following enrichment in SEL than in UPB. When the three bacterial species were inoculated (as a mixture) into ready-to-eat salami and into turkey meat samples, SEL led to the enrichment of all three species, permitting detection by pathogen-specific lateral flow assays or multiplex PCR. The authors concluded that SEL is a promising new multiplex selective enrichment broth for the detection of the three most prominent foodborne pathogens by antibody- or nucleic acid-based methods.

Others have investigated the use of UPB to simultaneously recover multiple bacterial species. Jiang et al. (1998) examined the use of UPB for the recovery of verotoxigenic *E. coli, Salmonella* spp., and *L. monocytogenes* from milk and cheese samples. UPB supported the growth of low inoculum levels (10 CFU/ml) of these organisms in pure cultures and in mixed cultures containing higher levels of other pathogens or bacterial flora from raw milk. UPB also supported the recovery and growth of heat-injured *Salmonella* spp., *L. monocytogenes*, and verotoxigenic *E. coli* at inoculum levels of 10^2 CFU/ml to yield concentrations of 10^8 CFU/ml in pure cultures and at least 10^5 CFU/ml in the presence of high levels of known competitive pathogens or microflora of cheese samples following 24 h of incubation.

UPB performed better than *Listeria* enrichment broth in supporting the recovery and growth of heat-injured *L. monocytogenes* and equally as well as BPW or TSB in supporting the growth of uninjured *L. monocytogenes, Salmonella* spp., and verotoxigenic *E. coli*. Coenrichment of these pathogens in UPB reduced the quantity of milk or cheese samples that was required for analysis and also reduced the cost and labor involved in preparing and processing separate preenrichment media.

In a similar study, Kanki and colleagues (2009) compared UPB to mEC+n for enrichment of Shiga toxin-producing *E. coli* O157 and O26 and to BPW for preenrichment of *S. enterica* in beef and radish sprouts following thermal, freezing, or no treatment. The results indicated that UPB effected better recovery of *E. coli* O157 and O26 in beef samples at 42°C than UPB at 35°C and from radish sprout samples than both UPB at 35°C and mEC+n. There were no significant differences between growth in UPB at 42°C and UPB at 35°C or BPW for the recovery of *Salmonella* from 205 naturally contaminated poultry samples. As with the other studies, these results show that savings in labor, time, samples, and space may be achieved if multiple pathogens are enriched simultaneously.

Salmonella, Staphylococcus, Listeria (SSL) broth was recently developed for simultaneous enrichment of *S. enterica* serovar Enteritidis, *Staphylococcus aureus*, and *L. monocytogenes* (Yu et al., 2010). This broth contains several selective agents and supplemental elements, including nalidixic acid, lithium chloride, potassium tellurite, sodium pyruvate, and mannitol. All three bacterial pathogens were capable of growing in SSL to as high as 10^8 CFU/ml after 24 h of incubation at 37°C when being inoculated at 50 to 100 CFU/ml, either individually or when simultaneously enriched. SSL broth led to the recovery of the three bacterial species from artificially contaminated raw beef and ready-to-eat chicken samples and also led to the recovery of the bacteria from 710 suspicious food samples by SSL with real-time PCR, yielding no false-positive or -negative results.

Finally, Sharma and Carlson (2000) developed a fluorogenic PCR assay for simultaneous detection of *Salmonella* spp. and *E. coli* O157:H7. As part of the assay, the bacteria were simultaneously enriched in gram-negative TSB broth (GNTSB), which consisted of a mixture of equal volumes of gram-negative broth and TSB. The bacteria were enriched for 18 h at 37°C with continuous agitation at 160 rpm. When tested with artificially inoculated meat or feces, the assay was able to detect less than 10 CFU/g of *S. enterica* serovar Typhimurium or *E. coli* O157: H7 when both pathogens were coenriched for 6 to 18 h.

CONCLUSION

The recovery of bacterial pathogens from food samples using cultural methods remains an integral part of food safety analysis. Regardless of the continued rise of advanced detection technologies, culture methods will continue to be vital during foodborne outbreak investigations, in which isolation of the pathogen from suspected food and infected populations is vital to understanding the source of the outbreak. Also, the need for isolated bacterial colonies during routine monitoring of the food processing environment remains a necessity. Furthermore, until rapid methods are developed with the sensitivity to allow detection of bacteria without enrichment, cultural methods will continue to be employed as an upfront technique to increase the concentrations of the target bacteria to a detectable level. However, many enrichment methods have been described in the scientific literature with respect to the recovery and isolation of bacterial pathogens from foods, with little consensus as to which methods are the most reliable or robust.

REFERENCES

Aragon-Alegro, L. C., D. C. Aragon, E. Z. Martinez, M. Landgraf, B. D. Gombossy de Melo Franco, and M. T. Destro. 2008. Performance of a chromogenic medium for the isolation of *Listeria monocytogenes* in food. *Food Control* **19:**483–486.

Becker, B., S. Schuler, M. Lohneis, A. Sabrowski, G. D. Curtis, and W. H. Holzapfel. 2006. Comparison of two chromogenic media for the detection of *Listeria monocytogenes* with the plating media recommended by EN/DIN 11290-1. *Int. J. Food Microbiol.* **109:**127–131.

Cassar, R., and P. Cuschieri. 2003. Comparison of *Salmonella* chromogenic medium with DCLS agar for isolation of *Salmonella* species from stool specimens. *J. Clin. Microbiol.* **41:**3229–3232.

Centers for Disease Control and Prevention. 1998. Outbreak of *Vibrio parahaemolyticus* infections associated with eating raw oysters—Pacific Northwest, 1997. *MMWR Morb. Mortal. Wkly. Rep.* **47:** 457–462.

Centers for Disease Control and Prevention. 1999. Outbreak of *Vibrio parahaemolyticus* infection associated with eating raw oysters and clams harvested from Long Island Sound—Connecticut, New Jersey, and New York, 1998. *MMWR Morb. Mortal. Wkly. Rep.* **48:**48–51.

Chain, V. S., and D. Y. C. Fung. 1991. Comparison of Redigel, Petrifilm, spiral plate system, Isogrid, and aerobic plate count for determining the numbers of aerobic bacteria in selected foods. *J. Food Prot.* **54:**208–211.

Church, D. L., D. Emshey, H. Semeniuk, T. Lloyd, and J. D. Pitout. 2007. Evaluation of BBL CHROMagar O157 versus sorbitol-MacConkey medium for routine detection of *Escherichia coli* O157 in a centralized regional clinical microbiology laboratory. *J. Clin. Microbiol.* **45:**3098–3100.

Cooke, V. M., R. J. Miles, R. G. Price, and A. C. Richardson. 1999. A novel chromogenic ester agar medium for detection of salmonellae. *Appl. Environ. Microbiol.* **65:**807–812.

Daniels, N. A., L. MacKinnon, R. Bishop, S. Altekruse, B. Ray, R. M. Hammond, S. Thompson, S. Wilson, N. H. Bean, P. M. Griffin, and L. Slutsker. 2000a. *Vibrio parahaemolyticus* infections in the United States, 1973–1998. *J. Infect. Dis.* **181:**1661–1666.

Daniels, N. A., B. Ray, A. Easton, N. Marano, E. Kahn, A. L. McShan II, L. Del Rosario, T. Baldwin, M. A. Kingsley, N. D. Puhr, J. G. Wells, and F. J. Angulo. 2000b. Emergence of a new *Vibrio parahaemolyticus* serotype in raw oysters. *JAMA* **284:**1541–1545.

de Boer, E., and A. E. Heuvelink. 2000. Methods for the detection and isolation of Shiga-toxin producing *Escherichia coli*. *J. Appl. Microbiol. Symp. Suppl.* **88:**133S–143S.

DePaola, A., J. Ulaszek, C. A. Kaysner, B. J. Tenge, J. L. Nordstrom, J. Wells, N. Puhr, and S. M. Gendel. 2003. Molecular, serological, and virulence characteristics of *Vibrio parahaemolyticus* isolated from environmental, food, and clinical sources in North America and Asia. *Appl. Environ. Microbiol.* **69:**3999–4005.

Duan, J., and Y. C. Su. 2005. Comparison of a chromogenic medium with thiosulfate–citrate–bile salts–sucrose agar for detecting *Vibrio parahaemolyticus*. *J. Food Sci.* **70:**125–128.

El Marrakchi, A., N. Boumhandi, and A. Hamama. 2005. Performance of a new chromogenic plating medium for isolation of *Listeria monocytogenes* from marine environments. *Lett. Appl. Microbiol.* **40:**87–91.

Food and Drug Administration. 2007. *Bacteriological Analytical Manual*, chapt. 5. *Salmonella*. (Online.) http: / / www.fda.gov / Food / ScienceResearch / LaboratoryMethods / BacteriologicalAnalytical ManualBAM/ucm070149.htm. Accessed 17 September 2010.

Funatsu, T., T. Taniyama, T. Tajima, H. Tadakuma, and H. Namiki. 2002. Rapid and sensitive detection method of a bacterium by using a GFP reporter phage. *Microbiol. Immunol.* **46:**365–369.

Gaillot, O., P. di Camillo, P. Berche, R. Courcol, and C. Savage. 1999. Comparison of CHROMagar *Salmonella* medium and Hektoen enteric agar for isolation of salmonellae from stool samples. *J. Clin. Microbiol.* **37:**762–765.

Gasanov, U., D. Hughes, and P. M. Hansbro. 2005. Methods for the isolation and identification of *Listeria* spp. and *Listeria monocytogenes*: a review. *FEMS Microbiol. Rev.* **29:**851–875.

Greenwood, M., C. Willis, P. Doswell, G. Allen, and K. Pathak. 2005. Evaluation of chromogenic media for the detection of *Listeria* species in food. *J. Appl. Microbiol.* **99:**1340–1345.

Hitchins, A. D. 2003. *Listeria monocytogenes*, p. 10.01–10.13. *In* Food and Drug Administration (ed.), *Bacteriological Analytical Manual*, 8th ed. AOAC International, Washington, DC.

Hoorfar, J., and D. Baggesen. 1998. Importance of pre-enrichment media for isolation of *Salmonella* spp. from swine and poultry. *FEMS Microbiol. Lett.* **169:**125–130.

Jenkins, C., J. Evans, H. Chart, G. A. Willshaw, and G. Frankel. 2008. *Escherichia coli* serogroup O26—a new look at an old adversary. *J. Appl. Microbiol.* **104:**14–25.

Jenkins, C., M. C. Pearce, A. W. Smith, H. I. Knight, D. J. Shaw, T. Cheasty, G. Foster, G. J. Gunn, G. Dougan, H. R. Smith, and G. Frankel. 2003. Detection of *Escherichia coli* serogroups O26, O103, O111, and O145 from bovine faeces using immunomagnetic separation and PCR/DNA probe techniques. *Lett. Appl. Microbiol.* **37:**207–212.

Jiang, J., C. Larkin, M. Steele, C. Poppe and J. A. Odumeru. 1998. Evaluation of universal preenrichment broth for the recovery of foodborne pathogens from milk and cheese. *J. Dairy Sci.* **81:** 2798–2803.

Kanki, M., K. Seto, J. Sakata, T. Harada, and Y. Kumeda. 2009. Simultaneous enrichment of Shiga toxin-producing *Escherichia coli* O157 and O26 and *Salmonella* in food samples using universal preenrichment broth. *J. Food Prot.* **72:**2065–2070.

Kim, H., and A. K. Bhunia. 2008. SEL, a selective enrichment broth for simultaneous growth of *Salmonella enterica, Escherichia coli* O157:H7, and *Listeria monocytogenes*. *Appl. Environ. Microbiol.* **74:**4853–4866.

Klein, E. J., J. R. Stapp, M. A. Neill, J. M. Besser, M. T. Osterholm, and P. I. Tarr. 2004. Shiga toxin antigen detection should not replace sorbitol MacConkey agar screening of stool specimens. *J. Clin. Microbiol.* **42:**4416–4417.

Kobayashi, T., S. Enomoto, R. Sakazaki, and S. Kuwahara. 1963. A new selective medium for pathogenic vibrios: T.C.B.S. agar (Modified Nakanishi's Agar). *Jpn. J. Bacteriol.* **18:**387–391.

Leimeister-Wächter, M., E. Domann, and T. Chakraborty. 1991. Detection of a gene encoding a phosphatidylinositol-specific phospholipase C that is co-ordinately expressed with listeriolysin in *Listeria monocytogenes*. *Mol. Microbiol.* **5:**361–366.

Line, J. E. 2001. Development of a selective differential agar for the isolation and enumeration of *Campylobacter* spp. *J. Food Prot.* **64:**1711–1715.

Manafi, M. 2000. New developments in chromogenic and fluorogenic culture media. *Int. J. Food Microbiol.* **60:**205–218.

Manafi, M., and B. Kremsmaier. 2001. Comparative evaluation of different chromogenic/fluorogenic media for detecting *Escherichia coli* O157:H7 in food. *Int. J. Food Microbiol.* **71:**257–262.

March, S. B., and S. Ratnam. 1986. Sorbitol-MacConkey medium for detection of *Escherichia coli* O157:H7 associated with hemorrhagic colitis. *J. Clin. Microbiol.* **23:**869–872.

Mead, P. S., L. Slutsker, V. Dietz, L. F. McCaig, J. S. Bresee, C. Shapiro, P. M. Griffin, and R. V. Tauxe. 1999. Food-related illness and death in the United States. *Emerg. Infect. Dis.* **5:**607–625.

Miller, R. G., and E. T. Mallinson. 2000. Improved detection of *Salmonella* of nontyphoid and typhoid salmonellae with balanced agar formulations. *J. Food Prot.* **63:**1443–1446.

Murray, P., E. J. Baron, J. H. Jorgensen, M. A. Pfaller, and R. H. Yolken (ed.). 2003. *Manual of Clinical Microbiology*, 8th ed. ASM Press, Washington, DC.

Nakanishi, Y. 1963. An isolation agar medium for cholerae and enteropathogenic halophilic vibrios. *Modern Media* **9:**246.

National Advisory Committee on Microbiological Criteria for Foods. 2007. Analytical utility of *Campylobacter* methodologies. *J. Food Prot.* **70:**241–250.

Notermans, S. H., J. Dufrenne, M. Leimeister-Wachter, E. Domann, and T. Chakraborty. 1991. Phosphatidylinositol-specific phospholipase C activity as a marker to distinguish between pathogenic and nonpathogenic *Listeria* species. *Appl. Environ. Microbiol.* **57:**2666–2670.

Novicki, T. J., J. A. Daly, S. L. Mottice, and K. C. Carroll. 2000. Comparison of sorbitol MacConkey agar and a two-step method which utilizes enzyme-linked immunosorbent assay toxin testing and a chromogenic agar to detect and isolate enterohemorrhagic *Escherichia coli*. *J. Clin. Microbiol.* **38:**547–551.

Oyarzabal, O. A., K. S. Macklin, J. M. Barbaree, and R. S. Miller. 2005. Evaluation of agar plates for direct enumeration of *Campylobacter* spp. from poultry carcass rinses. *Appl. Environ. Microbiol.* **71:** 3351–3354.

Perry, J. D., M. Ford, J. Taylor, A. L. Jones, R. Freeman, and F. K. Gould. 1999. ABC medium, a new chromogenic agar for selective isolation of *Salmonella* spp. *J. Clin. Microbiol.* **37:**766–768.

Pignato, S., G. Giammanco, and G. Giammanco. 1995. Rambach agar and SM- I D medium sensitivity for presumptive identification of *Salmonella* subspecies I-VI. *J. Med. Microbiol.* **43:**68–71.

Pontello, M., S. Russolo, F. Carozzi, and U. Bottiroli. 1987. Evaluation of a new, rapid method (MUCAP test) for the presumptive identification of *Salmonella* on primary isolation media. *Abstr. 5th Int. Symp. Rapid Methods and Automation Microbiol. Immunol.*, Florence, Italy, p. 339.

Poupart, M. C., M. Mounier, F. Denis, J. Sirot, C. Couturier, A. M. Freydiere, and Y. Gille. 1991. Detection of salmonellae by using Rambach agar and by a C8 esterase spot test. *J. Clin. Microbiol.* **29:**2357–2359.

Rambach, A. 1990. New plate medium for facilitated differentiation of *Salmonella* spp. from *Proteus* spp. and other enteric bacteria. *Appl. Environ. Microbiol.* **56:**301–303.

Reissbrodt, R. 2004. New chromogenic plating media for detection and enumeration of pathogenic *Listeria* spp.—an overview. *Int. J. Food Microbiol.* **95:**1–9.

Schönenbrücher, V., E. T. Mallinson, and M. Bulte. 2008. A comparison of standard cultural methods for the detection of foodborne *Salmonella* species including three new chromogenic plating media. *Int. J. Food Microbiol.* **123:**61–66.

Sharma, V. K., and S. A. Carlson. 2000. Simultaneous detection of *Salmonella* strains and *Escherichia coli* O157:H7 with fluorogenic PCR and single-enrichment-broth culture. *Appl. Environ. Microbiol.* **66:**5472–5476.

Silliker, J. H., and D. A. Gabis. 1973. ICMSF methods studies. I. Comparison of analytical schemes for detection of *Salmonella* in dried foods. *Can. J. Microbiol.* **19:**475–479.

Skirrow, M. B. 1977. *Campylobacter* enteritis: a new disease. *Br. Med. J.* **2:**9–11.

Sperber, W. A., M. A. Moorman, and T. A. Freier. 2001. Cultural methods for the enrichment and isolation of microorganisms, p. 45–51. *In* F. P. Downes and K. Ito (ed.), *Microbiological Examination of Foods*. American Public Health Association, Washington, DC.

Su, Y. C., J. Duan, and W. H. Wu. 2005. Selectivity and specificity of a chromogenic medium for detecting *Vibrio parahaemolyticus*. *J. Food Prot.* **68:** 1454–1456.

Taylor, W. I., and J. H. Silliker. 1961. Isolation of salmonellae from food samples. IV. Comparison of methods of enrichment. *Appl. Microbiol.* **9:**484–486.

van Dijk, S., M. J. Bruins, and G. J. H. M. Ruijs. 2009. Evaluation and implementation of a chromogenic agar medium for *Salmonella* detection in stool in routine laboratory diagnostics. *J. Clin. Microbiol.* **47:**456–458.

Willis, C., T. Baalham, M. Greenwood, and F. Presland. 2006. Evaluation of a new chromogenic agar for the detection of *Listeria* in food. *J. Appl. Microbiol.* **101:**711–717.

Wong, H. C. 2003. Detecting and molecular typing of *Vibrio parahaemolyticus*. *J. Food Drug Anal.* **11:** 100–107.

Yu, Y. G., H. Wu, Y. Y. Liu, S. L. Li, X. Q. Yang, and X. L. Xiao. 2010. A multipathogen selective enrichment broth for simultaneous growth of *Salmonella enterica* serovar Enteritidis, *Staphylococcus aureus*, and *Listeria monocytogenes*. *Can. J. Microbiol.* **56:**585–597.

Zadik, P. M., P. A. Chapman, and C. A. Siddons. 1993. Use of tellurite for the selection of verocytotoxigenic *Escherichia coli* O157. *J. Med. Microbiol.* **39:**155–158.

AUTOMATED AND LARGE-SCALE CHARACTERIZATION OF MICROBIAL COMMUNITIES IN FOOD PRODUCTION

E. Reynisson, K. Rudi, V. Þ. Marteinsson, J. Nakayama, N. Sakamoto, A. Rasooly, and J. Hoorfar

4

MICROBIAL COMMUNITIES IN FOOD

Microbial communities in food and food production establishments have highly versatile structures based on various extrinsic factors. These include food type, temperature, salinity, pH, water content, etc. (Olofsson et al., 2007a, 2007b; Reynisson et al., 2009, 2010). The influence of environmental factors on microbial population structure in food products is becoming more evident (Abriouel et al., 2008; Hovda et al., 2007a; Nakayama et al., 2007; Tanaka et al., 2010). The information gained by large-scale microbial community analysis not only can deepen the understanding of food microbiology in general but also can lead to improvement of food production systems for increased quality and safety and extension of shelf life (Rudi et al., 2004; Reynisson et al., 2010). In this chapter, the most recent methods applied to microbial communities are reviewed and examples of some cutting-edge technologies are presented.

Molecular methods have been developing rapidly in recent years both for specific detection of single species and for screening assays that allow the species composition of a given food sample to be unraveled. Examples of those methods are 16S rRNA clone analysis (Reynisson et al., 2010; Schmidt et al., 1991), fingerprinting methods such as terminal restriction fragment length polymorphism (t-RFLP) (Liu et al., 1997; Wilson et al., 2008), and denaturing gradient gel electrophoresis (DGGE) (Hovda et al., 2007a, 2007b; Muyzer et al., 1993), tag-encoded pyrosequencing (Huse et al., 2008), single-nucleotide primer extension (SNuPE) (Nikolausz et al., 2009a, 2009b), and microarrays (Wu et al., 2001; Rasooly and Herold, 2008). Furthermore, flow cytometry is also addressed, but this technique is based on single-cell analysis whereby a cell suspension is concentrated by nanofluidics and analyzed by laser technology (Veal et al., 2000; Comas-Riu and Rius, 2009).

E. Reynisson and V. Þ. Marteinsson, Matis—Icelandic Food Research and Biothech R&D, Vínlandsleið 12, 113 Reykjavík, Iceland. *K. Rudi,* Nofima Food, Osloveien 1, 1430 Ås (Nofima Food) and Hedmark University College, Holsetgt. 22, 2318 Hamar (Hedmark University College), Norway. *J. Nakayama and N. Sakamoto,* Department of Bioscience and Biotechnology, Faculty of Agriculture, Graduate School Kyushu University, 6-10-1 Hakozaki Higashi-ku Fukuoka, Japan. *A. Rasooly,* Division of Biology, Office of Science and Engineering, FDA, Silver Spring, MD 20993 U.S. and Cancer Diagnosis Program, National Cancer Institute, 6130 Executive Blvd., Rockville, MD 20852. *J. Hoorfar,* National Food Institute, Technical University of Denmark, Mørkhøj Bygade 19, DK-2860 Søborg, Denmark.

Rapid Detection, Identification, and Quantification of Foodborne Pathogens, ed. by J. Hoorfar

Why Analyze Bacterial Communities in Food?

The bacterial species present in food or in the processing environment are usually heterotrophs that multiply fast under favorable environmental conditions. Such an ideal environment is often humid, with excessive water flow containing nutrients and other important components that promote bacterial growth. The vast majority of the bacteria present in contact surfaces in processing lines originate from the raw material itself such as in fish and meat production. They are usually nonhazardous to humans but responsible for food deterioration and spoilage (Bagge-Ravn et al., 2003; Brightwell et al., 2006). However, there is also a high risk of contamination in foods from humans working at the processing line. The bacterial load in food and processing surfaces is dependent on the food type in question.

For example, it may take several days before caught fish is delivered to a processing plant, and further storage is often required when fish lots are queued for processing. During this time, bacteria present on the skin, gills, and digestive tract will grow in numbers and will diminish product quality if storage conditions are inadequate (Cakli et al., 2006; Lauzon et al., 2009). During this time, the microbial community composition is altered, and minor changes in environmental factors can have a great effect on the microbial community structure (Reynisson et al., 2010).

The production of other food types, such as meat, can face similar problems but with different members of the microbiota (Bagge-Ravn et al., 2003; Brightwell et al., 2006). The prevalence and growth of pathogenic bacteria in food can be influenced by interactions with the bacterial community, making it important to understand the interaction patterns in order to be able to control the risk of human infections.

The current focus is on developing predictive models to describe dynamic microbial community interactions, which can reliably describe microbial growth in food production (McMeekin and Ross, 1996; Kitts, 2001; Rudi et al., 2007a). The models are intended to predict either the shelf life of food products or the risk of pathogens entering and growing in different food types. A more comprehensive understanding of microbial population dynamics in food will certainly help to build more accurate models, but this requires more cost-effective and large-scale methods to produce more data.

Food and the Microbiota of the Human Gut

The human gut has through millions of years of evolution been adapted to commensal bacteria in food, water, and the environment (Ley et al., 2008). In the last century, the fight against foodborne pathogens through food preservation and hygiene, however, has dramatically altered the commensal microbiota we are exposed to, potentially leading to the emergence of diseases related to lack of bacterial exposure, such as allergy and other immunologically related diseases (Blaser and Falkow, 2009). The incidences of chronic immunological gut-related disease are increasing dramatically in the developed world, with an importance that by far exceeds that of traditional foodborne pathogens (Binder, 2004).

A major analytical challenge is the presence of a wide and undefined range of microorganisms that can potentially be harmful, and an obstacle is the lack of proper analytical tools to describe the complete microbial diversity and to distinguish beneficial from harmful bacteria (Rudi et al., 2007b). Obviously, we would not want to go back to a situation with high foodborne infection incidences, because we want to preserve the gut commensal microflora. A potential direction, however, could be to define the normal microflora in different types of food and to use deviations from the normal flora as a pathogen indicator (Rudi et al., 2007a). It may then be possible to detect long-term changes in the food microbiota that can have harmful consequences.

A new principle in analyses of potential chemical contamination of food is to determine the normal chemical profile and use deviation from this as an indicator of contamination. In principle, using this approach, a single analysis can be used to detect any chemical contamination. With the potential to screen the complete microbiota, we foresee that the detection of emerging pathogens can be accomplished through the analysis of deviations from the normal flora. Historically, it has proven very difficult to detect new pathogens. The reason for this is that pathogens can be of any phylotype (Marshall and Warren, 1984). Now, however, with the tools described, we should be able to define and survey the normal microbiota in different categories of food.

In the future, more important than the direct pathogen detection will be the monitoring of the complete microbiota in order to detect potential changes and to correlate historical changes with the emergence of new diseases. For instance, such data would have been very useful in identifying potential underlying microbial causes for the increase of immunologically related disorders seen today.

With emerging technologies becoming available at a lower price, it will be realistic to rapidly screen the complete microbiota (Brugere et al., 2009). This would enable us to preserve and define the beneficial commensal microflora, while combating pathogens.

This chapter gives an overview of established and newer methods available for rapid or large-scale microbial community screening. Most of them are based on 16S rRNA sequencing information. Ever-growing public databases containing sequencing information of various bacteria will keep supporting their application for a long time to come (Cole et al., 2009; Petrosino et al., 2009).

CURRENT METHODS

There are two basic concepts of microbiota profiling techniques. Either the analyses identify all bacteria present in a sample, or the analyses detect an array of bacteria that are expected to be present in a given matrix (Brugere et al., 2009). The benefit of detecting all bacteria is obviously the possibility to detect new and emerging bacteria. The disadvantages, however, are cost, lack of sensitivity, and complexity. Diagnostics of predefined groups, on the other hand, are obviously limited with respect to the detection of new and emerging bacterial groups, while the benefits are that relatively rapid and cost-efficient procedures can be developed.

Due to complementary properties, a relatively low frequency of use of techniques could cover the complete microbiota with high frequent analyses of predefined groups. In this way, potential new and emerging bacteria can be detected and included in a predefined group assay. An overview of current microbial community analysis methods is provided in Table 1.

16S Clone Analysis

It has been known for many years that it is not possible to cultivate and analyze all the microbes that can be seen under the microscope (Staley and Konopka, 1985). It has been estimated that it is possible to cultivate only about 0.1 to 5% of the microbes depending on the environment or their habitat, but with molecular methods it is now possible to analyze some of these uncultivated microbes (Amann et al., 1995; DeLong, 1997). Such estimations are more or less based on using the 16S rRNA molecule that is present in all prokaryotic cells. The first analyses were done by using the 5S rRNA molecule (Stahl et al., 1984; Lane et al., 1985). However, the larger ribosomal molecules, such as 16S rRNA and 23S rRNA, showed much higher sequence resolutions (Woese and Fox, 1977; Olsen et al., 1986).

Today, the 16S rRNA gene has become the gold standard of macromolecules used in taxonomic classification and evolutionary relations. The advantage of the 16S rRNA molecule is that it is present in all bacteria, and universal primers that bind to the majority of known taxa have been designed. It is well conserved but with hypervariable areas (V1 through V9) within the gene, which enables

TABLE 1 Summary of fast and/or large-scale methods used in microbial community analysis

Method	Applications	Supplies	Output	Advantages	Disadvantage(s)	Yr first cited	Reference
Cultivation	Various	Chemicals for medium preparation	Isolated strains and possibility of various downstream applications	Cultivated isolates accessible Simple equipment	Various growth conditions needed for different species Some bacteria uncultivable	1876	Koch, 1876 *B. anthracis*
16S rRNA clone analysis	Dominating population structure	Thermal cycler Sequencer (Sanger)	Sequences of 100–1,500 bp	Long sequences Genus or species classifications possible Community analysis	Few sequences (usually 50–100) Rare species inaccessible	1991	Schmidt et al., 1991
Massively parallel V6 tag sequencing	Detailed population structure	Thermal cycler Sequencer (pyrosequencer)	Sequences of 100–400 bp	Many sequences (10,000–30,000) Accessibility to rare species	Short sequence information Expensive instrumentation	2006	Sogin et al., 2006
t-RFLP	Population fingerprint	Thermal cycler DNA fragment analyzer	Peak profile	Semihigh throughput Rapid (1–2 days)	No species information except by parallel 16S clone library	1997	Liu et al., 1997
FACS	Cell sorting, detection, and enumeration	Flow cytometer	Isolated cells Enumeration of total bacteria or selected groups	Isolation of noncultivated species Fast counting of physiologically active cells	Expensive instrumentation	1996	Fuchs et al., 1996
Microarray	Population structure Strain gene expression profile	Microarray chip Scanning device	Positive or negative signal from each representative probe	Rapid, high throughput	Only previously known species detectable High detection limit Specificity and sensitivity issues	1995 2001	Schena et al., 1995 (gene expression) Wu et al., 2001 (community analysis)
SNuPE	Dominant population structure	Capillary gel electrophoresis Microarray chip Scanning device	Probe signals for relative quantification	Rapid, high througput, quantitative, specific	Only previously known species detectable	2000	Syvanen et al., 1990 (SNP detection) Rudi et al., 2000 (community analysis)

both distant and close determination of evolutionary relations. It is, however, often too conserved to discriminate between closely related species (Tringe and Hugenholtz, 2008).

The method has been used in two separate applications: to sequence the gene of pure culture strains and to analyze bacterial community structure and composition in a given environment. In both cases, the DNA has to be isolated from a pure strain or from a sample containing a mixture of DNA from all organisms present in the sample. The 16S gene is then amplified with universal primers, and the PCR product is directly sequenced from the pure strain, but the PCR product from the DNA mixture is cloned in a vector, typically TOPO cloning or similar applications. Each clone has therefore a 16S rRNA gene from a single bacterium in the community. Sequencing a sufficient number of clones can give information on species or genus composition in the environment in question. With this approach, both cultivable and noncultivable bacteria can be detected. In order to use the sequence for identification, it has to be run against a database containing 16S rRNA sequences and species information (Tringe and Hugenholtz, 2008). To date (as of December 2009), there are about 1.25 million sequences positioned in the 16S ribosomal database (rdp.cme.msu.edu) and the number is rapidly growing (Fig. 1).

Hypervariable Pyrosequencing

There have been major breakthroughs in high-throughput sequencing during the last years, which is now referred to as next-generation sequencing (Voelkerding et al., 2009). Currently, hypervariable V6-to-V8 pyrosequencing (sequencing by synthesis) is the only technology that is widely used for deep sequencing of the microbiota (Wommack et al., 2009). The reason is that this technology enables relatively long sequencing lengths,

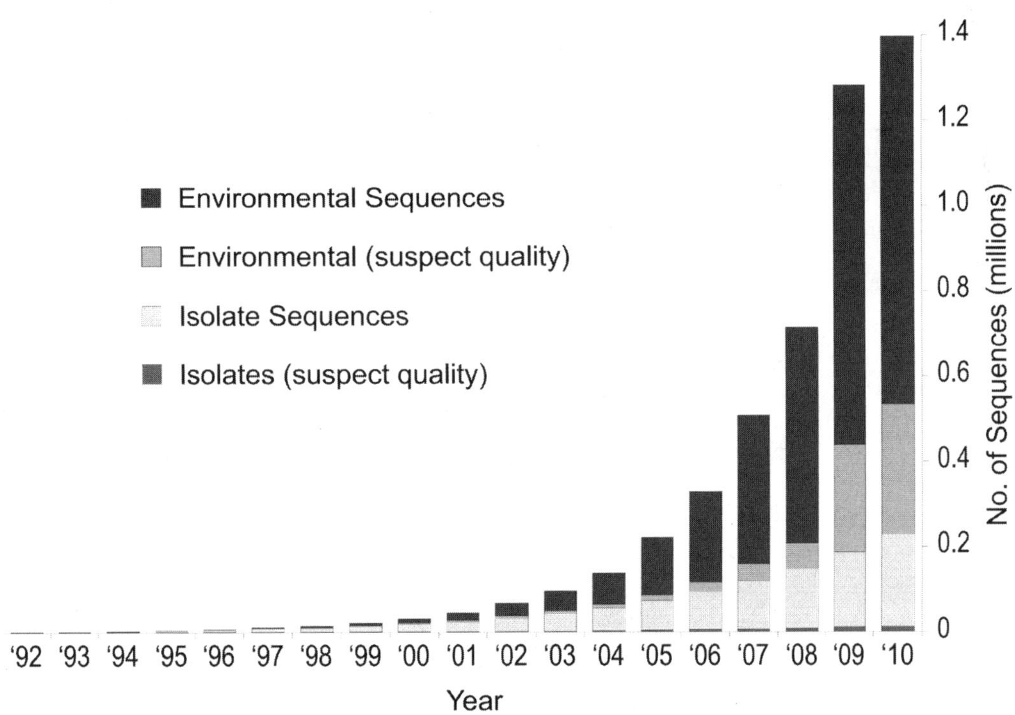

FIGURE 1 Growth of the 16S rRNA database since 1996 (Cole et al., 2009) (adapted from http://rdp.cme.msu.edu).

about 400 nucleotides. There are also developments of techniques that are based on single-molecule sequencing. It is, however, still unclear if these will reach the routine diagnostic market. It is likely that high-throughput sequencing will be implemented in the food industry as low-throughput surveillance techniques.

Recent progress in DNA sequencing technology is innovative and has stimulated research activities within microbial ecology as well as in other life science fields. Pyrosequencing-based 16S rRNA profiling is one of the most powerful techniques to explore microbial community structures in nature. Until now, this method has been used mainly for analysis of the environmental or human microbiome and only to a lesser extent in the food-associated microbiome (Huber et al., 2007; Roh et al., 2009; Zhang et al., 2009). However, with its accuracy, coverage, and the potential of in-depth analysis, it is just as applicable in the field of food microbiology.

As in some other molecular microbial community analyses, the process for V6-to-V8 pyrosequencing begins with isolation of total bacterial DNA from a sample (step 1). The V6-to-V8 region of the 16S rRNA gene, corresponding to *Escherichia coli* nucleotide positions 968 to 1390, is most commonly selected for the PCR amplification (step 2). The primers are linked with a bar code tag sequence in the 5′-end moiety, which enables assignment of mixed read sequences to the sample. The amplicons from different samples are mixed and then applied to emulsion PCR to generate clonal DNA molecules attached to a single bead (step 3). Finally, the pyrosequence reaction is performed in a picotiter plate containing a single bead in each well and is monitored by a charge-coupled-device camera over the plate (step 4).

The data handling can be a challenging task, but bioinformatics tools are available for this particular purpose. To process the pyrosequence data to bacterial composition data, "Pyrosequencing Pipeline" in the Ribosomal Database Project (RDP) website (http://pyro.cme.msu.edu/) is rather useful. The batch sequence data obtained by the instrument are subjected to the "pipeline initial process" available at the RDP website to sort each read into each piece of sample sequence data (step 5).

The sample sequence data are then converted to the bacterial population data by using "RDP classifier" or "RDP seqmatch" available at the Pyrosequencing Pipeline (step 6). RDP classifier uses a Bayesian algorithm to allow fast taxonomic classification of the 16S rRNA fragment sequences, in which batch sample sequence data can be converted to the bacterial population data automatically. The RDP Seqmatch outputs the phylogenetic information of the bacterium whose 16S rRNA sequence is found to be the closest to the query sequence. To illustrate the possibilities, we present an analysis scenario of 128 samples. For this, 128 bar code primers are needed, and by using only one-fourth of the area of a picotiter plate, approximately 100,000 reads are expected to be obtained in total; these are sorted into 128 sample sequence data, each containing approximately 1,000 reads, in which one read represents 0.1% of the total population. The method described here has also been applied for gastrointestinal tract bacterial composition analysis and reviewed in detail by Nakayama (2010).

Regarding food analysis, this application has recently been used in a study on "nukadoko," which is a fermented rice bran bed traditionally used for pickling vegetables in Japan (unpublished results). It is prepared by natural fermentation of salt water-paste rice bran and can be maintained, sometimes for many years, while refreshed by adding fresh rice bran every few weeks to few months. The microbial composition in the nukadoko has not yet been clarified in detail due to its complexity and variety depending on samples. For this type of naturally fermented food containing unexplored complex microbiota, pyrosequence-based 16S rRNA profiling was used to determine its bacterial community structure.

In this study, 16 different samples of nukadokos were analyzed and 18,268 valid reads including 6,519 different sequences were obtained. The data obtained enabled a detailed characterization of the bacterial composition, in which 489 operational taxonomic units were identified. As a result of bar code tag sorting, 685 to 1,525 reads were assigned to each sample (Fig. 2A), and bacterial composition was then obtained by a ribosomal database analysis tool (Fig. 2B).

The most abundant species was *Lactobacillus acetotolerans*, which was an uncultured and slowly growing lactic acid bacterium in nukadoko. *L. acetotolerans* was commonly detected as a predominant organism in some other nukadoko samples. The second predominant species, *Lactobacillus namurensis*, has been successfully cultured from this nukadoko. *Lactobacillus namurensis* was also commonly detected together with *L. acetotolerans* from some other nukadokos. The third predominant species was assigned to *Lactobacillus gallinarum*. Interestingly, however, the matching S_ab score was 0.871, corresponding to 97.8% identity between the sequences from the nukadoko and the database. This suggests that the corresponding bacteria in the nukadoko may be yet-to-be cultured species (Nakayama et al., 2007).

The benefit of pyrosequencing-based 16S rRNA profiling is its high resolution power and possibilities of multisample analysis, which is not possible to obtain with other methodologies.

Fingerprinting Microbial Communities

In some occasions, what is required is not knowledge of specific bacterial species composition but rather the fingerprint of the community, which can be used to evaluate changes in the microbial composition in relation with some intervening event, time in storage of food, depth of a water column, geographic location, etc. In all these cases, the rapid screening of the population structure is sometimes preferable over detailed information on species composition, which includes a greater workload and higher expenses. The most common method for this purpose is t-RFLP (Liu et al., 1997). The method depends on an amplification step with universal primers just as in 16S rRNA clone analysis but

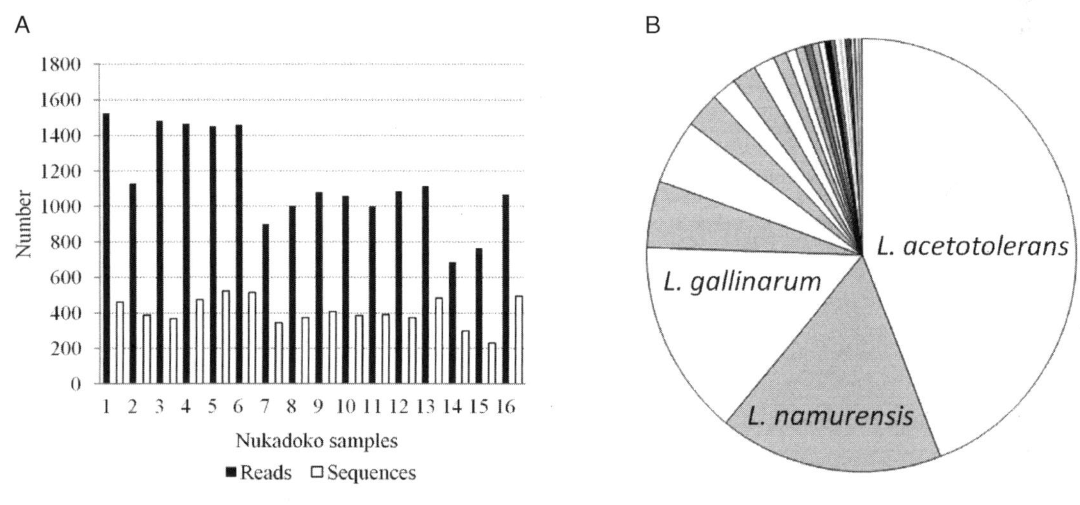

FIGURE 2 (Left) The number of reads (closed bar) and unique sequences (open bar) per nukadoko sample. A total of 18,268 valid reads including 6,519 different sequences were obtained. (Right) Circle graph representation of bacterial species composition in a long-aged nukadoko (sample no. 6).

with a different downstream application. One of the primers (or both) is labeled with a fluorescent dye giving each amplified DNA molecule a label on the corresponding end. Since various bacterial species contain different 16S rRNA nucleotide sequences, they also contain restriction sites at different positions. The PCR product is cleaved by restriction enzymes that recognize common sequences (usually four cutters), which results in labeled DNA fragments of different sizes according to the restriction site nearest to the terminus (Kitts, 2001). The sample is then run on a DNA analyzer in a fragment size mode along with an internal size standard (Fig. 3). Supporting databases for t-RFLP are available, containing a list of organisms and their terminal fragment size using specific primers and restriction enzymes (trflp.limnology.wisc.edu). The users can also upload their own data sets based on other primer-restriction enzyme combinations for phylogenetic assignment. However, such an assignment will never be as accurate as sequencing information, since some bacterial species can have differentiated restriction sites in different strains and unrelated bacteria can also share the same terminal restriction site (Kent et al., 2003). Reports on a specialized t-RFLP technique to screen a typical fish microbiota was recently described (Tanaka et al., 2010)

Another method, DGGE, can be used for the same purpose as the t-RFLP technique but cannot be considered as a high-throughput analysis. Instead of separation by terminal restriction sites, the PCR products originating from different species are separated on an agarose gel containing a gradient of a chemical denaturant, e.g., urea, which allows discrimination based on GC content (Muyzer et al., 1995). The main disadvantages of DGGE are that in complex communities, it is possible that one single band is composed of PCR products from more than one species and the method is low throughput compared to t-RFLP, which can be analyzed in a 96-well format.

Microarrays

Microarrays are spatially ordered arrays of recognition ligands (such as oligonucleotide, cDNA, aptamers, antibody, protein, peptide, or tissue) immobilized (chemically bonded) in discrete order on a solid matrix (e.g., glass or plastic). The technology is a high-throughput methodology capable of molecular identifica-

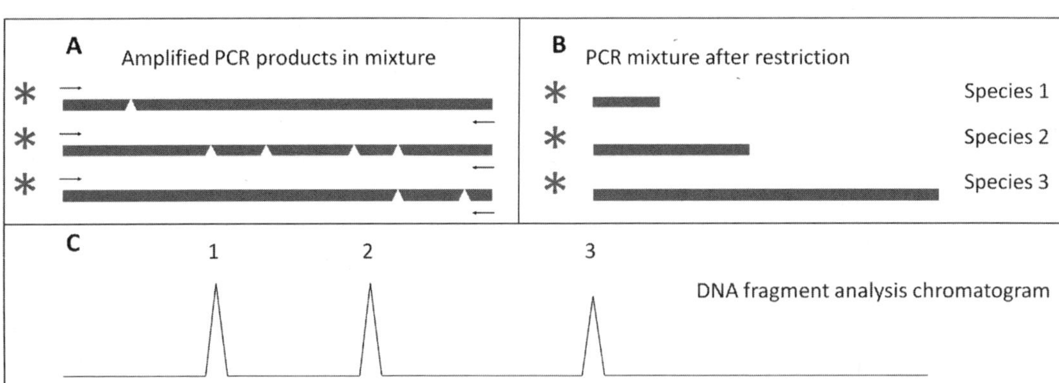

FIGURE 3 Flow sheet of the fundamentals of t-RFLP when used on a mixture of three cultures (species 1, 2, and 3). To start with, they are coamplified with PCR, resulting in a labeled PCR product (indicated by asterisk). The three species contain restriction sites on several locations within their 16S gene (A). Upon restriction, only the terminal fragment is detectable in a DNA fragment analyzer due to the fluorogenic labeling (B and C).

tion and characterization of multiple target analytes, mainly DNA sequences or proteins bound to the microarray in a single assay. Microarray platforms enable hundreds or thousands of parallel identification assays, each specific to a small section of a genome or a specific antigen.

The technology is mainly used for analysis of gene expression. Beyond that application, microarrays have a significantly wider potential for genomic analysis including microbial diagnostics enabled to obtain multiple detailed genomic or proteomic data regarding the pathogen, including identification of virulence factors or antibiotic resistance, and with the capability to analyze multiple organisms simultaneously.

Oligonucleotide Microarray Technology

One approach in microarray technology used for microbial analysis is oligonucleotide microarray, in which oligonucleotides are short (20 to 80 bases) single strands immobilized on glass slides. The DNA to be analyzed is labeled with a fluorescent dye (e.g., Cy3 or Cy5) and hybridized to the array. In such analysis, the target DNA hybridizes specifically to the complementary sequence on the array, and this annealing of the fluorescently labeled target can be detected by optical detectors such as a laser scanner, a charge-coupled-device camera, or even office scanners (Rasooly and Herold, 2008). The target's molecular label enables the identification of the binding position on the array, which identifies the sequence. The four main laboratory steps for using a microarray are (i) target preparation and labeling; (ii) hybridization of the labeled target to the array; (iii) washing of unbound material; and (iv) scanning, image acquisition, and data analysis.

Microarray Analysis of Microbial Populations

The ability for simultaneous analysis of a large number of specimens and a large number of molecular markers enables many potential applications including microbial gene expression, genotyping, microbial identification, DNA resequencing, mutational analysis, microbial community population analysis, microbial ecology analysis, antibiotic resistance determination, and virulence factor identification. For microbial microarray genotyping analysis, although it is feasible to probe every gene in the genome on a single array, more focused arrays are very useful to probe signature sequences from the genome. Several genomic characteristics (signature sequences) have been identified and detected including genes coding for rRNA (Edman et al., 2000; Small et al., 2001; Loy et al., 2002, 2005; Reyes-Lopez et al., 2003; Chandler et al., 2006; Peplies et al., 2006; Sanguin et al., 2006; Hoshino et al., 2007), virulence factors and antibiotic resistance potential (Chizhikov et al., 2001; Volokhov et al., 2002, 2003a, 2003b, 2004; Sergeev et al., 2004a, 2004b, 2006a, 2006b; Zhu et al., 2007), or shared conserved indels (insertions/deletions) (Gupta and Johari, 1998; Karlin and Brocchieri, 1998; Gupta and Griffiths, 2002; Gupta, 2004).

Microarray Analysis of the Physiology and Pathogenesis

Microarrays enable better understanding of microbial physiology and pathogenesis. For example, in *Salmonella enterica* serovar Typhimurium, the signaling of several critical cellular processes including metabolism, biofilm formation, transcription, translation, transport, and binding proteins, signal transduction, and regulatory functions were studied by microarray analysis, which identified two regulatory genes (*luxS* and *AI-2*) (Jesudhasan et al., 2009). Microarray analysis has shown that antagonistic flora members such as *Lactococcus lactis*, a lactic acid bacterium that is widely used in the dairy industry, can modulate *Staphylococcus aureus* proliferation and virulence expression (Even et al., 2009). Microarray analysis was used to decipher the regulation of sigma factor sigma(B), which has an important role in the acquisition of stress resistance and spore germination in *Bacillus cereus* and many other gram-positive bacteria (van Schaik et al., 2007).

Microarray-Based Identification and Genotyping

For simultaneous analysis of multiple bacterial virulence factors, oligonucleotide probes were integrated into a single DNA chip (Sergeev et al., 2004b) and used for identification of several food pathogens (*S. aureus* enterotoxin genes, *Listeria* spp., *Campylobacter* spp., and *Clostridium perfringens*). These bacteria represent the majority of food microbial pathogens. Using similar virulence factors, oligonucleotide-based microarrays were able to detect *E. coli* O157:H7, *S. enterica, Listeria monocytogenes,* and *Campylobacter jejuni* in meat (Suo et al., 2010). Genomic markers were used for a high-sensitivity pathogen detection microarray (10 fg of *Bacillus anthracis*). Target sequences were PCR amplified, targeting 18 potential biowarfare agents (Wilson et al., 2002). A microarray for broader species analysis targeting *E. coli, Shigella* spp., *Salmonella* spp., *Aeromonas hydrophila, Vibrio cholerae, Mycobacterium avium, Mycobacterium tuberculosis, Helicobacter pylori, Proteus mirabilis, Yersinia enterocolitica,* and *C. jejuni* was developed (Kostic et al., 2007). The introduction of competitive oligonucleotides in the labeling reaction successfully suppressed cross-reaction by closely related sequences, significantly improving the performance of the assay. Environmental performance was tested with environmental and veterinary samples harboring complex microbial communities. Detection sensitivity in the range of 0.1% has been demonstrated.

Other microarrays utilizing the 16S-23S rRNA gene internal transcribed spacer sequences and the *wzy* (O antigen polymerase) gene were used for detection of *Enterobacter sakazakii, Salmonella enterica, Klebsiella pneumoniae, Klebsiella oxytoca, Serratia marcescens, Acinetobacter baumannii, Bacillus cereus, Listeria monocytogenes, Staphylococcus aureus,* and *Escherichia coli* O157 in powdered infant formula (Wang et al., 2009).

Multiplex Diagnostics by Single Nucleotide Primer Extension

With the emerging extensive sequence database for foodborne pathogens, comprising both multilocus sequence typing (MLST) and 16S rRNA genes, it is now possible to target predefined mutations in characterizing the food microflora. There are now numerous techniques for mutation detection, including sequencing-, ligation- and hybridization-based techniques.

SNuPE, however, is currently the most accurate point mutation detection technique (Nikolausz et al., 2009b). The application range of SNuPE analyses is very wide, covering almost all areas of mutation analyses. In food microbiology, SNuPE analyses have been used to describe the microflora in ready-to-eat salads (Rudi et al., 2002) and in cheese ripening (Treimo et al., 2006), for discriminating pathogens and nonpathogens (Kostic et al., 2007), and in MLST (Rudi et al., 2003; Berget et al., 2007). Recently, it has also been shown that the sensitivity of SNuPE analyses enables PCR-independent point mutation detection (Nikolausz et al., 2009b).

SNuPE analyses have been adapted to a wide range of detection technologies such as capillary gel electrophoresis (Nikolausz et al., 2009a), mass spectroscopy (Tang et al., 1999), real-time PCR (Rudi and Holck, 2003; Rudi et al., 2006), and DNA array (Rudi et al., 1998). These detection platforms cover different ranges of mutations to be detected simultaneously and have potential for high-throughput analyses. The SNuPE detection adapted to real-time PCR instrumentation has the highest throughput potential but the lowest potential for multiplexing. The principle is that probes with a 5′-fluorescence label are SNuPE extended by a quencher. The signal is recorded in real time by the reduced report signal (Rudi and Holck, 2003). The multiplexing potential is limited to the number of overlapping fluorochromes (Rudi et al., 2006).

Mass spectroscopy is also a high-throughput technique whereby the increased mass of the probe is detected by a mass spectrophotometer (Tang et al., 1999). There are, however, several technical challenges with mass spectroscopy, since the technique detects all masses.

The most widely used SNuPE detection technology is capillary gel electrophoresis. Here, the principle is that different probes are detected by different sizes. Normally, a cytosine tail of different lengths is used for the size differentiation (Nikolausz et al., 2009a). The limitation of this technique is the synthesis of long oligonucleotide probes, limiting the technique to the detection of <20 probes simultaneously.

Microarray-based techniques represent the most complex techniques but have the possibility of detecting a much larger number of probes (>100) simultaneously. Here, there are three specific microarray platforms. One platform utilizes a universal tag (Bell et al., 2002) that enables a universal array. However, the disadvantage of this platform is the risk of cross-reactivity during labeling and hybridization. The second group of arrays is based on direct labeling of probes attached to the array (Pastinen et al., 2000). The challenge here is steric hindrance in the labeling step. The final group of array techniques is based on initial labeling of the probes, with subsequent array hybridization (Rudi et al., 1998). The challenge here is the compatibility of the probes in the multiplex labeling.

In food microbiota testing, different SNuPE approaches must be chosen based on application. We foresee nested approaches in which broad-range 16S rRNA gene arrays are used initially, with subsequent use of more specific SNuPE approaches based on, for example, pathogen- or virulence gene-specific single-nucleotide polymorphisms. The rationale is that there is no need to test for a certain pathogen if the bacterial group that the pathogen belongs to is not present in the sample at all. On the other hand, testing pathogens directly would require a very high number of runs, since there are a number of pathogens that can be expected to be present in foods.

A practical example of such an approach is the analysis of the microbiota in ready-to-eat food products. A 16S rRNA gene microbiota profiling would be sufficient to determine the presence of *Listeria* spp. in the food. If *Listeria* spp. are absent, then it is not necessary to further search for *L. monocytogenes*. On the other hand, detection of *Listeria* spp. would not necessarily mean that *L. monocytogenes* is present. However, in this case an additional SNupE MLST could be used, both to determine the presence of *L. monocytogenes* and to determine whether the virulent forms are present. Other pathogens can be analyzed in the same way. Together with the development of databases, this method using the SNuPE approach would enable global surveillance systems to cover most of the pathogens expected to be present in foods.

Flow Cytometry

Flow cytometry is a technique for counting, examining, and sorting microscopic particles and allows simultaneous multiparametric analysis of single cells flowing through an optical electronic detection apparatus. The method is based on the optical characterization of cells (and other particles) when they travel through a narrow fast-flowing fluid stream, introducing them to a focused beam. Three parameters are analyzed when the cells travel through the beam: forward scatter, side scatter, and fluorescence. It is a high-throughput technology that has mainly been used in medical cell immunology (Szczepanski et al., 2006). The application of this technique in microbiology has been increasing, especially in environmental microbiology fields (Bergquist et al., 2009) such as marine microbiology (Czechowska et al., 2008). Despite the potential, the major obstacles until now have been complicated instrumentation and the lack of technology to discriminate bacterial cells in complex samples such as mud and food (Veal et al., 2000). Recently, instrumentation has become more user-friendly, and sample preparation and labeling techniques are opening new fields of application for flow cytometry (Comas-Riu and Rius, 2009).

The real advantage of flow cytometry is the fluorescence-activated cell sorting (FACS), whereby fluorescent cells can be sorted for further analysis, e.g., genomic or physiological studies. Some distinctive bacterial species or groups can be differentiated by their natural differences in morphology, size, or fluorescence. The conventional way is, however, labeling by immunological or fluorescence probe-based methods. Fluorescent in situ hybridization (FISH) probes targeting the 16S gene of uncultivated strains have been used to sort the cells from an environmental sample, enabling whole-genome sequencing of the sorted cells. This was, for example, the case for the uncultivated TM7 phylum (Podar et al., 2007).

Microbial community analysis has mostly been restricted to marine micro- or phytoplankton containing autofluorogenic properties that are used as markers when studying community dynamics. Due to the small size of bacteria compared with mammalian cells, the use of flow cytometry to distinguish bacterial groups based solely on their size, shape, or morphology can be quite challenging. Nevertheless, the use of FISH probes targeted against the 16S rRNA molecule is the basis for further developments in this field.

Combining FISH strategies with FACS is an example of emerging technologies through which single bacterial cells can be sorted from a mixed population by their fluorescent signal (Czechowska et al., 2008). Determination of cell concentration is also achieved by flow cytometry, in which physiologically active cells can be counted separately from dead cells via staining by dyes that cannot penetrate the cell membrane of a living cell (Czechowska et al., 2008).

Determination of bacterial concentrations by flow cytometry has shown a high degree of correlation to the standard colony counting and has been suggested as a monitoring scheme for bacterial contamination in the food industry (Gunasekera et al., 2000; McHugh and Tucker, 2007).

CONCLUSION

High-throughput analysis of microbial populations in food products and food processing environments has revealed the existence of a higher complexity in the microbial world than previously expected. The new approaches provide opportunities for further understanding of the microbial developments that are initiated during food production and storage. Our current knowledge in food microbiology is largely still based on classical culture methods, leaving a large knowledge gap. It is expected that the use of molecular techniques for community analysis will narrow this gap. The key feature of most molecular methods is the 16S rRNA gene, which has been shown to be a highly suitable marker for estimation of evolutionary relations between organisms. At the same time, data banks with 16S rRNA gene sequences are rapidly expanding, which gives these methodologies a stronger backbone.

The future research trend in microbial community analysis in food and during food production will include further screening and data acquisition from different food types. Massive sequencing methodology based on pyrosequencing is becoming less expensive, with increased potential for in-depth analysis of microbial communities. The use of large-scale community analysis approaches in an applied sense is likely to include simpler, cheaper, and faster methods. Future examples of applied analysis of microbial communities in the food industry could be in monitoring of starter cultures, probiotic bacteria, and microbial contamination in food processing facilities; predictive food microbiology; and aquaculture, in which certain types of microbiota can provide optimal growth conditions for the cultured fish species.

ACKNOWLEDGMENTS

J. Hoorfar and K. Rudi were in part financially supported by the European Union project BIOTRACER (FOOD-2006-CT-036272).

REFERENCES

Abriouel, H., A. Martin-Platero, M. Maqueda, E. Valdivia, and M. Martinez-Bueno. 2008. Biodiversity of the microbial community in a Spanish farmhouse cheese as revealed by culture-dependent and culture-independent methods. *Int. J. Food Microbiol.* **127:**200–208.

Amann, R. I., W. Ludwig, and K. H. Schleifer. 1995. Phylogenetic identification and in situ detection of individual microbial cells without cultivation. *Microbiol. Rev.* **59:**143–169.

Bagge-Ravn, D., Y. Ng, M. Hjelm, J. N. Christiansen, C. Johansen, and L. Gram. 2003. The microbial ecology of processing equipment in different fish industries—analysis of the microflora during processing and following cleaning and disinfection. *Int. J. Food Microbiol.* **87:**239–250.

Bell, P. A., S. Chaturvedi, C. A. Gelfand, C. Y. Huang, M. Kochersperger, R. Kopla, F. Modica, M. Pohl, S. Varde, R. Zhao, X. Zhao, M. T. Boyce-Jacino, and A. Yassen. 2002. SNPstream UHT: ultra-high throughput SNP genotyping for pharmacogenomics and drug discovery. *BioTechniques* **2002**(Suppl.)**:**70–72, 74, 76–77.

Berget, I., E. Heir, J. Petcovic, and K. Rudi. 2007. Discriminatory power, typability, and accuracy of single nucleotide extension microarrays. *J. AOAC Int.* **90:**802–809.

Bergquist, P. L., E. M. Hardiman, B. C. Ferrari, and T. Winsley. 2009. Applications of flow cytometry in environmental microbiology and biotechnology. *Extremophiles* **13:**389–401.

Binder, V. 2004. Epidemiology of IBD during the twentieth century: an integrated view. *Best Pract. Res. Clin. Gastroenterol.* **18:**463–479.

Blaser, M. J., and S. Falkow. 2009. What are the consequences of the disappearing human microbiota? *Nat. Rev. Microbiol.* **7:**887–894.

Brightwell, G., J. Boerema, J. Mills, E. Mowat, and D. Pulford. 2006. Identifying the bacterial community on the surface of Intralox belting in a meat boning room by culture-dependent and culture-independent 16S rDNA sequence analysis. *Int. J. Food Microbiol.* **109:**47–53.

Brugere, J. F., A. Mihajlovski, M. Missaoui, and P. Peyret. 2009. Tools for stools: the challenge of assessing human intestinal microbiota using molecular diagnostics. *Expert Rev. Mol. Diagn.* **9:**353–365.

Cakli, S., B. Kilinc, T. Dincer, and S. Tolasa. 2006. Effects of using slurry ice during transportation on the microbiological, chemical, and sensory assessments of aquacultured sea bass (*Dicentrarchus labrax*) stored at 4 degrees C. *Crit. Rev. Food Sci. Nutr.* **46:**453–458.

Chandler, D. P., A. E. Jarrell, E. R. Roden, J. Golova, B. Chernov, M. J. Schipma, A. D. Peacock, and P. E. Long. 2006. Suspension array analysis of 16S rRNA from Fe- and SO(4)2-reducing bacteria in uranium-contaminated sediments undergoing bioremediation. *Appl. Environ. Microbiol.* **72:**4672–4687.

Chizhikov, V., A. Rasooly, K. Chumakov, and D. D. Levy. 2001. Microarray analysis of microbial virulence factors. *Appl. Environ. Microbiol.* **67:** 3258–3263.

Cole, J. R., Q. Wang, E. Cardenas, J. Fish, B. Chai, R. J. Farris, A. S. Kulam-Syed-Mohideen, D. M. McGarrell, T. Marsh, G. M. Garrity, and J. M. Tiedje. 2009. The Ribosomal Database Project: improved alignments and new tools for rRNA analysis. *Nucleic Acids Res.* **37:** D141–D145.

Comas-Riu, J., and N. Rius. 2009. Flow cytometry applications in the food industry. *J. Ind. Microbiol. Biotechnol.* **36:**999–1011.

Czechowska, K., D. R. Johnson, and J. R. van der Meer. 2008. Use of flow cytometric methods for single-cell analysis in environmental microbiology. *Curr. Opin. Microbiol.* **11:**205–212.

DeLong, E. F. 1997. Marine microbial diversity: the tip of the iceberg. *Trends Biotechnol.* **15:**203–207.

Edman, C. F., P. Mehta, R. Press, C. A. Spargo, G. T. Walker, and M. Nerenberg. 2000. Pathogen analysis and genetic predisposition testing using microelectronic arrays and isothermal amplification. *J. Investig. Med.* **48:**93–101.

Even, S., C. Charlier, S. Nouaille, N. L. Ben Zakour, M. Cretenet, F. J. Cousin, M. Gautier, M. Cocaign-Bousquet, P. Loubière, and Y. Le Loir. 2009. *Staphylococcus aureus* virulence expression is impaired by *Lactococcus lactis* in mixed cultures. *Appl. Environ. Microbiol.* **75:**4459–4472.

Fuchs, P., W. Weichel, S. Dubel, F. Breitling, and M. Little. 1996. Separation of *E. coli* expressing functional cell-wall bound antibody fragments by FACS. *Immunotechnology* **2:**97–102.

Gunasekera, T. S., P. V. Attfield, and D. A. Veal. 2000. A flow cytometry method for rapid detection and enumeration of total bacteria in milk. *Appl. Environ. Microbiol.* **66:**1228–1232.

Gupta, R. S. 2004. The phylogeny and signature sequences characteristics of Fibrobacteres, Chlorobi, and Bacteroidetes. *Crit. Rev. Microbiol.* **30:**123–143.

Gupta, R. S., and V. Johari. 1998. Signature sequences in diverse proteins provide evidence of a close evolutionary relationship between the *Deinococcus*-thermus group and cyanobacteria. *J. Mol. Evol.* **46:**716–720.

Gupta, R. S., and E. Griffiths. 2002. Critical issues in bacterial phylogeny. *Theor. Popul. Biol.* **61:**423–434.

Hoshino, T., K. Furukawa, S. Tsuneda, and Y. Inamori. 2007. RNA microarray for estimating relative abundance of 16S rRNA in microbial communities. *J. Microbiol. Methods* **69:**406–410.

Hovda, M. B., B. T. Lunestad, M. Sivertsvik, and J. T. Rosnes. 2007a. Characterisation of the bacterial flora of modified atmosphere packaged farmed Atlantic cod (*Gadus morhua*) by PCR-DGGE of conserved 16S rRNA gene regions. *Int. J. Microbiol.* **117:**68–75.

Hovda, M. B., M. Sivertsvik, B. T. Lunestad, G. Lorentzen, and J. T. Rosnes. 2007b. Characterisation of the dominant bacterial population in modified atmosphere packaged farmed halibut (*Hippoglossus hippoglossus*) based on 16S rDNA-DGGE. *Food Microbiol.* **24:**362–371.

Huber, J. A., D. B. M. Welch, H. G. Morrison, S. M. Huse, P. R. Neal, D. A Butterfield, and M. L. Sogin. 2007. Microbial population structures in the deep marine biosphere. *Science* **318:**97–100.

Huse, S. M., L. Dethlefsen, J. A. Huber, D. M. Welch, D. A. Relman, and M. L. Sogin. 2008. Exploring microbial diversity and taxonomy using SSU rRNA hypervariable tag sequencing. *PLoS Genet.* **4:**e1000255.

Jesudhasan, P. R., M. L. Cepeda, K. Widmer, S. E. Dowd, K. A. Soni, M. E. Hume, J. Zhu, and S. D. Pillai. 2009. Transcriptome analysis of genes controlled by luxS/aAutoinducer-2 in *Salmonella enterica* serovar Typhimurium. *Foodborne Pathog. Dis.* **7:**399–410.

Karlin, S., and L. Brocchieri. 1998. Heat shock protein 70 family: multiple sequence comparisons, function, and evolution. *J. Mol. Evol.* **47:**565–577.

Kent, A. D., D. J. Smith, B. J. Benson, and E. W. Triplett. 2003. Web-based phylogenetic assignment tool for analysis of terminal restriction fragment length polymorphism profiles of microbial communities. *Appl. Environ. Microbiol.* **69:**6768–6776.

Kitts, C. L. 2001. Terminal restriction fragment patterns: a tool for comparing microbial communities and assessing community dynamics. *Curr. Issues Intest. Microbiol.* **2:**17–25.

Koch, R. 1876. Untersuchungen über Bakterien. V. Die Ätiologie der Milzbrand-Krankheit, begründet auf die Entwicklungsgeschichte des *Bacillus anthracis* [Investigations into bacteria. V. The etiology of anthrax, based on the ontogenesis of *Bacillus anthracis*]. *Cohns Beitrage zur Biologie der Pflanzen* **2:**277–310.

Kostic, T., A. Weilharter, S. Rubino, G. Delogu, S. Uzzau, K. Rudi, A. Sessitsch, and L. Bodrossy. 2007. A microbial diagnostic microarray technique for the sensitive detection and identification of pathogenic bacteria in a background of nonpathogens. *Anal. Biochem.* **360:**244–254.

Lane, D. J., D. A. Stahl, G. J. Olsen, D. J. Heller, and N. R. Pace. 1985. Phylogenetic analysis of the genera *Thiobacillus* and *Thiomicrospira* by 5S rRNA sequences. *J. Bacteriol.* **163:**75–81.

Lauzon, H. L., H. Magnusson, K. Sveinsdottir, M. Gudjonsdottir, and E. Martinsdottir. 2009. Effect of brining, modified atmosphere packaging, and superchilling on the shelf life of cod (*Gadus morhua*) loins. *J. Food Sci.* **74:**M258–M267.

Ley, R. E., M. Hamady, C. Lozupone, P. J. Turnbaugh, R. R. Ramey, J. S. Bircher, M. L. Schlegel, T. A. Tucker, M. D. Schrenzel, R. Knight, and J. I. Gordon. 2008. Evolution of mammals and their gut microbes. *Science* **320:**1647–1651.

Liu, W. T., T. L. Marsh, H. Cheng, and L. J. Forney. 1997. Characterization of microbial diversity by determining terminal restriction fragment length polymorphisms of genes encoding 16S rRNA. *Appl. Environ. Microbiol.* **63:**4516–4522.

Loy, A., A. Lehner, N. Lee, J. Adamczyk, H. Meier, J. Ernst, K.-H. Schleifer, and M. Wagner. 2002. Oligonucleotide microarray for 16S rRNA gene-based detection of all recognized lineages of sulfate-reducing prokaryotes in the environment. *Appl. Environ. Microbiol.* **68:**5064–5081.

Loy, A., C. Schulz, S. Lucker, A. Schöpfer-Wendels, K. Stoecker, C. Baranyi, A. Lehner, and M. Wagner. 2005. 16S rRNA gene-based oligonucleotide microarray for environmental monitoring of the betaproteobacterial order "*Rhodocyclales*." *Appl. Environ. Microbiol.* **71:**1373–1386.

Marshall, B. J., and J. R Warren. 1984. Unidentified curved bacilli in the stomach of patients with gastritis and peptic ulceration. *Lancet* **i:**1311–1315.

McHugh, I. O., and A. L. Tucker. 2007. Flow cytometry for the rapid detection of bacteria in cell culture production medium. *Cytometry A* **71:**1019–1026.

McMeekin, T. A., and T. Ross. 1996. Shelf life prediction: status and future possibilities. *Int. J. Food Microbiol.* **33:**65–83.

Muyzer, G., E. C. de Waal, and A. G. Uitterlinden. 1993. Profiling of complex microbial populations by denaturing gradient gel electrophoresis analysis of polymerase chain reaction-amplified genes coding for 16S rRNA. *Appl. Environ. Microbiol.* **59:**695–700.

Muyzer, G., A. Teske, C. O. Wirsen, and H. W. Jannasch. 1995. Phylogenetic relationships of *Thiomicrospira* species and their identification in deep-sea hydrothermal vent samples by denaturing gradient gel electrophoresis of 16S rDNA fragments. *Arch. Microbiol.* **164:**165–172.

Nakayama, J. 2010. Pyrosequence-based 16S rRNA profiling of gastro-intestinal microbiota. *Biosci. Microflora* **29:**83–96.

Nakayama, J., H. Hoshiko, M. Fukuda, H. Tanaka, N. Sakamoto, S. Tanaka, K. Ohue, K. Sakai, and K. Sonomoto. 2007. Molecular monitoring of bacterial community structure in long-aged nukadoko: pickling bed of fermented rice bran dominated by slow-growing lactobacilli. *J. Biosci. Bioeng.* **104:**481–489.

Nikolausz, M., A. Chatzinotas, A. Tancsics, G. Imfeld, and M. Kastner. 2009a. Evaluation of single-nucleotide primer extension for detection and typing of phylogenetic markers used for investigation of microbial communities. *Appl. Environ. Microbiol.* **75:**2850–2860.

Nikolausz, M., A. Chatzinotas, A. Tancsics, G. Imfeld, and M. Kastner. 2009b. The single-nucleotide primer extension (SNuPE) method for the multiplex detection of various DNA sequences: from detection of point mutations to microbial ecology. *Biochem. Soc. Trans.* **37:**454–459.

Olofsson, T. C., S. Ahrne, and G. Molin. 2007a. The bacterial flora of vacuum-packed cold-smoked salmon stored at 7 degrees C, identified by direct 16S rRNA gene analysis and pure culture technique. *J. Appl. Microbiol.* **103:**109–119.

Olofsson, T. C., S. Ahrne, and G. Molin. 2007b. Composition of the bacterial population of refrigerated beef, identified with direct 16S rRNA gene analysis and pure culture technique. *Int. J. Food Microbiol.* **118:**233–240.

Olsen, G. J., D. J. Lane, S. J. Giovannoni, N. R. Pace, and D. A. Stahl. 1986. Microbial ecology and evolution: a ribosomal RNA approach. *Annu. Rev. Microbiol.* **40:**337–365.

Pastinen, T., M. Raitio, K. Lindroos, P. Tainola, L. Peltonen, and A. C. Syvanen. 2000. A system for specific, high-throughput genotyping by allele-specific primer extension on microarrays. *Genome Res.* **10:**1031–1042.

Peplies, J., C. Lachmund, F. O. Glockner, and W. Manz. 2006. A DNA microarray platform based on direct detection of rRNA for characterization of freshwater sediment-related prokaryotic communities. *Appl. Environ. Microbiol.* **72:**4829–4838.

Petrosino, J. F., S. Highlander, R. A. Luna, R. A. Gibbs, and J. Versalovic. 2009. Metagenomic pyrosequencing and microbial identification. *Clin. Chem.* **55:**856–866.

Podar, M., C. B. Abulencia, M. Walcher, D. Hutchison, K. Zengler, J. A. Garcia, T. Holland, D. Cotton, L. Hauser, and M. Keller. 2007. Targeted access to the genomes of low-abundance organisms in complex microbial communities. *Appl. Environ. Microbiol.* **73:**3205–3214.

Rasooly, A., and K. E. Herold. 2008. Food microbial pathogen detection and analysis using DNA microarray technologies. *Foodborne Pathog. Dis.* **5:**531–550.

Reyes-Lopez, M. A., A. Mendez-Tenorio, R. Maldonado-Rodriguez, M. J. Doktycz, J. T. Fleming, and K. L. Beattie. 2003. Fingerprinting of prokaryotic 16S rRNA genes using oligodeoxyribonucleotide microarrays and virtual hybridization. *Nucleic Acids Res.* **31:**779–789.

Reynisson, E., H. L. Lauzon, H. Magnusson, R. Jonsdottir, G. Olafsdottir, V. T. Marteinsson, and G. O. Hreggvidsson. 2009. Bacterial composition and succession during storage of North-Atlantic cod (*Gadus morhua*) at superchilled temperatures. *BMC Microbiol.* **9:**250.

Reynisson, E., H. L. Lauzon, L. Thorvaldsson, B. Margeirsson, A. R. Runarsson, V. Marteinsson, and E. Martinsdóttir. 2010. Effects of different cooling techniques on bacterial succession and other spoilage indicators during storage of whole, gutted haddock (*Melnogrammus aeglifinus*). *Eur. Food Res. Technol.* **231:**237–246.

Roh, S. W., K. H. Kim, Y. D. Nam, H. W. Chang, E. J. Park, and J. W. Bae. 2009. Investigation of archaeal and bacterial diversity in fermented seafood using barcoded pyrosequencing. *ISME J.* **4:**1–16.

Rudi, K., and A. L. Holck. 2003. Real-time closed tube single nucleotide polymorphism (SNP) quantification in pooled samples by quencher extension (QEXT). *Nucleic Acids Res.* **31:**e117.

Rudi, K., T. Katla, and K. Naterstad. 2003. Multi locus fingerprinting of *Listeria monocytogenes* by sequence-specific labeling of DNA probes combined with array hybridization. *FEMS Microbiol. Lett.* **220:**9–14.

Rudi, K., G. H. Kleiberg, R. Heiberg, and J. T. Rosnes. 2007a. Rapid identification and classification of bacteria by 16S rDNA restriction fragment melting curve analyses (RFMCA). *Food Microbiol.* **24:**474–481.

Rudi, K., O. M. Skulberg, F. Larsen, and K. S. Jakobsen. 1998. Quantification of toxic cyanobacteria in water by use of competitive PCR followed by sequence-specific labeling of oligonucleotide probes. *Appl. Environ. Microbiol.* **64:**2639–2643.

Rudi, K., O. M. Skulberg, R. Skulberg, and K. S. Jakobsen. 2000. Application of sequence-specific labeled 16S rRNA gene oligonucleotide probes for genetic profiling of cyanobacterial abundance and diversity by array hybridization. *Appl. Environ. Microbiol.* **66:**4004–4011.

Rudi, K., T. Maugesten, S. E. Hannevik, and H. Nissen. 2004. Explorative multivariate analyses of 16S rRNA gene data from microbial communities in modified-atmosphere-packed salmon and coalfish. *Appl. Environ. Microbiol.* **70:**5010–5018.

Rudi, K., M. Zimonja, S. E. Hannevik, and S. M. Dromtorp. 2006. Multiplex real-time single nucleotide polymorphism detection and quantification by quencher extension. *BioTechniques* **40:** 323–329.

Rudi, K., M. Zimonja, P. Trosvik, and T. Naes. 2007b. Use of multivariate statistics for 16S rRNA gene analysis of microbial communities. *Int. J. Food Microbiol.* **120:**95–99.

Rudi, K., S. L. Flateland, J. F. Hanssen, G. Bengtsson, and H. Nissen. 2002. Development and evaluation of a 16S rDNA array approach for describing complex microbial communities in ready-to-eat vegetable salads packed in modified atmosphere. *Appl. Environ. Microbiol.* **68:**1146–1156.

Sanguin, H., A. Herrera, C. Oger-Desfeux, A. Dechesne, P. Simonet, E. Navarro, T. M. Vogel, Y. Moënne-Loccoz, X. Nesme, and G. L. Grundmann. 2006. Development and validation of a prototype 16S rRNA-based taxonomic microarray for Alphaproteobacteria. *Environ. Microbiol.* **8:**289–307.

Schena, M., D. Shalon, R. W. Davis, and P. O. Brown. 1995. Quantitative monitoring of gene expression patterns with a complementary DNA microarray. *Science* **270:**467–470.

Schmidt, T. M., E. F. DeLong, and N. R. Pace. 1991. Analysis of a marine picoplankton community by 16S rRNA gene cloning and sequencing. *J. Bacteriol.* **173:**4371–4378.

Sergeev, N., K. E. Herold, and A. Rasooly. 2006a. Microarray analysis of microbial pathogens, p. 363–405. *In* M. J. Miller (ed.), *Encyclopedia of Rapid Microbiological Methods*, vol. 3. PDA, Bethesda, MD.

Sergeev, N., D. Volokhov, V. Chizhikov, and A. Rasooly. 2004a. Simultaneous analysis of multiple staphylococcal enterotoxin genes by an oligonucleotide microarray assay. *J. Clin. Microbiol.* **42:**2134–2143.

Sergeev, N., M. Distler, M. Vargas, V. Chizhikov, K. E. Herold, and A. Rasooly. 2006b. Microarray analysis of *Bacillus cereus* group virulence factors. *J. Microbiol. Methods* **65:**488–502.

Sergeev, N., M. Distler, S. Courtney, S. F. Al-Khaldi, D. Volokhov, V. Chizhikov, and A. Rasooly. 2004b. Multipathogen oligonucleotide microarray for environmental and biodefense applications. *Biosens. Bioelectron.* **20:**684–698.

Small, J., D. R. Call, F. J. Brockman, T. M. Straub, and D. P. Chandler. 2001. Direct detection of 16S rRNA in soil extracts by using oligonucleotide microarrays. *Appl. Environ. Microbiol.* **67:**4708–4716.

Sogin, M., H. G. Morrison, J. A. Huber, D. M. Welch, S. M. Huse, P. R. Neal, J. M. Arrieta, and G. J. Herndl. 2006. Microbial diversity in the deep sea and the underexplored "rare biosphere." *Proc. Natl. Acad. Sci. USA* **103:**12115–12120.

Stahl, D. A., D. J. Lane, G. J. Olsen, and N. R. Pace. 1984. Analysis of hydrothermal vent-associated symbionts by ribosomal RNA sequences. *Science* **224:**409–411.

Staley, J. T., and A. Konopka. 1985. Measurement of in situ activities of nonphotosynthetic microorganisms in aquatic and terrestrial habitats. *Annu. Rev. Microbiol.* **39:**321–346.

Suo, B., Y. He, G. Paoli, A. Gehring, S. I. Tu, and X. Shi. 2010. Development of an oligonucleotide-based microarray to detect multiple foodborne pathogens. *Mol. Cell. Probes* **24:**77–86.

Syvanen, A. C., K. Aalto-Setala, L. Harju, K. Kontula, and H. Söderlund. 1990. A primer-guided nucleotide incorporation assay in the genotyping of apolipoprotein-E. *Genomics* **8:**684–692.

Szczepanski, T., V. H. van der Velden, and J. J. van Dongen. 2006. Flow-cytometric immunophenotyping of normal and malignant lymphocytes. *Clin. Chem. Lab. Med.* **44:**775–796.

Tanaka, Y., H. Takahashi, N. Kitazawa, and B. Kimura. 2010. Rapid estimation of microbial populations in fish samples by using terminal restriction fragment length polymorphism analysis of 16S rDNA. *J. Food Prot.* **73:**104–113.

Tang, K., D.-J. Fu, D. Julien, A. Braun, C. R. Cantor, and H. Köster. 1999. Chip-based genotyping by mass spectrometry. *Proc. Natl. Acad. Sci. USA* **96:**10016–10020.

Treimo, J., G. Vegarud, T. Langsrud, and K. Rudi. 2006. Use of DNA quantification to measure growth and autolysis of *Lactococcus* and *Propionibacterium* spp. in mixed populations. *Appl. Environ. Microbiol.* **72:**6174–6182.

Tringe, S. G., and P. Hugenholtz. 2008. A renaissance for the pioneering 16S rRNA gene. *Curr. Opin. Microbiol.* **11:**442–446.

van Schaik, W., M. van der Voort, D. Molenaar, R. Moezelaar, W. M. de Vos, and T. Abee. 2007. Identification of the sigmaB regulon of *Bacillus cereus* and conservation of sigmaB-regulated genes in low-GC-content gram-positive bacteria. *J. Bacteriol.* **189:**4384–4390.

Veal, D. A., D. Deere, B. Ferrari, J. Piper, and P. V. Attfield. 2000. Fluorescence staining and flow cytometry for monitoring microbial cells. *J. Immunol. Methods* **243:**191–210.

Voelkerding, K. V., S. A. Dames, and J. D. Durtschi. 2009. Next-generation sequencing: from basic research to diagnostics. *Clin. Chem.* **55:** 641–658.

Volokhov, D., A. Rasooly, K. Chumakov, and V. Chizhikov. 2002. Identification of *Listeria* species by microarray-based assay. *J. Clin. Microbiol.* **40:**4720–4728.

Volokhov, D., V. Chizhikov, K. Chumakov, and A. Rasooly. 2003a. Microarray-based identification of thermophilic *Campylobacter jejuni, C.*

coli, *C. lari*, and *C. upsaliensis*. *J. Clin. Microbiol.* **41:** 4071–4080.

Volokhov, D., V. Chizhikov, K. Chumakov, and A. Rasooly. 2003b. Microarray analysis of erythromycin resistance determinants. *J. Appl. Microbiol.* **95:**787–798.

Volokhov, D., A. Pomerantsev, V. Kivovich, A. Rasooly, and V. Chizhikov. 2004. Identification of *Bacillus anthracis* by multiprobe microarray hybridization. *Diagn. Microbiol. Infect. Dis.* **49:**163–171.

Wang, M., B. Cao, Q. Gao, Y. Sun, P. Liu, L. Feng, and L. Wang. 2009. Detection of *Enterobacter sakazakii* and other pathogens associated with infant formula powder by use of a DNA microarray. *J. Clin. Microbiol.* **47:**3178–3184.

Wilson, B., B. S. Danilowicz, and W. G. Meijer. 2008. The diversity of bacterial communities associated with Atlantic cod *Gadus morhua*. *Microb. Ecol.* **55:**425–434.

Wilson, W. J., C. L. Strout, T. Z. DeSantis, J. L. Stilwell, A. V. Carrano, and G. L. Andersen. 2002. Sequence-specific identification of 18 pathogenic microorganisms using microarray technology. *Mol. Cell. Probes* **16:**119–127.

Woese, C. R., and G. E. Fox. 1977. Phylogenetic structure of the prokaryotic domain: the primary kingdoms. *Proc. Natl. Acad. Sci. USA* **74:**5088–5090.

Wommack, K. E., S. R. Bench, J. Bhavsar, D. Mead, and T. Hanson. 2009. Isolation independent methods of characterizing phage communities 2: characterizing a metagenome. *Methods Mol. Biol.* **502:**279–289.

Wu, L., D. K. Thompson, G. Li, R. A. Hurt, J. M. Tiedje, and J. Zhou. 2001. Development and evaluation of functional gene arrays for detection of selected genes in the environment. *Appl. Environ. Microbiol.* **67:**5780–5790.

Zhang, H., J. K. DiBaise, A. Zuccolo, D. Kudrna, M. Braidotti, Y. Yu, P. Parameswaran, M. D. Crowell, R. Wing, B. E. Rittmann, and R. Krajmalnik-Brown. 2009. Human gut microbiota in obesity and after gastric bypass. *Proc. Natl. Acad. Sci. USA* **106:**2365–2370.

Zhu, L. X., Z. W. Zhang, C. Wang, H. W. Yang, D. Jiang, Q. Zhang, K. Mitchelson, and J. Cheng. 2007. Use of a DNA microarray for simultaneous detection of antibiotic resistance genes among staphylococcal clinical isolates. *J. Clin. Microbiol.* **45:**3514–3521.

FAST AND HIGH-THROUGHPUT MOLECULAR TYPING METHODS

P. I. Fields, C. Fitzgerald, and J. R. McQuiston

5

As our understanding of pathogens has expanded and the world's food supply has become more interconnected, characterization and subtyping of foodborne pathogens have become critically important in monitoring specific foods, performing surveillance for foodborne pathogens, investigating foodborne outbreaks, and understanding the virulence properties of particular lineages or strains. The availability of highly discriminatory subtyping methods has greatly improved our ability to quickly identify contaminated foods, identify dispersed outbreaks, and trace pathogens to the offending lot of food.

WHICH METHOD TO USE?

A wide variety of subtyping methods have been described. It is important to consider the purpose for which the subtyping is being performed in order that the most appropriate method is selected and to ensure that the method provides meaningful results. Commonly, subtyping is performed for epidemiologic purposes (e.g., cluster detection, outbreak investigation, and tracing or attribution of specific subtypes to their source), in order to determine whether or not the isolates in question may be related to one another, indicating that they have a common origin and might have a common source. Factors to consider when developing or using a subtyping method include typeability, degree of resolution of subtypes, clinical or epidemiologic appropriateness of the subtypes determined, reproducibility and ease of interlaboratory comparisons, time to results, and cost of the method.

Of particular importance in selecting a subtyping method is standardization of the method. In order to compare subtyping results from one laboratory to the next, the labs need to use the same subtyping method and to perform it in a way that generates comparable results. The more successful subtyping schemes have been those that received widespread use, providing a mechanism to compare results between labs. For example, *Salmonella* serotyping, introduced in 1934 as a means of characterizing strains based on the immunoreactivity of two cell surface structures, identifies many specific pathotypes and epidemiologically relevant lineages and continues to provide valuable subtype information

P. I. Fields, C. Fitzgerald, and J. R. McQuiston, Enteric Diseases Laboratory Branch, Division of Foodborne, Waterborne, and Environmental Diseases, Centers for Disease Control and Prevention, 1600 Clifton Rd., Atlanta, GA 30333.

Rapid Detection, Identification, and Quantification of Foodborne Pathogens, ed. by J. Hoorfar

today. Molecular methods to replace traditional methods for serotype determination are being sought in order to maintain the serotype-based understanding of *Salmonella*. Similarly, phage typing, which characterizes strains based on their susceptibility to a panel of bacteriophages, has been widely used and standardized; it continues to provide important subtype information for organisms such as *Salmonella* serovar Enteritidis that seem to be very clonal and difficult to subtype by more modern methods. Standardized pulsed-field gel electrophoresis (PFGE) is the basis of national and international networks (e.g., PulseNet and PulseNet International); PFGE-based networks revolutionized cluster detection and outbreak investigations by identifying sometimes widely dispersed isolates as part of the same exposure or outbreak.

The earliest subtyping methods such as serotyping and phage typing relied on the characterization of phenotypic properties of a strain. Phenotypic subtyping methods have largely given way to molecular methods, which are easier to standardize and often faster to perform. This chapter will provide an overview of the molecular subtyping methods most commonly employed, in particular focusing on those methods providing fast and high-throughput analysis of strains. The heterogeneity that is targeted in molecular subtyping methods falls into three general categories. (i) Single-nucleotide polymorphisms (SNPs) are single-base changes in one sequence in comparison to another, related sequence. SNPs typically arise through sporadic mutation, so they represent a means of measuring random genetic drift. A large number of SNPs between two sequences may indicate a more distant genetic relationship; however, horizontal gene transfer can result in what appears to be multiple SNPs when a new genetic sequence is inserted into an otherwise homogenous genome. (ii) Second, some regions of the genome are more likely to change than others; they provide a means of differentiating closely related strains. Examples include variable number tandem repeats (described below) and highly polymorphic genes such as some surface structure and virulence genes. (iii) Finally, the presence/absence of specific loci can be found in both distantly and closely related strains. The presence of additional sequences in a strain can result from the insertion of a bacteriophage in the chromosome, acquisition of a plasmid, or other horizontal transfer events.

UNIVERSALLY APPLIED SUBTYPING METHODS

This section describes the most widely accepted genotyping methods that are used internationally for outbreak investigations, source attributions, and epidemiological surveillance.

Pulsed-Field Gel Electrophoresis

Macrorestriction digestion of total genomic DNA and analysis of fragment sizes by PFGE were first applied to subtyping bacterial pathogens in the early 1990s and initiated a new era in subtyping of bacteria. Prior to the use of PFGE, several genetic methods based on restriction of genomic DNA with frequently cutting restriction enzymes were introduced but typically produced complicated banding patterns that were difficult to standardize. These methods never received widespread use beyond the labs that introduced them. The combination of the universality of the PFGE method (i.e., its applicability to any bacterium), the relative simplicity and utility of the banding patterns produced, and the development of international subtyping networks based on standardized PFGE methods has contributed to the longevity of this subtyping method. Newer methods have improved the technical ease of the method and interpretation of the data, but PFGE remains the "gold standard" because no new method yet matches the combined discriminating power, epidemiologic concordance, and universality of PFGE.

The basic method for PFGE used in the PulseNet network was detailed by Ribot and colleagues (Ribot et al., 2006). Total genomic

DNA from a bacterial strain is purified in a gelatin plug such that the DNA can be recovered with a minimum of shearing. The DNA is restricted by using restriction enzymes that recognize low-frequency sequences within the genome of interest, and the DNA fragments, typically 50 to 600 kb, are separated by PFGE. The banding patterns, or DNA fingerprints, are analyzed in comparison to those produced by other strains. Differences in banding patterns arise from insertion or deletion of sequences that change the size of a particular fragment, genomic rearrangements, or nucleotide changes that produce or eliminate a restriction site.

Much like serotyping of enteric pathogens, the success of PFGE as a subtyping method can be attributed to the development of a standardized subtyping network based on the technique (Gerner-Smidt et al., 2006; Swaminathan et al., 2006). PulseNet was introduced in the United States in 1996 as a molecular surveillance network for foodborne disease and has expanded to include several international networks in 2010 (http://www.pulsenetinternational.org/). Initially, PulseNet was applied to the investigation of outbreaks detected by other methods. It quickly proved to be extremely efficient for cluster and outbreak detection, leading to the detection of dispersed, multistate outbreaks that had gone undetected under earlier surveillance systems. As PFGE databases grew, their utility increased; PFGE analysis has been applied to additional purposes such as routine surveillance, and comparison and tracking of subtypes across human, animal, and food isolates (Kerouanton et al., 2007; Soyer et al., 2010).

Despite the utility of PFGE-based subtyping, it does have drawbacks. Highly standardized methods and analysis algorithms are required in order to ensure comparability of results between laboratories, the method requires about 24 hours to perform, and only 10 to 20 isolates can be analyzed at a time. The analysis of banding patterns can be time-consuming, and the interpretation of banding patterns is somewhat subjective. The significance of minor differences between banding patterns is often not known, and interpretation of results beyond "match or no match" is often not possible. As a result, improved methods are being sought, but as yet no single method has been able to replace PFGE.

Multilocus Sequence Typing

Comparative sequencing of specific loci within the bacterial genome was first described as a method to help understand evolutionary relationships among bacteria and has since been applied to a wide range of research and public health purposes (Maiden, 2006). The genes targeted for multilocus sequence typing (MLST) are typically housekeeping genes because they are under more selective pressure and thus less likely to undergo random variation. MLST as it was originally conceived includes amplification of approximately seven housekeeping genes by PCR followed by DNA sequencing of the amplicons. The amplicons are typically 400 to 600 base pairs in length, such that they can be sequenced in a single sequencing reaction; both strands of DNA are typically sequenced. Each distinct gene sequence is considered a separate allele; alleles are assigned an allele number, and comparison of strains is done by analysis of allelic profiles. Publicly available databases have been established to provide information on the MLST schemes developed for different bacterial species, to house DNA sequences and allelic profiles for individual strains, and to provide sequence analysis tools (e.g., http://www.mlst.net/ and http://mlst.ucc.ie/). As with any subtyping method, standardization of the region of each gene sequenced and of the system and nomenclature for naming new alleles is critical in order to compare results from study to study (Maiden, 2006). This is not always the case. For example, several different schemes, all termed "MLST," have been reported for *Salmonella*; very different conclusions regarding serovar diversity of *Salmonella* and uses for the method were drawn depending on the MLST scheme used (Fakhr et al.,

2005; Kotetishvili et al., 2002; Sukhnanand et al., 2005; Torpdahl et al., 2005). Care must be taken to ensure that similar methods were used before trying to compare results from different studies.

MLST has the advantage of providing definitive information regarding relationships between strains. However, since housekeeping genes are typically conserved, they are often not variable enough to be used for subtyping. An alternative approach for subtyping is to target more-variable genes, instead of or in addition to housekeeping genes, in order to obtain the desired level of strain discrimination. *flaA* and *porA* have been targeted to improve the discriminatory power of MLST for *Campylobacter jejuni* and applied to outbreak investigations (Clark et al., 2005; Cody et al., 2009). Characterization of loci associated with prophages, both by presence/absence scoring and DNA sequencing, provided greater discrimination than PFGE in a study of *Salmonella* serovar Typhimurium (Ross and Heuzenroeder, 2005).

MLST has been particularly successful in subtyping *Campylobacter* strains, likely due to the fact that campylobacters are a genetically diverse group. MLST alone provided substantial strain discrimination in a study of *Campylobacter coli* epidemiology and showed that environmental isolates were distinct from geographically and temporally related human clinical strains (Sopwith et al., 2010). A study of almost 10,000 strains from human clinical specimens, food, and the environment showed specific associations between MLST type and the source of the *Campylobacter* isolate, indicating the potential for this method in studies of attribution of disease with specific genotypes (Sheppard et al., 2009). Further, MLST analysis of *C. jejuni* isolates from humans and chickens over a period of 11 years showed a decline in the number of strains with matching STs from these two sources, suggesting that human disease attributable to chicken had declined over the study period (de Haan et al., 2010).

Despite the proven utility of MLST and comparative gene sequencing for population and epidemiologic studies, it does require sequencing of multiple loci, typically covering several thousand nucleotides of sequence. Generating the data needed for MLST can be expensive and time-consuming, even with current improvements in technology. Alternatives have been sought, including targeting specific sequence differences among previously characterized alleles (Best et al., 2005); however, this approach does not allow for the discovery of new types. Mass spectrometry also holds promise for simplifying MLST analysis (see below).

Multilocus Variable-Number Tandem Repeat Analysis

Variable-number tandem repeats (VNTRs) are loci that contain short strings of nucleotides that are repeated a few to many times; the number of repeats can vary from strain to strain (Lindstedt, 2005). The availability of whole-genome sequences has facilitated the discovery of VNTRs, which are scattered throughout bacterial genomes, and has led to the birth of a new subtyping method, multilocus VNTR analysis (MLVA). MLVA is a subtyping method based on the characterization of multiple VNTR loci in a microbial genome. Five to 10 VNTR loci are targeted in a typical MLVA scheme, and the number of tandem repeats at each locus serves as a "fingerprint" to identify that strain.

The presence and nature of VNTR loci can be quite variable for different bacterial species and subtypes. As a result, the first step in developing an MLVA scheme is to analyze at least one and ideally more genome sequences from the bacteria of interest. The existence of multiple genomes allows comparison of the sequences to identify VNTR loci that might be more variable than others and thus potentially useful as a subtyping target. VNTR loci that were identified for one species or subtype are often useful for other subtypes (e.g., different serotypes of *Escherichia coli* or *Sal-*

monella). A panel of VNTRs useful for characterization of *E. coli* O157 was directly applicable to *E. coli* O26 strains (Miko et al., 2010). However, more commonly, additional VNTR loci must be sought for each new species or serotype to be characterized. For example, only two of seven VNTR loci used for characterization of *Salmonella* serotype Typhimurium were also useful for serovar Newport; the other five loci were either not present, not amplified, or not variable in serovar Newport (Davis et al., 2009).

Once appropriate VNTR loci are identified, the basic method for MLVA is amplification of each VNTR locus via PCR using primers flanking the VNTR site and determination of the size of the PCR fragments as a measure of the number of repeated units at each locus. Sizing of the PCR fragments was initially performed by using standard gel electrophoresis. Separation of the PCR fragments by capillary gel electrophoresis improved the accuracy of fragment size determinations but limited the use of VNTRs to those that are small enough to be resolved on a capillary gel. Analysis of fragment sizes with capillary gel electrophoresis is amenable to multiplexing by labeling a PCR primer for each locus with a different fluorescent dye.

MLVA protocols first began to appear in the early 2000s and have increased steadily since that time. In comparisons with one- or two-enzyme PFGE, MLVA based on 5 to 10 loci can provide similar or better levels of discrimination depending on the study and strain set. MLVA provided much better discrimination than PFGE for *Salmonella* serovar Typhimurium strains (Torpdahl et al., 2007). *Salmonella* serovar Enteritidis has been shown to be quite clonal by multiple subtyping methods; MLVA provided improved strain differentiation compared to both PFGE and phage typing (Beranek et al., 2009; Malorny et al., 2008). In a comparison of PFGE and MLVA for *E. coli* O157, MLVA seemed to provide better epidemiologic concordance for some outbreaks, whereas PFGE supported the epidemiologic data for others (Hyytia-Trees et al., 2006). MLVA and PFGE provided similar levels of discrimination for *Listeria monocytogenes* (Lindstedt et al., 2008; Sperry et al., 2008).

MLVA offers advantages over PFGE in that it is easier and faster to perform than PFGE and there is less need for excruciating attention to standardized protocols. However, analysis of banding patterns, albeit less complicated ones, is still required. Comparing the sizes of fragments on different capillary electrophoresis platforms is also difficult; Hyytia-Trees and colleagues have developed an algorithm that improved fragment sizing, and thus allele assignment, between two different capillary electrophoresis platforms (Hyytia-Trees et al., 2010). As discussed above, major drawbacks of MLVA are the variability of the VNTR loci themselves and the need for a customized set of loci for different species or subtypes in order to maximize strain discrimination. Also, MLVA has not been standardized through international networks as PFGE has. Four different MLVA protocols have been reported for *Salmonella* serovar Enteritidis (Beranek et al., 2009; Cho et al., 2010; Malorny et al., 2008; Ross and Heuzenroeder, 2009), and three protocols have been reported for *Listeria monocytogenes* (Lindstedt et al., 2008; Murphy et al., 2007; Sperry et al., 2008). Different protocols often target at least some of the same loci, but different sets of loci and different naming conventions can make comparisons between studies difficult.

DNA Microarrays

A DNA microarray is a high-throughput tool for determining the presence or absence of multiple DNA sequences in parallel. Microarrays are not a subtyping method per se but a technique to rapidly characterize sequence differences among DNA regions that have been characterized by other means.

Microarrays consist of a series of DNA molecules, typically oligonucleotides or PCR fragments, attached to a solid support such that the identity of the probe can be determined by its location on the solid support. Sequences in a biological sample that are homologous to a probe on the array are detected by hybridization of the test DNA to the probe, followed by a detection step that identifies the test DNA attached to the probe on the solid support (Miller and Tang, 2009). Early microarrays were typically formatted as a two-dimensional array of DNA probes on a glass slide. Depending on the technology used to produce the slide, an array can consist of hundreds to hundreds of thousands of DNA probes.

Whole-Genome Microarrays

As whole bacterial genome sequences became available in the early 2000s, microarrays representing the genome of the sequenced strain were used to characterize the gene content of additional strains. Probes were typically PCR fragments corresponding to open reading frames from the sequenced genomes; arrays consisting of several thousand PCR fragments were spotted onto glass slides, and the presence or absence of a homologous sequence in the test strain was determined by competitive hybridization between DNA from the sequenced and test strain (Garaizar et al., 2006). Whole-genome comparisons using microarrays have been applied to a variety of pathogens including *Salmonella, E. coli,* and *Campylobacter.* They expanded our understanding of genomic diversity within a species in an era when whole-genome sequencing was very expensive; however, the complexity of data produced does not lend itself to a high-throughput analysis.

Low-Density Microarrays

An improved understanding of strain diversity from whole-genome microarray analysis and other genetic comparisons, combined with the availability of an increasing number of whole-genome sequences, has led to a second generation of DNA microarrays that focus on a smaller number of more informative genetic targets. Low-density microarrays consisting of a few to several hundred probes that detect specific genes or loci have been developed for several foodborne pathogens. Several different arrays employing oligonucleotides 50 to 70 residues in length and targeting genes related to virulence, prophages, plasmids, antimicrobial resistance determinants, and other variable elements have been described. Total genomic DNA from the strain of interest is labeled with a fluorescent dye and hybridized to the array; the analysis of total genomic DNA allows the characterization of strains without preconceived knowledge of the gene content of the strain.

Huehn and colleagues developed a microarray that represented 282 genetic loci from *Salmonella* (Huehn and Malorny, 2009). Probes were 40- to 60-mers and detected genes associated with serotype, virulence, prophages, fimbria production, and antimicrobial resistance. The microarrays have been successfully used to understand the relationships between and virulence of specific *Salmonella* serotypes (Huehn et al., 2009a, 2009b) and identified diversity among *Salmonella* serovar Typhimurium strains possessing the same PFGE pattern (Litrup et al., 2010). Serovar Typhimurium is known to be quite diverse; the discriminatory power of the microarray for more clonal serovars remains to be determined.

Bruant and colleagues developed a microarray to detect 189 genes that have been associated with virulence in *E. coli* and 30 antimicrobial resistance genes, represented by 315 70-mer oligonucleotides (Bruant et al., 2006). The array was used to characterize *E. coli* isolates from surface waters, showing that 24% of the isolates possessed a virulence gene profile suggestive of specific *E. coli* pathotypes (Hamelin et al., 2007). Additionally, the array was used in combination with other subtyping methods to characterize atypical enteropathogenic *E. coli* strains from children with and without diarrhea and showed that specific virulence gene profiles were associated with strains from children with diarrhea (Afset et al., 2008).

Frye and colleagues developed a microarray based on all antimicrobial resistance gene sequences that were available in the GenBank database (Frye et al., 2010). The array consisted of 70-mer oligonucleotides homologous to 775 antimicrobial resistance genes. Since the array detects the presence or absence of specific genes, it does not detect antimicrobial resistance determinants that result from mutation; and only gene presence, not gene expression, is determined, so the result may or may not correspond to the resistance properties of the strain. The array has been used to characterize *Salmonella* strains from poultry and determine relationships between strains (Zou et al., 2009).

Low-density microarrays using 50–70-mer oligonucleotides produce simpler, more easily analyzed data sets than higher-density arrays, and oligonucleotide probe-based microarrays are easier to control for quality and have higher specificity than PCR probes. To date, low-density arrays have been used primarily to characterize strains; they have to some extent been standardized across multiple laboratories (Grønlund et al., 2010), but it remains to be seen whether they will have utility for the routine subtyping of strains.

Liquid Microarrays

Studies employing microarray slides predominate in the literature, but a tube-based microarray format provides substantial improvements in flexibility and throughput. The Luminex platform utilizes differentially labeled microspheres in liquid suspension, and such assays are thus often referred to as "liquid" or "suspension" microarrays (Dunbar, 2006). Each microsphere set, or bead set, is labeled with a variable proportion of two fluorescent dyes that is used to differentiate one bead set from the next. Individual bead sets are labeled with DNA, antibody, or other probes, and different bead sets are mixed to produce multiplex bead pools. Labeled targets, typically PCR fragments, are hybridized with a bead pool. Hybridization of the target to a particular bead is detected by passing the target-bead reaction mixture through a flow cytometer; the fluorescent signature of a bead and the hybridized target is detected as individual beads pass through a detector. The Luminex platform has been used to identify serotypes in *Salmonella* by detecting the genes responsible for the antigens that are detected in traditional serotyping (Fitzgerald et al., 2007; McQuiston et al., 2010). Multiplex assays based on SNPs that differentiated the four lineages of *Listeria monocytogenes* were developed on the Luminex platform and used to characterize strains from ready-to-eat foods; however, the assay did not provide as much strain discrimination as PFGE (Ward et al., 2008).

Summary

Microarrays have proven useful for characterizing unknown strains for whole-genome content or specific gene subsets, using well-defined probe sets. However, the start-up costs for microarrays can be high, and standardization and quality control are critical. Since the contents of a slide microarray must be defined and made in batch in advance of any experimentation, they are not easily adaptable when additional probes are desired. Also, processing and analyzing one glass slide per specimen limits throughput and thus may not be appropriate for large-scale screening. In order to maximize the utility of microarrays, the development of public databases cataloging useful targets and defining specific probes to detect those targets will be an important next step.

Mass Spectrometry

Mass spectrometry is an analytical technique for characterizing the chemical composition of a sample. Mass spectrometry can be divided into three steps: the sample is ionized by any of several techniques, the ions produced from the sample are separated based on their mass-to-charge ratio, and the nature and quantity of the individual ions are determined. Mass spectrometry has been used primarily in two ways for the characterization of microorganisms: characterization of whole cells and characterization of DNA from a single cell or from a complex sample. Characterization of whole bacteria by mass spectrometry has proven to be highly reproducible and useful for the identification of a particular organism by comparison of the mass spectrum of an unknown organism to a library of spectra from known organisms (Mandrell et al., 2005; Mellmann et al., 2009). However, analysis of whole organisms is typically not sensitive enough to provide strain discrimination. A variety of approaches using mass spectrometry for the characterization of DNA molecules or PCR fragments have been reported and may prove more appropriate for subtyping.

Mass spectrometry of biological samples typically employs one of two approaches to ionize the sample (Sauer and Kliem, 2010). In matrix-assisted laser desorption ionization, the sample to be analyzed is embedded in a matrix that protects fragile biomolecules. A spot on the sample-matrix crystal is irradiated with a laser, which ablates a portion of the crystal and volatilizes and ionizes the sample molecules. In electrospray ionization, the sample is dissolved in an organic solvent and applied to a capillary. A voltage is applied to the capillary, which produces an aerosol of droplets containing the sample; the sample is ionized as the droplets pass through a series of vacuum chambers. The mass of the ions produced by either method is typically determined by using time of flight analyzers that accelerate the ions via an electric field. The velocity of each ion is measured as the time it takes for the ion to reach the detector and used to calculate its mass-to-charge ratio.

Application of mass spectrometry to bacterial subtyping is still in its infancy. Methods reported to date take advantage of the understanding of bacterial diversity obtained from other subtyping methods. DNA sequence differences can be detected as mass differences in PCR fragments generated from appropriate DNA targets, such as MLST loci (Hall et al., 2009). This approach requires shorter PCR fragments than traditional MLST in order to accurately determine the weights of the fragments and provides only the nucleotide composition of the fragment, not the nucleotide sequence. However, with careful selection of target sequences, levels of discrimination can be comparable to those of traditional MLST. And as with other subtyping methods, inclusion of additional, species-specific and/or more variable targets can improve discrimination (Wolk et al., 2009).

An alternate method for characterization of sequence diversity is to amplify the region of interest by PCR, convert the PCR fragment to RNA via transcription, cleave the RNA at specific nucleotides in four separate reactions that correspond to the four bases making up the RNA sequence, and determine the masses of the cleavage products in each reaction (Honisch et al., 2007). The specific sequence of the test DNA can be deduced from the masses of the cleavage products from the test DNA in comparison to those of cleavage products produced by a reference sequence. Thus, this approach allows both the discovery of new sequence polymorphisms and the characterization of previously recognized ones. A third approach has been used to detect previously characterized SNPs. The region surrounding the SNP is amplified by PCR, a single nucleotide opposite the SNP is added to an oligonucleotide targeting the SNP region, the mass of the modified oligonucleotide is determined by mass spectrometry, and the nucleotide that was added is deduced from the mass of the modified oligonucleotide (Bishop et al., 2010; Honisch et al., 2010). In practice, up to 40 SNPs can be detected in a single spectrum/reaction.

FUTURE TRENDS

It is safe to say that whole-genome sequencing will play a bigger and bigger role in bacterial subtyping, but we can only imagine how that technology will improve and evolve so that it can be easily applied to routine problems. Whole-genome analysis of 95 strains of group A *Streptococcus* and characterization of SNPs by mass spectrometry in an additional 250 strains has already been used to describe the evolution of an epidemic serotype M3 strain over a period of 15 years (Beres et al., 2010). Whole-genome sequencing, completed in less than a week, has been applied to outbreak investigations (Gilmour et al., 2010; Lewis et al., 2010). Currently, the major limitations in applying whole-genome sequencing to subtyping and other biological questions are the need for improved computational power to analyze the data and the need for a means to store the massive data sets that whole-genome sequencing approaches produce. Both of these limitations will need to be addressed before whole-genome sequencing can be applied widely. Also, as with any subtyping method, whole-genome sequencing must be validated in the context of the biological questions it attempts to answer. It remains to be seen whether the straightforward concordance between whole-genome sequences and epidemiologic data, as in the Gilmour report of a *Listeria* outbreak investigation (Gilmour et al., 2010), is the exception or the rule.

While whole-genome sequencing may be beyond the capacity of many laboratories right now, the technology will undoubtedly become easier and more available in the future. In the meantime, the cost of whole-genome sequencing has decreased substantially, and comparative genome sequencing of multiple strains of a single species or subtype can be used to identify SNPs and other genetic heterogeneity, which can be applied to the improvement of existing subtyping targets and methods.

Currently, the ability to subtype microorganisms is dependent on obtaining the isolate in pure culture. Culture-independent diagnostic tests are becoming available and increasingly used, particularly in clinical medicine. However, bacterial isolates are critical in order to confirm the presence of a pathogen so that regulatory or other preventive actions can be taken and to subtype the isolates for monitoring, surveillance, outbreak investigations, and the like. New approaches to subtyping and surveillance will be required in order to adapt new diagnostic practices that do not result in a bacterial isolate.

REFERENCES

Afset, J. E., E. Anderssen, G. Bruant, J. Harel, L. Wieler, and K. Bergh. 2008. Phylogenetic backgrounds and virulence profiles of atypical enteropathogenic *Escherichia coli* strains from a case-control study using multilocus sequence typing and DNA microarray analysis. *J. Clin. Microbiol.* **46:** 2280–2290.

Beranek, A., C. Mikula, P. Rabold, D. Arnhold, C. Berghold, I. Lederer, F. Allerberger, and C. Kornschober. 2009. Multiple-locus variable-number tandem repeat analysis for subtyping of *Salmonella enterica* subsp. *enterica* serovar Enteritidis. *Int. J. Med. Microbiol.* **299:**43–51.

Beres, S. B., R. K. Carroll, P. R. Shea, I. Sitkiewicz, J. C. Martinez-Gutierrez, D. E. Low, A. McGeer, B. M. Willey, K. Green, G. J. Tyrrell, T. D. Goldman, M. Feldgarden, B.W. Birren, Y. Fofanov, J. Boos, W. D. Wheaton, C. Honisch, and J. M. Musser. 2010. Molecular complexity of successive bacterial epidemics deconvoluted by comparative pathogenomics. *Proc. Natl. Acad. Sci. USA* **107:**4371–4376.

Best, E. L., A. J. Fox, J. A. Frost, and F. J. Bolton. 2005. Real-time single-nucleotide polymorphism profiling using Taqman technology for rapid recognition of *Campylobacter jejuni* clonal complexes. *J. Med. Microbiol.* **54**(Pt. 10):919–925.

Bishop, C., T. Callman, and J. Wain. 2010. Characterization of SNPs in *Salmonella* Paratyphi A using MALDI-TOF MS, p. 65–67. *In* P. Colin and G. Clement (ed.), *International Symposium on Salmonella and Salmonellosis.* Zoopole Development-ISPAIA, Saint-Malo, France.

Bruant, G., C. Maynard, S. Bekal, I. Gaucher, L. Masson, R. Brousseau, and J. Harel. 2006. Development and validation of an oligonucleotide microarray for detection of multiple virulence and antimicrobial resistance genes in *Escherichia coli. Appl. Environ. Microbiol.* **72:**3780–3784.

Cho, S., T. S. Whittam, D. J. Boxrud, J. M. Bartkus, S. C. Rankin, M. J. Wilkins, P. Somsel, F. P. Downes, K. A. Musser, T. P. Root, L. D. Warnick, M. Wiedmann, and A. M. Saeed. 2010. Use of multiple-locus variable number tandem repeat analysis and phage typing for subtyping of *Salmonella* Enteritidis from sporadic human cases in the United States. *J. Appl. Microbiol.* **108:**859–867.

Clark, C. G., L. Bryden, W. R. Cuff, P. L. Johnson, F. Jamieson, B. Ciebin, and G. Wang.

2005. Use of the Oxford multilocus sequence typing protocol and sequencing of the flagellin short variable region to characterize isolates from a large outbreak of waterborne *Campylobacter* sp. strains in Walkerton, Ontario, Canada. *J. Clin. Microbiol.* **43:** 2080–2091.

Cody, A. J., M. J. Maiden, and K. E. Dingle. 2009. Genetic diversity and stability of the porA allele as a genetic marker in human *Campylobacter* infection. *Microbiology* **155**(Pt. 12)**:**4145–4154.

Davis, M. A., K. N. Baker, D. R. Call, L. D. Warnick, Y. Soyer, M. Wiedmann, Y. Grohn, P. L. McDonough, D. D. Hancock, and T. E. Besser. 2009. Multilocus variable-number tandem-repeat method for typing *Salmonella enterica* serovar Newport. *J. Clin. Microbiol.* **47:**1934–1938.

de Haan, C. P., R. Kivisto, M. Hakkinen, H. Rautelin, and M. L. Hanninen. 2010. Decreasing trend of overlapping MLST genotypes between human and chicken *Campylobacter jejuni* isolates over a decade in Finland. *Appl. Environ. Microbiol.* **76:**5228–5236.

Dunbar, S. A. 2006. Applications of Luminex xMAP technology for rapid, high-throughput multiplexed nucleic acid detection. *Clin. Chim. Acta* **363:**71–82.

Fakhr, M. K., L. K. Nolan, and C. M. Logue. 2005. Multilocus sequence typing lacks the discriminatory ability of pulsed-field gel electrophoresis for typing *Salmonella enterica* serovar Typhimurium. *J. Clin. Microbiol.* **43:**2215–2219.

Fitzgerald, C., M. Collins, S. van Duyne, M. Mikoleit, T. Brown, and P. Fields. 2007. Multiplex, bead-based suspension array for molecular determination of common *Salmonella* serogroups. *J. Clin. Microbiol.* **45:**3323–3334.

Frye, J. G., R. L. Lindsey, G. Rondeau, S. Porwollik, F. Long, M. McClelland, C. R. Jackson, M. D. Englen, R. J. Meinersmann, M. E. Berrang, J. A. Davis, J. B. Barrett, J. B. Turpin, S. N. Thitaram, and P. J. Fedorka-Cray. 2010. Development of a DNA microarray to detect antimicrobial resistance genes identified in the National Center for Biotechnology Information database. *Microb. Drug Resist.* **16:**9–19.

Garaizar, J., A. Rementeria, and S. Porwollik. 2006. DNA microarray technology: a new tool for the epidemiological typing of bacterial pathogens? *FEMS Immunol. Med. Microbiol.* **47:**178–189.

Gerner-Smidt, P., K. Hise, J. Kincaid, S. Hunter, S. Rolando, E. Hyytia-Trees, E. M. Ribot, and B. Swaminathan. 2006. PulseNet USA: a five-year update. *Foodborne Pathog. Dis.* **3:**9–19.

Gilmour, M. W., M. Graham, G. Van Domselaar, S. Tyler, H. Kent, K. M. Trout-Yakel, O. Larios, V. Allen, B. Lee, and C. Nadon. 2010. High-throughput genome sequencing of two *Listeria monocytogenes* clinical isolates during a large foodborne outbreak. *BMC Genomics* **11:**120.

Grønlund, H., L. Riber, H. Vigre, C. Löfström, L. Folling, S. Huehn, B. Malorny, P. Rådström, K. Rudi, and J. Hoorfar. 2010. Microarray-based genotyping of *Salmonella*: interlaboratory evaluation of reproducibility and standardization potential. *Int. J. Food Microbiol.* Epub ahead of print. doi:10.1016/j.ijfoodmicro.2010.08.007.

Hall, T. A., R. Sampath, L. B. Blyn, R. Ranken, C. Ivy, R. Melton, H. Matthews, N. White, F. Li, V. Harpin, D. J. Ecker, L. K. McDougal, B. Limbago, T. Ross, D. M. Wolk, V. Wysocki, and K. C. Carroll. 2009. Rapid molecular genotyping and clonal complex assignment of *Staphylococcus aureus* isolates by PCR coupled to electrospray ionization-mass spectrometry. *J. Clin. Microbiol.* **47:**1733–1741.

Hamelin, K., G. Bruant, A. El-Shaarawi, S. Hill, T. A. Edge, J. Fairbrother, J. Harel, C. Maynard, L. Masson, and R. Brousseau. 2007. Occurrence of virulence and antimicrobial resistance genes in *Escherichia coli* isolates from different aquatic ecosystems within the St. Clair River and Detroit River areas. *Appl. Environ. Microbiol.* **73:** 477–484.

Honisch, C., Y. Chen, C. Mortimer, C. Arnold, O. Schmidt, D. van den Boom, C. R. Cantor, H. N. Shah, and S. E. Gharbia. 2007. Automated comparative sequence analysis by base-specific cleavage and mass spectrometry for nucleic acid-based microbial typing. *Proc. Natl. Acad. Sci. USA* **104:**10649–10654.

Honisch, C., M. Mosko, C. Arnold, S. E. Gharbia, R. Diel, and S. Niemann. 2010. Replacing reverse line blot hybridization spoligotyping of the *Mycobacterium tuberculosis* complex. *J. Clin. Microbiol.* **48:**1520–1526.

Huehn, S., C. Bunge, E. Junker, R. Helmuth, and B. Malorny. 2009a. Poultry-associated *Salmonella enterica* subsp. *enterica* serovar 4,12:d:- reveals high clonality and a distinct pathogenicity gene repertoire. *Appl. Environ. Microbiol.* **75:**1011–1020.

Huehn, S., R. Helmuth, C. Bunge, B. Guerra, E. Junker, R. H. Davies, P. Wattiau, W. van Pelt, and B. Malorny. 2009b. Characterization of pathogenic and resistant genome repertoire reveals two clonal lines in *Salmonella enterica* subsp. *enterica* serovar Paratyphi B (+)-tartrate positive. *Foodborne Pathog. Dis.* **6:**431–443.

Huehn, S., and B. Malorny. 2009. DNA microarray for molecular epidemiology of *Salmonella*. *Methods Mol. Biol.* **551:**249–285.

Hyytia-Trees, E., P. Lafon, P. Vauterin, and E. M. Ribot. 2010. Multilaboratory validation study of standardized multiple-locus variable-number tandem repeat analysis protocol for Shiga toxin-producing *Escherichia coli* O157: a novel approach to normalize fragment size data between capillary electrophoresis platforms. *Foodborne Pathog. Dis.* **7:**129–136.

Hyytia-Trees, E., S. C. Smole, P. A. Fields, B. Swaminathan, and E. M. Ribot. 2006. Second generation subtyping: a proposed PulseNet protocol for multiple-locus variable-number tandem repeat analysis of Shiga toxin-producing *Escherichia coli* O157 (STEC O157). *Foodborne Pathog. Dis.* **3:** 118–131.

Kerouanton, A., M. Marault, R. Lailler, F. X. Weill, C. Feurer, E. Espie, and A. Brisabois. 2007. Pulsed-field gel electrophoresis subtyping database for foodborne *Salmonella enterica* serotype discrimination. *Foodborne Pathog. Dis.* **4:**293–303.

Kotetishvili, M., O. C. Stine, A. Kreger, J. G. Morris, Jr., and A. Sulakvelidze. 2002. Multilocus sequence typing for characterization of clinical and environmental *Salmonella* strains. *J. Clin. Microbiol.* **40:**1626–1635.

Lewis, T., N. J. Loman, L. Bingle, P. Jumaa, G. M. Weinstock, D. Mortiboy, and M. J. Pallen. 2010. High-throughput whole-genome sequencing to dissect the epidemiology of *Acinetobacter baumannii* isolates from a hospital outbreak. *J. Hosp. Infect.* **75:**37–41.

Lindstedt, B. A. 2005. Multiple-locus variable number tandem repeats analysis for genetic fingerprinting of pathogenic bacteria. *Electrophoresis* **26:**2567–2582.

Lindstedt, B. A., W. Tham, M. L. Danielsson-Tham, T. Vardund, S. Helmersson, and G. Kapperud. 2008. Multiple-locus variable-number tandem-repeats analysis of *Listeria monocytogenes* using multicolour capillary electrophoresis and comparison with pulsed-field gel electrophoresis typing. *J. Microbiol. Methods* **72:**141–148.

Litrup, E., M. Torpdahl, B. Malorny, S. Huehn, M. Helms, H. Christensen, and E. M. Nielsen. 2010. DNA microarray analysis of *Salmonella* serotype Typhimurium strains causing different symptoms of disease. *BMC Microbiol.* **10:**96.

Maiden, M. C. 2006. Multilocus sequence typing of bacteria. *Annu. Rev. Microbiol.* **60:**561–588.

Malorny, B., E. Junker, and R. Helmuth. 2008. Multi-locus variable-number tandem repeat analysis for outbreak studies of *Salmonella enterica* serotype Enteritidis. *BMC Microbiol.* **8:**84.

Mandrell, R. E., L. A. Harden, A. Bates, W. G. Miller, W. F. Haddon, and C. K. Fagerquist. 2005. Speciation of *Campylobacter coli, C. jejuni, C. helveticus, C. lari, C. sputorum,* and *C. upsaliensis* by matrix-assisted laser desorption ionization-time of flight mass spectrometry. *Appl. Environ. Microbiol.* **71:**6292–6307.

McQuiston, J. R., R. J. Waters, B. Dinsmore, M. Mikoleit, and P. I. Fields. 15 December 2010. Molecular determination of H-antigens of *Salmonella* using a microsphere-based liquid array. *J. Clin. Microbiol.* (Epub ahead of print.)

Mellmann, A., F. Bimet, C. Bizet, A. D. Borovskaya, R. R. Drake, U. Eigner, A. M. Fahr, Y. He, E. N. Ilina, M. Kostrzewa, T. Maier, L. Mancinellie, W. Moussaoui, G. Prévost, L. Putignani, C. L. Seachord, Y. W. Tang, and D. Harmsen. 2009. High interlaboratory reproducibility of matrix-assisted laser desorption ionization-time of flight mass spectrometry-based species identification of nonfermenting bacteria. *J. Clin. Microbiol.* **47:**3732–3734.

Miko, A., B. A. Lindstedt, L. T. Brandal, I. Lobersli, and L. Beutin. 2010. Evaluation of multiple-locus variable number of tandem-repeats analysis (MLVA) as a method for identification of clonal groups among enteropathogenic, enterohaemorrhagic and avirulent *Escherichia coli* O26 strains. *FEMS Microbiol. Lett.* **303:**137–146.

Miller, M. B., and Y. W. Tang. 2009. Basic concepts of microarrays and potential applications in clinical microbiology. *Clin. Microbiol. Rev.* **22:**611–633.

Murphy, M., D. Corcoran, J. F. Buckley, M. O'Mahony, P. Whyte, and S. Fanning. 2007. Development and application of multiple-locus variable number of tandem repeat analysis (MLVA) to subtype a collection of *Listeria monocytogenes*. *Int. J. Food Microbiol.* **115:**187–194.

Ribot, E. M., M. A. Fair, R. Gautom, D. N. Cameron, S. B. Hunter, B. Swaminathan, and T. J. Barrett. 2006. Standardization of pulsed-field gel electrophoresis protocols for the subtyping of *Escherichia coli* O157:H7, *Salmonella*, and *Shigella* for PulseNet. *Foodborne Pathog. Dis.* **3:**59–67.

Ross, I. L., and M. W. Heuzenroeder. 2005. Discrimination within phenotypically closely related definitive types of *Salmonella enterica* serovar Typhimurium by the multiple amplification of phage locus typing technique. *J. Clin. Microbiol.* **43:**1604–1611.

Ross, I. L., and M. W. Heuzenroeder. 2009. A comparison of two PCR-based typing methods with pulsed-field gel electrophoresis in *Salmonella enterica* serovar Enteritidis. *Int. J. Med. Microbiol.* **299:**410–420.

Sauer, S., and M. Kliem. 2010. Mass spectrometry tools for the classification and identification of bacteria. *Nat. Rev. Microbiol.* **8:**74–82.

Sheppard, S. K., J. F. Dallas, M. MacRae, N. D. McCarthy, E. L. Sproston, F. J. Gormley, N. J. Strachan, I. D. Ogden, M. C. Maiden, and K. J. Forbes. 2009. *Campylobacter* genotypes from food animals, environmental sources and clinical disease in Scotland 2005/6. *Int. J. Food Microbiol.* **134:**96–103.

Sopwith, W., A. Birtles, M. Matthews, A. Fox, S. Gee, S. James, J. Kempster, M. Painter, V. Edwards-Jones, K. Osborn, M. Regan, Q. Syed, and E. Bolton. 2010. Investigation of food and environmental exposures relating to the epidemiology of *Campylobacter coli* in humans in Northwest England. *Appl. Environ. Microbiol.* **76:** 129–135.

Soyer, Y., S. D. Alcaine, D. J. Schoonmaker-Bopp, T. P. Root, L. D. Warnick, P. L. McDonough, N. B. Dumas, Y. T. Grohn, and M. Wiedmann. 2010. Pulsed-field gel electrophoresis diversity of human and bovine clinical *Salmonella* isolates. *Foodborne Pathog. Dis.* **7:**707–717.

Sperry, K. E., S. Kathariou, J. S. Edwards, and L. A. Wolf. 2008. Multiple-locus variable-number tandem-repeat analysis as a tool for subtyping *Listeria monocytogenes* strains. *J. Clin. Microbiol.* **46:** 1435–1450.

Sukhnanand, S., S. Alcaine, L. D. Warnick, W. L. Su, J. Hof, M. P. Craver, P. McDonough, K. J. Boor, and M. Wiedmann. 2005. DNA sequence-based subtyping and evolutionary analysis of selected *Salmonella enterica* serotypes. *J. Clin. Microbiol.* **43:**3688–3698.

Swaminathan, B., P. Gerner-Smidt, L. K. Ng, S. Lukinmaa, K. M. Kam, S. Rolando, E. P. Gutierrez, and N. Binsztein. 2006. Building PulseNet International: an interconnected system of laboratory networks to facilitate timely public health recognition and response to foodborne disease outbreaks and emerging foodborne diseases. *Foodborne Pathog. Dis.* **3:**36–50.

Torpdahl, M., M. N. Skov, D. Sandvang, and D. L. Baggesen. 2005. Genotypic characterization of *Salmonella* by multilocus sequence typing, pulsed-field gel electrophoresis and amplified fragment length polymorphism. *J. Microbiol. Methods* **63:**173–184.

Torpdahl, M., G. Sorensen, B. A. Lindstedt, and E. M. Nielsen. 2007. Tandem repeat analysis for surveillance of human *Salmonella* Typhimurium infections. *Emerg. Infect. Dis.* **13:**388–395.

Ward, T. J., T. F. Ducey, T. Usgaard, K. A. Dunn, and J. P. Bielawski. 2008. Multilocus genotyping assays for single nucleotide polymorphism-based subtyping of *Listeria monocytogenes* isolates. *Appl. Environ. Microbiol.* **74:**7629–7642.

Wolk, D. M., L. B. Blyn, T. A. Hall, R. Sampath, R. Ranken, C. Ivy, R. Melton, H. Matthews, N. White, F. Li, V. Harpin, D. J. Ecker, B. Limbago, L. K. McDougal, V. H. Wysocki, M. Cai, and K. C. Carroll. 2009. Pathogen profiling: rapid molecular characterization of *Staphylococcus aureus* by PCR/electrospray ionization-mass spectrometry and correlation with phenotype. *J. Clin. Microbiol.* **47:**3129–3137.

Zou, W., J. G. Frye, C. W. Chang, J. Liu, C. E. Cerniglia, and R. Nayak. 2009. Microarray analysis of antimicrobial resistance genes in *Salmonella enterica* from preharvest poultry environment. *J. Appl. Microbiol.* **107:**906–914.

CRITICAL CONSIDERATIONS BEFORE SETTING UP RAPID METHODS

II

SAMPLING, TRANSPORT, AND SAMPLE PREPARATION IN EMERGENCY SITUATIONS AND RAPID RESPONSE

S. O'Brien, P. Whyte, C. Iversen, and S. Fanning

6

The nature of the modern food industry has changed in recent times, with many food producers now exporting their products worldwide. This global distribution means that outbreaks of foodborne illness can occur over large geographical regions, often affecting people in a number of countries simultaneously. As a result, robust and integrated food chain surveillance is an essential risk management component to ensure food safety hazards are effectively controlled and noncompliances are quickly identified when they arise. Surveillance combined with the implementation of suitable traceability systems in the food industry facilitates timely intervention and limits the risk to consumers and damage to food businesses.

The means by which samples for microbiological analysis are collected, transported, and prepared for confirmatory testing have a profound effect on the speed and accuracy of results obtained. The quality of all diagnostic data produced in the laboratory relies on the quality of the initial sample generated. Consequences of poorly collected, incorrectly transported, and inadequately prepared samples include failure to recover and identify the foodborne pathogens or toxins present and the misidentification of potential hazards or sources of contamination. These failures or delays in identifying a causative agent can inadvertently facilitate the dissemination of the outbreak, contributing to an increasing number of cases (Kretzer et al., 2008).

For this reason it is vital that the sample procurement process take account of a number of crucial factors that could influence the efficacy of downstream detection methods. These are likely to include the following (Kretzer et al., 2008).

1. Target pathogen may be present in low numbers (e.g., verocytotoxigenic *Escherichia coli* in minced beef).

2. Food/feces may contain physical/chemical components that may inhibit the detection method employed.

3. Pathogens may be spread heterogeneously through sample matrices (e.g., *Cronobacter sakazakii* in powdered infant formula).

For the purpose of this chapter, the abbreviation ST (sample treatment) will be used to describe the three-stage process of sampling, sample transport, and sample preparation. Ide-

S. O'Brien, P. Whyte, C. Iversen, and S. Fanning, Centres for Food Safety & Foodborne Zoonomics, UCD Veterinary Sciences Centre, University College Dublin, Belfield, Dublin 4, Ireland.

Rapid Detection, Identification, and Quantification of Foodborne Pathogens, ed. by J. Hoorfar

ally, the ST process should be rapid and provide a working sample that is likely to contain the highest concentration of the target pathogen, which should be free of inhibitors to enhance sensitivity. This process should also preserve the number(s) of target organisms in cases where quantification will be necessary. Having an accurate estimate of the level of contamination in a food product is a key to the subsequent hazard analysis/dose-response stage of a risk assessment (Cummins et al., 2008).

Whether using serotyping or molecular subtyping methods to identify a cluster of related cases, outbreak investigations are increasingly more complex, with many food products being comprised of ingredients originating from multiple suppliers often in different geographical regions. Contamination of food products (e.g., ready-to-eat meats) while in the postprocessing environment is a common cause of product recalls and a significant source of food poisoning events and shelf life problems (Kornacki, 2000) in developing countries.

As many foodborne pathogens have a low infectious dose (estimated to be between 10 and 1,000 viable cells), the correct food samples and laboratory specimens must be acquired for analysis. Available epidemiologic data should guide this selection so that the most likely sources are correctly sampled. Investigations should also extend beyond the immediate preparation of the implicated food and investigate the processing chain environment from farm to fork including sources of ingredients, processing and storage environments, and transportation.

Food microbiologists typically sample foods for four main reasons, as follows:

1. Suspicion of a source of foodborne illness directly or following a customer complaint.

2. Quality control and shelf life stability studies.

3. Investigation of process hygiene and examination of food handling techniques.

4. To verify compliance with current regulatory requirements (e.g., official testing for food hygiene and safety).

OUTBREAK INVESTIGATIONS

The successful identification of a foodborne outbreak increasingly requires local public health laboratories to carry out molecular characterization by subtyping of pathogenic bacteria recovered from suspect samples. In these cases, surveillance data must be rapidly generated and compared over increasingly broad regions (e.g., regionally, nationally, and internationally). The PulseNet USA system was originally designed and set up to support this requirement (Gerner-Smidt et al., 2006). Once a foodborne outbreak is confirmed, its containment becomes a priority. This is achieved by identifying the source(s) of the outbreak, tracing and removing contaminated product from the market, generating public awareness if necessary, minimizing secondary spread where relevant, and implementing procedures to limit/prevent future incidents.

In this chapter, the focus will be on the foodborne illness outbreak scenario, and much of it will be applicable to validation procedures relevant to hazard analysis critical control points (HACCP) or to situations in which a customer complaint is received.

Primarily, when investigating potential cases of food poisoning, every effort should be made to recover the remains of the suspected food. In addition, suspect food from the same meal should be sought, even if it is not the suspect ingredient, followed by other batches of food from the same supplier or caterer on either side of the production cycle under investigation. Furthermore, information on the preparation, cooking, and storage of the suspect food should be acquired as soon as possible. This may provide vital clues to the epidemiological investigation team.

As it is rarely feasible for a laboratory to test for all potential pathogens, it is important that as much detail as possible on the symptoms of the illness be gathered. Detailed methods for the recovery and enumeration of specific pathogenic organisms are given elsewhere in this book. Such an approach will allow the most relevant tests to be selected and reduce the amount of sample needed for testing. Once the etiological agent has been identified, the point of entry into the food chain should be ascertained and strategies should be developed for the control/elimination of the pathogen.

Finally, advice must be given to food producers on how to avoid similar food safety failures in the future. Figure 1 shows the steps required to deal with a foodborne illness outbreak.

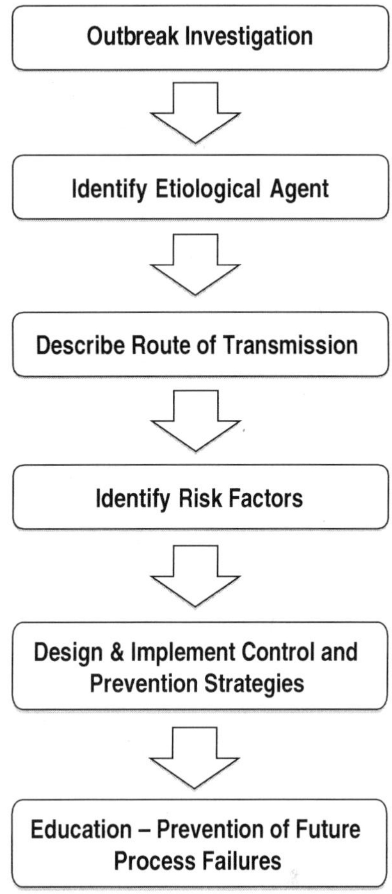

FIGURE 1 Handling a foodborne illness outbreak.

BEFORE SAMPLING

Before discussing the various sample types, it is important to first define what constitutes a sample. In terms of food safety microbiology, a sample may be considered to be a portion of food, surface, or air considered representative of a larger matrix. It is therefore vital that such samples give a representative profile of all possible microbes present in the complete matrix and that they allow for an accurate estimate of the level of contamination to be deduced should it be required.

Furthermore, it is important to take a sample of sufficient size so as to obtain a detectable number of pathogenic cells. For example, 25 g of minced beef may contain a clinically significant level of verocytotoxigenic *E. coli* (e.g., approximately 10 CFU), which may be below the limit of detection (LOD) of some microbiological detection methods.

When developing standard operating procedures for ST, it is important to be mindful of the capabilities of the testing laboratory that will analyze the sample(s). Furthermore, sample requirements may vary for different analytes such as microbes, nucleic acids, and toxins and for different detection methods including qualitative, quantitative, or subtyping methods. Therefore, consultation with the laboratory will enable the development of best-practice protocols to be established and validated.

SAMPLE TYPES

Sample collection methods include contact plates, destructive samples (food/excision), food rinsates, cotton swabs, premoistened sponge, wipes, sterile tongue depressors, and air sampling vacuums (agar plate/filter). When deciding which of these approaches to use, there are two important considerations. First, it is appropriate to have an indication of the suspect pathogen under investigation. In cases where outbreaks have occurred, medical reports and the food matrix type in question should give an indication of the likely biological agent(s). The importance of this "short listing" cannot be overstated.

The diversity that exists among bacteria, viruses, and fungi in terms of their stability is enormous. The stability of other target analytes such as nucleic acids and toxins can also vary widely, depending on various environmental and storage conditions. For example, a sample type that would be appropriate for detecting an aerobic microbe may adversely affect an anaerobic organism. In cases where there is a degree of uncertainty concerning the nature of the causative agent, it may be prudent to acquire a number of samples by using a variety of preservation media in order to accommodate the most likely targets/agents.

Another consideration when planning for the direct analysis of DNA in a sample is the possible presence in the swab of inhibitory chemicals that may interfere with the nucleic acid detection method or of environmental impurities such as metal ions, which may adversely affect the performance of a PCR method (Greenfield and White, 1993).

The second consideration is to ensure that the sampling method used has been validated. It is essential to use a sampling method that will give the most efficient recovery as well as provide the best chance of identifying the target analyte.

Sampling of Foods

When examining a food processing plant, sampling should initially be based on the HACCP plan. Often, foodborne outbreaks can occur when critical points such as heating or cooling events have deviated from the required specification. While frequent sampling will increase the reliability of the data obtained, sample pooling can also allow for a higher throughput (Kretzer et al., 2008).

In the European Union, HACCP systems were mandated through the "Food Hygiene Package" (Commission Regulations EC 852, 853, 854, and 2073 and formal Directive EC 471), which requires the use of all seven HACCP principles (O'Brien et al., 2007). One of the requirements of the HACCP system is the implementation of a set of process hygiene procedures that must be undertaken to verify that the system is being operated properly by food business operators.

The microbiological criteria contained within EC 2073 rely on the use of indicator organisms such as total viable counts and *Enterobacteriaceae* as indicators of process hygiene, and the results of these analyses are expressed as mean $\log_{10}$ CFU per square centimeter. In the meat processing industry, samples are collected by excising small pieces of meat at four sites on the animal carcass or by swabbing a defined surface area at these sites and pooling the samples for analysis. The pooled sample is then examined to determine carcass hygiene levels during slaughter.

While sample sizes are clearly specified in EC 2073, in other situations the determination of sample sizes may be more arbitrary and can have a significant impact on the sensitivity of detection. The main considerations should include (i) the likely prevalence and numbers of the target organism in the process and (ii) the LOD for the analytical method to be used.

The majority of standard operating procedures for foodstuffs recommend the portions to be of 10 to 25 g/ml of foodstuffs as a representative sample. In cases where bovine, ovine, or porcine carcasses are being examined, pooled swab or excision samples are recommended. For example, when testing porcine carcasses, a swab is taken from a 100-cm^2 area of the ham, back, belly, and jowl (O'Brien et al., 2007). The four swabs are pooled and examined as a composite sample. However, it should be noted that swab samples are considered representative of only 20% or less of the microbial counts obtained by excision.

While some foods only have microbial flora on the surface (e.g., beefsteak), others have contaminants distributed throughout (e.g., cheese and milk). It is thus impossible to make a single sampling protocol applicable to all foodstuffs. Sampling procedures should be optimized to take account of the matrix being investigated, the pathogen that may be present, and the detection method to be used.

Sampling Surfaces

Another source of contamination in the food processing plant environment can be the niche growth of a pathogen in a biofilm. Ineffective cleaning and sanitizing of factory pipes, surfaces, and equipment can facilitate biofilm formation, with cells in some biofilms displaying 1,000-fold or more increased resistance to destruction when compared to planktonic cells (Kornacki, 2010).

The main problem with sampling surfaces has been highlighted in the section above, where the swab method may only remove a fraction of the microbial flora from a carcass when compared to an excision sample. It is important therefore to understand that any given surface sampling method may not recover all of the microbes on that surface. Nonetheless, if the same method is consistently used to sample the same areas in a processing plant and/or the method's LOD is adequate, it may still prove to be a valuable sampling tool.

The swab method is perhaps the most commonly used investigative tool for the examination of surfaces. Use of a premoistened sponge is generally regarded to be more efficient than the traditional (dry) cotton swab (Silliker and Gabis, 1975). While the cotton swab is an effective means for sampling inaccessible areas (e.g., cracks in stone surfaces), more downward pressure can be applied to a premoistened sponge, making it more efficient at removing biofilms and covering/sampling a larger surface area.

Usually, swabbing will require the use of a sterile template with openings generally ranging from 1 to 100 cm^2 depending on the surface area represented. The sterile template is placed on the area to be sampled and swabbed in a uniform fashion. The swab (e.g., cotton, polyurethane sponge, etc.) is then returned to its original containment vessel, which should also contain a known amount of transport/neutralizer medium depending on the target organism and the nature of the surface sampled.

The contact plate method uses a petri dish containing the appropriate agar so that the medium has a rounded convex surface raised above the walls of the dish. The plate is inverted and pressed against the test surface (such as a wall or a floor) in a rolling motion, with a light uniform pressure. Plates are capped with a lid, inverted, and incubated at an appropriate temperature to facilitate the growth of the target organism(s) under investigation. Following incubation, characteristic colonies are counted, and the results are compared to appropriate microbiological standards.

The use of presterilized tongue depressors has been recommended for dry scraping or sampling between sandwiched areas of equipment (Kornacki, 2010). The contact area of the depressor, used to sample the surface, can be broken off into a sterile container and diluent/transport media added. Where samples are taken from delineated areas, it may be possible to obtain reliable quantitative results as well as qualitative results. The tongue depressor may also be used for taking samples of freshly voided animal feces from the lairage or transport vehicles when sampling at an abattoir.

Air Sampling

Monitoring the hygiene of the processing environment is important as a means of reducing the threat posed by airborne microbial contaminants. Control of environments surrounding sensitive processing and packing operations has become an increasingly important part of hygienic food manufacturing in recent years, especially in the pharma-food industry. In the absence of mandatory standards, food producers should monitor the quality of the manufacturing plant air within a HACCP framework, first identifying an acceptable level of contamination, followed by routine testing of high-risk areas of the processing line. Air sampling may also play an important role in identifying the source of microbial contamination in food processing plants during an outbreak.

Several methods are available for air sampling, ranging from simple qualitative methods (such as gravity settling plate) to the more complex quantitative methods (including the Anderson six-stage impactor). Informative and detailed analyses of filtration samplers, slit and sieve impingers, electrostatic precipitation impactors, and centrifugal samplers have been published previously by Kretzer et al. (2008) and Evancho et al. (2001).

Sample Transport and Storage

The next key stage in the sampling chain is the transport of the acquired sample under appropriate conditions. The nature by which samples are expedited will have a potentially profound effect on the final results obtained by subsequent analysis. The three main problems that arise during transport are as follows:

1. Growth of nontarget or background microflora making the detection of the target organism more difficult.
2. Death of injured target microbes by outgrowth of nontarget organisms or unfavorable transport conditions (e.g., anaerobes exposed to aerobic conditions).
3. Cross-contamination of samples with nontarget microbes. This is often caused by poor aseptic technique during sample collection or storage and can have a major impact on final results.

To reduce the risk posed by these problems, samples should be stored at 0 to 4°C and transported to the laboratory as soon as possible.

FUTURE TRENDS

Rapid identification of an outbreak and the corresponding etiological agent is crucial in preventing the spread of disease. A number of pathogens including verocytotoxigenic *E. coli* and some *Salmonella* strains can have a low infectious dose, posing a serious threat to public health even when present in foods at low levels. For this reason, it is critical that professionals involved in public health have well-planned strategies to efficiently deal with foodborne outbreak scenarios including rapid sampling techniques, sampling plans, rapid detection, and molecular subtyping protocols for the major foodborne pathogens.

Furthermore, this process can be underpinned by ongoing monitoring and collection of microbiological data, relevant not only to clinical isolates but also to isolates of veterinary and food origin. Many countries have compiled valuable data on foodborne pathogens contained in various databases and formats, e.g., food safety authorities, universities, research institutes, and agricultural and industrial organizations. A key challenge in the future will be the integration of these data streams into online databases that can be accessed by relevant stakeholders. While linking these databases may be challenging on both a practical and a political level, there can be no doubt as to the benefit of such a resource to enhance the protection of public health. Databases containing information on the microbial ecology of particular foodstuffs could be connected to a database in which the kinetic properties of microorganisms are stored, and these could be in turn linked to databases with processing and epidemiological data. Such a system of linked databases would act not only as a powerful tool to facilitate rapid responses to foodborne outbreaks but also as a tool for prevention, acting as a resource for microbial risk assessments.

REFERENCES

Cummins, E., P. Nally, F. Butler, G. Duffy, and S. O'Brien. 2008. Development and validation of a probabilistic second-order exposure assessment model for *Escherichia coli* O157:H7 contamination of beef trimmings from Irish meat plants. *Meat Sci.* **79:**139–154.

Evancho, G. M., W. H. Sveum, L. J. Moberg, and J. F. Frank. 2001. Microbiological monitoring of the food processing environment, p. 25–35. *In* F. P. Downes and K. Ito (ed.), *Compendium of Methods for the Microbiological Examination of Foods.* American Public Health Association, Washington, DC.

Gerner-Smidt, P., K. Hise, J. Kincaid, S. Hunter, S. Rolando, E. Hyytiä-Trees, E. M. Ribot, and B. Swaminathan. 2006. PulseNet USA: a five-year update. *Foodborne Pathog.* **3:**9–19.

Greenfield, L., and T. White. 1993. Sample preparation methods, p. 122–137. *In* D. Persing, S. Smith, F. Tenover, and T. White (ed.), *Diagnostic Molecular Microbiology: Principles and Applications.* ASM Press, Washington, DC.

Kornacki, J. L. 2000. The nuts and bolts of food safety. *Food Testing Anal.* **40:**18–22.

Kornacki, J. L. 2010. How do I sample the environment and equipment?, p. 125–136. *In* J. L. Kornacki (ed.), *Principles of Microbiological Trouble-*

shooting in the Industrial Food Processing Environment. Springer Press, New York, NY.

Kretzer, J. W., M. Biebl, and S. Miller. 2008. Sample preparation: an essential prerequisite for high-quality bacteria detection, p. 15–23. *In* M. Zourob (ed.), *Principles of Bacterial Detection: Biosensors, Recognition Receptors and Microsystems*. Springer Press, New York, NY.

O'Brien, S., M. Lenahan, T. Sweeney, and J. Sheridan. 2007. Assessing the hygiene of pig carcasses using whole body carcass swabs compared to the four site method in EU Decision 471. *J. Food Prot.* **70:**432–439.

Silliker, J. H., and D. A. Gabis. 1975. A cellulose sponge sampling technique for surfaces. *J. Milk Food Technol.* **38:**504.

STATISTICS OF SAMPLING FOR MICROBIOLOGICAL TESTING OF FOODBORNE PATHOGENS

T. Ross, P. M. Fratamico, L. Jaykus, and M. H. Zwietering

7

SAMPLING PLANS

A sampling plan is part of a process to assess the acceptability of a batch of product against some criterion. At a minimum, a sampling plan involves specification of the number of samples to be drawn and tested from a lot and a criterion (or criteria) that must be satisfied to consider the entire batch as acceptable for sale or other specified uses.

In this text, the main interest is the assessment of the microbiological safety of foods, but the same considerations discussed in this chapter apply equally to assessment of microbiological quality. Even with the widespread implementation of preventative strategies such as hazard analysis and critical control point (HACCP) and related food safety management strategies, there are still many situations in which food safety assurance mainly relies on microbiological testing. End product testing might be needed in some circumstances, for example, when there are no critical control points in a process (e.g., raw or minimally processed ready-to-eat foods) or when the history of a product is unknown. Equally, food safety objectives and performance objectives (International Commission on Microbiological Specifications for Foods, 2002; Anonymous, 2005) may nominate microbiological criteria to be met.

When Is Testing Appropriate?

The Codex Alimentarius Commission (CAC) (1997) considers that microbiological testing of foods should only be required when

- There is good evidence that the product or product group has been implicated in foodborne disease or when the food has inadequate shelf life or if other microbiological issues will arise if controls are not applied.
- There is good evidence or reason to believe that the application of testing will reduce the health risk, or quality defect, associated with the food or will effectively assess adherence to microbiological control measures or process controls.

T. Ross, Tasmanian Institute of Agricultural Research, School of Agricultural Science, University of Tasmania, Private Bag 54, Hobart 7001, Tasmania, Australia. *P. M. Fratamico*, U.S. Department of Agriculture—Agricultural Research Service, Eastern Regional Research Center, Wyndmoor, PA 19038. *L. Jaykus*, Department of Food, Bioprocessing and Nutrition Sciences, North Carolina State University, Raleigh, NC 27695-7624. *M. H. Zwietering*, Laboratory of Food Microbiology, Agrotechnology & Food Sciences Group, Wageningen Agricultural University, Postbus 8129, 6700EV, Wageningen, The Netherlands.

Rapid Detection, Identification, and Quantification of Foodborne Pathogens, ed. by J. Hoorfar

Expanding on this theme, the International Commission on Microbiological Specifications for Foods (ICMSF) identifies five situations in which microbiological testing of products may be relevant and necessary (International Commission on Microbiological Specifications for Foods, 2002), relating to the following:

- Primary products or ingredients for which no subsequent step in processing could be considered to provide microbiological control (e.g., minimally processed salad vegetables or cocoa powder dusted over chocolates) or processes that could be overwhelmed by unusually high contamination levels.
- In-process testing to verify a kill step or to assess the potential for cross-contamination to products after a critical control point.
- Monitoring of the processing environment if that environment is a potential source of microbial contaminants (e.g., *Listeria monocytogenes* in plants manufacturing lightly preserved refrigerated products with extended shelf life).
- Shelf life testing, for products whose shelf life may be limited by microbial spoilage and where shelf life validation or accelerated shelf life testing might identify potential problems before they become apparent in the market place.
- End product testing when no other information is available to assess the status of the product, or when end product testing is used to verify that the *process* is under control, or as noted above, where there is no other critical control point available for the product/process.

Limitations of Testing

Microbiological criteria related to food safety are usually inherently, if not explicitly, numerical and specify tolerable limits or frequencies of microbiological contamination beyond which a batch of production is not acceptable for sale or consumption.

In this context, the integrity of methods for detection and enumeration of microorganisms or their toxic products of metabolism is of paramount concern not only for protection of consumer health but also for trade.

Baird-Parker (2000), in presenting a succinct history of the evolution of microbial food safety legislation, noted that the ad hoc manner in which most microbiological criteria were developed created inconsistencies that were problematic for international trade in foods, viz., a product that is microbiologically acceptable in one nation is not considered acceptable in another, or the methods used to assess compliance in one nation produce different results from those mandated in another. Consequently, an understanding of method reliability and sensitivity in relation to microbiological criteria, including the efficacy of the sampling plan, is needed to establish consistent criteria both for trade and public health protection.

At the outset it must be emphasized that no sampling process can ever guarantee the microbiological safety of a food because, at present, there are no established methods for microbiological analysis of foods that are nondestructive. A sampling plan can only ever provide a certain degree of confidence that the lot does, or does not, comply with some microbiological criterion that the sampling plan is designed to assess. To implement a sampling plan appropriate to a specific food safety criterion, it is necessary to know the degree of confidence the plan provides and, accordingly, to understand the statistics of sampling plans.

There are numerous sources of guidance on the statistics, appropriate use, and interpretation of sampling plans (detailed below). In this chapter, basic concepts of sampling plan nomenclature, design, and interpretation are presented to complement the main focus of this text, namely, rapid microbiological detection, identification, and enumeration technologies. Additionally, recently introduced concepts concerning the numerical interpretation of microbiological presence/absence testing and a discussion of sample compositing are also presented.

TERMINOLOGY AND TYPES OF SAMPLING PLANS

There is a series of ISO standards (ISO 2589; ISO 3591, ISO 8422, ISO 8423, etc.) related to sampling plans including, but not restricted to, microorganisms in foods. The CAC has prepared extensive guidelines (Codex Alimentarius Commission, 2004) on appropriate procedures for sampling and design and use of sampling plans for food safety in general, including food microbiology. Additionally, the ICMSF has published a number of texts (International Commission on Microbiological Specifications for Foods, 1986, 2002, 2011) that provide background and specific guidance for microbiological assessment of particular types of food commodities, including types of microbiological tests and sampling plans appropriate to those foods and food processes. The Codex guidelines (Codex Alimentarius Commission, 2004) also provide specific advice on randomization of sampling (so that samples are representative of the lot) and for transport to the testing laboratory. The ICMSF (2002) also provides specific advice on sampling procedures, sample handling and transport, and similar topics.

The main issues are that samples analyzed must be representative of the lot, that sample integrity and identity must be preserved, and that the sample must be in the same (microbiological) condition when it reaches the laboratory for testing as when it was originally taken.

Two main types of sampling plan are recognized: "attributes" and "variables" sampling plans. In general, most of the interest in food safety management is with attributes sampling plans, because they test foods for compliance with a fixed microbiological criterion related to the safety of the product. Attributes plans are often employed for assessment of "zero tolerance" approaches, a topic that will be returned to in "Presence/Absence Sampling and Zero Tolerance" below.

Variables Sampling Plans

Variables plans use the full range of numerical data describing microbial loads on the foods of interest and are based on the known mean and standard deviation of log-transformed counts of the product being tested. Consequently, they can only be applied to situations in which the distributions of microbial loads are known (e.g., when the logarithm of the cell concentrations follows a normal distribution). Variables plans are more appropriate to producers who regularly perform microbiological testing of their products. Such information is rarely available to a regulatory authority or a purchaser, who consequently will usually default to attributes plans.

Attributes Sampling Plans

Attributes sampling plans test against a single criterion or attribute, such as the presence of *Salmonella* in 25 g or the proportion of sample units that contain >100 CFU $\cdot$ g^{-1} of the target organism. Unlike what occurs in the variables sampling plans, the magnitude of the deviation between the number of microorganisms in the sample and that specified in the criterion is not considered: samples simply pass or fail, i.e., they either do, or do not, have the attribute being assessed. Attributes plans are described as two-class or three-class plans. Both types are characterized by three elements: (i) n, the number of samples taken from the batch of product being tested; (ii) m, the attribute or condition that is being assessed; (iii) c, the maximum tolerable proportion (or number) of the samples that do not satisfy the attribute m, but for which the batch is still considered to be acceptable. In three-class sampling plans, however, a fourth characteristic, (iv) M, an attribute or condition that, if exceeded in *any* sample, is completely unacceptable and leads to rejection of the entire batch, is also specified.

There is specific terminology associated with the application of sampling plans. Some relevant terms and their definitions are presented in Table 1. A more complete list is

TABLE 1 Terminology associated with sampling plans used in food microbiology

Term	Definition
Lot	A specified quantity of some commodity manufactured or produced under conditions which are presumed to be, or have been, uniform for the purpose of application of sampling plans
Consignment	A quantity of some commodity delivered at one time: it may consist of either a portion of a lot or a set of several lots
Batch	A term sometimes used interchangeably for "lot" or for "consignment." (For the purposes of this chapter, a batch is considered to be a subset of a lot.)
Sample	One or several units, or portions, of the material comprising the lot and selected randomly (by appropriate means)
Representative sample	A sample that maintains the characteristics of the lot from which it is drawn
Producer's risk	The probability that a satisfactory lot is rejected by the sampling plan, because of chance, i.e., that an unrepresentatively *high* number of positive units/portions are sampled
Consumer's risk	The probability that an *unsatisfactory* lot is accepted by the sampling plan, because of chance, and sold/consumed, i.e., that an unrepresentatively *low* number of positive units/portions are sampled
Characteristic	A property that helps to identify, or differentiate (e.g., their acceptability) between items within a given lot
Sampling plan	A procedure that enables selection of samples from a lot, in order to obtain information needed to decide whether a lot complies with some specified microbiological criterion
Homogenous	When the characteristic of interest is uniformly distributed throughout the lot
OC curve	A graph that relates the probability of acceptance of a lot as a function of its actual quality/rate of defects. The shape of the curve depends on the stringency of the sampling plan (no. of samples, no. of positive samples acceptable [Fig. 1])
Acceptable quality level (AQL)	The rate of nonconforming items in a lot that can be tolerated before the lot is rejected: the AQL is linked with a certain level of confidence that depends on the stringency of the sampling plan. Typically, the confidence is set to 95%.
Analytical unit	The actual volume/mass/area, etc., of a sample that is tested in the specified method: an analytical unit may comprise the entire sample, or may be an aliquot of the sample.
Compositing	In presence/absence testing, the combining of samples/analytical units into a larger size to minimize testing effort by reducing the number of separate tests that need to be performed and followed
Microbiological criterion	A statement or specification that defines the acceptability of a product or a food lot, based on the absence/presence or number of specified microorganisms including parasites, and/or quantity of their toxins/metabolites, per unit(s) of mass, volume, area, or lot

presented by the CAC (Codex Alimentarius Commission, 2004).

Attributes sampling plans can be specified in terms of these three or four values. For example, an $n = 10$, $m = 100$ CFU · g^{-1}, $c = 2$, two-class plan involves testing 10 samples from a lot. The lot is acceptable if at least 8 of 10 samples have <100 CFU · g^{-1}. If the results for 10 samples were <10, <10, <10, 20, 20, 80, 50, 70, 150, and 45,000 CFU · g^{-1}, the two-class sampling scheme presented would lead to acceptance of the batch because no more than two samples exceeded *m* (i.e., 100 CFU · g^{-1}). If, however, *m* were reduced to 50 CFU · g^{-1}, or *c* were reduced to 1, or the sampling plan were extended to a three-class plan with $M = 1{,}000$ CFU · g^{-1}, then the same batch would be considered defective and rejected. Thus, the specification of a sampling plan dictates the probability of acceptance of a lot and also dictates the stringency of a criterion.

STATISTICS OF SAMPLING

As noted in the section above ("Terminology and Types of Sampling Plans"), there are numerous sources of information on the statistics of sampling plans. An overview of fundamental concepts, however, is presented below.

Basic Concepts

In essence, the greater the quantity of material sampled and tested, the more sensitive the detection limit of the sampling plan. For example, if 1 kilogram of product were tested by enrichment and no pathogens were detected, it could be inferred, by simple logic, that the level of contamination in the product was probably less than one cell per kilogram. One could not be sure, however, because pathogens are not usually homogenously distributed in foods. Thus, a single sample taken at random may not, by chance, be representative of the lot from which it was drawn. Additional 1-kilogram samples could be tested to increase confidence that the sample, as well as the result, *was* representative.

If another four samples of 1 kilogram each were tested and each yielded similar results, there would be more confidence that the original sample was representative of the entire lot. There would be greater confidence that, on average, the contamination in the lot is less than 1 cell per kilogram. But how much confidence would there be in the representativeness of the results? What is the chance that five completely unrepresentative samples were taken and that the lot, as a whole, has an average contamination level that is much higher (or lower) than the results of the five samples suggest? Fortunately, the statistics to enable calculation of the confidence associated with the results of sampling plans are well established and readily accessible.

The Binomial Distribution

In probability theory and statistics, the binomial distribution describes the chance of finding a positive when drawing samples from a lot and applying a test that has only two possible outcomes, e.g., a positive or a negative (or yes/no). In microbiological testing for pathogens, in particular, this is frequently the case because it is hoped that the contaminant will be present only at low levels, if at all, and either is present in the sample or is not. This kind of testing is often termed "presence/absence" testing and used for so-called zero-tolerance regulations (see also "Presence/Absence Sampling and Zero Tolerance" below).

Returning to the binomial distribution, consider, for example, a 1,000-liter vat of milk in which the average *Escherichia coli* contamination level is one cell per liter. It might be expected (by simple logic, but see further discussion below) that if a 100-ml sample were chosen at random there would be a 1-in-10 chance that the sample contains an *E. coli* cell. Conversely, there is a 90% chance that the contaminant would *not* be present in the sample. If two samples are randomly drawn, each has a 1/10 chance of containing an *E. coli* cell, so the probability of *not* detecting the contaminant is 90% × 90%, i.e., 81%. If three samples are taken and a positive in any sample causes rejection, the probability of not detecting the contamination is 72%, and so on (see also the discussion on the hypergeometric distribution below). Thus, the probability of *not* detecting the contamination, if present, decreases as more samples are taken: confidence in the correctness of the results of the sampling plan increases as more samples are taken and analyzed.

These relationships can be summarized mathematically: for a given number of samples (n), and the true prevalence of contamination in the lot (p), the probability (P) of *not* detecting the contamination, and thereby accepting the lot as satisfactory, is given by the following equation:

$$P_{(acceptance)} = (1 - p)^{\text{number of samples}} \quad \textbf{(1)}$$

since all samples are required to be negative. If a contamination frequency of 10% were considered unacceptable, however, an obvious question is to ask "how many samples would have to be tested to be confident that contamination if present at that level, or higher, would be detected?" To answer this question, it would be hoped that the probability of acceptance of a lot with greater than 10% defects is low. Thus, the left-hand side of equation 1 might be set to 5% (i.e., 0.05), meaning that we require 95% confidence that if ≥10% of the units were contaminated, such contamination would be detected. The equation would then be solved as follows:

$$P_{(acceptance)} = (1 - p)^{\text{number of samples}}$$

By substituting in the relevant values, we obtain:

$$0.05 = (1 - 0.1)^{\text{number of samples}}$$

and transforming both sides of the equation:

$$\log(0.05) = \log[(0.9)^{\text{number of samples}}]$$

and rearranging:

$$\log(0.05) = \text{number of samples} \times \log(0.9)$$

Thus, the number of samples required

$$= \log(0.05)/\log(0.9)$$

$$= 29$$

More generally:

$$\text{samples required} = \log(\text{required level of confidence})/\log(1 - \text{prevalence of interest}) \quad \textbf{(2)}$$

In the above example, 29 *negative* samples are required to determine, with 95% confidence, that the target contamination level in the batch is <10%. While that is a large number of samples, the contamination frequency is rather high in terms of modern expectations of food quality and safety. If the acceptable contamination frequency were ≤1% of products, equation 2 shows that 299 negative samples are required to be 95% confident that the required upper frequency of contamination (i.e., 1%) is not exceeded. If lower frequencies of noncompliance are required, or higher levels of confidence, even greater numbers of samples are required. The relationship between contamination level and probability of detection for a given number of samples has traditionally been summarized in a graph called an "operating characteristic" curve, often abbreviated as "OC" curve, from which sampling plans can be designed to meet particular levels of confidence or from which the efficacy of a specified sampling plan can be rapidly assessed without recourse to a series of calculations using equation 1.

Equation 1 is appropriate in cases where detection in any one of the samples will cause rejection of the batch, i.e., it can be used to interpret the reliability of sampling plans for which $c = 0$. Sometimes, however, sampling plans allow one or more of the samples to be positive, i.e., to exceed the criterion that specifies the attribute. In this case a more complex equation is required to determine the probability of detection for a given number of samples and a given frequency of contamination (equation 3).

$$P_{(acceptance)} = 1 - \sum_{i=0}^{c} \frac{n!}{c!(n-c)!} p^{c}(1-p)^{(n-c)} \quad \textbf{(3)}$$

where *n*, *c*, and *p* are as previously defined, and which may be abbreviated as:

$$P_{(acceptance)} = 1 - \sum \text{Binomial}(c; n, p)$$

(Note that in the case of $c = 0$, equation 3 simplifies to equation 1).

Clearly, this is an even more complex equation to calculate, and one that must be calculated for each integer value up to the value *c*. As for the case of $c = 0$ plans, OC curves are presented to simplify the task of analysis of sampling plan performance. Fortunately, common spreadsheet software applications also include equation 3 as a predefined function and can be used to calculate confidence levels for different sampling plans. For example, in Microsoft® Excel, the function is "Binomdist," and for a given value of *c*, given the number of samples (*n*) and the true or maximal tolerable defect rate (*p*), it can be used to calculate the probability of detection. Alternatively, it can be used (by iterative methods, using functions such as "Solver" or "Goal Seek" in Microsoft® Excel) to determine the number of samples required for a specific level of confidence in the result.

Examples of OC curves for various sampling plans with different *c* and *n* values are shown in Fig. 1.

It is also illustrative to consider the limits of detection at 95% confidence for a given number of samples. Thus, by rearranging equation 1, for a $c = 0$ sampling plan, five negative samples provide 95% confidence only that the true prevalence of contaminated units is <45%. For 10 samples, contamination prevalence of >26% can be detected with 95% confidence, while for 20 negative samples, 95% confidence is obtained that the true prevalence of contaminated units in a batch is <14%. Clearly, most practical presence/absence sampling plans do not provide a high level of assurance of zero tolerance.

Hypergeometric Distribution

In fact, the calculations and equations presented above are strictly true only if the amount of material sampled is relatively small compared to the total amount of material in the lot. This is because in food microbiology we sample without replacement because every test is destructive (of the sample). As a consequence, the total sample volume is reduced with every subsequent sample taken, and accordingly, the probability of detecting the contaminant increases slightly each time a negative sample is taken. This situation is described by the hypergeometric distribution, which can also be evaluated to calculate the probability of detecting one, or multiple, positive samples when sampling without replacement and given (i) the number of samples taken, (ii) the true prevalence of positives in the total sample volume, and (iii) the total number of samples of that size that can be removed from the batch (i.e., batch size divided by sample size). However, provided that the total sample volume (i.e., sample size multiplied by the number of samples taken for the test) is a relatively small proportion of the total batch size, the binomial distribution and calculations described offer a good approximation.

Compositing

For those microbiological hazards that typically (i) lead to infection even when only low numbers (e.g., a few thousands or less) of cells are ingested, or (ii) for which food processing and preparation may be inadequate, or unreliable, to eliminate the pathogen, there is a need to have confidence that the pathogen, if present, is present only at very low levels. Note, however, that very low contamination levels, e.g., less than one cell per sample unit, tend to be regarded and discussed as *frequencies* of contamination, because multiple samples have to be analyzed to detect and enumerate them. In reality, however, they may form part of a continuum of contamination levels from high to very low.

The presence of enterohemorrhagic *E. coli* in ground meat intended for beefburgers is a relevant example of an organism for which very low concentrations in foods are known to have caused outbreaks. The potential presence of *Salmonella* or *Cronobacter* spp. in pow-

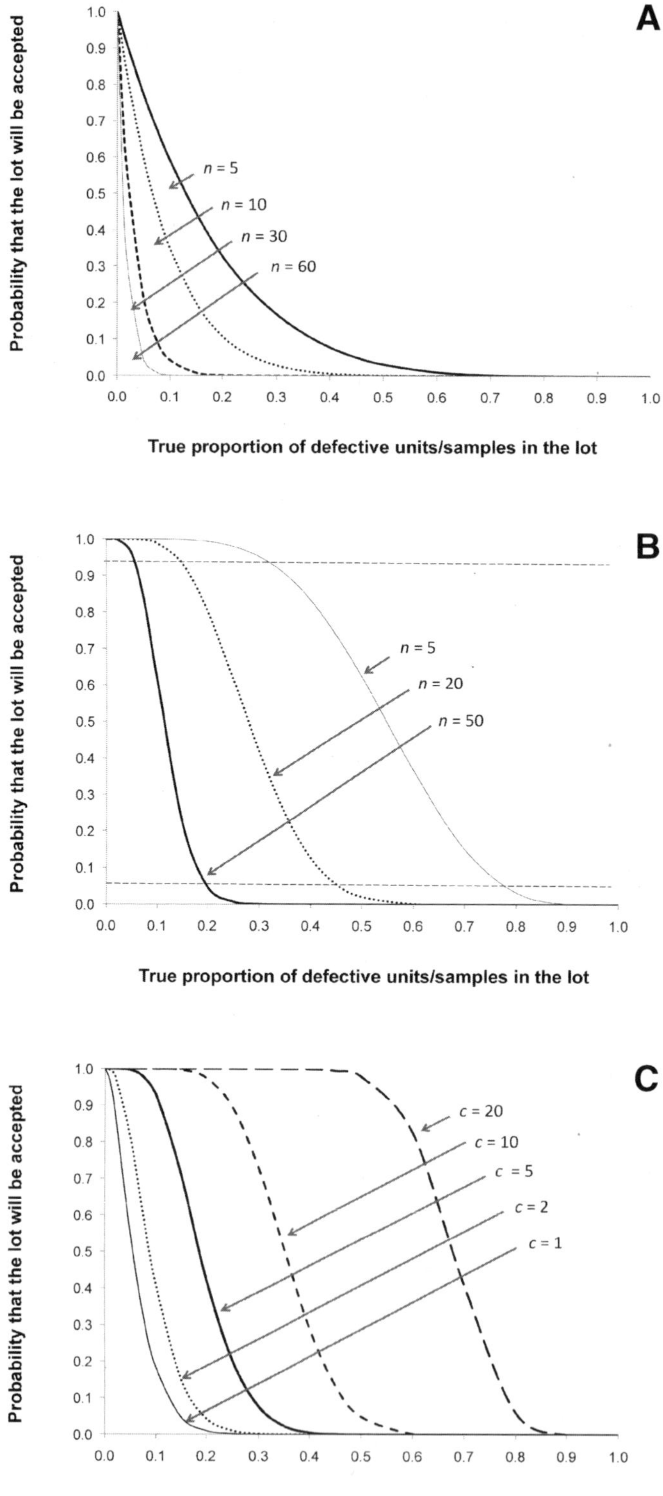
A
Probability that the lot will be accepted
True proportion of defective units/samples in the lot
n = 5
n = 10
n = 30
n = 60
B
Probability that the lot will be accepted
True proportion of defective units/samples in the lot
n = 5
n = 20
n = 50
C
Probability that the lot will be accepted
True proportion of defective units/samples in the lot
c = 20
c = 10
c = 5
c = 2
c = 1

dered infant milk formula is another example, not only because of the susceptibility of the target population for these products but also because of the potential for growth of the pathogen in the reconstituted milk/formula prior to consumption. The threat from these hazards requires a very high level of confidence that lots are free of these pathogens, leading in turn to large numbers of samples having to be tested on a frequent basis.

The detection of pathogens in the presence of other members of the microbiota (such as in raw meat), or in processed foods in which cells may have become sublethally injured, requires multiple steps to demonstrate that the pathogen is *not present* rather than merely *not detected*. Thus, typical procedures for pathogen detection in foods involve a resuscitation, or preenrichment, step to allow injured cells to recover and be able to endure the more stringent selective enrichment processes that usually follow to increase the numbers of the target organisms relative to other organisms that may have been present in the sample initially. After that stage, further selective media followed by biochemical and serological tests may be applied, or, increasingly, identification can be based on detection of genetic markers (e.g., specific 16S RNA genes by PCR or other methods). In either case, a potentially large number of analytical units, typically of 25 g and each added to 225 ml of preenrichment broth, have to be tested for every lot. This requires time, labor, space, and reagents and has led to renewed interest in the idea of compositing of samples, i.e., combining the individual analytical units into fewer, larger units to reduce the number of tests per lot that have to be performed. If the result from the composite sample is negative, as would usually be expected for a well-controlled production system operating under good agricultural practices, good manufacturing practices, and HACCP, etc., much time and expense can potentially be saved by compositing. If the sampling plan is a $c = 0$ scheme, a positive result for a composite sample has the same consequence as detection in any single sample from the lot, i.e., the lot is noncompliant.

Similarly, for any sampling scheme that involves enrichment procedures, if the composite sample is negative, each individual sample in the composite sample can be assumed to be negative (subject to the caveats discussed below). If the sampling plan allows for some positives (i.e., $c > 0$), retained material from each of the individual samples can be retested individually to determine what proportion of samples from the lot are noncompliant.

Validity of Compositing

Jarvis (2007) provided a useful analysis of the strengths and potential pitfalls of compositing of samples. Provided that the method sensitivity allows detection of a single cell in the larger volume of the composite sample, then there is no difference in the confidence of test results derived from material tested as multiple small analytical units or as an equivalent amount composited into fewer, larger, analytical units.

For compositing to be valid, however, it must be demonstrated that the methods used will enable detection of a single contaminant with equal confidence in the larger (composited) volume and in the (smaller) individual analytical unit. This will require validation that a single cell present in the preenrichment will grow to a sufficient concentration so that

FIGURE 1 OC curves. (A) Effect of the number of samples on the probability of detecting a defective unit (e.g., a contaminant) as a function of the prevalence of defective units in the lot for a scheme in which any positive leads to rejection of that batch (i.e., $c = 0$). (B) Sampling plan with $c = 5$ and $n = 10$, 20, or 50 showing the influence of the number of samples on the difference between the producer's risk (shown as 95% probability of acceptance, upper dashed line) and the consumer's risk (95% probability of rejection, lower dashed line). As n increases, the difference between the consumer's risk and the producer's risk is reduced. (C) Influence of the number of positive results permitted by the sampling scheme on the probability of acceptance of the lot, using a sampling plan based on $n = 30$ samples.

a 1-ml (or other size) aliquot of the preenrichment media will contain at least one viable cell for transfer to the next step in the isolation/identification process. In cultural methods, typically 1 ml or 0.1 ml is subcultured, while in molecular approaches, even smaller volumes are used (e.g., 20 to 50 μl). Thus, the method must allow a single cell in the composite to grow to the same concentration as would be achieved in an enrichment of the individual analytical unit.

For a $c = 0$ scheme, if the composite sample is 10 times larger, then the cell effectively needs to grow 10 times more to achieve the same final concentration in the composite because the initial concentration is 10-fold lower (assuming that only one of the original samples is contaminated). If the composite were composed of 100 individual analytical units, a cell initially present would be required to grow 100 times more to achieve the same final concentration, i.e., an additional 2 log of growth. For example, growth from one cell in 2.5 liters of resuscitation culture to one cell per ml depends on the lag time (including injury recovery time) and the generation time of the target organism in the enrichment broth. Growth from one cell per 2.5 liters to one cell per ml equates to ~12 generation times, in addition to the lag time. If the aliquot taken for subsequent testing is 0.1 ml, an additional three generation times would be needed (i.e., 3.3 doublings equate approximately to a 10-fold increase).

Moreover, even at one cell per 0.1 ml, there is no certainty that a 0.1-ml aliquot would contain a cell, because the cells will be randomly distributed. The probability of sampling a cell in 0.1 ml when the concentration is 0.1 cell per ml can be estimated from the Poisson distribution:

$$P_{\text{(detection of 1 cell)}} = 1 - e^{\text{aliquot volume} \times \text{concentration of cells in sample}} \quad \textbf{(4)}$$

from which it can be determined that the mean concentration needs to be >47 cells per ml to have >99% confidence that every 0.1-ml aliquot drawn at random will contain at least one cell. Thus, a further three generations of growth are required if a smaller volume is used for the next stage of the procedure.

The above is intended to demonstrate that compositing can influence the results of enrichments and decrease detection sensitivity, particularly in cases where enrichment processes have been optimized to the minimum enrichment time needed for a smaller volume of material.

Jarvis (2007) identifies three possible modes of compositing. The first is to combine all samples and take a subsample of the composite and process that as a normal (e.g., 25-g) enrichment sample. Clearly, this approach decreases the detection sensitivity because the chance of including a contaminant from any of the samples in the new subsample is reduced in proportion to the number of samples that are combined. This approach is clearly invalid.

The second option is to combine all the material into a single large unit and combine that with the total volume of enrichment media that would be used if all analytical units were processed individually. The approach is subject to the caveats described above but, if validated, is effectively the same as processing each sample individually.

The third option, described by Jarvis as "wet compositing," can be used when the preenrichment of a composite does not give individual cells the same probability of being amplified sufficiently so that they, or their progeny, are certain to be passed on (in the chosen aliquot) to the next stage of the process. In this case, the individual analytical units are preenriched separately. Appropriate aliquots from each preenrichment are then composited into the enrichment step. Preenrichment cultures can also be preserved and retested individually if a positive result ensues and if the sampling plan allows for one or more positives (i.e., $c > 0$).

MICROBIOLOGICAL CRITERIA FOR FOODS

Inherent in any sampling plan is some microbiological criterion for which the sampling plan is designed to assess compliance. While the above discussion has enabled evaluation of the confidence one can have in the reliability of a sampling plan to assess compliance with a specific microbiological criterion, it has not considered how those criteria, or attributes, are established. Before commencing a discussion on that topic, it is useful to introduce some basic ideas about microbiological criteria.

Definition of a Microbiological Criterion

Several organizations have considered the appropriate specification of microbiological criteria. Codex guidelines (Codex Alimentarius Commission, 1997) are representative of those deliberations and state that a microbiological criterion must include the following.

1. A statement of the microorganisms of concern and/or their toxins/metabolites and the reason for that concern
2. The analytical methods for their detection and/or quantification
3. A plan defining the number of field samples to be taken and the size of the analytical unit
4. Microbiological limits considered appropriate to the food at the specified point(s) of the food chain
5. The number of analytical units that should conform to these limits

The Codex further adds that a microbiological criterion should state the following.

6. The food to which the criterion applies
7. The point(s) in the food chain where the criterion applies
8. Any actions to be taken when the criterion is not met

As discussed throughout this text, not all microbiological methods produce equivalent results, and results may depend on the food product, the strain of the target organism, the history of the organism, and other factors. Equally, the tolerable microbiological limit may depend on the food. For example, in foods that support growth of *L. monocytogenes*, absence in multiple 25-g samples is usually specified; but for foods that do *not* support its growth, up to 100 CFU · g^{-1} is considered tolerable, based on the relatively high 50% infective dose (ID_{50}) for this pathogen (Food and Agriculture Organization/World Health Organization, 2004). Test sensitivity also varies according to the amount of material sampled or included in the test. For these reasons, when microbiological criteria are specified and used for regulatory purposes or commercial agreements, unambiguous specification of the criterion, method, food, etc., are all important to ensure that the criterion is used for the purpose and in the manner that were originally intended.

Establishing Microbiological Criteria—Role of Risk Assessment

In "Sampling Plans" above, it was noted that the ad hoc development of microbiological criteria led to problems in international trade in foods and that this, in turn, had resulted in the introduction of risk assessment to enable relevant and "defensible" criteria to be established.

Risk assessment is being adopted in many businesses, organizations, and institutions as a basis for supporting decision making. In general, it offers a structured approach to synthesizing information relevant to some undesirable event (i.e., a hazard), including the likelihood that it will occur and the severity of the consequences if it did. These three elements (hazard, probability, and severity) constitute "risk."

Microbial food safety risk assessment is a relatively new activity internationally. It was given impetus by the resolution of the Uruguay Round of the General Agreement on Tariffs and Trade in 1993–1994, which led to the creation of the World Trade Organization. Also included in the resolution of the Uruguay Round were decisions agreeing that the only justification for a nation to prevent or limit the import of foods from another nation was by demonstration that the imported product posed an unacceptable risk to human, animal, or plant life or health in the importing nation. These decisions are found in (i) the

"Agreement on the Application of Sanitary and Phytosanitary Measures" (also known as the SPS Agreement) and (ii) the Agreement on Technical Barriers to Trade (also known as the TBT Agreement).

Demonstration of unacceptable risk was to be by transparent and scientifically based risk assessments. Through the CAC, the Food and Agriculture Organization and World Health Organization were charged with development of methods for such risk assessments, a process that began in 2000 with the instigation of the Joint Expert Meetings on Microbiological Risk Assessment. The outcomes of that work, which is ongoing, are presented in a series of publications available in print or online from those two organizations.

In microbiological risk assessment, four elements are considered. "Hazard identification" concerns the identification of microorganisms capable of causing adverse health effects and which may be present in a particular food or group of foods. Dose-response assessment ("Hazard characterization") seeks to establish the relationship between the number of pathogen cells (or their toxins) ingested and the consequent probability of illness, i.e., to define the "infectiousness" of the organism. Hazard characterization identifies factors that affect the risk, such as the severity of disease caused by the pathogen of interest, and factors that affect its potential for growth or survival or inactivation in the host. Exposure assessment seeks to identify the frequency with which consumers are exposed to the pathogen in specific foods and the levels found in those foods, or in the food supply in general.

Predictive modeling of the microbial ecology of foods is used to estimate doses ingested based on survey data obtained at retail, or at the time of production, and knowledge of times and temperatures of handling and product formulation. "Risk characterization" attempts to synthesize all these data and relevant knowledge to identify points at which control can best be exercised, as well as to be able to define tolerable levels and frequencies of exposure. Dose-response information can be used to establish microbiological criteria that are appropriate at the point of consumption. Using exposure assessment models, such levels can be translated into criteria relevant at the point of manufacture, where testing is more likely to be applied as a control. This has been discussed and illustrated in the context of food safety objectives and of process and performance criteria by ICMSF (2002), CAC (2004), Whiting et al. (2006), and van Schothorst et al. (2009) and has been included in Codex procedural manuals since 2004. (Note that the procedural manuals are available for download from http://www.codexalimentarius.net/web/procedural_manual.jsp).

A detailed discussion of the role and application of risk assessment methods in the establishment of science-based microbiological criteria for foods in international trade is beyond the scope of this chapter. Ross and Chan (2002) and Havelaar et al. (2005) do, however, provide useful discussions of these topics, while Uyttendaele et al. (2006), Membré et al. (2007), and Nauta and Havelaar (2008) present practical examples of how this can be achieved.

Risk-Based Sampling—the ICMSF Approach

Despite the advent of risk assessment thinking, there are only a few structured approaches to development of sampling plans that are overtly based on risk. Among those that do exist is that presented by ICMSF in 1986 and revised in 2002 (International Commission on Microbiological Specifications for Foods, 1986, 2002). That approach identifies a number of factors affecting foodborne microbial risks including the severity of illness caused by the organism of concern, the sensitivity of the population for whom the product is intended, and the potential for growth (or inactivation) of pathogens or other microorganisms of concern in the food, which lead to increased (or decreased) consumer risk. Using a matrix of those factors guides the selection of sampling plans of different stringency, appropriate to the relative risk determined by those factors. The sampling plan matrix, including the risk-based criteria for sampling plan selection, is shown in Table 2.

The stringency of all sampling plans can be increased or decreased by modification of the microbiological criterion that is the attribute being assessed. For example, the criterion for any sampling plan could be that the average concentration of the target organism is ≤100 CFU g^{-1}, or absence in 1 g, or absence in 25 g. The mean contamination level detected by a sampling plan for which the criterion is absent in 25 g is much lower than that of the same plan based on a criterion of ≤100 CFU g^{-1}. While sampling plans have usually been interpreted in terms of the confidence provided that the attribute is consistently satisfied throughout the lot, even for presence/absence testing, the size of the sample unit can convey information about the mean concentration of samples drawn from the lot. At the simplest level, absence in 1 g can be interpreted as a <1 CFU g^{-1} and absence in 25 g as <0.04 CFU g^{-1}. This was alluded to in "Basic Concepts," above, but the interpretation of presence/absence sampling in terms of target concentrations of pathogens has been rigorously investigated over recent years and is discussed in the next section.

PRESENCE/ABSENCE SAMPLING AND ZERO TOLERANCE

Presence/Absence Testing as One-Dilution Most Probable Number

As demonstrated above, no sampling plan can ever demonstrate that a lot is free of contamination, a point made emphatically by ICMSF

TABLE 2 Example of a risk-based sampling plan scheme[a]

Type of hazard	Conditions in which the food is expected to be handled and consumed after sampling:		
	Reduce risk	Cause no change in risk	May increase risk
Utility (general contamination, reduced shelf life, and spoilage)	Case 1 3-class, $n = 5$, $c = 3$	Case 2 3-class, $n = 5$, $c = 2$	Case 3 3-class, $n = 5$, $c = 1$
Indirect: low, indirect hazard	Case 4 3-class, $n = 5$, $c = 3$	Case 5 3-class, $n = 5$, $c = 2$	Case 6 3-class, $n = 5$, $c = 1$
Moderate: not usually life threatening, usually no sequelae, normally of short duration, symptoms self-limiting, can cause severe discomfort	Case 7 3-class, $n = 5$, $c = 2$	Case 8 3-class, $n = 5$, $c = 1$	Case 9 3-class, $n = 10$, $c = 1$
Serious: incapaciting but usually not life threatening, sequelae rare, direct, of moderate duration	Case 10 2-class, $n = 5$, $c = 0$	Case 11 2-class, $n = 10$, $c = 0$	Case 12 2-class, $n = 20$, $c = 0$
Serious: (i) for the general population or (ii) for restricted populations, causing life-threatening illness or substantial chronic sequelae, of long duration	Case 13 2-class, $n = 15$, $c = 0$	Case 14 2-class, $n = 30$, $c = 0$	Case 15 2-class, $n = 60$, $c = 0$

[a] Data from International Commission on Microbiological Specifications for Foods, 2002.

(2002) in advocating that more realistic microbiological criteria be developed and implemented through Food Safety Objectives and Process and Performance Criteria (Codex Alimentarius Commission, 2007). Nondetection of pathogens in a number of replicate samples of a specified size can only ever provide confidence, at some defined probability, that if contamination *is* present in the lot, the concentration of contaminants is below some (definable) level.

In effect, a series of presence/absence tests can be regarded as a one-dilution most-probable-number method. For example, if 1 in 10 samples of 25 grams of food are found to contain the target organism, this might be expressed as a 1/10 chance of detecting a cell in 25 grams, i.e., approximately one cell per 250 g. If the size of the analytical unit is 10 g, and 1 in 10 samples are found to contain the target organism, it could be inferred that the average contamination level is in the region of one cell per 100 grams.

Jarvis (1989) presents a similar analysis of contamination levels that can be inferred for specific sampling plans, taking into consideration binomial sampling statistics. This was extended by Legan et al. (2001) to be able to interpret the sensitivity of sampling plans to target *concentration* levels, rather than prevalence of contamination.

Importantly, all the statistics presented above are based on the assumption that the contaminants are homogenously distributed in the lot. Legan et al. (2001) presented an approach to estimate the performance of sampling plans when the contaminants are not homogenously distributed. While that approach provides a good approximation for target levels that can be assessed by quantitative methods (e.g., by viable count or real-time PCR), the influence of type 1 errors (false positives) can significantly increase the probability of lot rejection when presence/absence methods are used. The Legan et al. (2001) approach was further refined by van Schothorst et al. (2009) to take this effect into account.

The first step in that approach involves translating the concentration of contaminants in the lot to an equivalent expected prevalence for *the size of the analytical unit* that is used to represent the sample in the presence/absence protocol. From this expected prevalence, the probability of detecting the contamination can be determined by using equation 3 for the number of samples (of that size) defined by the sampling plan. If this process is repeated for a range of concentrations of contaminants, the probability of detection can be related to the *concentration of contaminants* to generate an OC curve that is specific to the sampling plan but is a *function of concentration* rather than of prevalence (Fig. 2A).

By using this approach, the detection limit of any sampling plan at a specified level of confidence can be determined. Analogous calculations can be performed for plans with $c > 0$ and for three-class plans. Thus, any attributes sampling scheme can be interpreted quantitatively, for specified levels of confidence.

The above treatment assumes that the contaminants are homogenously distributed throughout the lot, but in real systems, even if the process is consistent, contaminants will not usually be distributed evenly throughout the lot. There is evidence, however, that the distribution of the logarithm of concentration of contamination levels in a lot is well described by a normal distribution (note that the counts themselves are described by a log-normal distribution).

Assuming a log-normal distribution, the calculations described above can be repeated for lots that are characterized by a *distribution* of concentrations. The probability of detection for a given sampling plan is, however, now also a function of the *variability* of the log (pathogen concentration) throughout the lot. The greater the variability and the greater the mean log (concentration), the greater the proportion of samples from that lot that will exceed the criterion (Fig. 2B).

In essence, in this approach the probability of detection is equated to the proportion of samples in the distribution that would be expected to exceed the target value. This can be determined for any combination of mean,

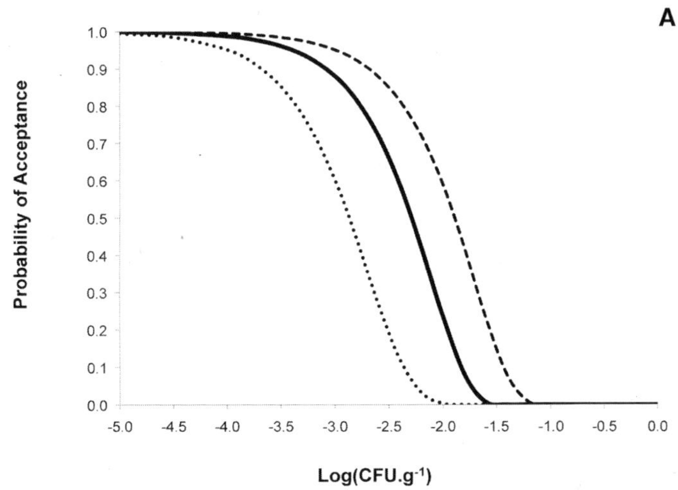

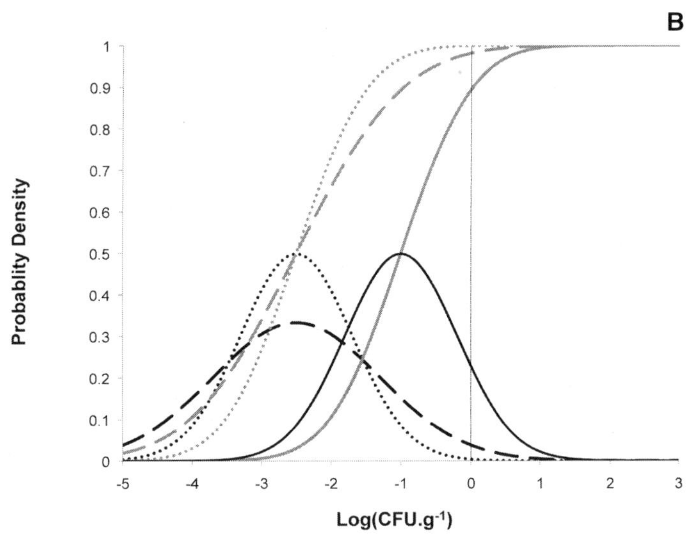

FIGURE 2 OC curves for selected sampling plans. (A) Probability of batch acceptance as a function of contamination levels, for different numbers of samples and sizes of analytical units for $c = 0$ (i.e., no positives permitted) sampling plans. The dotted line represents a plan comprising 20 25-g samples. The solid line represents a sampling plan comprising five 25-g samples. The dashed line represents a sampling plan comprising five 10-g samples. Though not shown, a plan with 50 10-g samples has the same OC curve as that for 20 25-g samples (i.e., the dotted-line OC curve), showing that compositing has no effect on plan sensitivity (subject to the caveats described in the text). (B) Effect of the mean and standard deviation of the distribution on the expected prevalence of defects determined as a proportion of samples above the criterion. In the plot, an arbitrary threshold of 1 CFU $\cdot$ g^{-1} (0 log CFU $\cdot$ g^{-1}) is shown. Three distributions are shown as probability density (black) and corresponding cumulative probability curves (grey). The distributions shown (mean $\pm$ standard deviation in log CFU $\cdot$ g^{-1}) are as follows: solid lines, -1 ± 0.8; dotted lines, -2.5 ± 0.8; dashed lines, -2.5 ± 1.2. The proportions that have unacceptably high microbial loads are represented by that part of the distribution to the right of the vertical line representing the threshold. This proportion can be deduced most easily from the cumulative probability curves (grey lines). The point at which cumulative probability curves cross the threshold value gives the proportion of samples below the threshold and, by inference ($1 -$ proportion below the threshold), the proportion above the threshold. It can be seen that the proportion of defectives is a function of both the mean and standard deviation of the distribution.

standard deviation, and target value by using a table of z-values or appropriate functions in spreadsheet software (e.g., the function "Normdist" in Microsoft® Excel). (The z-value is the number of standard deviations between the mean and some target value. It is calculated as the difference between the target value and the mean value, divided by the standard deviation. The proportion of the distribution that is above this target value can be determined from the number of z-values.) Thus, the proportion of samples that would be expected to become positive for a given log-normal distribution and a given detection limit per sample can be estimated.

Consequences of the Poisson Distribution

The approach described is a reasonable approximation for target levels that can be determined by quantitative methods. Where the quantification itself is based on the *proportion* of samples of a specified size that become positive, there are further considerations related to the probability of detection, including the chance described, e.g., by the Poisson distribution, that a sample will contain a target cell even when the concentration is below the target value (Fig. 3). For this reason, an additional set of calculations is required, as described by van Schothorst et al. (2009), to account for the chance that a positive sample will be detected and the lot rejected, even though the average concentration in the sample is below the target level, i.e., due to type 1 errors or false positives. The closer the average log(concentration) is to the target value, the greater the probability of false positives. To account for their contribution, this probability is determined for all concentrations

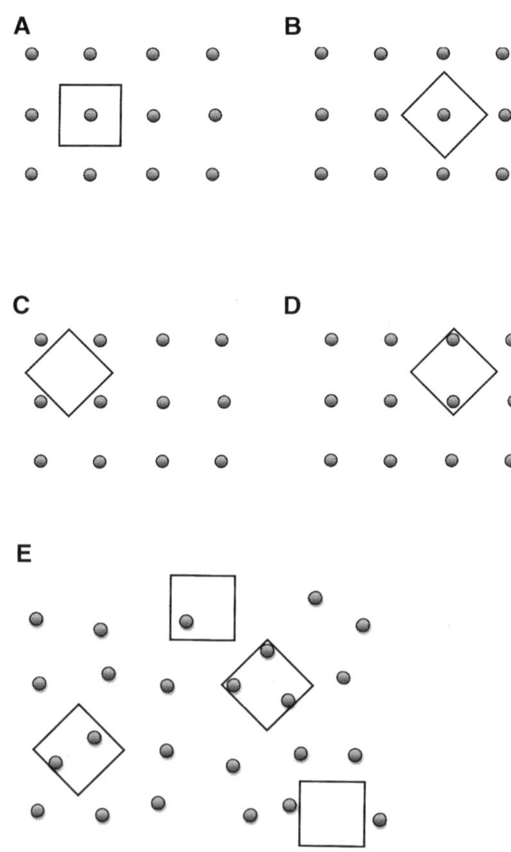

FIGURE 3 Illustration of sampling errors in microbiology, showing a two-dimensional depiction of a food sample contaminated with microbial cells, shown as shaded circles. Each square or diamond represents a sample drawn from the food. The cells are evenly distributed at a concentration of 1 cell per square centimeter, and the sample size is exactly 1 square centimeter, so that it would be expected that every sample would capture one cell, as is shown in panels A and B. However, it is easy to see that by chance a sample might not capture a cell, as shown in panel C. Similarly, even though the cells are perfectly homogenously distributed at a concentration equivalent to one per sample, more than one cell could be captured in a sample (panel D). In reality, cells would be randomly distributed as shown in panel E, and this would exacerbate the situation depicted in panels A through D so that even if the *average* concentration is one cell per sample, some samples will contain no cells while others will contain multiple cells.

within the distribution and the cumulative probability of false positives is included in the estimate of the probability of detection. The net effect of this is to reduce the number of negative samples required compared to that estimated by the binomial distribution (equation 3) for a specified level of confidence.

Thus, to determine the overall performance of a sampling plan, i.e., the concentration of contaminants that it can detect with a specified level of confidence, it is necessary to consider (i) the number of samples, (ii) the size of the analytical units, (iii) the mean concentration in the lot, (iv) the standard deviation of distribution of contamination levels [log(CFU · g^{-1})], and (v) the required confidence in the reliability of the test.

Equally, the calculations can be rearranged to design sampling plans with required performance, by manipulating sample size, sample number, and confidence required, given the mean concentration and standard deviation of the log(concentration) of organisms within the lot.

The calculations required for the approach described are complex and repetitive and are best undertaken by using software to automate the calculations. The ICMSF has developed a spreadsheet that undertakes these calculations and provides guidance and interpretation. It is available, without cost, for download from the ICMSF website (http://www.icmsf.iit.edu/main/home.html).

CONCLUSIONS AND FUTURE TRENDS

In general, the microbiological safety of food cannot be effectively ensured by testing because microorganisms are not completely or evenly spread throughout the food, as might be the case for a chemical contaminant. As such, the safety of food lots, assumed to be produced under consistent conditions and resulting in food with consistent quality, is often assessed by microbiological analysis of a series of samples drawn from that lot. Understanding the reliability of sampling plans is, therefore, important for food safety management. At acceptable frequencies of contamination, the probability of detecting unacceptable levels of contamination is low unless large numbers of samples are analyzed. For this reason, preventive food safety management based on HACCP, good agricultural practices, and good hygiene practices, etc., are preferred to, and more reliable than, end product testing. Nonetheless, in some situations, testing for microbiological quality and safety is necessary. Strategies such as compositing can in some circumstances reduce the associated workload.

When the target level to be measured in the sample is high, the sampling error is relatively small, but when contamination frequency is low and the test is by presence/absence, the error can be relatively large and affect the interpretation of results. Recent advances in analysis of microbiological sampling plans have suggested analytical approaches to accommodate the influence of such sampling error. Those approaches assume certain forms of distribution of cells within foods and within samples, and these assumptions remain to be rigorously assessed. Equally, the quantitative interpretation of the performance of presence/absence plans presented above relies on knowledge of the known or assumed standard deviations of concentrations (log CFU) in the lot, but these data are rarely available.

Finally, the interpretations of sampling plan performance presented above assume that methods used for detection of microorganisms in foods are completely reliable, i.e., that if the organism is present, it will be detected.

REFERENCES

Baird-Parker, A. C. 2000. The production of microbiologically safe and stable food, p. 3–18. *In* B. M. Lund, A. C. Baird-Parker, and G. W. Gould (ed.), *The Microbiological Safety and Quality of Food*. Aspen Publishers Inc., Gaithersburg, MD.

Codex Alimentarius Commission. 1997. *Principles for the Establishment and Application of Microbiological Criteria for Foods (CAC/GL 21-1997)*. http://www.codexalimentarius.net/download/standards/394/CXG_021e.pdf. Accessed 2 September 2010.

Codex Alimentarius Commission. 2004. *General Guidelines on Sampling (CAC/GL 50-2004)*. http://

www.codexalimentarius.net/download/standards/10141/CXG_050e.pdf. Accessed 2 May 2010.

Codex Alimentarius Commission. 2007. *Principles and Guidelines for the Conduct of Microbiological Risk Management (MRM).* (*CAC/GL 63-2007*). http://www.codexalimentarius.net/download/standards/10741/cxg_063e.pdf. Accessed 6 September 2010.

Food and Agriculture Organization/World Health Organization. 2004. *Risk Assessment of* Listeria monocytogenes *in Ready-to-Eat Foods. Technical Report. Microbiological Risk Assessment Series 5.* Food and Agriculture Organization, Rome, Italy.

Havelaar, A., M. J. Nauta, and J. T. Jansen. 2005. Fine-tuning Food Safety Objectives and risk assessment. *Int. J. Food Microbiol.* **93:**11–29.

International Commission on Microbiological Specifications for Foods. 1986. *Microorganisms in Foods 2. Sampling for Microbiological Analysis: Principles and Specific Applications*, 2nd ed. Blackwell Scientific, Oxford, United Kingdom.

International Commission on Microbiological Specifications for Foods. 2002. *Microorganisms in Foods 7. Microbiological testing in food safety management.* Kluwer Academic/Plenum Publishers, New York, NY.

International Commission on Microbiological Specifications for Foods. 2011. *Microorganisms in Foods: Use of Data for Assessing Process Control and Product Acceptance.* Kluwer Academic/Plenum Publishers, New York, NY.

Jarvis, B. 1989. *Statistical aspects of the microbiological analysis of foods. Progress in Industrial Microbiology*, vol. 21. Elsevier, Amsterdam, The Netherlands.

Jarvis, B. 2007. On the compositing of samples for qualitative microbiological testing. *Lett. Appl. Microbiol.* **45:**592–598.

Legan, J. D., M. H. Vandeven, S. Dahms, and M. B. Cole. 2001. Determining the concentration of microorganisms controlled by attributes sampling plans. *Food Control* **12:**137–147.

Membré, J.-M., J. Bassettand, and L. M. Gorris. 2007. Applying the food safety objective and related standards to thermal inactivation of *Salmonella* in poultry meat. *J. Food Prot.* **70:**2036–2044.

Nauta, M. J., and A. H. Havelaar. 2008. Risk-based standards for *Campylobacter* in the broiler meat chain. *Food Control* **19:**372–381.

Ross, T., and C. Chan. 2002. Microbiological criteria and microbiological risk assessment, p. 214–247. *In* M. Brown and M. Stringer (ed.), *Microbiological Risk Assessment in Food Processing.* Woodhead Publishing, Cambridge, United Kingdom.

Stringer, M. 2005. Summary report: food safety objectives—role in microbiological food safety management. *Food Control* **16:**775–794.

Uyttendaele, M., K. Baert, Y. Ghafir, G. Daube, L. De Zutter, L. Herman, K. Dierick, D. Pierard, J. J. Dubois, B. Horion, and J. Debevere. 2006. Quantitative risk assessment of *Campylobacter* spp. in poultry based meat preparations as one of the factors to support the development of risk-based microbiological criteria in Belgium. *Int. J. Food Microbiol.* **111:**149–163.

van Schothorst, M., M. H. Zwietering, T. Ross, R. L. Buchanan, M. B. Cole, and the International Commission on Microbiological Specifications for Foods. 2009. Relating microbiological criteria to food safety objectives and performance objectives. *Food Control* **20:**967–979.

Whiting, R. C., A. Rainosek, R. L. Buchanan, M. Miliotis, D. LaBarre, W. Long, A. Ruple, and S. Schaub. 2006. Determining the microbiological criteria for lot rejection from the performance objective or food safety objective. *Int. J. Food Microbiol.* **110:**263–267.

PREANALYTICAL SAMPLE PREPARATION AND ANALYTE EXTRACTION

P. Rossmanith, J. Hedman, P. Rådström, J. Hoorfar, and M. Wagner

8

OVERVIEW OF SAMPLE PREPARATION ISSUES

In physical and chemical terms "food" represents a large and heterogeneous group of very different matrices. This influences the quality of a sample in terms of purity and can lead to difficulties in analyzing food samples by molecular methods, such as PCR. These methods often require appropriate reaction conditions (Al Soud and Rådström, 2001; Rådström et al., 2004; Rossen et al., 1992; Wilson, 1997). These challenges can be addressed by the development of either dedicated detection methods for separate food categories or universal sample preparation methods to produce the required sample purity and volume.

Sample preparation methods transform various heterogeneous inputs into one homogeneous output that can be processed by a subsequent sensitive method (Stevens and Jaykus, 2004a; Brehm-Stecher et al., 2009). Thus, the pathogen detection is a multistep process: single-step applications, such as those offered by, e.g., culture enrichment of pathogens or plate counts, are less practical if quantitative data are to be obtained.

Given the historical background of this field, most molecular methods were developed mainly for in vitro applications on purified analytes such as DNA, which means their use for food diagnostics requires some adjustments. Therefore, enzymatic assays, such as real-time PCR, must be preceded by sample preparation steps and then followed by DNA isolation and purification methods, in order to establish a standard of purity suitable for reliable quantitative measurement. A simple crude boiling is simply not sufficient to obtain quantitative results. The entirety of these processes comprises a detection chain that is defined by numerous elements connected in a logical sequence.

What Is Sample Preparation?

In short, sample preparation methods aim at separation of the target cells or virus particles from the surrounding food matrix. Down-

P. Rossmanith, Christian Doppler Laboratory for Molecular Food Analytics, University of Veterinary Medicine, Veterinärplatz 1, A-1210 Vienna, Austria. *J. Hedman*, Department of Biology, Swedish National Laboratory of Forensic Science (SKL), SE-581 94 Linköping, Sweden. *P. Rådström*, Department of Applied Microbiology, Lund University, Veterinärplatz 1, SE-221 00 Lund, Sweden. *J. Hoorfar*, National Food Institute, Technical University of Denmark (DTU), Mørkhøj Bygade 28, DK-2860 Søborg, Denmark. *M. Wagner*, Institute of Milk Hygiene, Milk Technology and Food Science, Department of Veterinary Public Health and Food Science, University of Veterinary Medicine, Veterinärplatz 1 A-1210 Vienna, Austria.

Rapid Detection, Identification, and Quantification of Foodborne Pathogens, ed. by J. Hoorfar

stream methods, such as cell disruption, analyte purification, and detection via, for example, real-time PCR, have to be taken into consideration individually when designing or choosing a suitable preparation method. Sample treatment is defined as the preanalytical step in the method protocol, which is also necessary for reduction of the sample volume while maintaining the initial target number, as much as possible. This step can concentrate the target cells and remove inhibitory substances that would otherwise hinder the subsequent detection methods. To achieve these objectives, one or more consecutive methodological steps may be necessary.

Optimal Performance Characteristics of Sample Preparation

In summary, sample preparation should meet the following essential requirements:

- Reduction of the sample size to volumes that can easily be handled in downstream steps, such as analyte isolation/purification from the target organism
- Concentration of the target organism
- First removal of possible inhibitory substances that affect subsequent molecular methods
- Providing consistent sample quality that is adequate for subsequent methods in the detection chain, independent of the original nature or composition of the respective food matrices
- Providing maximum recovery, which is a maximum throughput rate with minimal loss of targets during the process
- Providing linear recovery over the whole range of target concentrations down to low contamination levels, which is especially important for the subsequent use of quantitative methods
- Providing a low detection limit
- Preservation of the integrity of the targets (bacteria/viruses), to support subsequent quantitative recovery of the analyte (e.g., DNA molecule for real-time PCR). At best, the viability of the target organisms should be conserved if it was given in the sample
- Providing simplified handling that supports reproducibility
- Cost-effectiveness, as the sample preparation step can increase the overall cost of a test
- Noncomplex and time-saving

Separation as the Basis of Sample Preparation

Separation processes are based on difference (Geankopolis, 2003). In analyzing foods, the target pathogens must be extracted from very diverse matrices. The given physical, chemical, and biological differences can then be exploited to separate these targets. Separation parameters are set according to the characteristics of these differentiating factors (Stieβ, 1997).

The underlying principles of separating the target cells from the food matrix are chemical, physical, physicochemical, or biochemical-biological (Stevens and Jaykus, 2004a). Every separation process exploits features of the target that are distinct from the physical or biochemical properties of the food matrix or nontarget organisms. This could rely on recognition of molecules on cell walls of target cells that are not present in the competing flora, or separation on the basis of size, hydrophobicity, conductivity, and many other principles. Separation parameters are defined on the basis of the characteristics of these differentiating factors, and separation efficiency is achieved by the extent of significance of the differentiating factors.

The separation principle for a specific sample treatment method has to be chosen according to the proposed downstream detection method. For example, in the case of immunoaffinity-based detection of bacterial cells, chemical degradation of the bacterial cell wall beforehand would not be appropriate; it might lead to the destruction of the very same epitopes that are necessary for the paratopes in the binding sites of the antibody.

Classification of Food Matrices in Connection with Sample Preparation

The topic of food matrix classification in connection with sample preparation is covered in various textbooks on food chemistry, technology, and processing (Berlitz et al., 2009). Nevertheless, the classification of foodstuffs into food category, chemical composition, structure and physical properties, and grade of processing helps to appreciate the task of separating bacterial or viral targets from food matrices.

Traditionally, food is classified as follows: dairy products; eggs; meat; fish and seafood; edible fat and oil; cereals and cereal products; pulses (beans, lentils, etc.); vegetables; fruit; sugar, alcohol, and honey; spices and coffee; vinegar, salt, and water (Berlitz et al, 2009). This classification gives a good overview of the dimension of the field.

The chemical composition and properties of food can be classified as follows: water content; amino acids, peptides, and proteins; lipids; carbohydrates; food additives; and pH value (Berlitz et al., 2009).

The variety of different chemical compounds and molecules in foodstuffs is comparable to that offered by organic chemistry itself. Most foodstuffs can be classified in the aforementioned categories according to their chemical content. Not only does the influence of the chemical composition of the food on the performance of the sample preparation method have to be taken into account, but so must also the influence of food additives or aberrant pH values. The pH value of a certain food matrix can also affect the sample preparation method itself, as most chemical preseparation methods require a stable pH value (Rossmanith et al., 2007; Stevens and Jaykus, 2004a); and this also applies to methods utilizing hydrogen or other weak bonds such as antigen-antibody-related methods.

The structure and physical properties of food with importance for sample preparation are cell walls, tissue cohesion, solubility, and viscosity; the state of aggregation and food can be existent as emulsion, solution, and solid mixture. This is because the underlying separation methods generally depend on physical separation or must be performed in aqueous solution (Brehm-Stecher et al., 2009; Stevens and Jaykus, 2004a). Presample preparation methods, such as enzymatic or chemical treatment for solubilization of the foodstuff, help to circumvent this problem.

Food can be mainly processed as follows: raw, cooked (heat treatment), matured (cheese), fermented, smoked, salted, dried, subjected to irradiation, and subjected to freezing. Food processing confers chemical or physical changes to a given foodstuff at a secondary level. The type and grade of processing play a major role, as these different treatments affect the structure of the food matrices as well as their chemical composition. Additional characteristics have an impact on the performance of the sample preparation method, regardless of the original composition of the food matrix. The following similar foodstuffs can demonstrate some different characteristics depending on the type of processing.

PRACTICAL SOLUTIONS

Physical Separation Methods

Filtration and Dielectrophoresis/Ultrasound

The difference in particle size of target cells and the food matrix is the basis for filtration as a separation method (Geankopolis, 2003; Stieß, 1997). Filtration often involves sieving, which is related to but still different from the physical separation method, being rapid, simple, and inexpensive. The recoveries of target cells demonstrated in recent publications using filtration for sample treatment vary from 10 to 100% (Besse et al., 2004; Fernandez-Astorga et al., 1996; Wang et al., 1992). These recoveries have been obtained by filtration of beer, wine, and other liquid and low-content beverages (Thomas, 1988). D'Urso et al. (2009) introduced a combined filtration method involving a discrimination buffer in a prefiltration step (Fig. 1). The major advantage of filtration is the nearly unlimited sample volume that can be processed.

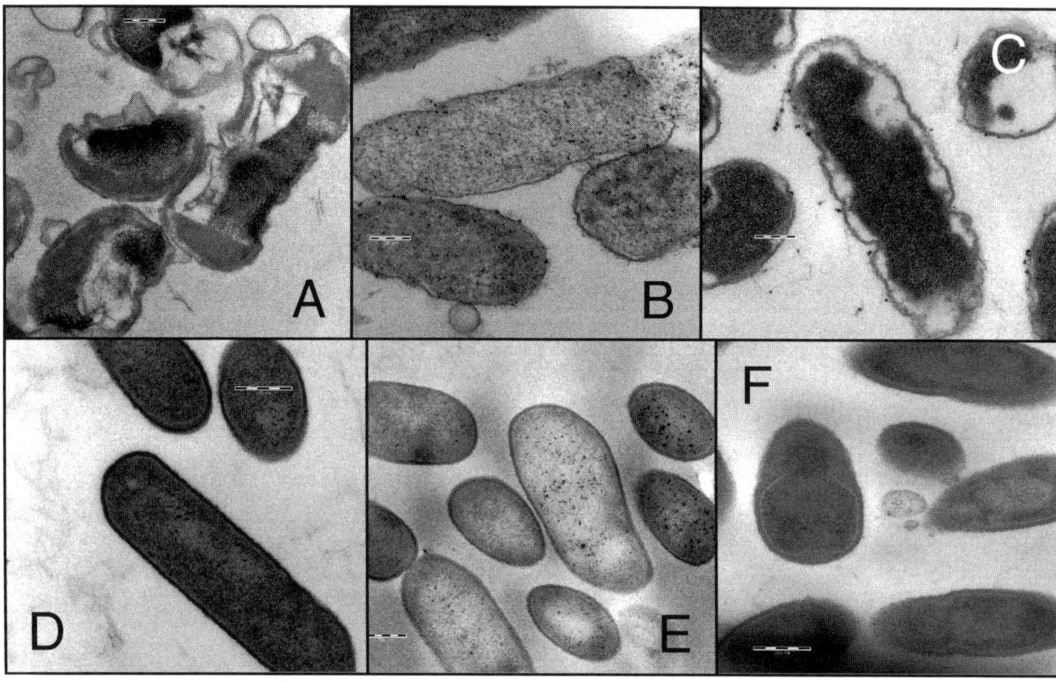

FIGURE 1 Transmission electron microscopic pictures of *S. enterica* serovar Typhimurium and *L. monocytogenes* after using "First generation" chemical treatment in sample preparation for solubilization of food matrices. (A and F) Exposure to 4 M guanidine-isothiocyanate and 2 M NaCl according to D'Urso et al. (2009). (B and E) Treatment with 0.6% NaCl and 6% sodium citrate according to Stevens and Jaykus (2004b). (C and D) 2% Sodium citrate treatment according to Ulve et al. (2008).

Still, solid foods are incompatible with this method, as the filter easily clogs (Bylund, 1995). Also, the method is likely to be influenced by bacterial targets absorbing to the filter or capturing of foodstuff particles: therefore, Duodu et al. (2009) developed a combined filtration and immunomagnetic separation method.

Dielectrophoresis separates targets from matrices by generating a high-frequency electric field (0.1 to 10 MHz) in which targets move according to their charges. Given a minimum concentration of 10^7 CFU per ml, 50% recovery from pure bacterial culture could be obtained (Markx et al., 1996; Stevens and Jaykus, 2004a). The question remains whether these targets are viable. Additionally, the method is not suitable for direct application to food due to numerous inherent chemical reactions in food samples, e.g., oxidation and reduction at both electrodes (Hamann and Vielstich, 2005).

Sonication of samples leads to concentration of particles in resonance nodes dependent on the wavelength and energy used (Stevens and Jaykus, 2004a; Limaye and Coakley, 1998). For this, 10^7 cells are the minimum amount necessary to achieve reasonable results; recoveries from 72% to 96% were reported for $>10^7$ cells (Limaye and Coakley, 1998).

Centrifugation

Centrifugation is the method of choice for separating particles from liquid media. The underlying mechanism of sedimentation depends on particle density and diameter, viscosity of the liquid, and the relative centrifugation force applied (Fliss et al., 1991; Tjhie et al., 1994). Differential centrifugation works by a stepwise increase of centrifugation speed, which removes the heavier food particles right at the beginning. The increased speed eventually leads to pelleting of the sample itself (Meyer et al., 1991; Neiderhauser et al., 1992).

Apart from the one-time costs for a suitable centrifuge, the method is fast, easy, and inexpensive. Also, detection limits of 10^3 to 10^4 CFU per milliliter were shown (Stevens and Jaykus, 2004a). Still, the adhesion of target cells to the food matrices and cosedimentation of the targets with the food particles both can lead to the loss of targets and thereby false results.

Density gradient centrifugation takes advantage of the targets seeking the equilibrate portion of the sample tube while continuously decreasing the density of the suspending solution during centrifugation. The constant change in the system and some food components, e.g., high fat contents, make this method difficult to standardize (Stevens and Jaykus, 2004a). Therefore, this method is currently limited to liquid samples. Apart from this, density gradient centrifugation has been broadly used for the last couple of decades (Wolffs et al., 2004; Lindqvist, 1997; Lindqvist et al., 1997; Fukushima et al., 2007). Recovery rates were shown to vary between 11% and 45% with detection limits between 10^3 and 10^4 cells (Stevens and Jaykus, 2004a, 2004b).

Adsorption

Different adsorption effects like van der Waals forces, electrostatic interactions, hydrophobic interactions, and hydrogen bonding mediate nonspecific adsorptions to the surface of solid materials: metal hydroxides, ion-exchange resins, and lectins as well as targets are forced to matching liquid phases (Bailon et al., 2000). These reversible interactions are also valid in immunological binding of epitope and antibody or viral binding (Carta and Jungbauer, 2010).

A representative for liquid systems is the aqueous two-phase system. Recovery rates from 1% to 50% with a detection limit of 10^4 CFU per ml were reported (Lantz et al., 1994; Pedersen et al., 1998). Recoveries in both phases change with varying pH values and polymers. Two-phase systems are also sensitive to temperature and chemical composition of samples; partitioning may be impaired by the presence of food components. The resulting difficulties in standardization as well as high detection limits and heterogeneous results are major disadvantages of this method. This can only be partially compensated by the ability to process up to 4 g of sample volumes.

Several nonspecific solid-phase adsorption methods such as metal hydroxide coatings and ion-exchange resins were shown to separate bacterial cells from sample matrices with varying recoveries of 9.5% to 99.0% (Berry and Siragusa, 1997; Cullison and Jaykus, 2002; Lucore et al., 2000; Stevens and Jaykus, 2004a). A drawback is the inactivation of cells by pH value manipulations, which dissociate ionic interactions that are necessary for recovering the targets.

Another rapid and simple method involves lectins, proteins binding the *N*-acetylglucosamine residue of the bacterial cell wall. Lectins in turn can be bound to agarose beads and used in affinity columns or in conjunction with magnetic vehicles such as paramagnetic beads. Recoveries from 23% to 50% could be obtained (Patchett et al., 1991; Payne et al., 1992; Stevens and Jaykus, 2004a), but the requirements of presample treatment, the lack of efficient dissociation of bound cells, and high costs make this method unfavorable.

Fumian et al. (2009) developed a combined adsorption-filtration method, resulting in recoveries of norovirus particles from 5.2% to 72.3%, but due to large variability of the results, further development is necessary.

Biochemical and Biological Separation Methods

The high affinity of antibodies and viral binding proteins for their targets can be exploited to separate bacterial target cells from food matrices (Carta and Jungbauer, 2010; Stevens and Jaykus, 2004a). This conveys the benefit of a preselection step, which concentrates the targets and separates them from their matrices at the same time. Other recently introduced methods make use of aptamers and antimicrobial peptides, both related to lectins (see "Adsorption," above) in their nonspecific binding capacity.

Overall, publications regarding affinity bindings and bead technology have increased over the past few years (Brehm-Stecher et al., 2009). Antibodies, viral binding proteins, and antimicrobial peptides (AMP) from higher plants, such as magainin I, all bind to the surface of bacteria. They have been used in combination with paramagnetic nanobeads (Amagliani et al., 2006; Cudjoe and Krona, 1997; Nogva et al., 2000; Jung et al., 2003; Kretzer et al., 2007; Loessner et al., 2002; Hallier-Soulier and Guillot, 1999; Rådström et al., 2004; and Niederhauser et al., 1994), with silanized glass slides (Kulagina et al., 2005), and with direct colony blotting (Belyi et al., 1995).

For enhanced recovery, nonspecific DNA-binding paramagnetic beads were combined with real-time PCR for specific target identification (Nogva et al., 2000; Josefsen et al., 2004, 2007). The crucial parameters when using magnetic beads are the spherical shape, uniformity of the beads, and correct binding and dissociation buffer conditions. Magnetic beads can have the advantage of consistency and reproducibility due to the controlled amount of beads and buffer volumes used in a protocol.

Using paramagnetic beads, Josefsen et al. (2004) have developed a semiquantitative strategy to quantify low numbers of *Campylobacter* from chicken rinse samples. They showed that even after a 12-hour selective enrichment phase under standardized conditions, the initial amount of cells in carcass rinse was inversely correlated to threshold cycle values. Thus, a higher initial amount of target bacteria may result in lower threshold cycle values. However, the precision of the method needs to be elucidated in more detail. Generally, this strategy was also applicable to *Salmonella* by using adapted enrichment media and times followed by real-time PCR quantitation (Krämer et al., 2010). Thus, careful consideration should be given to the enrichment strategies for *Salmonella* cells in combination with subsequent quantitative real-time PCR analysis.

Antibodies

Skjerve et al. (1990) were the first to report the separation of *L. monocytogenes* from culture media and heterogeneous suspensions by means of monoclonal antibodies. Jung et al. (2003) combined an immunogenetic separation method with flow cytometry and obtained recovery rates from 7% to 23% per ml sample of culture medium. In general, the detection limits of immunoassay separation methods are 10^2 to 10^3 CFU per milliliter (Stevens and Jaykus, 2004a; Nogva et al., 2000; Belyi et al., 1995; Fluit et al., 1993; Uyttendaele et al., 2000; Yu and Bruno, 1996).

Viral Binding Proteins

Viral binding proteins in combination with paramagnetic beads were used for the separation of bacterial cells by Kretzer et al. (2007), although this was performed with culture media only. The detection limit of this approach is reported to be lower than for antibody-covered beads (<10 CFU) (Kretzer et al., 2007). Promising results for *L. monocytogenes* were achieved with listeriolysin O-coated beads and the following phage endolysin-derived cell wall binding domain-based magnetic separation (Amagliani et al., 2006; Kretzer et al., 2007). Recovery rates were not determined, but a detection limit of less than 10 CFU per ml for all artificially contaminated foods was established. In the previous report, however, the artificial contaminations did not affect the whole sample, since solid foodstuffs were not homogenized prior to analysis.

Aptamers and AMP

The use of aptamers coated to magnetic beads was demonstrated with a detection limit range of 10^2 to 10^3 CFU (Joshi et al., 2009). Another AMP-based method developed by Kulagina et al. (2005) determined a detection limit of 6.5×10^4 and 6.8×10^5 CFU. This was achieved by using an array-based biosensor without determination of the recovery rate.

So far, these methods have been limited to pure cultures of bacteria or liquid matrices such as blood, skim milk, or water, and their application to solid foods remains problematic (Stevens and Jaykus, 2004a). Heterogeneous detection limits and recoveries, insufficient processed volumes, and overall high costs are the common disadvantages of this category of methods. Additionally, the range of bacterial targets is limited to matching epitopes and specific separation carriers. Cross-linking and other side effects dependent on food composition and washing/blocking steps also influence recovery rates (Bathen and Breitbach, 2001; Stevens and Jaykus, 2004a).

CHEMICAL AND ENZYMATIC DIGESTION FOR PRESEPARATION

The methods described below are defined as preseparation methods because in all cases digestion of the food matrix is followed by a physical separation method such as centrifugation.

Historic Development: Enzymatic Digestion of the Food Sample Matrix

Foods can be degraded by enzymatic digestion through specific cleavage of macromolecules. Only a few investigators have worked on enzymatic digestion methods for dairy products by using pronase or an enzymatic system combining pronase with lysozyme and proteinase K (Wegmuller et al., 1993; Allmann et al., 1995). In one case a detection limit of 50 CFU per g following conventional PCR was published (Wegmuller et al., 1993).

However, enzymatic digestion is a widespread molecular biological method but quickly reaches its limits in terms of sample numbers and volumes due to the high costs of the enzymes. Only a few foodstuffs can be processed by this method, as each class of chemical compound needs a matching enzyme to cut it. Moreover, most of the applications are time-consuming, as overnight incubation is usually necessary to achieve acceptable recoveries. However, this approach marks the first attempt at easing the separation process by removal of the gross food particles.

Chemical Digestion of the Food Sample Matrix

Chemical digestion or solubilization of the food matrices is executed by chemical cleavage and solubilization of the foodstuffs in aqueous solutions. The reaction conditions have to be selected specifically, utilizing the cell wall or other target-specific characteristics (Rossmanith et al., 2007, 2010). To set up one universal protocol for a wide range of heterogeneous food categories is difficult, as the reactions are influenced by pH value, temperature, salt concentration, viscosity, or the main component of each foodstuff such as fat, protein, or carbohydrates. Therefore, the food composition is more important for this sample preparation approach than in separation techniques that are not based on digestion of the sample matrix.

Mechanisms that can be used to bring the foodstuff into aqueous solution are micelle formation with integration of water-insoluble components within the micelle, chaotropic effects, change of pH value, salting in, salting out, and heat (Mester et al., 2010; Rossmanith et al., 2007, 2010).

For the use of micelle formation, detergents such as sodium dodecyl sulfate (SDS) can be used; this process occurs in everyday life during the cleaning of dishes or clothes with similar chemical agents. The formation of micelles is based on the amphiphilic properties of the detergent molecules with lipophilic and hydrophobic components (Bhairi, 2001; Rosen, 2004). The lipophilic compounds of the food matrix, such as fat, are then enclosed in the micelle, whereas the hydrophilic parts of the molecules build a surface on the outside of the vesicle, which is in contact with the aqueous solution via hydrogen bonds. Due to their molecular structure, detergents are the most relevant compounds if used in chemical solubilization, in terms of negative effects on the integrity and viability of the targets (Mayrl et al., 2009; Mester et al., 2010; Rossmanith et al., 2010). This molecular structure is highly similar to the structure of the cell membrane of bacterial cells, therefore structurally affecting this membrane, which is crucial for the condition of the cell (Bhairi, 2001; Rosen, 2004).

Chaotropic substances such as urea change the surface tension of aqueous solutions and thus also facilitate the solubilization of various food compounds, but mainly the protein fraction (Mester et al., 2010; Rosen, 2004).

The role of the change in pH value is not fully understood yet, but again, proteins are mainly affected. The folding of the proteins is changed according to their isoelectric point; therefore, hydrophilic properties of the molecule are brought to the surface, which then facilitates the solubilization of these proteins (Rosen, 2004).

The refolding of proteins is also the mechanism of action of salting in/out. Salting in/out also helps to disrupt tissue coherence, such as releasing myofibrils from the muscle fibers (Mayrl et al., 2009).

Heat helps to accelerate every chemical reaction and brings fat into a liquid state, which then accumulates in a liquid phase independent of the aqueous phase. This phase settles on top of the sample after centrifugation and can easily be discarded (Rossmanith et al., 2007).

"Classical" Chemical Solubilization Approaches

Choi (2000, 2001) and Choi and Hong (2003) developed a guanidine thiocyanate/phenol/chloroform extraction method for direct preparation of DNA from *Listeria monocytogenes, Yersinia enterocolitica*, and *Salmonella enterica* serovar Enteritidis. They were able to detect up to 10^3 CFU of *L. monocytogenes* per ml after conventional PCR (Choi and Hong, 2003). Another group used a diethyl ether/chloroform extraction method and performed an incubation step in 2 M urea and 1% SDS (Ramesh et al., 2002).

Stevens and Jaykus (2004b) could reduce 11 g of plain nonfat yogurt or cheddar cheese to a pellet of 500 μl by combining sodium citrate in a protocol with centrifugation and the DNAzol BD reagent (Fig. 1). The resulting pellet size from the first step in sodium citrate was 1 to 5 g. Subsequently, the cheese was blended and filtered, and the yogurt was strained through sterile cheesecloth. This removed a major part of the food matrix. Following the centrifugation, the bacterial recoveries varied from 53% to 143%. These data were obtained by directly plating the pellets of *L. monocytogenes* and *S. enterica* serovar Enteritidis from both foodstuffs. Depending on the food matrix and the bacterial species analyzed, detection limits from 10^0 and 10^3 CFU per g were established by conventional PCR and following Southern hybridization (Stevens and Jaykus, 2004b).

Matrix Lysis: a Modular Solubilization System

So far, it appears that there are two main approaches for sample preparation: separating the targets from the more or less unchanged food matrix and solubilization of foodstuff with subsequent harvesting of the remaining targets. The benefit of the latter approach is the simplification of the separation process, which can be easily performed by centrifugation.

Centrifugation can be advantageous with regard to recovery rate, since if performed accurately in a liquid of low viscosity, the theoretical recovery of 100% is almost achievable in practice (Koch and Blumberg, 1976; Rossmanith et al., 2010). Once the problem of accurate lysis of the foodstuff is solved, the resulting advantages of the food solubilization approach are simple handling steps, low costs, maximal recovery, unlimited sample volume, maximal rate of target concentration/volume reduction, minimal interaction with the food matrix during separation, unspecific application for every target, which has a higher specific weight than the buffer solution, and a broad range of applicable foodstuffs. The enzymatic approach (see also "Historic Development: Enzymatic Digestion of the Food Sample Matrix," above) would also carry these advantages, but it involves long incubation times and high costs. The "classical" chemical solubilization approaches either were using chemistry as an additional step to enhance the performance of the following DNA isolation or were using a buffer chemistry, which was not capable of fully lysing great amounts of foodstuff (Choi and Hong, 2003; Stevens and Jaykus, 2004b).

Over the past few years, a modular system for bacterial separation covering all relevant species and foodstuffs has been developed. This method, called "matrix lysis," is capable of reducing 6- to 12-g samples of the foodstuff in 50-ml volumes to a size that can be processed in commercial isolation kits for subsequent DNA isolation (<200 μl). The first proof of principle was demonstrated by processing *L. monocytogenes* (Rossmanith et al., 2007). A buffer system containing the chaotrope urea (8 M) and SDS (1%) as detergent was used for solubilization of the food matrix. Separation of the targets from the resulting solution was achieved by centrifugation at 3,220 × *g*. A detection limit of 7.8 CFU per gram was obtained after real-time PCR, with a recovery of 39.5%. Nevertheless, the harsh incubation conditions, originated by the use of SDS as detergent, affect the overall condition of the bacterial cells, especially the cell walls and membranes (Fig. 2 and 3). This results in total lysis of gram-negative bacteria such as *Salmonella* and *Escherichia coli*.

In a second attempt, the application was broadened to include gram-negative species (Mayrl et al., 2009). The degradation of all sorts of foodstuffs was obtained by a buffer system containing Lutensol, a nonionic detergent, instead of SDS. However, Lutensol creates milder incubation conditions. This is most likely based on the molecular structure of Lutensol, suggesting that this detergent does not interfere with the bacterial cell membrane. The mean recovery was 49.9% for *S. enterica* serovar Typhimurium, *Staphylococcus aureus*, and *L. monocytogenes* with an average detection limit of 6.3 CFU/g.

Both studies included the testing and demonstration of the linearity of the recovery over the range of contaminations from the detection limit up to 10^9 initial cells per gram. Also included in the second study was the demonstration of the removal of 5-log scales of free target DNA during the protocol, which was shown by real-time PCR analysis. Consequently, contamination with free DNA derived from dead bacterial cells does not significantly impair quantification. Qualitative results will be impaired only with a presumptive contamination of over 10^6 dead cells, which would not be associated with growth contamination.

The strategy was further advanced in a third attempt to allow for the separation of living cells, both gram positive and gram negative, from the food matrices (Mester et al., 2010). Preserving viability can facilitate the interpretation of PCR results, since such a signal would then indicate the occurrence of potentially infective cells.

The separation of serovar Typhimurium and *L. monocytogenes* from dairy products has been demonstrated with a respective recovery of ±100% and 66% after real-time PCR following the use of ionic liquids (ILs) in the lysis buffer, a completely new family of chemicals that have not yet been used in food science. The recovery of living cells obtained by plate count was between 36% and 60% on selective agar and ±100% on unselective agar. ILs are low-melting-point organic salts that are liquid below the boiling point of water. Of special interest have been ILs based on the 1-alkyl-3-methylimidazolium cation $[C_n mim]^+$. 1-Ethyl-3-methylimidazolium thiocyanate was introduced to the matrix lysis protocol as a main reactive buffer compound (Fig. 4.).

One of the interesting aspects of these ILs is the amphiphilic character of the cation, comparable with conventional detergents. This provides these ILs with surfactant-like properties, which led to their increased use in biochemistry (Liu et al., 2009). Additionally, the thiocyanate anion of 1-ethyl-3-methylimidazolium thiocyanate is chaotropic, according to Hoffmeister's series, and there-

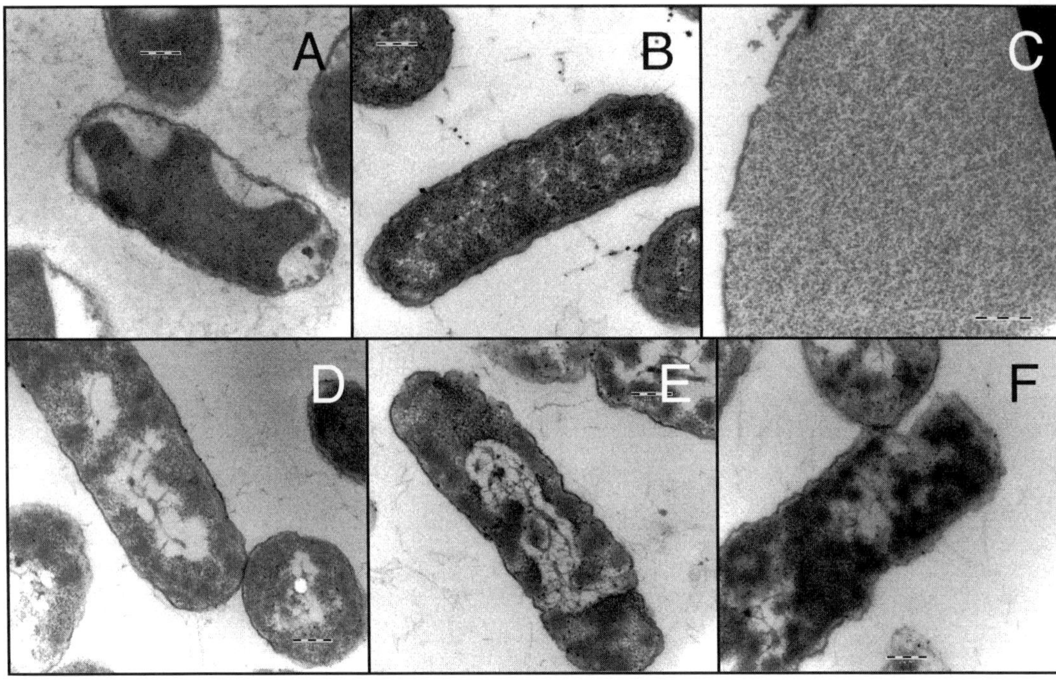

FIGURE 2 Transmission electron microscopic pictures of *S. enterica* serovar Typhimurium after exposure to chemical treatments used in sample preparation for solubilization of food matrices (matrix lysis). (A) Control group, an overnight culture of untreated serovar Typhimurium cells. (B) A penicillin G-treated culture for comparison of cell wall damage. (C) Treatment with 8 M urea and 1% SDS. The visible secondary structure was generated from the remnants of several lysed cells. (D) Exposure to 1 M $MgCl_2$. The cellular integrity is intact; DNA seems to reversibly precipitate within the cells. (E) Treatment with 7.5% $[Emim]^{-}SCN^{+}$. The cells show appearance comparable to that of cells treated with 1 M $MgCl_2$ with more effect of the chemical towards the cell wall; also, the precipitation seems more advanced. (F) Treatment with 8 M urea and 1% Lutensol^{TM}. The cellular integrity is mostly intact, but the shape of the cells is changed and the cell wall is affected.

fore suspected to enhance the solubilization properties of the buffer system (Kunz et al., 2004).

In preliminary experiments, Mester et al. (2010) were able to demonstrate the use of $MgCl_2$ as a practical solvent for dairy products and meat. This inexpensive method allows for recovery of ~100% of living bacterial cells, both gram-positive and gram-negative (Patent application, EPA09007959.1: Method for isolating viable cells).

This method comprises a standardized handling protocol with a modular system of different buffers for several applications (Mayrl et al., 2009; Mester et al., 2010; Rossmanith et al., 2007, 2010). This enables standardized handling and simplifies routine application of the method. Sample volumes solely depend on laboratory equipment and could theoretically be extended to 50 g to 100 g, and the incubation steps do not exceed 3 hours. The protocol may also be applied as a washing protocol to detect surface-attached targets, e.g., bacterial pathogens on carcasses (Rossmanith et al., 2007).

ANALYTE EXTRACTION

Analyte extraction is the subsequent step following sample preparation. It is necessary to obtain access to the molecules that are the tar-

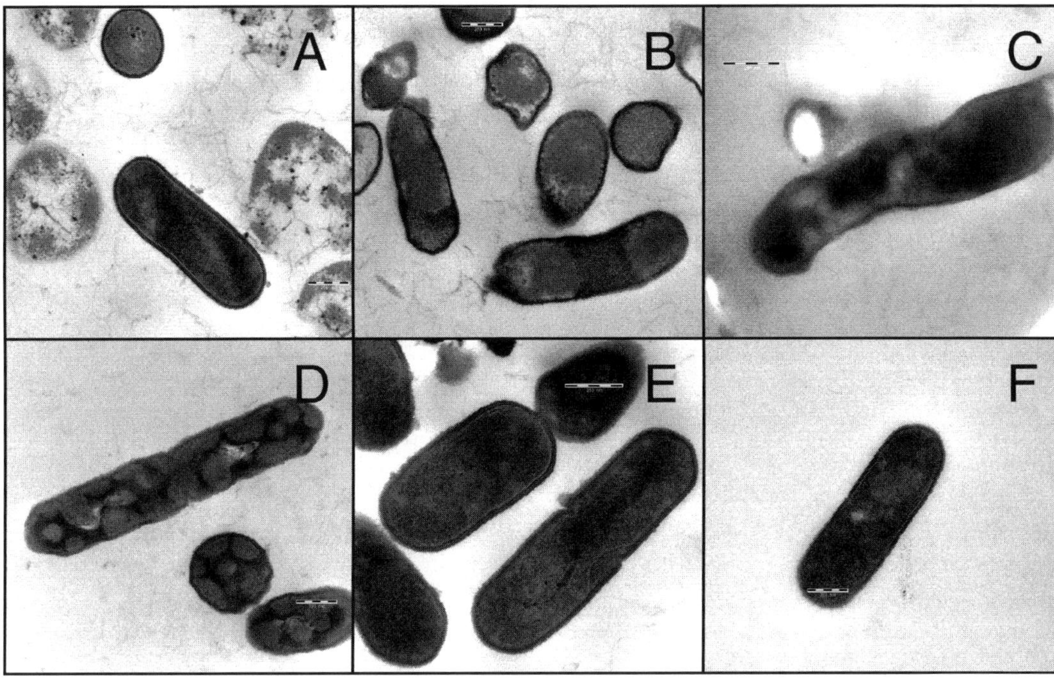

FIGURE 3 Transmission electron microscopic pictures of *L. monocytogenes* after exposure to chemical treatments used in sample preparation for solubilization of food matrices (matrix lysis). (A) Control group, an overnight culture of *L. monocytogenes* cells including storage for 4 weeks at 4°C to demonstrate the natural degradation of the cells. (B) A penicillin G-treated culture for comparison of cell wall damage. (C) Treatment with 8 M urea and 1% SDS. The cellular appearance reflects the harsh chemical stress that affected the cells. (D) Treatment with 8 M urea and 1% Lutensol™. The cells show significant influence of the reagent, with nevertheless intact cell walls. (E) Exposure to 1 M $MgCl_2$. The cells remain seemingly unaffected by the treatment. (F) Treatment with 7.5% $[Emim]^{-}SCN^{+}$. The cellular appearance also reflects the high rate of recovery obtained by the sample preparation method using this reagent.

get of molecular detection assays such as real-time PCR. Another aspect is the purification of the analyte from inhibitory compounds and perturbing remnants of the targets or food-

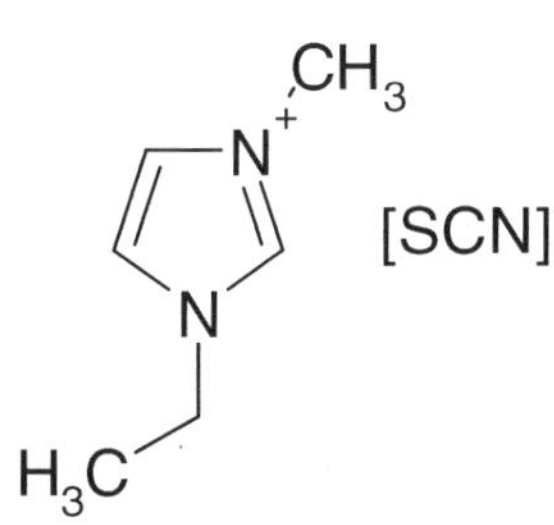

FIGURE 4 Schematic chemical structure of 1-ethyl-3-methylimidazolium thiocyanate.

stuff. In order to gain access to the analyte molecules, nucleic acids, or proteins, the integrity of the target cells has to be destroyed in most cases.

Within the detection sequence, the step that is often called in a demotic way "isolation" of either nucleic acid or protein actually comprises several independent but reasonably progressively ordered procedures:

- Disintegration of the cell wall/capsid
- Separation of the cell/capsid remnants
- Purification of the analyte (molecule)

Since analyte extraction as a part of the detection chain in food pathogen detection is similar to the application in basic research, a brief summary is given below. Further information is given in the specialized books and articles covering this field (Sambrook et al., 1989).

Disruption of the Cell Wall

The main difference in analyte extraction between research settings and routine end use laboratories is the time factor (Brehm-Stecher et al., 2009). The first step of analyte extraction is the destruction of the bacterial cell wall. This process requires some incubation time, especially for gram-positive bacteria. In a research setting, enzymatic digestion of the cell wall by overnight incubation is not problematic, while in a food control laboratory time is crucial and therefore disruption of the bacterial cell wall will delay the test response time. In the subsequent steps, analyte separation and purification are less time-consuming. Figure 5 gives an overview of the several established cell disruption methods (Rossmanith et al., 2006).

Purification of Nucleic Acids

Among the variety of methods available, there are sufficient differences in the grade of purity of the analyte after purification. Those providing the most reliable purification of the analyte are favorable (Flekna et al., 2007). However, the factors of time and ease of handling should be taken into consideration. For quantitative detection using real-time PCR, solid-phase extraction and silica matrices are favorable. These methods are available on a commercial basis and correspondingly well validated. Crude methods can be used if qualitative results should be obtained by PCR. However, in this case a higher detection limit has to be expected.

A short overview of popular combined isolation/purification methods is given below.

- Guanidinium-thiocyanate-phenol-chloroform extraction: this method works especially well for gram-negative bacteria. For gram-positive bacteria, an additional subsequent enzymatic digest of the cell wall is necessary. It includes cell disruption, as well as separation of the analyte and purification. Purification mainly targets salt impurities.
- Alkaline extraction method: this classic method works best for plasmid extraction and purification.

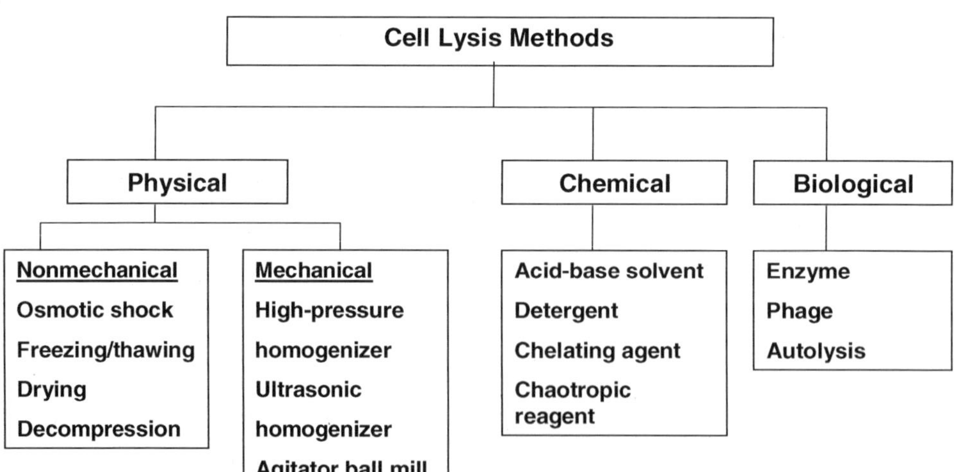

FIGURE 5 Schematic overview of preanalytical sample preparation methods that are usually used to disrupt the cellular structure of target bacteria in food samples.

- CTAB extraction: this method can be used for gram-negative strains and is based on precipitation of nucleic acids.
- Chelex-based DNA extraction: this is a short and reliable protocol, providing DNA in a quality that supports qualitative PCR.
- Nucleic acid purification methods: cesium chloride (CsCl) gradient centrifugation: this complex method is nevertheless superior with regard to purity of nucleic acid samples.
- Solid-phase nucleic acid purification: this approach allows for fast and efficient purification and is widely used in commercial kits.
- Silica matrix-based methods: this matrix specifically binds DNA, which is also the underlying principle of several commercial kits.
- Purification by oligo(dT)-cellulose chromatography, diatomaceous earth, magnetic beads, and anion-exchange material: these methods are of less relevance in the field of food pathogen detection.

Optimization of PCR Chemistry

Apart from removing inhibiting compounds, the chemistry of the downstream analysis may be optimized in order to neutralize the inhibition. In PCR, the DNA polymerase is a key component. DNA polymerases from different sources have different abilities to withstand various inhibitors (Rådström et al., 2004). Mixing two commercial DNA polymerases has also been shown to neutralize the inhibition, compared to using each polymerase separately (Hedman et al., 2010). Adding components such as bovine serum albumin or Tween 20 or elevating the pH value of the PCR buffer may also relieve inhibition effects (Rådström et al., 2004). In order to enhance diagnostic PCR in general, an integrated pre-PCR processing approach has been suggested (Rådström et al., 2004). The aim of this approach is to generate PCR-amplifiable samples by integrating sampling, sample preparation, and amplification chemistry.

Purification Methods for Proteins

Although a major topic in protein biochemistry, extraction and purification of proteins are not so important in the field of food pathogen detection. Most protein-dependent detection methods, such as enzyme-linked immunosorbent assays or imunofluorescence, rely on the surface properties of the target cells rather than on cellular proteins (Walsh, 2002). Therefore, the target cell is also the analyte. Here, the purification capacities of the sample preparation methods are crucial for the purity of the analyte. Otherwise, additional purification steps have to be included after sample preparation, when the purity of the target cells for use as analyte is not sufficient for downstream detection methods.

Moreover, most of the protein purification methods, such as several chromatographic methods, gel electrophoresis, dialysis, selective precipitation, isoelectric focusing, and some automated extraction systems, are complex and therefore are not useful in large-scale diagnosis (Carta and Jungbauer, 2010; Walsh, 2002).

Further investigations, following detection of a pathogenic organism in food by using proteomics or other continuative experiments for more detailed investigation of a specific outbreak or basic research, may then benefit from extensive purification of proteins.

ACKNOWLEDGMENTS

This work was supported in part by the Christian Doppler Society and in part by EU FP6 project BIOTRACER.

REFERENCES

Allmann, M., C. Hofelein, E. Koppel, J. Luthy, R. Meyer, C. Niederhauser, B. Wegmuller, and U. Candrian. 1995. Polymerase chain reaction (PCR) for detection of pathogenic microorganisms in bacteriological monitoring of dairy products. *Res. Microbiol.* **146:**85–97.

Al Soud, W. A., and P. Rådström. 2001. Purification and characterization of PCR-inhibitory components in blood cells. *J. Clin. Microbiol.* **39:** 485–493.

Amagliani, G., E. Omiccioli, A. Campo, I. J. Bruce, G. Brandi, and M. Magnani. 2006. Development of a magnetic capture hybridization-PCR assay for *Listeria monocytogenes* direct detection in milk samples. *J. Appl. Microbiol.* **100:**375–383.

Bailon, P., G. K. Ehrlich, W. J. Fung, and W. Berthold. 2000. *An Overview of Affinity Chromatography.* Humana Press, New York, NY.

Bathen, D., and M. Breitbach. 2001. *Adsorptionstechnik.* Springer, Berlin, Germany.

Belyi, Y. F., N. A. Varfolomeeva, and I. S. Tartakovskii. 1995. A simple colony-blot method for identification of *Listeria* in food samples. *Med. Microbiol. Immunol.* **184:**105–108.

Berlitz, H. D., W. Grosch, and P. Schieberle. 2009. *Food Chemistry,* 4th ed. Springer, Heidelberg, Germany.

Berry, E. D., and G. R. Siragusa. 1997. Hydroxyapatite adherence as a means to concentrate bacteria. *Appl. Environ. Microbiol.* **63:**4069–4074.

Besse, N. G., N. Audinet, A. Beaufort, P. Colin, M. Cornu, and B. Lombard. 2004. A contribution to the improvement of *Listeria monocytogenes* enumeration in cold-smoked salmon. *Int. J. Food Microbiol.* **91:**119–127.

Bhairi, S. M. 2001. *Detergents—a Guide to the Properties and Uses of Detergents in Biological Systems.* Calbiochem-Novabiochem, San Diego, CA.

Brehm-Stecher, B., C. Young, L. A. Jaykus, and M. L. Tortorello. 2009. Sample preparation: the forgotten beginning. *J. Food Prot.* **72:**1774–1789.

Bylund, G. 1995. *Dairy Processing Handbook.* Tetra Pak Processing Systems AB, Lund, Sweden.

Carta, G., and A. Jungbauer. 2010. *Protein Chromatography.* Wiley-VCH, Weinheim, Germany.

Choi, W. S. 2000. Detection of *Salmonella* in milk by polymerase chain reaction. *J. Food Hyg. Saf.* **15:** 262–266.

Choi, W. S. 2001. Detection of *Yersinia enterocolitica* in milk by polymerase chain reaction. *Food Sci. Biotechnol.* **10:**451–454.

Choi, W. S., and C. H. Hong. 2003. Rapid enumeration of *Listeria monocytogenes* in milk using competitive PCR. *Int. J. Food Microbiol.* **84:**79–85.

Cudjoe, K. S., and R. Krona. 1997. Detection of *Salmonella* from raw food samples using Dynabeads(R) anti-Salmonella and a conventional reference method. *Int. J. Food Microbiol.* **37:**55–62.

Cullison, M. A., and L. A. Jaykus. 2002. Magnetized carbonyl iron and insoluble zirconium hydroxide mixture facilitates bacterial concentration and separation from nonfat dry milk. *J. Food Prot.* **65:**1806–1810.

Duodu, S., I. Mehmeti, A. Holst-Jensen, and S. Loncarevic. 2009. Improved sample preparation for real-time PCR detection of *Listeria monocytogenes* in hot-smoked salmon using filtering and immunomagnetic separation techniques. *Food Anal. Methods* **2:**23–29.

D'Urso, O. F., P. Poltronieri, S. Marsigliante, C. Storelli, M. Hernandez, and D. Rodriguez-Lazaro. 2009. A filtration-based real-time PCR method for the quantitative detection of viable *Salmonella enterica* and *Listeria monocytogenes* in food samples. *Food Microbiol.* **26:**311–316.

Fernandez-Astorga, A., M. J. Hijarrubia, B. Lazaro, and I. Barcina. 1996. Effect of the pretreatments for milk samples filtration on direct viable cell counts. *J. Appl. Bacteriol.* **80:**511–516.

Flekna, G., W. Schneeweiss, F. J. Smulders, M. Wagner, and I. Hein. 2007. Real-time PCR method with statistical analysis to compare the potential of DNA isolation methods to remove PCR inhibitors from samples for diagnostic PCR. *Mol. Cell. Probes* **21:**282–287.

Fliss, I., E. Emond, R. E. Simard, and S. Pandian. 1991. A rapid and efficient method of lysis of *Listeria* and other gram-positive bacteria using mutanolysin. *BioTechniques* **11:**453, 456–457.

Fluit, A. C., R. Torensma, M. J. Visser, C. J. Aarsman, M. J. Poppelier, B. H. Keller, P. Klapwijk, and J. Verhoef. 1993. Detection of *Listeria monocytogenes* in cheese with the magnetic immuno-polymerase chain reaction assay. *Appl. Environ. Microbiol.* **59:**1289–1293.

Fukushima, H., K. Katsube, Y. Hata, R. Kishi, and S. Fujiwara. 2007. Rapid separation and concentration of food-borne pathogens in food samples prior to quantification by viable-cell counting and real-time PCR. *Appl. Environ. Microbiol.* **73:**92–100.

Fumian, T. M., J. P. Leite, V. A. Marin, and M. P. Miagostovich. 2009. A rapid procedure for detecting noroviruses from cheese and fresh lettuce. *J. Virol. Methods* **155:**39–43.

Geankopolis, G. 2003. *Transport Processes and Separation Process Principles,* 4th ed. Prentice-Hall, Lebanon, IN.

Hallier-Soulier, S., and E. Guillot. 1999. An immunomagnetic separation polymerase chain reaction assay for rapid and ultra-sensitive detection of *Cryptosporidium parvum* in drinking water. *FEMS Microbiol. Lett.* **176:**285–289.

Hamann, C. H., and W. Vielstich. 2005. *Elektrochemie.* Wiley-VCH, Weinheim, Germany.

Hedman, J., A. Nordgaard, C. Dufva, B. Rasmusson, R. Ansell, and P. Rådström. 2010. Synergy between DNA polymerases increases polymerase chain reaction inhibitor tolerance in forensic DNA analysis. *Anal. Biochem.* **405:**192–200.

Josefsen, M. H., P. S. Lübeck, F. Hansen, and J. Hoorfar. 2004. Toward an international standard for PCR-based detection of foodborne thermotolerant *Campylobacter*: interaction of enrichment media and pre-PCR treatment on carcass rinse samples. *J. Microbiol. Methods* **58:**39–48.

Josefsen, M. H., M. Krause, F. Hansen, and J. Hoorfar. 2007. Optimization of a 12-hour TaqMan PCR-based method for detection of *Salmonella* in meat. *Appl. Environ. Microbiol.* **73:**3040–3048.

Joshi, R., H. Janagama, H. P. Dwivedi, T. M. Senthil Kumar, L. A. Jaykus, J. Schefers, and S. Sreevatsan. 2009. Selection, characterization, and application of DNA aptamers for the capture and detection of *Salmonella enterica* serovars. *Mol. Cell. Probes* **23:**20–28.

Jung, Y. S., J. F. Frank, and R. E. Brackett. 2003. Evaluation of antibodies for immunomagnetic separation combined with flow cytometry detection of *Listeria monocytogenes*. *J. Food Prot.* **66:** 1283–1287.

Koch, A. L., and G. Blumberg. 1976. Distribution of bacteria in the velocity gradient centrifuge. *Biophys. J.* **16:**389–405.

Krämer, N., C. Löfström, H. Vigre, J. Hoorfar, C. Bunge, and B. Malorny. 2010. A novel strategy to obtain quantitative data for modelling: combined enrichment and real-time PCR for enumeration of salmonellae from pig carcasses. *Int. J. Food Microbiol.* Epub ahead of print.

Kretzer, J. W., R. Lehmann, M. Schmelcher, M. Banz, K. P. Kim, C. Korn, and M. J. Loessner. 2007. Use of high-affinity cell wall-binding domains of bacteriophage endolysins for immobilization and separation of bacterial cells. *Appl. Environ. Microbiol.* **73:**1992–2000.

Kulagina, N. V., M. E. Lassman, F. S. Ligler, and C. R. Taitt. 2005. Antimicrobial peptides for detection of bacteria in biosensor assays. *Anal. Chem.* **77:**6504–6508.

Kunz, W., P. Lo Nostro, and B. W. Ninham. 2004. The present state of affairs with Hoffmeister effects. *Curr. Opin. Colloid Interface Sci.* **9:**1–18.

Lantz, P. G., F. Tjerneld, E. Borch, B. Hahn-Hagerdal, and P. Rådström. 1994. Enhanced sensitivity in PCR detection of *Listeria monocytogenes* in soft cheese through use of an aqueous two-phase system as a sample preparation method. *Appl. Environ. Microbiol.* **60:**3416–3418.

Limaye, M. S., and W. T. Coakley. 1998. Clarification of small volume microbial suspensions in an ultrasonic standing wave. *J. Appl. Microbiol.* **84:** 1035–1042.

Lindqvist, R. 1997. Preparation of PCR samples from food by a rapid and simple centrifugation technique evaluated by detection of *Escherichia coli* O157:H7. *Int. J. Food Microbiol.* **37:**73–82.

Lindqvist, R., B. Norling, and S. T. Lambertz. 1997. A rapid sample preparation method for PCR detection of food pathogens based on buoyant density centrifugation. *Lett. Appl. Microbiol.* **24:**306–310.

Liu, R., J. F. Liu, Y. G. Yin, X. L. Hu, and G. B. Jiang. 2009. Ionic liquids in sample preparation. *Anal. Bioanal. Chem.* **393:**871–883.

Loessner, M. J., K. Kramer, F. Ebel, and S. Scherer. 2002. C-terminal domains of *Listeria monocytogenes* bacteriophage murein hydrolases determine specific recognition and high-affinity binding to bacterial cell wall carbohydrates. *Mol. Microbiol.* **44:**335–349.

Lucore, L. A., M. A. Cullison, and L. A. Jaykus. 2000. Immobilization with metal hydroxides as a means to concentrate food-borne bacteria for detection by cultural and molecular methods. *Appl. Environ. Microbiol.* **66:**1769–1776.

Markx, G. H., P. A. Dyda, and R. Pethig. 1996. Dielectrophoretic separation of bacteria using a conductivity gradient. *J. Biotechnol.* **51:**175–180.

Mayrl, E., B. Roeder, P. Mester, M. Wagner, and P. Rossmanith. 2009. Broad range evaluation of the matrix solubilization (matrix lysis) strategy for direct enumeration of foodborne pathogens by nucleic acids technologies. *J. Food Prot.* **72:** 1225–1233.

Mester, P., M. Wagner, and P. Rossmanith. 2010. Use of ionic liquid-based extraction for recovery of *Salmonella* Typhimurium and *Listeria monocytogenes* from food matrices. *J. Food Prot.* **73:** 680–687.

Meyer, R., J. Luthy, and U. Candrian. 1991. Direct detection of polymerase chain reaction (PCR) of *Escherichia coli* in water and soft cheese and identification of enterotoxigenic strains. *Lett. Appl. Microbiol.* **13:**268.

Neiderhauser, C., U. Candrian, C. Hofelein, M. Jermini, H. P. Buhler, and J. Luthy. 1992. Use of polymerase chain reaction for detection of *Listeria monocytogenes* in food. *Appl. Environ. Microbiol.* **58:**1564.

Niederhauser, C., C. Hofelein, B. Wegmuller, J. Luthy, and U. Candrian. 1994. Reliability of PCR decontamination systems. *PCR Methods Appl.* **4:**117–123.

Nogva, H. K., K. Rudi, K. Naterstad, A. Holck, and D. Lillehaug. 2000. Application of 5′-nuclease PCR for quantitative detection of *Listeria monocytogenes* in pure cultures, water, skim milk, and unpasteurized whole milk. *Appl. Environ. Microbiol.* **66:**4266–4271.

Patchett, R. A., A. F. Kelly, and R. G. Kroll. 1991. The adsorption of bacteria to immobilized lectins. *J. Appl. Bacteriol.* **71:**277–284.

Payne, M. J., S. Campbell, R. A. Patchett, and R. G. Kroll. 1992. The use of immobilized lectins in the separation of *Staphylococcus aureus, Escherichia coli, Listeria* and *Salmonella spp.* from pure cultures and foods. *J. Appl. Bacteriol.* **73:**41–52.

Pedersen, L. H., P. Skouboe, L. Rossen, and O. F. Rasmussen. 1998. Separation of *Listeria monocytogenes* and *Salmonella berta* from a complex food matrix by aqueous polymer two-phase partitioning. *Lett. Appl. Microbiol.* **26:**47–50.

Rådström, P., R. Knutsson, P. Wolffs, M. Lovenklev, and C. Lofstrom. 2004. Pre-PCR processing: strategies to generate PCR-compatible samples. *Mol. Biotechnol.* **26:**133–146.

Ramesh, A., B. P. Padmapriya, A. Chrashekar, and M. C. Varadaraj. 2002. Application of a convenient DNA extraction method and multiplex PCR for the direct detection of *Staphylococcus aureus* and *Yersinia enterocolitica* in milk samples. *Mol. Cell. Probes* **16:**307–314.

Rosen, M. J. 2004. *Surfactants and Interfacial Phenomena*, 3rd ed. John Wiley & Sons, Hoboken, NJ.

Rossen, L., P. Norskov, K. Holmstrom, and O. F. Rasmussen. 1992. Inhibition of PCR by components of food samples, microbial diagnostic assays and DNA-extraction solutions. *Int. J. Food Microbiol.* **17:**37–45.

Rossmanith, P., M. Krassnig, M. Wagner, and I. Hein. 2006. Detection of *Listeria monocytogenes* in food using a combined enrichment/real-time PCR method targeting the *prfA* gene. *Res. Microbiol.* **157:**763–771.

Rossmanith, P., B. Suss, M. Wagner, and I. Hein. 2007. Development of matrix lysis for concentration of gram positive bacteria from food and blood. *J. Microbiol. Methods* **69:**504–511.

Rossmanith, P., K. Frühwirth, B. Süß, E. Schopf, and M. Wagner. 2010. The use of chromogenic bacteria as coloured substitutes for pathogens: a simple strategy during design and development of a new method for sample pretreatment. *Lett. Appl. Microbiol.* **50:**230–233.

Sambrook, J., T. Maniatis, and E. F. Fritsch. 1989. *Molecular Cloning: a Laboratory Manual*, 2nd ed. Cold Spring Harbor Laboratory Press, Cold Spring Harbor, NY.

Skjerve, E., L. M. Rorvik, and O. Olsvik. 1990. Detection of *Listeria monocytogenes* in foods by immunomagnetic separation. *Appl. Environ. Microbiol.* **56:**3478–3481.

Stevens, K. A., and L. A. Jaykus. 2004a. Bacterial separation and concentration from complex sample matrices: a review. *Crit. Rev. Microbiol.* **30:**7–24.

Stevens, K. A., and L. A. Jaykus. 2004b. Direct detection of bacterial pathogens in representative dairy products using a combined bacterial concentration-PCR approach. *J. Appl. Microbiol.* **97:**1115–1122.

Stieß, M. 1997. *Mechanische Verfahrenstechnik 2.* Springer, Heidelberg, Germany.

Thomas, D. S. 1988. Electropositively charged filters for the recovery of yeasts and bacteria from beverages. *J. Appl. Bacteriol.* **65:**35–41.

Tjhie, J. H., F. J. van Kuppeveld, R. Roosendaal, W. J. Melchers, R. Gordijn, D. M. MacLaren, J. M. Walboomers, C. J. Meijer, and A. J. van den Brule. 1994. Direct PCR enables detection of *Mycoplasma pneumoniae* in patients with respiratory tract infections. *J. Clin. Microbiol.* **32:**11–16.

Ulve, V. M., C. Monnet, F. Valence, J. Fauquant, H. Falentin, and S. Lortal. 2008. RNA extraction from cheese for analysis of in *situ* gene expression of *Lactococcus lactis. J. Appl. Microbiol.* **105:**1327–1333.

Uyttendaele, M., I. Van Hoorde, and J. Debevere. 2000. The use of immuno-magnetic separation (IMS) as a tool in a sample preparation method for direct detection of *L. monocytogenes* in cheese. *Int. J. Food Microbiol.* **54:**205–212.

Walsh, G. 2002. *Biochemistry and Biotechnology.* John Wiley & Sons Ltd., Chichester, West Sussex, England.

Wang, R. F., W. W. Cao, and M. G. Johnson. 1992. 16S rRNA-based probes and polymerase chain reaction method to detect *Listeria monocytogenes* cells added to foods. *Appl. Environ. Microbiol.* **58:**2827–2831.

Wegmuller, B., J. Luthy, and U. Candrian. 1993. Direct polymerase chain reaction detection of *Campylobacter jejuni* and *Campylobacter coli* in raw milk and dairy products. *Appl. Environ. Microbiol.* **59:**2161–2165.

Wilson, I. G. 1997. Inhibition and facilitation of nucleic acid amplification. *Appl. Environ. Microbiol.* **63:** 3741–3751.

Wolffs, P., R. Knutsson, B. Norling, and P. Rådström. 2004. Rapid quantification of *Yersinia enterocolitica* in pork samples by a novel sample preparation method, flotation, prior to real-time PCR. *J. Clin. Microbiol.* **42:**1042–1047.

Yu, H., and J. G. Bruno. 1996. Immunomagnetic-electrochemiluminescent detection of *Escherichia coli* O157 and *Salmonella typhimurium* in foods and environmental water samples. *Appl. Environ. Microbiol.* **62:**587–592.

CRITERIA FOR CHOOSING THE RIGHT RAPID METHOD

H. Joosten and J. Marugg

9

SELECTION CRITERIA

Microbiological food safety is achieved by strict adherence to Good Manufacturing Practices and implementation of adequate control measures in the production process to prevent the entry, survival, and/or growth of pathogens. Due to the nonhomogeneous distribution of microorganisms and the destructive nature of analytical methods, testing alone cannot guarantee the absence of pathogens. Nevertheless, analytical methods are important tools for food safety management and applied for various purposes such as the ones indicated below.

- Testing of finished products to verify compliance with regulations or to verify that control measures are properly implemented
- Monitoring production line samples and environmental samples to verify the efficacy of Good Hygienic Practices and Good Manufacturing Practices
- Checking raw materials to confirm the compliance with criteria that have been agreed upon with suppliers
- Investigations into the cause of a positive test result of a finished product, line sample, or environmental sample and assessment of the effect of corrective actions
- Outbreak investigations and epidemiological tracing
- Determining the prevalence of pathogens in raw materials or finished products in the framework of a hazard analysis and critical control point or risk assessment study

Conventional microbiological methods are generally time-consuming, requiring typically 1 to 3 days to obtain a result for direct enumeration assays, 3 to 4 days for a negative result in presence/absence tests, and 5 days or more for a confirmed positive result. Identification and typing of pathogens can cause a further delay in obtaining the test results. The application of rapid methods can therefore be a great advantage to respond quickly to outbreaks, reject contaminated raw materials before they enter the production chain, rapidly identify process failures to take corrective measures, etc.

It is fairly easy to find rapid methods for almost any purpose by consulting the scientific literature (Patel, 1994; Stannard et al., 1990; Clark et al., 1999; Downes and Ito, 2001; Levin, 2009), searching the Internet, or con-

H. Joosten and J. Marugg, Nestlé Research Center, Vers-chez-les-Blanc, 1000 Lausanne 26, Switzerland.

Rapid Detection, Identification, and Quantification of Foodborne Pathogens, ed. by J. Hoorfar

tacting commercial method suppliers. The choice can be overwhelming, making the comparison of the pros and cons of the different alternatives a challenging task. However, depending on the importance of the results and the number of tests that need to be done, it is crucial to make this effort, especially because of the potential public health impact and financial implications of choosing an inadequate or poorly documented method.

To facilitate making this comparison, an overview of criteria that can be useful to judge the suitability of a method is given. Table 1 provides some general examples of the relative importance of these criteria for different applications, but it is up to the user to define which criteria are really essential and which are only desirable.

RELIABILITY

Obviously, reliability is a key requirement for any method and must be evaluated carefully.

Detection and Enumeration Methods

For detection and enumeration methods, the following performance indicators should be checked.

Analytical Sensitivity and Specificity

Does the method correctly recognize the target if it is present, even if it only represents a small fraction of the total microbial population in the sample (inclusivity), and is it able to discriminate closely related species or genera as nontarget (exclusivity)?

Detection Limit

What is the minimum amount of target pathogen that needs to be present in a sample to allow for a positive test result? The detection limit is often established by analyzing artificially contaminated samples, but it should be realized that it may also depend on the physiological state, the strain, and the presence of competitors. This implies that the spiking

TABLE 1 Typical examples of the relative importance of selection criteria for analytical methods

User	Application	Selection criteria	
		Critical	Less relevant
Food manufacturer	Routine testing of ingredients and finished products	Reliability Validation Time to (negative) result Costs Acceptance	Shelf life of reagents
	Environmental monitoring	Costs	Acceptance
	Outbreak investigations	Time to (positive) result Reliability Sample throughput Supply chain confidence	Costs
	Critical control point validation study	Reliability	Time to result
Public health authority	Routine testing of finished products	Reliability Validation Costs	Time to result
	Outbreak investigations	Time to (positive) result, including (sub-)typing Reliability Acceptance Throughput Supply chain confidence	Costs
	Microbiological risk assessment	Reliability	Time to result

protocol that is used to obtain the artificially contaminated samples should be adapted to what can be expected to be present in "real" samples, for example, by selecting different target strains, by applying stress on the spiking material (e.g., dry or heat stress), and by adding competing microorganisms as well. The detection limit can be a key criterion in case "pooling" or "compositing" of samples is envisaged. This cost-saving practice is very common, especially if only negative results are expected. However, many validation protocols do not specifically cover this aspect; all too often, it is simply assumed that if pooling is allowed for the standard protocol, it is also permitted for an alternative method (Jarvis, 2008; Miled, 2010).

Compatibility with the Matrix

Depending on the analytical procedure, the chemical composition of the sample may interfere with detection or enumeration. Examples are PCR inhibitors, colored ingredients, or growth inhibitors. The physical characteristics may also be important, for example, if the sample needs to be filtered as part of the analytical protocol.

The performance of the method may also depend on the microbiological composition of the matrix, which includes not only contaminants but also deliberately added microorganisms such as starter bacteria and probiotics. As they are often present in high levels, they can easily interfere with the detection of undesirable microorganisms (Joosten et al., 2006).

Repeatability and Reproducibility

Repeatability is a measure of the capability of the method to give the same result if the test is repeated with the same sample by the same technician in the same laboratory. Reproducibility measures this capability if the test is done with the same sample in another lab by another technician.

Measurement Uncertainty

The measurement uncertainty parameter indicates the extent to which the measured value can differ from the true value. However, as the true value is usually not known, it is difficult to determine the measurement uncertainty directly, nor is it feasible to estimate it through adding up the individual error sources. Therefore, preference is usually given to a procedure that determines the standard deviation of reproducibility to estimate the measurement uncertainty of a method (Lombard, 2006).

Robustness

The robustness of a method relates to the extent to which the analytical result is influenced by small deviations in the procedure or differences in the chemical composition of the reagents that are likely to occur under normal circumstances.

Identification Methods

Methods to identify foodborne pathogens are usually applied on isolated colonies, thereby excluding the matrix and the presence of competitors as error sources. For such methods, reliability depends entirely on whether it consistently gives the correct result with individual strains, e.g., identification at genus or species level, serotype, pulse type, or ribotype. In many cases it is important to consider the discriminatory power of the technique as well, to determine if it is capable of distinguishing between closely related strains and/or if it is capable of revealing taxonomic relationships between different isolates.

VALIDATION

A validated method is a method that has been proven to be fit for purpose, and it is evident that validation is an important criterion in choosing a method. Validation usually consists of generating experimental evidence that the performance of a method meets the requirements, but as it is difficult to define unambiguous performance requirements for a method, validation is usually limited to comparing the performance of an alternative method with the reference method according to a defined protocol (ISO16140 [Anonymous, 2003] or AOAC [Feldsine et al., 2002]).

If an alternative method passes this test, it will be granted the status of a validated method by the supervising organization. However, it should be realized that the reference method may not be very reliable itself and alternative methods that are successfully evaluated do not necessarily perform much better. This is a particular concern for methods used to detect pathogens that have emerged recently such as *Cronobacter* spp. (Guillaume-Gentil et al., 2005; Iversen et al., 2008).

It should also be noted that differences exist between the various validation protocols, which is reflected in the existence of expressions such as "in-house validated," "performance tested validated," "fully validated," etc. (Feldsine et al., 2002). Often it is not clear what the consequences are in terms of confidence: if a method is validated, it is per definition proven to be fit for purpose, and in theory the validation procedure should be subordinate, but in practice there is a difference, at least in perception, in the confidence that can be given to a method that has been superficially tested by a laboratory that has commercial interests in selling the method and the confidence given to a method that has undergone an independent evaluation according to an internationally accepted protocol.

As there are also differences in the criteria that are used to judge if the observed performance meets the expectations, it is recommended to always critically evaluate the validation protocol and the results obtained.

Whereas there is a wealth of information on validation of enumeration and detection methods for food pathogens, it is often more difficult to find validation data for identification and typing methods. Internationally accepted validation procedures for these methods do not exist, nor is there an independent entity that organizes validation trials and provides certificates of approved methods. The interested users therefore have to rely on the reputation of the supplier, on scientific publications (provided that they exist), or on their own results.

ACCEPTANCE BY PUBLIC HEALTH AUTHORITIES AND TRADE PARTNERS

If analytical results are to be shared with other parties, it is important to verify beforehand that there is agreement about the analytical method. In most cases, microbiological criteria not only define the sampling plan and maximum tolerable limits but also prescribe the method that has to be used to verify compliance. It cannot be taken for granted that rapid or alternative methods will be accepted by regulators or control authorities, even if they are claimed to be fully validated.

TIME TO RESULT

One of the principal reasons for the long delays when applying conventional microbiological methods is the fact that these methods depend on growth of the target microorganism, which must reach a level high enough for the organism to be visible as an individual colony or to generate biochemical changes of the growth medium that can be easily detected such as acidification, gas production, etc. Under ideal circumstances, microbial growth can be relatively rapid, and an increase of 1 log per hour is not unusual for most pathogens, which would imply that even for methods with a very low detection limit (e.g., 1 cell per liter), an overnight incubation would be more than sufficient to reach the threshold. However, depending on the physiological state of the individual cells, additional time may be required to overcome the lag phase (Stephens et al., 1997). Furthermore, growth may come to a halt prematurely due to the presence of competing microorganisms (Miled, 2010), in which case an additional enrichment may be necessary before plating on a selective or elective isolation agar. Further delay may be caused by the fact that the procedures are adapted to the availability of personnel in the laboratory, which normally imposes incubation periods of ca. 24 h.

Obtaining results faster, with rapid methods, allows one to respond promptly to outbreaks, reject raw materials, identify process failures, and take corrective measures, etc., but the degree to which this is important for the user may vary. In this respect, it is also important to consider if at the moment at which a result becomes available it is possible to take action immediately. This may not always be the case; for instance, if a method gives a result at the end of the afternoon, it may not be until the following morning before the product can be released and shipped.

COSTS

For most users, the cost of a method is a very important criterion, which merits to be calculated precisely, taking into account all the contributing elements, including the equipment, reagents and kits, laboratory space, waste disposal, maintenance, labor, etc. As the impact of the labor costs is often relatively high, it can be necessary to perform a detailed analysis of how the workload for the technicians is distributed and whether the tasks can be easily combined with other activities during a normal working day. Obviously, high labor costs favor the selection of automated methods that require very little hands-on time.

Renting equipment from the supplier instead of buying it can be an interesting alternative if the procurement costs are high, for example, to have more flexibility to change to another method or to discontinue the analysis without having to write off the investment. When calculating the test costs, it may also be appropriate to consider the cost savings that can be achieved if a rapid method is implemented, for example, if this allows for a more rapid release of raw materials or finished products.

LABORATORY SPACE

The equipment needed for rapid microbiological methods can be quite voluminous, and depending on the available space, this may be a decisive factor when choosing the best alternative. In particular, molecular methods often need specialized laboratories with separated operational units to avoid contamination of reagents, to operate sensitive detection instruments, and so on. Computers have also become an integrated part of almost any detection equipment for digital data generation, but they take up some space. The possibility of sharing one computer for several pieces of equipment should be explored at the time of procurement.

SAMPLE THROUGHPUT

Most microbiological laboratories experience fluctuating demands for analytical support, which should be taken into account when choosing an alternative rapid method. The capacity of the equipment should be large enough to respond to peak demands during outbreak situations, whereas equipment may also stay idle for large parts of the year but still need service and maintenance to ensure that it can be used immediately when it is suddenly needed. A solution to this problem may be to obtain multipurpose platforms that can support different tests.

SHELF LIFE OF REAGENTS

Whereas the chemicals used for traditional microbiological methods are usually relatively cheap, this cannot be said of the reagents and kits that are necessary for alternative methods, rendering it important to consider their shelf life, in particular if the analytical throughput is variable. Transportation of temperature-sensitive reagents can be quite costly, which is why an increasing number of kit manufacturers formulate lyophilized reagents and leave the addition of water to their reagents to the receiving laboratory.

For high-throughput routine laboratories that have to perform several types of analyses, it may be efficient to have rapid methods that are multifunctional, allowing the detection and/or enumeration of more than one parameter on the same equipment, or even in the same run. Some new reverse transcriptase PCR methods even have the capacity to detect multiple targets from the same sample.

COMPETENCY LEVEL AND WORKFLOW COMPATIBILITY

Before rapid methods can be successfully used, it may be necessary to provide specific training to the technicians that will use the equipment, so it is important to determine beforehand how this will be achieved: will it be provided by the supplier, are the technicians willing and able to undergo the training and use the equipment, and what should be done about technicians that become redundant if they do not have the required competency level?

Another relevant question is whether the procedure is compatible with the normal working hours of the technicians. There may be peaks in the workload or activities that need to be carried out at unusual times, which can make an alternative method less attractive.

DATA MANAGEMENT AND INTERPRETATION

Reliable and efficient data management is a key requirement for a modern food microbiology laboratory, and this aspect should therefore not be overlooked when assessing alternative methods. There are several pertinent questions that can be raised, such as whether it is possible to introduce information about the samples and the corresponding results through connection with the Laboratory Information Management System. Is the information easily traceable and protected against undesirable manipulation? Are the analytical data reported as unambiguous and actionable results, or do they first require interpretation by specialists?

SUPPLY CHAIN CONFIDENCE

When choosing a rapid method, it is advisable to consider the probability and consequences of a scenario in which the method is suddenly not available anymore; for example, because the reagents are not received or if the equipment is out of order.

It can also happen that new validation information becomes available showing that the method is not as reliable as previously thought. If the consequences are serious, it becomes even more important to ensure that the validation data do not leave any room for doubt and also that only vendors that can guarantee swift delivery of reagents, technical service, spare parts, and backup equipment are chosen, even if there is a sudden worldwide increase in demand, such as during a serious foodborne outbreak. This aspect is less critical if a suitable fall-back option is readily available, for example, using the conventional method or an external laboratory.

To further strengthen the supply chain confidence, it may be necessary to evaluate the financial strength and the quality system of the supplier, in particular to assess the procedures that ensure the consistent quality of reagents, equipment, and service.

MOTIVATION AND REPUTATION

Using fancy equipment can be motivating for technicians, and this can be a legitimate reason to choose a specific rapid method, even though it is not likely to be given the highest priority. Likewise, it can be beneficial for the reputation of a laboratory, a food producer, or a regulatory authority if it can claim that it uses state-of-the-art methods.

CONCLUSION AND FINAL REMARKS

There are many criteria that can be used to select the right rapid method. The degree to which they are relevant will differ depending on the user and the specific application (Table 1). The present overview will hopefully make it easier to set priorities and to avoid overlooking important features.

Once the selection criteria are established, the next challenge is to find reliable information on how candidate methods score with respect to the requirements.

The sources for this information are identical to those mentioned in "Selection Criteria" above, such as textbooks, scientific publications, the Internet, and method suppliers, but it is also recommended to consult validation bodies like Microval (http://www.microval.org/), Nordval (http://www.nmkl.org/Engelsk/index.htm), AOAC (http://

www.aoac.org/), etc., and of course to discuss with colleagues in the field. It is also hoped that this overview helps method developers and suppliers to better understand the needs of the method users and allows them to focus on developing superior methods with clear added value rather than providing solutions for nonexisting problems.

REFERENCES

Anonymous. 2003. *Microbiology of Food and Animal Feeding Stuffs—Protocol for the Validation of Alternative Methods. (EN ISO 16140)*. European Committee for Standardization, AFNOR, Paris, France.

Clark, S. A., K. C. Thompson, C. W. Keevil, and M. S. Smith (ed.). 1999. *Rapid Detection Assays for Food and Water.* The Royal Society of Chemistry, Cambridge, United Kingdom.

Downes, F. P., and K. Ito (ed.). 2001. *Compendium of Methods for the Microbiological Examination of Foods*, 4th ed. American Public Health Association, Washington, DC.

Feldsine, P., C. Abeyta, and W. Andrews. 2002. AOAC International Methods Committee guidelines for validation of qualitative and quantitative food microbiological official methods of analysis. *J. AOAC Int.* **85:**1187–1200.

Guillaume-Gentil, O., V. Sonnard, M. C. Kandhai, J. D. Marugg, and H. Joosten. 2005. A simple and rapid cultural method for detection of *Enterobacter sakazakii* in environmental samples. *J. Food Prot.* **68:**64–69.

Iversen, C., P. Druggan, S. Schumacher, A. Lehner, C. Feer, K. Gschwend, H. Joosten, and R. Stephan. 2008. Development of a novel screening method for the isolation of *Cronobacter* spp. (*Enterobacter sakazakii*). *Appl. Environ. Microbiol.* **74:**2550–2553.

Jarvis, B. 2008. On the compositing of samples for qualitative microbiological testing. *Lett. Appl. Microbiol.* **45:**592–598.

Joosten, H., E. Bidlas, and N. Garofalo. 2006. *Salmonella* detection in probiotic products. *Int. J. Food Microbiol.* **110:**104–107.

Levin, R. E. 2009. *Rapid Detection and Characterization of Foodborne Pathogens by Molecular Techniques.* CRC Press, Boca Raton, FL.

Lombard, B. 2006. Estimation of measurement uncertainty in food microbiology: the ISO approach. *Accred. Qual. Assur.* **11:**94–100.

Miled, R. B., S. Neves, N. Baudouin, B. Lombard, V. Deperrois, P. Colin, and N. G. Besse. 2010. Impact of pooling powdered infant formula samples on bacterial evolution and *Cronobacter* detection. *Int. J. Food Microbiol.* **138:**250–259.

Patel, P. D. (ed.). 1994. Rapid analysis techniques in food microbiology. Blackwell Academic and Professional, Glasgow, United Kingdom.

Stannard, C., S. Petitt, and F. Skinner (ed.). 1990. *Rapid Microbiological Methods for Foods, Beverages and Pharmaceuticals.* Wiley-Blackwell, Chichester, United Kingdom.

Stephens, P. J., J. A. Joynson, K. W. Davies, R. Holbrook, H. M. Lappin-Scott, and T. J. Humphrey. 1997. The use of an automated growth analyser to measure recovery times of single heat-injured *Salmonella* cells. *J. Appl. Microbiol.* **83:** 445–455.

YOUR RESULTS ARE YOUR CONTROLS: INCLUSION OF CRITICAL TEST CONTROLS

C. Löfström and J. Hoorfar

10

Molecular methods, including nucleic acid-based and antibody-based techniques, are the most commonly used alternative methods for routine analysis of foodborne pathogens. In many cases, these methods have also been to some extent validated against culture-based methods by international organizations. Although using molecular methods, such as PCR and enzyme-linked immunosorbent assay (ELISA), appears to be straightforward, the importance of including proper test controls may be overlooked.

Results of tests developed or published by one laboratory can sometimes be difficult to reproduce by other laboratories. Although this relates to most laboratory techniques, lack of reproducibility is more pronounced in molecular techniques due to sensitive reagents, complex equipment, and the need for specialized personnel. Therefore, inclusion of proper analytical controls is an absolute prerequisite for successful implementation of molecular methods for diagnostic purposes.

C. Löfström and J. Hoorfar, National Food Institute, Technical University of Denmark, Mørkhøj Bygade 19, DK-2860 Søborg, Denmark.

In the case of PCR, it has been suggested in several ISO standards that a PCR-based method cannot be given diagnostic status, no matter how limited the application, before it includes, as a minimum, an internal amplification control (IAC) and positive and negative process controls, as well as positive and negative PCR controls (Table 1) (International Organization for Standardization, 2005, 2006a, 2006b, 2006c, 2007). Furthermore, to enable correct evaluation of results, the interpretation of controls should be clearly specified in the method descriptions (International Organization for Standardization, 2005). However, analytical controls cannot be a substitute for a thorough quality assurance program in the laboratory to ensure proper handling of samples.

To obtain reliable results, the correct functioning of all steps of the analysis needs to be controlled, including the use of high-quality reagents and careful sample preparation steps. In this chapter, the selection and inclusion of the analytical control(s), its use, interpretation, applications, and limitations are described, using PCR and ELISA techniques as examples. The inclusion of controls associated with biosensors and lab-on-a-chip is also touched upon.

Rapid Detection, Identification, and Quantification of Foodborne Pathogens, ed. by J. Hoorfar

TABLE 1 Controls necessary to assess the performance of molecular methods

Molecular method	Control	Abbreviation	Content
PCR	Internal amplification control	IAC	Chimeric nonrelevant DNA added to master mixture amplified by the same primer set as the target DNA or another set of primer pairs. The amplicon size should be visually distinguishable from the target amplicon, or a probe binding site for a probe that fluoresces at a different wavelength should be used.
	Negative process control	NPC	A negative sample inoculated with a sufficient amount of a closely related but nontarget strain processed throughout the entire protocol
	Positive process control	PPC	A negative sample inoculated with a sufficient amount of the pathogen and processed throughout the entire protocol
	Negative extraction control/extraction blank		A control carried through all steps of the DNA extraction procedure in the absence of a test sample
	Positive PCR control		All reagents together with the target DNA in a defined amount or copy number
	Negative PCR control/ negative template control	NTC	All reagents, but no nucleic acid apart from the primers and probes
	Premise control		A tube containing the master mixture left open in the PCR setup room to detect possible contaminating DNA in the environment (to be done in certain intervals as part of the quality assurance program)
	Standard control (only for quantification)		Three to four samples containing 10-fold dilution of known number of target DNA copies in a range above the detection limit (necessary only for quantitative PCR
	Nonamplification control	NAC	Optionally applicable in real-time PCR assays. All PCR reagents except DNA polymerase. Increase in fluorescence indicates a bad batch of probe.
ELISA	Blank		Background absorbance levels in the absence of a test sample. Could be measured using buffer or distilled water.
	Positive ELISA control		All reagents together with the target in a defined amount
	Negative ELISA control		All reagents, but no target added
	Negative process control	NPC	A negative sample inoculated with a sufficient amount of a closely related, but nontarget strain processed throughout the entire protocol

TABLE 1 (*Continued*)

Molecular method	Control	Abbreviation	Content
	Positive process control	PPC	A negative sample inoculated with a sufficient amount of the pathogen and processed throughout the entire protocol
	Negative extraction control/extraction blank		A control carried through all steps of the sample preparation procedure in the absence of a test sample
	Standard control (only for quantification)		Three to four samples containing a 10-fold dilution of a known number of target molecules in a range above the detection limit
	Conjugate control		A control where the coating antibody is left out, to ensure that there is no nonspecific binding of the conjugate to well surfaces or to the coating antibody
	Substrate/chromogen control		A control leaving out the coating antibody and conjugate. Test for nonspecific production of the color reaction.

DIFFERENT TYPES OF CONTROLS

Negative and positive controls are generally included in most molecular methods and are used to verify that the assay used is able to detect the target and to rule out the risk of cross-contamination during the analysis step. Apart from these controls, controls for the preanalytical steps also need to be included. This is to ascertain that the sample preparation steps have functioned as intended with a high probability of concentrating the target without any interfering substances. The specific controls needed for PCR and antibody-based methods are described in separate sections below.

Another important type of controls includes process controls, with the purpose of verifying that all steps in the detection process work as intended. These steps can include sample treatment and nucleic acid extraction, together with amplification and detection (International Organization for Standardization, 2005). Process controls should be included at regular time intervals as laid down by the quality assurance program of the testing laboratory (International Organization for Standardization, 2005).

Furthermore, process controls should be included if the positive and negative controls are not giving the expected results, with the aim of identifying the source(s) of error in the analysis. There are two types of process controls: positive process control and negative process control. In addition, the positive process control can be combined with the IAC by addition of a suitable strain or DNA at the beginning of the analysis chain.

PCR CONTROLS

To ensure that the amplification and detection steps are working properly, negative and positive controls are to be included in every batch of samples analyzed (International Organization for Standardization, 2005). Apart from this, an IAC should be included with every single PCR to make sure that it was possible to detect the target DNA sequence in each individual sample. Furthermore, process controls should be included at regular time intervals. Details on the content of these controls and the reason for their inclusion are discussed below.

Negative Controls

The negative control consists of a sample in which the target DNA sequence is absent (International Organization for Standardization, 2005). In this case, all reagents are added to the PCR tube, but the sample is replaced with the same volume of DNA-free water without any PCR inhibitors. This control is also referred to as negative template control (NTC). The result of this control should always be negative, thereby verifying the accuracy of the amplification and detection steps. It also indicates that no cross-contamination with PCR products and/or target organisms has taken place before the PCR, e.g., by improper handling of reagents or samples.

An important aspect of using PCR, due to the possibility of amplifying and detecting only one copy of the target gene if present, is the control of workflow in the laboratory. NTC should be present in all PCR runs (International Organization for Standardization, 2005). In real-time PCR, some software programs use the NTCs to assign the threshold used for calculating the cycle threshold (C_T) values. In this case, it is recommended that at least three NTCs be included, in order to ensure a reproducible threshold value. As an alternative, the NTC can also include nontarget DNA instead of water, e.g., *Escherichia coli* DNA if *Salmonella* is the target organism.

Another type of negative control in real-time PCR is the nonamplification control. This type of control is especially useful if it is suspected that there is an increase in fluorescence that is not the result of amplification, i.e., degradation of a poorly made fluorescent probe. The control consists of all the reagents except for the DNA polymerase. If an increase in fluorescence is detected in this control, a new batch of probe should be tested.

Positive Control

The positive control consists of a reaction containing the target molecule in a defined amount or copy number (International Organization for Standardization, 2005). Usually, a specified amount of purified DNA or a cell suspension with a known amount of target organisms or molecules is used for this purpose. This control should be included in every run and should always be positive for the target. If it is negative, a possible failure of analysis has occurred and all samples tested in the same setup should be reanalyzed.

For PCR, the positive control is added to the PCR tube containing all other reagents. The number of freeze-thaw cycles of the positive PCR control should be minimized, since it might lead to breakdown of the DNA. Care should also be taken to use an appropriate number of DNA copies; the number should be high enough to ensure consistent detection but as low as possible to minimize the risk of cross-contamination.

Internal Amplification Control

In order to ensure that no false-negative results are obtained due to PCR inhibition, a defect thermocycler, or incorrect PCR mixture, the use of an IAC is mandatory in diagnostic PCR (Hoorfar et al., 2004a, 2004b; International Organization for Standardization, 2005).

The IAC consists of a defined amount or copy numbers of DNA added to the reaction mixture (International Organization for Standardization, 2005). For real-time PCR assays employing probes such as molecular beacons, TaqMan probes, or hybridization probes, the IAC also consists of a probe specific for the IAC amplicon with a reporter that fluoresces at a wavelength different from that of the target probe. Depending on the channels available on the thermocycler, one example of a suitable combination of dyes could be 6-carboxyfluorescein for the target probe and HEX/VIC/JOE for the IAC (Malorny et al., 2004). A multiple IAC containing several primer binding sites specific for five different targets but with a unique probe binding site in the middle region of the IAC has been constructed (Maaroufi et al., 2006). This approach facilitates the incorporation of IAC at laboratories performing routine analysis, as it means that only one IAC reagent is necessary when performing analyses of multiple organisms.

In a PCR without an IAC, a negative response (no band or signal) could mean that there was no target sequence present in the reaction. However, it could also mean that the reaction was inhibited, due to a malfunction of the thermal cycler, incorrect PCR mixture, poor DNA polymerase activity, or not least the presence of inhibitory substances in the sample matrix (Rådström et al., 2003). Conversely, in a PCR with an IAC, a control signal (from the IAC and/or from the target amplicon) should always be produced, even though there is no target sequence present.

There are numerous reports about the use of IACs in diagnostic PCR assays both for bacteria (Josefsen et al., 2007; Lund et al., 2004; Lund and Madsen, 2006) and for viruses (Maaroufi et al., 2006; Tang et al., 2005). Furthermore, IACs have also been successfully implemented into several quantitative real-time PCR assays, giving the possibility to control the underestimation of contamination loads due to PCR failure (Rodríguez-Lázaro et al., 2005). Unfortunately, very few PCR methods for detection of foodborne fungi incorporate IACs (Paterson, 2007).

For real-time PCR assays, it is important that the IAC have an amplification efficiency similar to that of the target amplicon in order to avoid a biased detection level (International Organization for Standardization, 2006c). To achieve this, it is recommended that the homologous IAC (the same oligonucleotide primer set used for both target and IAC) have a slightly lower amplification efficiency than the target DNA. The heterologous IAC (different primer sets for the target and for the IAC) should use a reduced concentration of primers compared to the target assay in order to support a better amplification of the target.

Due to competition, if the target DNA is amplified but the IAC is not, it is assumed that the target DNA is present in a proportionally greater amount. When this occurs, the positive PCR result is still valid, since the IAC amplification is not necessary. If neither the IAC nor the target DNA is amplified, it is assumed that inhibition of the PCR has occurred and the test results for that particular sample are not valid. However, the drawback of this approach could be a less sensitive limit of detection of the target pathogen due to the competition by the IAC.

Combination of IAC and Positive PCR Control

Some recent reports suggest the addition of the IAC before DNA extraction in order to control the pre-PCR sample preparation steps, including the extraction process (positive process control) as well as the PCR step (positive PCR control) (Long et al., 2008; Murphy et al., 2007; Sohni et al., 2008). For this purpose, whole mutated cells added at the beginning of the analysis chain can be used instead of DNA (see below).

Positive Process Control

The positive process control is defined, according to ISO 22174:2005, as a "sample spiked with the target organism, which should be treated in the same way as the test samples" (International Organization for Standardization, 2005). The target organism is added in sufficient quantities to the sample at the beginning of the analysis. However, the number of cells or copies of DNA added should not be too large, as this increases the risk of cross-contamination to the samples. As an example, in an assay developed in our laboratory for detection of *Salmonella* in meat, an amount of approximately 40,000 CFU was analyzed in parallel in the DNA extraction step as a positive process control (Josefsen et al., 2007). A negative response from the positive process control, while the positive PCR control is positive, indicates a failure of the steps somewhere upstream of the PCR testing. In this case, the samples need to be reexamined, including a new purification of DNA and a new PCR run. As an alternative, the process control can be combined with the positive PCR control.

In a recent study, a *Listeria monocytogenes* mutant strain that was used as an IAC was developed (Long et al., 2008). A computational DNA random shuffling method was used to create an IAC sequence that has the same length and GC% content as the *hly* gene with the primer sequences flanking the shuffled fragment. By using this method, similar DNA extraction and amplification efficiencies were obtained for the IAC mutant strain and for the wild-type *L. monocytogenes* strain. However, to avoid the use of possibly pathogenic organisms in the laboratory, it might be useful to use this approach on a closely related nonpathogenic species instead.

In another study, the *gfp* gene was cloned into the chromosome of a nonpathogenic strain of *E. coli* and used as an internal control (Murphy et al., 2007). The IAC was duplexed with assays for the *iroB* gene of *Salmonella* and the *hlyA* gene of *L. monocytogenes* and optimized to obtain sensitivities for the respective assays similar to those achieved for single reactions. A similar approach was used for detection of the *rpoB* gene of *Bacillus anthracis*, where a mimic of the amplicon was integrated into a host strain of nonpathogenic *Bacillus subtilis* (Sohni et al., 2008).

Negative Process Control

Similar to the positive process control, a negative process control is, according to ISO 22174:2005, a "target pathogen-free sample of the food matrix which is run through all stages of the analytical process" (International Organization for Standardization, 2005). The purpose of this control is to verify that there is no cross-contamination from other positive samples analyzed at the same time; for instance, in the same centrifuge or DNA extraction robot. Moreover, it serves as a general control of the lack of cross-contamination at the laboratory.

This has also been mentioned as a negative extraction control (International Organization for Standardization, 2005) with the purpose to control the method from the extraction of DNA/RNA, to the amplification and final detection steps. If the negative process control results in a positive signal, it is recommended that samples be retested by PCR. If so, results of the other samples should be disregarded and the test repeated from the beginning of the analysis chain.

Quantification Controls Used in Real-Time PCR

Real-time PCR offers the possibility to enumerate bacteria and other microorganisms. However, inclusion of a standard curve is important for the enumeration purpose. In real-time PCR, the quantification is not based on the end point signal but rather on the exponential increase of the initial DNA amount in regard to the performed PCR cycles (Mackay, 2004). By using this characteristic, the initial amount of DNA in a PCR can be quantified relative to another parameter or as absolute numbers.

Relative Quantification

Relative quantification is often used when studying expression of specific genes and when we are only interested in up- or down-regulation of the gene(s) of interest. The same DNA sample is analyzed for both the target gene and the reference gene, for which the C_T values are compared to obtain a relative expression of the target gene(s). The amplification efficiencies of the target and reference genes are also taken into account when calculating the relative expression. A more detailed description of relative quantification is beyond the scope of this chapter and has been published elsewhere (Bustin, 2002).

Absolute Quantification

In absolute quantification, the C_T values of known amounts of target are compared directly to the C_T values obtained for the unknown samples. Absolute quantification needs a standard curve design with known amounts of the target. This is the type of quantification that is most common when analyzing pathogens in food on a routine basis. Generally, it is possible to define DNA genome equivalents based on the measurement of DNA spectrophotometrically or fluorometrically or to define CFU or cell equivalents based on plating and cell counts. The selection of DNA or cell equivalents must be chosen with care to avoid variation in the interpretation of results (for a review, see Malorny et al., 2008).

The correlation between the number of DNA equivalents and the number of CFU depends to a large extent on the matrix and the target organism. Previous reports are contradictory; some report that the numbers of target organisms are overestimated when using PCR in comparison with culture-based methods (Löfström et al., 2010; Reichert-Schwillinsky et al., 2009); some find an underestimation (Pennacchia et al., 2009); or no significant difference is found between the methods (Martín et al., 2006). Overestimating target cell numbers has been explained by the presence of extracellular target DNA or nonculturable or dead target organisms.

According to the draft ISO standard (ISO/DIS 22119), any stable DNA or RNA stock can be used to prepare a standard curve (International Organization for Standardization, 2006c), using serial dilutions of the stock solution. Commonly used standards include plasmid DNA and in vitro-transcribed RNA. It is important that the target concentration of the unknown samples fall within the linear range of the standard curve and that an appropriate number of calibration points and replicates for the quantification range be included. It is recommended to include at least four calibration points with two replicates or six calibration points with one measurement at each point (International Organization for Standardization, 2006c).

Preparation of the Standard Curve

The preparation of the standard curve should take into consideration the loss of nucleic acids due to sample preparation. That means samples for the standard curve should be processed in the same way as the test samples. Furthermore, it is important to take into account the DNA recovery rate of a pre-PCR sample treatment. Recovery rates of the cells could be different for each food item and should therefore be determined for the food of interest. One approach is to artificially contaminate the food in replicates with 10-fold serially diluted cell suspensions ranging from approximately 10^7 per g food to 1 cell per g food (Malorny et al., 2008) or within the range that is relevant for the specific target organism in the specific food item (International Organization for Standardization, 2007).

Thereafter, the samples are treated by using the same sample preparation and PCR method as for the test samples. The recovery of all the steps and the amplification efficiency of the target in a specific matrix are then taken into account when the standard curve is generated. The number of CFU in the cell suspensions is determined by plating on an appropriate agar, and CFU is used as the unit for the standard curve.

ANTIBODY-BASED METHODS

There are many examples of antibody-based methods for detection and quantification of foodborne pathogens. These include, for example, ELISA, enzyme-linked fluorescence assay, lateral flow devices, and the more recently developed antibody-based biosensors and lab-on-a-chip systems. In this chapter, ELISA is used as an example of antibody-based methods. For all these systems, different types of controls are needed, but they all have in common the need to include positive and negative controls and process controls as well as quantification standards. It has been recommended to include the assay blank, the sample blank, and the positive standards in each run and that the unknown samples should be run in duplicates in each setup (International Organization for Standardization, 2004).

Positive and Negative Controls

In ELISA and some other antibody-based analysis methods, there are two types of negative controls used: an assay blank and a sample blank consisting of the same type as the one being analyzed. The assay blank is used to check that no cross-contamination has taken place during the analysis and to correct for the background noise. The average absorbance values of the assay blanks are subtracted from the average absorbance values of the unknown samples in order to obtain a normalized value that can be compared between different runs (International Organization for Standardization, 2004). This approach is particularly valuable if no standard curve is included and when a particular cutoff value is used to determine if a sample is positive or negative for the selected analyte.

One example of this approach is the serological determination of *Salmonella* in pigs or cattle (Nielsen et al., 1995). It is recommended to include several negative controls in every setup, including agreed acceptance criteria, which are determined either on the basis of a specific value on the absorbance threshold (International Organization for Standardization, 2004) or as a percentage of a positive control in standard sample (Anonymous, 2008). Furthermore, the samples and controls should be loaded onto the ELISA plate in a randomized manner to avoid position effects in the plate (International Organization for Standardization, 2004).

At least one positive control is included to check for false-negative results. The positive control can consist of the target molecule, e.g., a toxin, or the whole organism, e.g., *Salmonella* cells. In some standardized assays, a strong positive control is used together with the weak positive control (Anonymous, 2008). The positive controls often consist of standardized samples that can be obtained from reference laboratories or other sources. Standardized positive controls facilitate intralaboratory comparisons of results. This could help, for instance, when a sample needs to be declared free of a specific agent. Acceptance criteria often applied to positive controls include that the absorbance should be above a specified absorbance threshold (Anonymous, 2008; International Organization for Standardization, 2004) and that replicates should not have a too-high standard variation (International Organization for Standardization, 2004).

Process Controls

Also for ELISA and other antibody-based analysis methods, process controls should be included to make sure that the target is extracted from the sample matrix in a reproducible way. Similar to DNA methods, the target can be added at the start of the analysis chain to a negative sample of the same type(s) that is included for the unknown samples and/or using an artificially contaminated water sample. It can be recommended to include process controls consisting of the same matrices that are to be analyzed, because the recovery of the target can vary consistently between different matrices.

For ELISA, it is also important to include at regular intervals controls of the ELISA reagents, including conjugate, substrate, and chromogen. This is done to ensure that the absorbance change is specific to the target and that the detection system is working properly.

Controls for Quantitative Antibody-Based Methods

ELISAs can also be used to generate quantitative data. Reference samples with known amounts of the target organism are analyzed together with the unknown samples to generate a standard curve. The values from the assay blank (negative control) should be subtracted from all values for diluted test sample solutions and reference standards. The standards are run at least in duplicate, and the average values could be used to generate a standard curve. The number of target organisms/molecules can then be obtained by interpolation of the concentrations from this standard curve. For some assays, limitations have been set on when results are acceptable. For example, in the EN ISO 21572:2004 standard for an ELISA method, it is recommended that the confidence variation of the optical density values obtained for any of the reference standards should be $\leq$15% and that the 2.5% reference standard should have an optical density value (at 450 nm) of $\geq$0.8 (International Organization for Standardization,

2004). These values are specific for each assay and should be carefully selected and validated.

INTERPRETATION OF CONTROLS

Depending on the results of controls, the interpretation of the results of molecular tests varies (Tables 2 and 3). It is important to include in the quality assurance program of the laboratory what actions should be taken when controls are not as expected. Depending on the outcome of controls, specific problems with the testing steps can be identified and corrective actions implemented. For instance, if the positive PCR control is positive, despite all other controls (including IAC) being negative, it means a possible inhibition of the PCR has occurred, and this prompts for a reexamination of the sample, starting from the DNA extraction.

Another possibility in this case is to dilute the extracted DNA and perform the PCR assay again. If the sample is still positive, it can be judged as positive. However, a negative PCR result should be confirmed by reextracting DNA from the original sample. On the other hand, if all controls are positive, it can indicate a possible contamination at some step along the testing. Negative samples can still be judged as negative if the internal control is positive for that particular sample, but positive

TABLE 2 Evaluation of results from diagnostic PCR depending on PCR result and results of controls[a,b]

Test sample	Positive PCR control	Negative PCR control	Positive process control	Negative process control	IAC	Interpretation of results	Actions taken
+	+	−	+	−	+/−	Positive	Analysis is working properly, no actions needed
	+	+	+	+	+/−	Inconclusive	Possibility of contamination. Rerun samples with new master mix, and/or make new DNA extraction.
	−	−	+/−	−	+/−	Positive	Make new positive control for the next analysis run.
−	+	−	+	−	+	Negative	Analysis is working properly, no actions needed.
	+	−	+	−	−	Inconclusive	Possibility of PCR inhibition, rerun samples with new master mix.
	+	−	−	−	−	Inconclusive	Possibility of failure of process/extraction. Rerun extraction of samples and PCR.
	−	−	−	−	−	Inconclusive	Possibility of failure of PCR. Make new PCR master mix and rerun samples.

[a] Modified from International Organisation for Standardization, 2005.
[b] +, PCR product detectable; −, PCR product not detectable.

TABLE 3 Evaluation of results from diagnostic ELISA depending on ELISA result and results of controls[a]

Test sample	Positive control	Negative control/ blank	Positive process control	Negative process control	Conjugate/ substrate/ chromogen control	Interpretation of results	Actions taken
+	+	−	+	−	−	Positive	Analysis is working properly, no actions needed.
	+	+	+	+	−	Inconclusive	Possibility of contamination. Make new sample preparation and rerun samples.
	+	−	+	+	+	Inconclusive	Possibility of contamination. Rerun samples with new ELISA reagents.
	−	−	+/−	−	+/−	Positive	Make new positive control for the next analysis run.
	+	−	+/−	−	+	Inconclusive	Analysis not working properly. Make new reagents and rerun samples.
−	+	−	+	−	−	Negative	Analysis is working properly, no actions needed
	+	−	−	−	−	Inconclusive	Possibility of failure of process/extraction. Rerun preparation of samples and ELISA.
	−	−	−	−	−	Inconclusive	Possibility of failure of ELISA. Make new ELISA reagents and rerun samples.

[a] +, ELISA product detectable; −, ELISA product not detectable.

samples need to be reanalyzed starting from the DNA extraction step.

CONCLUSIONS AND FUTURE DIRECTIONS

Despite the strength of rapid methods, there are some limitations with these technologies that need to be controlled in order to obtain reliable results. First, due to the highly sensitive detection limit of PCR especially, there is a risk of false-positive results. This issue can be resolved by addition of negative and process controls to check the entire process at all testing steps. Second, there might be problems with amplification and detection steps caused, for instance, by the fact that the capacity of PCR can be altered due to the presence of PCR-interfering substances common in biological samples, a malfunction of the PCR thermocycler, and/or improper reagents. All these possible negative factors influencing PCR can be controlled by the addition of an amplification control that acts as an independent control of each separate sample. However, analytical controls alone cannot ensure proper results when performing molecular methods for routine applications. They need to be supported with proper quality assurance and quality control programs, including proper handling of reagents and adequate laboratory procedures.

Novel methods for detection and quantification of foodborne pathogens are constantly being developed, and among the new techniques are biosensors and lab-on-a-chip methods (for a review, see Velusamy et al., 2010). Promising results have been gained by using these approaches for pathogen detection. However, proper analytical controls are crucial also in sophisticated methods, to ensure that the methods are working properly and that reliable results are obtained. For these techniques, positive and negative controls should be included, similar to the ones described for nucleic acid-based and antibody-based methods in the previous sections. Also, method-specific controls are needed, depending on the analysis technique employed.

ACKNOWLEDGMENT

The authors were in part financially supported by the European Union project BIOTRACER (FOOD-2006-CT-036272).

REFERENCES

Anonymous. 2008. Enzootic bovine leukosis, p. 733–736. *In OIE Terrestrial Manual 2008.* World Organisation for Animal Health, Paris, France.

Bustin, S. A. 2002. Quantification of mRNA using real-time reverse transcription PCR (RT-PCR): trends and problems. *J. Mol. Endocrinol.* **29:**23–39.

Hoorfar, J., N. Cook, B. Malorny, M. Wagner, D. De Medici, A. Abdulmawjood, and P. Fach. 2004a. Diagnostic PCR: making internal amplification control mandatory. *Lett. Appl. Microbiol.* **38:**79–80.

Hoorfar, J., B. Malorny, A. Abdulmawjood, N. Cook, M. Wagner, and P. Fach. 2004b. Practical considerations in design of internal amplification controls for diagnostic PCR assays. *J. Clin. Microbiol.* **42:**1863–1868.

International Organization for Standardization. 2004. *EN ISO 21572:2004. Foodstuffs—Methods for the Detection of Genetically Modified Organisms and Derived Products—Protein Based Methods.* International Organization for Standardization, Geneva, Switzerland.

International Organization for Standardization. 2005. *ISO 22174:2005. Microbiology of Food and Animal Feeding Stuffs—Polymerase Chain Reaction (PCR) for the Detection of Food-Borne Pathogens—General Requirements and Definitions.* International Organization for Standardization, Geneva, Switzerland.

International Organization for Standardization. 2006a. *ISO/DIS 20837:2006. Microbiology of Food and Animal Feeding Stuffs—Polymerase Chain Reaction (PCR) for the Detection of Food Borne Pathogens—Requirements for Sample Preparation for Qualitative Detection.* International Organization for Standardization, Geneva, Switzerland.

International Organization for Standardization. 2006b. *ISO/DIS 20838:2006. Microbiology of Food and Animal Feeding Stuffs—Polymerase Chain Reaction (PCR) for the Detection of Food Borne Pathogens—Requirements for Amplification and Detection for Qualitative Methods.* International Organization for Standardization, Geneva, Switzerland.

International Organization for Standardization. 2006c. *ISO/DIS 22119:2007. Draft International Standard. Microbiology of Food and Animal Feeding Stuffs—Real-Time Polymerase Chain Reaction (PCR) for the Detection of Food-Borne Pathogens—General Requirements and Definitions.* International Organization for Standardization, Geneva, Switzerland.

International Organization for Standardization. 2007. *ISO/DIS 22118. Draft International Standard. Microbiology of Food and Animal Feeding Stuffs—Polymerase Chain Reaction (PCR) for the Detection of Food-Borne Pathogens—Performance Characteristics of Molecular Detection Methods.* International Organization for Standardization, Geneva, Switzerland.

Josefsen, M. H., M. Krause, F. Hansen, and J. Hoorfar. 2007. Optimization of a 12-hour TaqMan PCR-based method for detection of *Salmonella* bacteria in meat. *Appl. Environ. Microbiol.* **73:**3040–3048.

Löfström, C., J. Schelin, B. Norling, H. Vigre, J. Hoorfar, and P. Rådström. 2010. Culture-independent quantification of *Salmonella enterica* in carcass gauze swabs by flotation prior to real-time PCR. *Int. J. Food Microbiol.* Epub ahead of print. doi:10.1016.j.ijfoodmicro.2010.03.042.

Long, F., X. N. Zhu, Z. M. Zhang, and X. M. Shi. 2008. Development of a quantitative polymerase chain reaction method using a live bacterium as internal control for the detection of *Listeria monocytogenes. Diagn. Microbiol Infect. Dis.* **62:**374–381.

Lund, M., and M. Madsen. 2006. Strategies for the inclusion of an internal amplification control in conventional and real time PCR detection of *Campylobacter* spp. in chicken fecal samples. *Mol. Cell. Probes* **20:**92–99.

Lund, M., S. Nordentoft, K. Pedersen, and M. Madsen. 2004. Detection of *Campylobacter* spp. in chicken fecal samples by real-time PCR. *J. Clin. Microbiol.* **42:**5125–5132.

Maaroufi, Y., J. M. de Bruyne, V. Duchateau, R. Scheen, and F. Crokaert. 2006. Development of a multiple internal control for clinical diagnostic real-time amplification assays. *FEMS Immunol. Med. Microbiol.* **48:**183–191.

Mackay, I. M. 2004. Real-time PCR in the microbiology laboratory. *Clin. Microbiol. Infect.* **10:**190–212.

Malorny, B., C. Löfström, M. Wagner, N. Krämer, and J. Hoorfar. 2008. Enumeration of *Salmonella* bacteria in food and feed samples by real-time PCR for quantitative microbial risk assessment. *Appl. Environ. Microbiol.* **74:**1299–1304.

Malorny, B., E. Paccassoni, P. Fach, C. Bunge, A. Martin, and R. Helmuth. 2004. Diagnostic real-time PCR for detection of *Salmonella* in food. *Appl. Environ. Microbiol.* **70:**7046–7052.

Martín, B., A. Jofré, M. Garriga, M. Pla, and T. Aymerich. 2006. Rapid quantitative detection of *Lactobacillus sakei* in meat and fermented sausages by real-time PCR. *Appl. Environ. Microbiol.* **72:**6040–6048.

Murphy, N. M., J. McLauchlin, C. Ohai, and K. A. Grant. 2007. Construction and evaluation of a microbiological positive process internal control for PCR-based examination of food samples for *Listeria monocytogenes* and *Salmonella enterica*. *Int. J. Food Microbiol.* **120:**110–119.

Nielsen, B., D. Baggesen, F. Bager, J. Haugegaard, and P. Lind. 1995. The serological response to *Salmonella* serovars *typhimurium* and *infantis* in experimentally infected pigs. The time course followed with an indirect anti-LPS ELISA and bacteriological examinations. *Vet. Microbiol.* **47:** 205–218.

Paterson, R. R. 2007. Internal amplification controls have not been employed in fungal PCR hence potential false negative results. *J. Appl. Microbiol.* **102:** 1–10.

Pennacchia, C., D. Ercolini, and F. Villani. 2009. Development of a real-time PCR assay for the specific detection of *Brochothrix thermosphacta* in fresh and spoiled raw meat. *Int. J. Food Microbiol.* **134:**230–236.

Rådström, P., C. Löfström, M. Lövenklev, R. Knutsson, and P. Wolffs. 2003. Optimisation strategies for overcoming PCR inhibition. *In* C. W. Dieffenbach and G. S. Dveksler (ed.), *PCR Primer: a Laboratory Manual.* Cold Spring Harbor Laboratory Press, Cold Spring Harbor, NY.

Reichert-Schwillinsky, F., C. Pin, M. Dzieciol, M. Wagner, and I. Hein. 2009. Stress- and growth rate-related differences between plate count and real-time PCR data during growth of *Listeria monocytogenes*. *Appl. Environ. Microbiol.* **75:**2132–2138.

Rodríguez-Lázaro, D., M. Pla, M. Scortti, H. J. Monzó, and J. A. Vázquez-Boland. 2005. A novel real-time PCR for *Listeria monocytogenes* that monitors analytical performance via an internal amplification control. *Appl. Environ. Microbiol.* **71:** 9008–9012.

Sohni, Y., S. Kanjilal, and V. Kapur. 2008. Cloning and development of synthetic internal amplification control for *Bacillus anthracis* real-time polymerase chain reaction assays. *Diagn. Microbiol. Infect. Dis.* **61:**471–475.

Tang, Y., Q. Wang, and Y. M. Saif. 2005. Development of a ssRNA internal control template reagent for a multiplex RT-PCR to detect turkey astroviruses. *J. Virol. Methods* **126:**81–86.

Velusamy, V., K. Arshak, O. Korostynska, K. Oliwa, and C. Adley. 2010. An overview of foodborne pathogen detection: in the perspective of biosensors. *Biotechnol. Adv.* **28:**232–254.

INTERNATIONAL VALIDATION, RING TRIAL, AND STANDARDIZATION OF RAPID METHODS

S. Qvist

11

The need for the food industry to rapidly assess the microbiological quality of raw materials and finished products and the microbiological status of manufacturing procedures has led to the development and refinement of alternative methods that are faster and easier to perform than the corresponding culture-based (hereafter called reference) methods.

In order to ensure the reliability of the proprietary methods, the food and drink industry, accredited laboratories, public health services, and other authorities demand agreed validation protocols. The validation data generated can constitute the basis for certification by an independent organization or company, which declares the method performance equivalent to official reference methods.

During the last decade, different validation organizations have provided detailed protocols and guidelines for the validation of alternative microbiological methods. In order to perform such validations, expertise on statistical design, sampling schemes, and data analysis is required (see chapter 12, this volume).

S. Qvist, NordVal, c/o National Veterinary Institute, P.O. Box 750, Sentrum N-0106, Oslo, Norway.

In the following, a general outlook of the validation procedures and the implementation of their phases are outlined. Detailed information on the various validation organizations and their validation protocols are beyond the scope of this article. Interested parties are advised to consult the references given at the end of this chapter.

VALIDATION PROCESS

The validation of qualitative and quantitative methods comprises two phases. Phase A, "Comparative study by an expert laboratory of the alternative method against a reference method," is sometimes referred to as in-house validation by an expert laboratory. Phase B, "Collaborative study of the alternative method," is carried out in a ring trial organized by the same expert laboratory from Phase A.

The test characteristics for alternative methods are shown in Table 1. The comparative and collaborative studies and calculations and interpretations of obtained data are core elements of any validation program (Anonymous, 2007, 2008). Detailed descriptions of these procedures are beyond the scope of this paper. Information can be studied in the appropriate validation protocols issued by the

Rapid Detection, Identification, and Quantification of Foodborne Pathogens, ed. by J. Hoorfar

TABLE 1 Test characteristics that are validated for alternative (rapid) methods

Qualitative (end point or yes/no) methods	Quantitative (enumeration) methods
1. Selectivity (inclusivity/exclusivity)	1. Selectivity (inclusivity/exclusivity)
2. Relative accuracy	2. Lowest validated detection level with satisfactory precision
3. Detection limit	3. Repeatability
4. Relative sensitivity	4. Reproducibility
5. Relative specificity	5. Method uncertainty
6. Agreement between the methods (κ)	

different standardization and validation organizations.

Alternative methods are always compared to reference culture methods. Most commonly used reference methods are methods published by official recognized standard organizations: International Organization for Standardization/European Commission for Standardization (ISO/CEN) methods, FDA/*Bacteriological Analytical Manual* (FDA/BAM) methods, and Nordic Committee for Food Analysis (NMKL) methods.

VALIDATION ORGANIZATIONS

As of 2010, the following international validation organizations are in function: AFNOR (Association Française de Normalisation), AOAC (Association of Analytical Communities), MicroVal, and NordVal. Individual validation protocols have been in existence for each organization, but during the last decade considerable harmonization work has been carried out. Special efforts in harmonizing the American and European systems are being exerted. This work has been promoted and facilitated by publishing EN ISO Standard 16140 (Anonymous, 2003). However, since this standard has shown to have several drawbacks, the decision has been made to put it through a revision for improvements.

During the period of revision, AFNOR, MicroVal, and NordVal are using the ISO 16140 protocol as the basis for their validations, including corrections and modifications considered necessary in order to achieve satisfactory results according to these organizations' interpretations. For NordVal, the differences are specified in a NordVal validation protocol (Anonymous, 2009).

In the United States, a validation protocol having many similarities with ISO 16140 is in use by AOAC International, which is responsible for the AOAC *Official Methods* Program. Interestingly, the AOAC subsidiary, the AOAC Research Institute, is managing another validation procedure, referred to as the AOAC Performance Tested Methods Program (AOAC International, 1989; AOAC Research Institute, 1989). This program provides an independent third-party review of proprietary methods, diagnostic kits, and commercial reagents and tests in order to confirm the manufacturers' performance claims. This program does not include a collaborative study, and therefore, methods approved according to this program cannot be considered fully validated.

SELECTION OF SAMPLE CATEGORIES

The selection of food categories and types used within the validation will depend on the type or group of microorganisms and the scope of the validation. Validation can be carried out for a restricted number of food categories, e.g., meat products and milk and dairy products. Then, only these categories require to be studied.

In addition to food categories, feed and environmental samples and samples from feces and primary production can be included.

When selecting samples for the study, it is important to find those that are naturally contaminated. If it is not possible to acquire a sufficient number of naturally contaminated samples, artificial contamination of samples is permissible. It is desirable that food samples come from as wide a distribution as possible in order to minimize any bias from local food specialties and in order to broaden the validation spectrum.

It must be ensured that with the selection of the different food types both high and low amounts of background microflora and different types of stresses due to processing are considered. In addition, raw (unprocessed) food items must be included. The international validation protocol EN ISO 16140:2003 shows an example illustrating the validation of a method for detection of *Listeria monocytogenes* (Anonymous, 2003).

For the food category "milk products," the following submatrices are selected: (i) raw milk as high background flora, unstressed; (ii) milk powder as low background flora, under drying stress; and (iii) pasteurized milk as low background flora, under heat stress.

For the food category "meat products," the following food types are selected: (i) cooked ham as lower background flora, under heat stress; (ii) fermented salami as high background flora, under pH and water activity stress; and (iii) raw cured pork filet, lower background flora, under water activity stress.

TEST SAMPLES

The contamination levels of both naturally and artificially contaminated samples should be estimated. Some naturally contaminated samples may contain a high number of the target analyte. In such cases, a naturally contaminated sample can be diluted with uncontaminated material of the same food item to achieve a lower level of contamination. The reference and proprietary methods shall be performed with, as far as possible, exactly the same sample type.

In general, the data shall be presented in order to give an overview of the raw data obtained, and information shall be given on the type of contamination (naturally contaminated or artificially contaminated) of the samples used. In case naturally contaminated samples are used, the level of contamination shall be determined. This information is of importance for optimal interpretation of the results. For artificially contaminated samples, the inoculation procedure shall be specified.

REVISION OF EN ISO 1640:2003

A comprehensive discussion of the revision items is beyond the scope of this article. However, a short presentation of some of the differences under discussion seems relevant in order to explain special requirements and anticipations for improvements of the document. The following aspects represent the author's point of view, seen from the chair of NordVal.

Relative inclusivity is defined as the ability to detect the target microorganism from a wide range of strains. In the NordVal protocol, 50 pure cultures of target organisms relevant to the alternative method and the food matrices should be selected for the study of inclusivity. This is the same for the ISO protocol except for *Salmonella*, for which only 30 pure cultures are selected. NordVal finds this exception for *Salmonella* in the ISO document inappropriate, since more than 2,500 *Salmonella* strains, belonging to different somatic groups, are known. There seems now to be a general agreement that the number of *Salmonella* strains tested should be at least 50 and preferably be increased to 100 strains. In addition, to cover the epidemiologically prevalent *Salmonella* serotypes, it is recommended to include a minimum number of serotypes that represent all serogroups (Malorny et al., 2003).

As far as relative accuracy, sensitivity, and specificity are concerned, NordVal has in its protocol tables for calculations. This is important not only for screening results (as is the case with the ISO standard) but also for confirmation. The tables for screening results are used for calculation of positive and negative deviations. However, in order to demonstrate the true performance of the alternative methods, it is necessary to divide the values for positive deviation into values for true positives and false positives (Löfström et al., 2009). NordVal finds that confirmation tables are of great value for such calculations, although the necessity of confirmation can be explained in the text.

Furthermore, the ISO standard does not specify any acceptance criteria for results obtained in the validation. NordVal has established an acceptance criterion of 95% for sensitivity. This criterion indicates how many false negatives could be accepted by using an alternative microbiological method. In addition, NordVal has acceptance criteria for the degree of agreement quantified by the kappa value (κ). Kappa values above 0.80 are interpreted as "very good agreement."

Regarding food matrices to be tested, many similarities exist between the ISO and the NordVal protocols. However, the NordVal protocol in addition includes animal feeds in its matrix lists. In the Nordic countries, feed is considered an important matrix in programs covering the farm-to-table control policy. Hence, special requirements are introduced for analytical methods intended for feed analyses.

As far as the collaborative studies are concerned, the number of samples to be tested is under discussion. In the past, problems have been encountered with sample shipment, sample homogeneity and stability, and the lack of sufficient qualified laboratories with the necessary statistical knowledge.

NordVal has questioned the value of using extended collaborative studies with numerous results by using only one matrix such as pasteurized milk. The outcome of the revision work on this issue is awaited with great interest.

EU LEGISLATION

In recent years, the impact of European Union (EU) legislations on the acceptance of alternative methods for food control has considerably increased in Europe. Despite the fact that the availability of rapid analytical results is of paramount importance for the investigation of foodborne outbreaks and for the development in fresh meat trade in the internal market, the EU system has been rather slow to accept the use of alternative microbiological methods for food control.

The interest for the use of alternative methods in the EU came on the agenda in relation to the *Salmonella* guarantees requested by Finland and Sweden when these countries entered the EU in 1995. In the EU legislation (Anonymous, 2005a, 2005b), only classical methods were allowed for *Salmonella* control of meat at that time. Since these methods take 4 to 5 days for a negative result, meat could be exported to Finland and Sweden only if it was either frozen or vacuum-packed.

Therefore, meat exporters, kit manufacturers, and authorities in Denmark requested the Commission to accept the use of alternative methods that could release meat for export within 24 hours. In this process, NordVal played an active role as adviser—highly recognized by meat exporters and kit producers.

The EU Commission responded by seeking advice from the Scientific Committee on Veterinary Measures relating to Public Health (SCVPH). In June 2002, the SCVPH issued an opinion on criteria for evaluation of methods for *Salmonella* detection; and consequently, in June 2003 the EU Commission issued a decision on authorization of certain alternative methods to be used in microbiological testing of meat intended for Finland and Sweden.

The requirements for approval of the alternative methods for *Salmonella* control of meat for Finland and Sweden are still based on advice from the SCVPH but are more specific in the Commission Regulation (EC) No 2073/2005 on microbiological criteria for foodstuffs, which is an implementing rule of the recent hygiene legislation, entered into force in January 2006. It is here stated that alternative methods to be authorized for special guarantees for *Salmonella* control must have been validated against the most recent editions of the ISO 6579 or NMKL 71 and, if a proprietary method, certified by a third party in accordance with the protocol set out in Standard EN ISO 16140 or other international similar accepted protocols. AFNOR, AOAC-Official Methods of Analysis, MicroVal, and NordVal are considered to be using such similar international protocols, even though they are not mentioned in the EU regulation.

CONCLUSION

Producers and users of alternative methods appreciate the ongoing harmonization of validation protocols, but disagreement exists regarding centralizing all validations within one organization, because of the fear of monopoly followed by increased costs.

An interesting aspect is the increasing understanding that the performance of the alternative methods, due to higher sensitivity, can produce more reliable results than the reference method. However, conservatism by lawmakers is generally considerable, claiming that the results obtained by the reference method (the gold standard) are the true results and that only these results can be used by regulators in exercising food control.

This position might change in future, so that alternative methods will be accepted not only to be used for official food control but also as official standard methods. In fact, this is already the case for a commercial *Salmonella* PCR method.

REFERENCES

Anonymous. 2003. *ISO 16140. Microbiology of Food and Animal Feeding Stuffs. Protocol for the Validation of Alternative Methods, ISO TC 34/SC.* International Organization for Standardization, Geneva, Switzerland.

Anonymous. 2005a. Commission Regulation (EC) No 1688/2005 of 14th October 2005 implementing Regulation (EC) No 853/2004 of the European Parliament and of the Council as regards special guarantees concerning salmonella for consignments to Finland and Sweden of certain meat and eggs. *Official Journal of the European Union* L **271:**17–19.

Anonymous. 2005b. Commission Regulation (EC) No 2073/2005 of 15th November 2005 on microbiological criteria for foodstuffs. Amended by Commission Regulation (EC) No 1441/2007 of 5th December 2007. *Official Journal of the European Union* L **338:**1–29.

Anonymous. 2007. *NMKL Procedure No 20: Evaluation of Results from Qualitative Methods.* NMKL, Oslo, Norway.

Anonymous. 2008. *NMKL Procedure No 8: Measurements of Uncertainty in Quantitative Microbiological Examination of Foods.* NMKL, Oslo, Norway.

Anonymous. 2009. *Protocol for the Validation of Alternative Microbiological Methods.* NordVal, Oslo, Norway.

AOAC International. 1989. *Handbook for AOAC Members,* 6th ed. AOAC International, Gaithersburg, MD.

AOAC Research Institute. 1989. AOAC *Performance Tested Methods. Program.* AOAC International, Gaithersburg, MD.

Löfström, C., M. Krause, M. H. Josefsen, F. Hansen, and J. Hoorfar. 2009. Validation of a same-day real-time PCR method for screening of meat and carcass swabs for Salmonella. *BMC Microbiol.* **9:**85.

Malorny, B., J. Hoorfar, C. Bunge, and R. Helmuth. 2003. Multi-center validation of the analytical accuracy of Salmonella-PCR: towards an international standard. *J. Appl. Environ. Microbiol.* **69:**290–296.

STATISTICAL DATA ANALYSIS OF RESULTS BASED ON ALTERNATIVE DETECTION AND ENUMERATION METHODS

M. Greiner, H. Vigre, and I. Gardner

12

Microbiological data on foodborne pathogens used in research for risk assessment or risk mitigation should be accurate and reliable to the specified level of the underlying measurement performance characterization. The performance requirements are purpose specific. For example, the monitoring of prevalence trends and comparison among countries should be based on well-standardized methods with high or at least known diagnostic sensitivity and specificity to allow correct inferences. On the other hand, quantitative microbiological risk assessments require quantitative data on the occurrence and concentration, e.g., the number of CFU per gram, of foodborne pathogens, whereby variability (e.g., differences among batches and sampling sites and over time) and uncertainty (e.g., standard errors of population parameters) are accounted for. The measurement uncertainty in the enumeration of foodborne pathogens has been reviewed by Corry et al. (2007).

Culture-based and molecular microbiological methods may be used for both detection (classification of sample status) and enumeration. Application areas for detection include the determination of the proportion of positive samples in the context of hygiene process control, monitoring and control programs, outbreak investigations, and many others. Application areas for enumeration, i.e., quantification of CFU per weight or volume unit of the sample, may be relevant for the same areas as for detection. Furthermore, enumeration is usually required to assess the efficiency of intervention measures, for estimation of the growth characteristics, or for method comparison. Some microbiological methods provide only qualitative results (presence/absence of the target analyte) and thus are only applicable for detection.

However, when used on a dilution series of a positive sample, a semiquantitative measure of the concentration may be obtained. On the other hand, enumeration methods provide quantitative results, i.e., estimates of the number of CFU per sample unit, and thereby support both detection and enumeration at the same time. The latter data type is especially useful for risk assessors. Detection and enu-

M. Greiner, Federal Institute for Risk Assessment, Scientific Services, Epidemiology, Biostatistics and Mathematical Modelling, D-14195 Berlin, Germany. *H. Vigre*, Department of Microbiology and Risk Assessment, National Food Institute, Technical University of Denmark, DK-2860 Søborg, Denmark. *I. Gardner*, Department of Medicine and Epidemiology, School of Veterinary Medicine, University of California, Davis, CA 95616.

Rapid Detection, Identification, and Quantification of Foodborne Pathogens, ed. by J. Hoorfar

meration methods should be developed, validated, applied, and interpreted based on consideration of the necessary taxonomic level of the pathogen and the relevant features of the sample such as unit of sampling, food type, matrix, point of sampling along the food chain, and environmental conditions.

Alternative methods for detection and enumeration cover a wide range of technical approaches, some of which aim at improved performance and operational characteristics such as the limit of detection (LOD), inclusivity, exclusivity, diagnostic sensitivity, diagnostic specificity, detection range, sample throughput, standardization, or automation, while others provide additional information or allow detection and enumeration at higher degrees of taxonomical or phylogenetic resolution.

The present chapter reviews standard statistical approaches to performance characterization for detection and enumeration with emphasis on validity aspects. The performance characterization with regard to precision (interlaboratory reproducibility) is briefly described. Relevant for both classical and alternative methods for detection and enumeration are issues related to the LOD, sometimes referred to as analytical sensitivity. Data generated by using methods with an LOD include nondetects, i.e., false-negative results, and are referred to as censored data, because only values that are greater than the LOD can be reliably observed. Ignoring the censoring problem and also the substitution of nondetects with an arbitrary value selected from a range between zero and the LOD leads to biases.

We describe procedures for establishing an LOD as well as statistical approaches for dealing with censored data for enumeration (qualitative) methods. Nowadays, newer statistical methodologies are being used to classify samples as contaminated or not and allow more refined identification of the microbial contamination status based on multiple diagnostic indicators. Although not the focus of this review, these novel procedures and applications will be briefly covered. Finally, we review specialized statistical approaches in the context of alternative detection and enumeration methods. Guidelines for reporting standards are available for diagnostic test evaluation and for many alternative methods (see http://www.mibbi.org/index.php/Main_Page).

INTERNATIONAL STANDARDS

The performance characteristics related to the validity of a diagnostic method are derived from the deviation of the results from the true status of the sample. Valid qualitative and quantitative diagnostic assays should theoretically yield results in accordance with true presence (qualitative method) or number (quantitative) of microbes in a sample or subsample of a specified food matrix. According to the International Organization for Standardization (2003) (ISO 16140), validation of an alternative method requires demonstration of adequate confidence that the results obtained by the alternative method are comparable to the true presence or number of microorganisms in the sample. The true presence or number of microbes in a sample can be determined by different approaches. The use of artificially contaminated samples or certified reference material makes it possible to evaluate the alternative method, not relying on other diagnostic methods.

When the presence or number of microorganisms in a sample is unknown (which typically is the case when working with real samples), the true status of the sample is not known exactly and can only be estimated. This estimation can be done by using a reference method, under the assumption that the result of the reference method is equivalent to the true result. Recently, statistical methods have been developed that can estimate measures of validity without known presence or numbers of microorganisms in the sample.

The estimation of the measurement uncertainty of a microbial test result is affected by a number of method-related criteria that must be considered in the validation of a test (Corry et al., 2007): namely, the sensitivity and the specificity of the method, the recovery, and the LOD.

The ISO standard is continuously proposing the general principles for deriving measures for the validation of alternative methods in the field of microbiological analysis of food. In this section, we describe the most important statistical measures described by ISO 16140 for validation of qualitative and quantitative methods of accuracy used for alternative methods. We also discuss the limitations of this approach related to the assumptions used to define the true status of the sample.

Accordingly, a validation protocol comprises two phases: comparison of the method against other methods and comparison of the performance of the method in different laboratories (interlaboratory reproducibility study). The measures of accuracy depend on whether the method is qualitative or quantitative.

Statistical Measures Described by the ISO for Validation of Qualitative Methods

In this context, the statistical measures that must be considered are diagnostic sensitivity (the probability that a truly positive sample containing the target analyte gives a positive test result) and diagnostic specificity (the probability that a truly negative sample without the target analyte gives a negative test result). Traditionally, diagnostic sensitivity and specificity of a qualitative test are obtained by comparison of the sample results using the diagnostic test of interest and a reference diagnostic method (which is assumed to give the true status of the sample). The sensitivity is calculated as the proportion of positive results in the reference method that also were positive in the method under validation. The specificity is calculated as the proportion of negative results in the reference method that also were negative in the method under validation.

The diagnostic sensitivity of a qualitative method depends on the LOD of the method. In general, the smaller the value of the LOD, the higher the probability that a truly positive sample gives a positive test result (sensitivity), especially when the concentration of the target analyte is low in naturally contaminated samples. Theoretically, the diagnostic sensitivity is positively related to the concentration of the analyte in the sample. Because the concentration of the analyte might change between animal populations or between food items, sensitivity and specificity should be defined with regard to the animal population from which the raw materials were sourced and the point at which the sample was taken along the food processing chain.

Many factors such as those related to the actual sampling process may lead to variability of the diagnostic sensitivity and specificity among slaughterhouses or food processing sites. A valid estimate of sensitivity and specificity is important for the design and interpretation of surveillance programs based on a new test. Underestimation of the true sensitivity of a new test may hinder the implementation of more sensitive alternative methods in surveillance programs because we do not have true knowledge of the performance of alternative methods.

Most scientific papers reporting the validation of qualitative alternative methods have compared the results of the alternative method to results obtained using bacterial cultures of naturally contaminated samples (Bohaychuk et al., 2007; Mäde et al., 2004) or artificially contaminated (laboratory-spiked) samples (Hu and Arsov, 2009; Bohaychuk et al., 2005; Mäde et al., 2004).

In many situations with naturally contaminated samples, the presence and amount of a pathogen in a sample are not known, which strictly means that there is a lack of a perfect reference method (sometimes referred to as gold standard; see Wilk, 2001, for objections against using this term) to determine the true status of the sample, which is needed to validate the diagnostic validity of the alternative method. Two approaches for the estimation of the validity of an alternative diagnostic method using naturally contaminated samples in the absence of a perfect reference method are (i) the use of a relative (imperfect) standard and (ii) the use of a latent-class model. In latent-class models, the true status is a latent (unknown) variable and no assumptions are made about the accuracy of the new test or the reference method (see "Validation in the Absence of a Perfect Reference Method," below, for details).

Agent isolation methods (e.g., bacterial cultivation and virus isolation) often are used as reference methods for obtaining the true status of naturally contaminated samples. On the assumption that the result of the agent isolation methods is the true status of the sample, the sensitivity and the specificity (and associated 95% confidence intervals [CI]) of the alternative method can be calculated. The advantage of this approach is that it is easy to perform the calculation and communicate the result even with a limited knowledge of statistical methods. The major disadvantage of this approach is that culture and isolation can be imperfect protocols. The direction and magnitude of the bias caused by use of a relative standard depend on how the samples are misclassified by the imperfect reference method (either as false positives or false negatives).

When the assumption that the test used as standard is perfect is not valid, the estimated value of sensitivity and specificity of the alternative method can never be higher than the true (but unknown) sensitivity and specificity of the standard (which is assumed to have a sensitivity and specificity equal to 100%).

The specificity of isolation methods is often assumed to be 100% (no false-positive samples), because false isolation of an agent is not possible with the possible exception of cross-contamination of samples or laboratory errors. In contrast, the sensitivity of isolation methods very often is considerably lower than 100% due to false-negative samples. When assuming that the results of the isolation method are perfect, some of the samples classified by the isolation method as negative can be truly positive. In the alternative method, there are some of these negatively classified samples that are truly positive which might be classified as positive samples, and consequently, the calculated specificity of the alternative method will be underestimated. Because the isolation methods do not produce false-positive results, the samples used for obtaining the probability that the alternative method yields a positive result given a positive sample (sensitivity) do not include truly negative samples, and thereby the risk of underestimation is eliminated. However, because some of the truly positive samples are not classified as positive by the reference method, the samples used for evaluating the sensitivity of the alternative method only represent a fraction of the truly positive samples, and thereby the sensitivity likely will be overestimated.

To avoid the assumption that the reference method gives perfect results, an advanced statistical technique has been developed. This technique, latent-class analysis (LCA), makes no assumption about the accuracy of either the alternative method or the reference culture method. LCA allows the evaluation of the accuracy of two or more qualitative methods when there is no perfect reference method and may yield valid estimates subject to inherent model assumptions (see "Validation in the Absence of a Perfect Reference Method," below, for details).

The use of artificially contaminated (spiked) samples is valuable in the initial phases of development of alternative methods because the concentration of the pathogen in the sample and subsequent dilution series is known (with allowance for error created by dilution). However, artificially contaminated samples cannot be used to evaluate the diagnostic accuracy of alternative methods. The diagnostic accuracy of a method is influenced by the concentration and distribution of the target pathogen and nontarget microorganisms (competing flora). Also, the concentration of the pathogen in naturally contaminated samples might vary more than the range of concentration of the pathogen in the artificially contaminated samples. Therefore, caution is needed when interpreting estimates of diagnostic sensitivity and specificity obtained using artificially contaminated samples because the estimated accuracy with artificially contaminated samples is typically biased upwards.

Statistical Measures for Validation of Quantitative Methods

Colony counts from a given sample are the most common reference value used for validation of quantitative alternative methods. Depending on the alternative method, relevant validation criteria are the relationship and the agreement between the alternative method and the number of pathogens in the sample. Because no reference material with a known number of pathogens is available, "known values" of pathogens in samples can be obtained either by using artificially contaminated (spiked) samples or by performing many replicated measurements using a reference method.

Depending on whether the alternative method is measuring the same quantity as the reference method (e.g., number of CFU) or another quantity than that used by the reference method (cycle threshold [C_T] values obtained in a quantitative PCR), the relationship should be measured as agreement and association, respectively.

The assessment of association between the concentration of a pathogen in a sample based on the reference method and by the alternative method can be done using linear regression. In linear regression, the association between the true number (which is assumed to be perfect) and the value measured by the alternative method is expressed mathematically by a line defined by an intercept and slope. A significant association (statistically significant slope) between the reference and the alternative method shows that when the quantity measured by the reference method increases, the quantity measured by the alternative method also increases. A situation where this is applied is at the preparation of PCR standard curves, where the association between the C_T value before a positive signal in a quantitative real-time PCR and known concentrations of pathogens in the sample are estimated by linear regression. Linear regression to estimate the relationship between C_T values and the concentration of pathogens has been applied in several studies (Yang et al., 2003; Park et al., 2008; Löfström et al., 2010).

When the alternative and reference methods are measuring the same quantity, it is interesting to observe not only the association but also the *agreement* between the methods. Very often, however, we do not know the true quantity. Bland and Altman (1986) assume that the mean of the two measures obtained by the reference and the alternative methods is the best estimate of the quantity. Based on this assumption, they present a statistical method focusing on the quantitative deviation of the reference and the alternative method according to the mean. The advantage of this method is that the underlying assumption that one of the methods is measured without error (an assumption that is implicit in linear regression) is not necessary.

Intralaboratory Repeatability and Interlaboratory Reproducibility

The variability of test results from the same laboratory (method applied in constant conditions with the same apparatus, same operator, etc.), often termed repeatability, is generally lower than the variability of test results from different laboratories (method applied in various conditions with different apparatus, different operator, etc.), often termed reproducibility.

To assess variability of the result obtained by a detection or enumeration method, repeated measurements on a series of common samples should be performed either at different times (to assess the within-laboratory variability) or at different laboratories (to assess the variability among laboratories). A study of the variability within the same laboratory often is referred to as "proficiency tests," and a study of the variability among different laboratories often is referred to as "ring trials."

For qualitative methods, the variability is expressed as the probability of finding different results (i.e., one negative and one positive) from two identical samples analyzed using the same method in the same laboratory (accordance) or in two different laboratories (concordance). Formulas for calculation are presented in the ISO 16140 (International Organization for Standardization, 2003). The values of accordance and concordance should only be interpreted in the context of variability of results—these measures do not indicate the accuracy of the method. If the concordance is smaller than the accordance, this suggests that there is more variability among laboratories. However, a certain amount of variability among laboratories can always be expected, and numerical assessment of the result is as important as a significance test for whether or not the variation between laboratories is important.

For quantitative methods, the variability of repeated results on the same sample (often termed precision) is measured either as standard deviations or coefficients of variation (standard deviation divided by the mean of the replicates). The statistical methods for analysis of studies of assay precision among laboratories are similar to those used for assessment of within-laboratory precision. However, as part of an interlaboratory study, it might be considered important to assess and rank variability in test results from multiple sources.

For example, if a study was designed to test an assay in three laboratories each using two highly trained technicians and running the samples in duplicate, each test sample would be tested 12 times. The selected factors (laboratory, technician, and replicate result) can be considered to be fixed or random, depending on how they are selected and whether they are representative of the target population. For this experimental design, variance components can be estimated and the intraclass (intracluster) correlation coefficient can be estimated as a measure of the similarity of sample results (Bartlett and Frost, 2008).

DEALING WITH CENSORED DATA

Problem Formulation

Enumeration methods can be characterized by their LOD, which is the lowest detectable number of CFU per sample unit with a specified probability (e.g., 50% or 95%). The LOD is one important performance criterion and has been studied extensively in the validation of alternative enumeration methods (Leonard et al., 2003). Censored observations (nondetects) are those that are below the LOD. Such results are often reported as zero or missing or are substituted with the value of the LOD or some other arbitrary value below the LOD. For quantitative real-time PCR, if a sample is not positive at the end of the maximum number of temperature cycles (e.g., 40), the result might be reported as either negative or >40 if a quantitative result was needed.

In this section, the term "censored data" refers to a data set in which censored observations occur. Shorten et al. (2006) and Lorimer and Kiermeier (2007) have shown that the substitution of censored observations results in biased statistical summaries of censored microbiological data. In the presence of censoring, the population mean value is often overestimated and the variance is underestimated. The degree of bias increases if the LOD is shifted towards higher concentration values. The issue of nondetects can occur also in data generated by qualitative methods. False-negative results obtained with qualitative methods are dealt with in the context of method validation through estimation of the diagnostic sensitivity. Qualitative microbiological tests are often applied on pooled samples or other complex sampling procedures. The estimation of the diagnostic sensitivity and specificity is then based on mathematical models that reflect the sampling process (Jordan, 2005; Arnold et al., 2009, 2010).

Definition of LOD

If false-positive signals are known to occur with the method under consideration, the LOD can be derived from replicated measurement obtained with known negative samples with a true concentration of zero. In the environmental sciences, where this approach is used for analysis of contaminants or toxins, the measurement error at zero concentration is often approximated by replicate measurements of a standard solution with low concentration. Helsel (2005) uses an example in which LOD = 3.14 · s, where s is the standard deviation. The value 3.14 is the one-sided t statistic for a sample size of 7 and type I error (false positive) of 1 percent ($\alpha = 0.01$). The standard deviation, s, is estimated by using seven replicate measurements at low concentration.

However, this approach is not applicable for microbiological enumeration techniques if repeated measurements of samples with true concentration of zero always yield negative test results. In this case, the LOD may be determined as the lowest possible dilution of a known concentration standard that yields reliable positive measurements (e.g., in at least 50% of replicate samples).

Storing Censored Observations in Databases

Database entries should indicate observations that are known to be censored. The storage format should allow different LOD values to be used. It is strongly advised not to use text strings such as "<LOD" to indicate a censored observation because this would change the data type of the variable to text. Different LOD values may occur for data generated in different laboratories or using different methods with different LOD values. Helsel (2005) presents an example of three measurements, two nondetects with different LODs (1 and 5) and one measurement above LOD, x_i = <1, <5, and 10 with some arbitrary unit of concentration for each element in the sample (Table 1). He recommends two alternative data formats for storing censored data, one using an indicator and another using interval endpoints, respectively. According to the indicator format, two variables are needed. The first variable denotes a concentration value, which is either the LOD value in the case of a censored observation or a measured concentration above the LOD. The second variable is an indicator identifying whether the concentration value is censored (1) or not (0).

According to the interval endpoint format, two variables refer to the left and right ends of the interval in which the observation occurs. The two endpoints are identical for noncensored observations. Depending on the software used to store the data, "NA" or another entry may be used to indicate a missing value.

The names of the two variables ("Left" and "Right") and the use of "NA" as missing value follow the convention for using the program package "fitdistrplus" (see "Observations from One Population," below). Using either of the two formats, it is easy to transform the data as needed for the software program chosen for analysis.

Statistical Approaches for Dealing with Censored Observations

One important distinction should be made when dealing with censored data. Either the data can be assumed to have been sampled

TABLE 1 Obsolete and two alternative recommended formats to store censored data in a database

Obsolete notation	Censoring	Indicator format		Interval endpoint format	
		Concn	Censored	Left	Right
<1	Yes (LOD of 1)	1	1	NA	1
<5	Yes (LOD of 5)	5	1	NA	5
10	No	10	0	10	10

from a single population with an unknown but nonzero true concentration, or the data are assumed to be from a mixed population with two subpopulations: one contaminated subpopulation with a nonzero true concentration and another free subpopulation with a true concentration of zero. "Population" in this context refers to a subset of either truly positive or truly negative elements from which a sample was selected. The distinction about which of these two different scenarios applies for a given data set should be based on microbiological expert judgment so that the appropriate statistical analysis is used.

The statistical methods mentioned below, if not stated otherwise, presuppose a normal distribution model. That means that the "true" measurement values in the absence of a LOD would follow a normal distribution. Microbial counts are usually $\log_{10}$ transformed to achieve approximate normality. We focus below on methods that are applicable to microbiological data. Reviews in the context of other application areas are found elsewhere (Helsel, 2005; European Food Safety Authority, 2010). The methods for dealing with mixed populations are described in more detail below, as this topic is not well covered in the microbiological literature.

Observations from One Population

The scenario of "one population" is mainly relevant in experimental settings, e.g., when comparing the effect of inactivation conducted on samples with known starting concentrations. For example, the effect of a new disinfectant X versus comparator disinfectant Y is investigated, using 20 replicate treatments with either disinfectant of bacterial suspensions with defined CFU/ml. The anticipated effect of any treatment is a reduction of the bacterial concentration. The scenario applies also to quantification of preselected "field samples" if the presence of foodborne pathogens has been confirmed by using a method with high specificity. For example, a quantitative (enumeration) method is applied to a subset of fresh table eggs that were confirmed as *Salmonella* positive by a highly specific qualitative method. Under the assumption of one population, nondetects have to be interpreted as false negatives. In practice, situations may be encountered where a treatment effect may actually lead to a reduction of the true concentration to zero in some of the study units. It must be decided on a case-by-case basis whether the one-population scenario applies or whether two endpoints of treatment effects, i.e., reduction to nonzero concentration and the proportion of reduction to zero, should be considered. In the latter case, the mixed-population scenario applies.

Statistical methods have been developed to provide unbiased estimation of the population mean value and standard deviation by using censored data. Shorten et al. (2006) provide closed formulas for estimation of the mean and standard deviation adjusted for censoring by use of the methods of moments. Maximum likelihood (ML) estimators of the mean and standard deviation for censored data are known to have superior statistical properties, especially for large sample sizes. The ML point estimates for mean and standard deviation can be obtained by numerical methods, i.e., no closed formula exists. The variance and covariance of the ML estimates can be calculated as shown by Shorten et al. (2006).

The same reference provides also statistical methods for the comparison of mean values (*t* test) and variances (*F* test) and for the estimation of microbial growth using censored data. Lorimer and Kiermeier (2006) provide technical guidance on the use of the free statistical software R (R Development Core Team, 2009) for the analyses of microbiological data, including censored data. Delignette-Muller et al. (2009) have developed the R package "fitdistrplus," in which method of moments and ML methods are implemented for estimating the mean and standard deviation of censored data. Further details of these methods are obtained by typing "?fitdistcens" at the R-prompt.

Observations from a Mixed Population

The scenario of a mixed population applies under the assumption of a true nonzero concentration for a subpopulation of contaminated (or infected) elements and a true concentration of zero for the free subpopulation. The proportions of contaminated and free subpopulations are p and $1 - p$, which can be interpreted as prevalence and (1 − prevalence), respectively. The methods described in "Observations from One Population," above are not applicable under the assumption of a mixed population. As an example for this scenario, let the variable Y denote the counts of CFU/25 g (without log transformation) obtained from a sample of 20 batches of 25 g of a food product, drawn at random at one stage of the production process,

$$Y = (0, 0, 1, 0, 0, 3, 0, 0, 0, 2, 0, 0, 3, 0, 0, 0, 2, 0, 0, 0)$$

No bacteria were detected in 15 batches, whereas 5 batches were positive with counts between 1 and 3. It is plausible to assume that many, if not all, of the negative batches are actually not contaminated. However, due to a limited analytical sensitivity of the enumeration method used, some of the zeros may in fact be false-negative results. There is no way of evaluating whether a given negative count value is a true negative observation from the free subpopulation or a false negative (censored) observation from the contaminated subpopulation. Consequently, the data storage formats discussed in "Storing Censored Observations in Databases," above are not applicable.

When censoring is ignored, the naïve population estimates for the prevalence ($5/20 = 0.25$) and for the mean count [$(1 + 3 + 2 + 3 + 2)/5 = 2.2$] are biased. However, the observed positive counts, along with distributional assumptions for the counts in the contaminated subpopulation, enable the analyst to estimate these parameters with adjustment for censoring. Assume that the count values are Poisson distributed (mean equal to the variance of the distribution), given that they are from the contaminated subpopulation and that they are equal to zero for the free subpopulation, i.e.,

$$Y|\text{contaminated} \sim \text{Pois}(\lambda)$$
$$Y|\text{free} = 0$$

where λ denotes the parameter of the Poisson density, i.e., the mean count number per sample unit. The distribution of Y, given that the true subpopulation is not known, is a mixture distribution with the two distribution components combined with weights given by the unknown prevalence parameter, p,

$$Y \sim f(y|p, \lambda) = (1 - p)\text{Pois}(y|0) + p\ \text{Pois}(y|\lambda)$$

This type of model is also known as a zero-inflated Poisson (ZIP) model. Because of the free subpopulation component, there is an excess of zeros compared to what would be expected if all data were observations from the Poisson distribution. Various R packages support ZIP models based on Poisson regression models. The command when using the "pscl" package for the count values (Y) is simply "zeroinfl(Y~1)" and returns the intercept term for the count model (Poisson with log link) and the intercept term for the zero-inflation model (binomial with logit link), here denoted a_C and a_Z, respectively. The inverse link functions must then be used to obtain the parameters:

$$p = 1 - \exp(a_Z)/[1 + \exp(a_Z)]$$
$$\lambda = \exp(a_C)$$

Note that the zero-inflation parameter a_Z refers to the zeros, whereas the prevalence, p, based on our notation refers to the nonzeros. Also note the regression framework of this approach. Hence, explanatory covariates such as inactivation protocols as well as experimental conditions (e.g., time and temperature) affecting p and/or λ may be included easily (see details by typing "?zeroinfl" at the R prompt). Using the "zeroinfl" command, we obtain estimates for p of 0.3 (95% CI = 0.1 to 0.6) and estimate λ as 1.8 (95% CI = 0.9 to 3.8) for the example data. The naïve, nonadjusted results underestimate the true prevalence and overestimate the true Poisson parameter, while these biases are mitigated by use of the ZIP model.

VALIDATION IN THE ABSENCE OF A PERFECT REFERENCE METHOD

In their review of measurement of uncertainty, Corry et al. (2007) pointed out that the diagnostic sensitivity and specificity of a new alternative microbiological detection method cannot be estimated if the standard comparative method is less sensitive. This commonly occurs when the standard method is bacterial culture and the alternative method is nucleic acid detection. This general problem in the validation of diagnostic tests has been addressed by using LCA. A detailed description of LCA statistical methods is outside the scope of this chapter, and hence, we refer readers to Enøe et al. (2000) for a review of maximum likelihood and Bayesian methods.

Briefly, LCA is applicable for cross-tabulated dichotomous (positive/negative) results of two or more diagnostic tests for samples from one or more populations under the assumption that (i) the (unknown) prevalences in the study populations (if more than one was used) differ, (ii) the diagnostic sensitivity and specificity of the tests are constant across the study populations (if more than one was used), and (iii) the tests are conditionally independent (Gardner et al., 2000) given the contamination status. The latter assumption suggests that diagnostic tests included in LCA should be based on different diagnostic principles, such as culture-based, DNA-based, or immunologically based principles.

LCA has mainly been applied to the validation of tests conducted on samples from individual animals. For example, Nielsen et al. (2004) considered enzyme-linked immunosorbent assays (ELISAs) and fecal culture for the diagnosis of *Salmonella* serotype Dublin in cattle, and Mainar-Jaime et al. (2008) investigated the modified agglutination test and an indirect ELISA for the detection of serum antibodies against *Toxoplasma gondii* in sheep. Table 2 contains the data reported by Nielsen et al. (2004) for cattle aged between 0 and 99 days and an ELISA cutoff value of 25 ODC% (corrected optical density percent). Note that this format of reporting cross-tabulated diagnostic test results is suitable for extension of the study through inclusion of more tests or more populations. Under the assumption of a diagnostic specificity of 100% of the culture method, the authors obtained 0.46 (95% CI = 0.38 to 0.53), 0.89 (95% CI = 0.87 to 0.91), and 0.11 (95% CI = 0.03 to 0.19) as estimates of the diagnostic sensitivity and specificity of the ELISA and the sensitivity of fecal culture, respectively. It is noted that the data would allow also the estimation of the specificity of culture. However, if the assumption of 100% specificity of the culture method holds, one degree of freedom in the data can be used for checking the goodness of fit of the model.

MULTIVARIATE STATISTICAL APPROACHES FOR ALTERNATIVE METHODS

New technologies have emerged in response to the need for fast, cost-efficient, and high-throughput devices for the detection and classification of foodborne microorganisms.

TABLE 2 Cross-tabulated results of ELISA and fecal culture for detection of *Salmonella enterica* serotype Dublin in cattle aged 0 to 99 days[a]

ELISA	Fecal culture	Population 1	Population 2
Negative	Negative	397	517
Negative	Positive	11	7
Positive	Negative	117	99
Positive	Positive	11	4

[a] ELISA cutoff, 25 ODC%. Values are count numbers of observations. Data from Nielsen et al., 2004.

Biosensor technologies are part of this development (Velusamy et al., 2010). A variety of traits such as genetic, morphological, immunological, biochemical, physical, or functional are studied in this context. The measurement principles for such traits often give rise to sets of observations rather than single-valued measurements for each isolate in question. Such data, in statistical jargon, are referred to as multivariate data.

The treatment of multivariate data, which typically show a high degree of correlation, is well established in mathematical statistics. However, the appropriate choice of one method from a suite of statistical (multivariate) approaches requires a clear problem formulation, appropriate data, and specialized statistical training. Here we review some applications of statistical methods and provide references for further reading. We note that the statistical methods for analysis of biosensor data reach far into the area of bioinformatics and undergo a dynamic evolution along with the technical developments.

Purpose of the Analysis and Choice of Statistical Method

Mariey et al. (2001) reviewed the application of the so-called Fourier transform infrared photoacoustic spectroscopy method for the differentiation, classification, and identification of microorganisms. In their paper, the authors distinguish between so-called supervised and unsupervised statistical methods. The terminology points to the area of artificial intelligence, where a similar distinction is made between unsupervised and supervised learning algorithms. "Supervised learning algorithms" refer to statistical classification approaches that require the known true classification of at least a subset of the samples. Classification rules can then be found by using the "learning data" and applied for classification of an unknown sample. In contrast, "unsupervised learning algorithms" allow statistical classification of a sample without prior knowledge of any classification. In this case, the classification rules are solely based on similarity patterns observed in the learning sample. Both approaches can be applied for taxonomical classification and identification. This important distinction also applies to other sensor methods.

Supervised methods are applicable if the true characteristic (e.g., classification of species or subtype) is known for each isolate in the study. It is then possible to develop and optimize an algorithm aiming at correct classification of all isolates. Using the optimal classification rule, a new sample of defined isolates can be typed in order to characterize the classification performance of the method.

In principle, this supervised approach is similar to the optimization (selection of cutoff value) and validation of a diagnostic test (estimation of diagnostic sensitivity and specificity). The ultimate goal is the application of the method for classification of unknown samples. Mariey et al. (2001) and Savić et al. (2008) list in their reviews as examples for supervised methods the discriminant analysis, K-nearest neighbor, soft independent modeling of class analogy, partial least squares regression, and artificial neural networks.

The choice of statistical approach can be considered to be part of the optimization process. The investigator may have little choice if the analysis method is implemented in the laboratory software. Further multivariate statistical methods that are applied in environmental microbiology have been reviewed by Ramette (2007), and a genetic algorithm has been applied by Li et al. (2009).

In contrast, unsupervised methods are applicable if the true classification is not known for the study sample and only explorative analysis of the similarities among isolates is required. Mariey et al. (2001) and Savić et al. (2008) consider in this category the principle component analysis, hierarchical cluster analysis using the Pearson's product moment correlation or Euclidian distance with Ward's and unweighted-pair group method using average linkages algorithm, canonical variate analysis, and multidimensional scaling.

Optimization and validation studies (supervised approaches in the terminology above) should be conducted with samples that are representative for the target application. If it is necessary to work with selected or spiked samples, the pattern of strains or subtypes should reflect their empirical occurrence. The sample size should be large enough to allow for using robust statistical techniques such as bootstrap or sample splitting and cross-validation. In application studies (unsupervised methods), the selection of samples should follow good statistical practices.

CONCLUSIONS AND FUTURE TRENDS

Scientists working in the area of foodborne pathogens may benefit from procedures for assay performance characterization which are established in the general diagnostic arena including techniques for estimating sensitivity and specificity without a perfect reference method and adjustment methods for the limit of detection. On the other hand, the nature of data generated by alternative detection methods has led to the development of highly specialized statistical methodologies. In the future, we shall see more sophisticated laboratory and statistical method developments going hand in hand. This will provide both opportunities for more reliable data analyses (e.g., better source attribution) and challenges in terms of training of specialists.

REFERENCES

Arnold, M., J. J. Carrique-Mas, and R. H. Davies. 2010. Sensitivity of environmental sampling methods for detecting *Salmonella* Enteritidis in commercial laying flocks relative to the within-flock prevalence. *Epidemiol. Infect.* **138:**330–339.

Arnold, M. E., D. Mueller-Doblies, J. J. Carrique-Mas, and R. H. Davies. 2009. The estimation of pooled-sample sensitivity for detection of *Salmonella* in turkey flocks. *J. Appl. Microbiol.* **107:**936–943.

Bartlett, J. W., and C. Frost. 2008. Reliability, repeatability and reproducibility: analysis of measurement errors in continuous variables. *Ultrasound Obstet. Gynecol.* **31:**466–475.

Bland, J. M., and D. G. Altman. 1986. Statistical method for assessing agreement between two methods of clinical measurement. *Lancet* **i:**307–310.

Bohaychuk, V. M., G. E. Gensler, R. K. King, J. T. Wu, and L. M. McMullen. 2005. Evaluation of detection methods for screening meat and poultry products for the presence of foodborne pathogens. *J. Food Prot.* **68:**2637–2647.

Bohaychuk, V. M., G. E. Gensler, M. E. McFall, R. K. King, and D. G. Renter. 2007. A real-time PCR assay for the detection of *Salmonella* in a wide variety of food and food-animal matrices. *J. Food Prot.* **70:**1080–1087.

Corry, J. E. L., B. Jarvis, S. Passmorec, and A. Hedges. 2007. A critical review of measurement uncertainty in the enumeration of food micro-organisms. *Food Microbiol.* **24:**230–253.

Delignette-Muller, M. L., R. Pouillot, J. B. Denis, and C. Dutang. 2009. Use of the package fitdistrplus to specify a distribution from non-censored or censored data. http://riskassessment.r-forge.r-project.org/intro2fitdistrplus.pdf.

Enøe, C., M. P. Georgiadis, and W. O. Johnson. 2000. Estimation of sensitivity and specificity of diagnostic tests and disease prevalence when the true disease state is unknown. *Prev. Vet. Med.* **45:**61–81.

European Food Safety Authority. 2010. Management of left-censored data in dietary exposure assessment of chemical substances. European Food Safety Authority, Parma, Italy. http://www.efsa.europa.eu/en/efsajournal/doc/1557.pdf.

Gardner, I. A., H. Stryhn, P. Lind, and M. T. Collins. 2000. Conditional dependence between tests affects the diagnosis and surveillance of animal diseases. *Prev. Vet. Med.* **45:**107–122.

Helsel, D. R. 2005. *Nondetects and Data Analysis. Statistics for Censored Environmental Data.* John Wiley & Sons, Chichester, United Kingdom.

Hu, Y., and I. Arsov. 2009. Nested real-time PCR for hepatitis A detection. *Lett. Appl. Microbiol.* **49:** 615–619.

International Organization for Standardization. 2003. *ISO 16140:2003. Microbiology of Food and Animal Feeding Stuffs—Protocol for the Validation of Alternative Methods.* International Organization for Standardization, Geneva, Switzerland.

Jordan, D. 2005. Simulating the sensitivity of pooled-sample herd tests for fecal *Salmonella* in cattle. *Prev. Vet. Med.* **70:**59–73.

Leonard, P., S. Hearty, J. Brennan, L. Dunne, J. Quinn, T. Chakraborty, and R. O'Kennedy. 2003. Advances in biosensors for detection of pathogens in food and water. *Enzyme Microb. Technol.* **32:**3–13.

Li, Y., Y. Li, B. Zheng, L. Qu, and C. Li. 2009. Determination of foodborne pathogenic bacteria by multiplex PCR-microchip capillary electrophoresis with genetic algorithm-support vector regression optimization. *Anal. Chim. Acta* **643:**100–107.

Löfström, C., J. Schelin, B. Norling, H. Vigre, J. Hoorfar, and P. Rådström. 2010. Culture-independent quantification of *Salmonella enterica* in carcass gauze swabs by flotation prior to real-time PCR. *Int. J. Food Microbiol.* Epub ahead of print. doi:10.1016.j.ijfoodmicro.2010.03.042.

Lorimer, M. F., and A. Kiermeier. 2006. A guide for using R for analysing microbiological data. SARDI Food Safety, Adelaide, Australia. http://www.sardi.sa.gov.au.

Lorimer, M. F., and A. Kiermeier. 2007. Analysing microbiological data: Tobit or not Tobit? *Int. J. Food Microbiol.* **116:**313–318.

Mäde, D., R. Petersen, K.Trümper, R. Stark, and L. Grohmann. 2004. In-house validation of a real-time PCR method for rapid detection of *Salmonella* ssp. in food products. *Eur. Food Res. Technol.* **219:**171–177.

Mainar-Jaime, R. C., N. Atashparvar, and M. Chirino-Trejo. 2008. Estimation of the diagnostic accuracy of the *invA*-gene-based PCR technique and a bacteriological culture for the detection of *Salmonella* spp. in caecal content from slaughtered pigs using Bayesian analysis. *Zoonoses Public Health* **55:**112–118.

Mariey, L., J. P. Signolle, C. Amiel, and J. Travert. 2001. Discrimination, classification, identification of microorganisms using FTIR spectroscopy and chemometrics. *Vib. Spectrosc.* **26:**151–159.

Nielsen, L. R., N. Toft, and A. K. Ersbøll. 2004. Evaluation of an indirect serum ELISA and a bacteriological faecal culture test for diagnosis of *Salmonella* serotype Dublin in cattle using latent class models. *J. Appl. Microbiol.* **96:**311–319.

Park, H. J., H. J. Kim, S. H. Park, E. G. Shin, J. H. Kim, and H. Y. Kim. 2008. Direct and quantitative analysis of *Salmonella enterica* serovar Typhimurium using real-time PCR from artificially contaminated chicken meat. *J. Microbiol. Biotechnol.* **18:**1453–1458.

Ramette, A. 2007. Multivariate analyses in microbial ecology. *FEMS Microbiol. Ecol.* **62:**142–160.

R Development Core Team. 2009. *R: A Language and Environment for Statistical Computing.* R Foundation for Statistical Computing, Vienna, Austria. http://www.R-project.org.

Savić, D., N. Joković, and L. Topisirović. 2008. Multivariate statistical methods for discrimination of lactobacilli based on their FTIR spectra. *Dairy Sci. Technol.* **88:**273–290.

Shorten, P. R., A. B. Pleasants, and T. K. Soboleva. 2006. Estimation of microbial growth using population measurements subject to a detection limit. *Int. J. Food Microbiol.* **108:**369–375.

Velusamy, V., K. Arshak, O. Korostynska, K. Oliwa, and C. Adley. 2010. An overview of foodborne pathogen detection: in the perspective of biosensors. *Biotechnol. Adv.* **28:**232–254.

Wilk, C. 2001. Gold standards and fool's gold. *Aust. Vet. J.* **79:**115.

Yang, C., Y. Jiang, K. Huang, C. Zhu, and Y. Yin. 2003. Application of real-time PCR for quantitative detection of *Campylobacter jejuni* in poultry, milk and environmental water. *FEMS Immunol. Med. Microbiol.* **38:**265–271.

MEAT PRODUCTION CHAIN

III

SALMONELLA IN PORK, BEEF, POULTRY, AND EGG

B. Malorny, A. Bhunia, H. J. M. Aarts, C. Löfström, and J. Hoorfar

13

PREVALENCE IN MEAT AND EGGS

The contribution of the various foods and their categories to the occurrence of foodborne cases of human salmonellosis varies between countries depending on the prevalence of different *Salmonella* serovars in food animals and in their various food production chains, as well as consumption habits and food preparation practices. In the European Union and many other countries, eggs and egg products are the foods most frequently implicated in human salmonellosis (European Food Safety Authority, 2009a; Centers for Disease Control and Prevention, 2006; Hald et al., 2004). Most contaminated eggs have *Salmonella* on the shell surface only, and therefore, eggs are usually sanitized by a variety of methods and agents (Berrang et al., 2000).

B. Malorny, Federal Institute for Risk Assessment, National Reference Laboratory for Salmonella, Diedersdorfer Weg 1, D-12277 Berlin, Germany. *A. Bhunia*, Department of Food Science, Purdue University, 745 Agriculture Mall Dr., West Lafayette, IN 47907-1160. *H. J. M. Aarts*, Laboratory for Zoonoses and Environmental Microbiology (LZO), National Institute for Public Health (RIVM)—Centre for Infectious Disease Control, P.O. Box 1, 3720 BA, Bilthoven, The Netherlands. *C. Löfström and J. Hoorfar*, National Food Institute, Technical University of Denmark, Mørkhøj Bygade 19, DK-2860 Søborg, Denmark.

Meat is another important source of foodborne salmonellosis, with poultry and pork implicated more often than beef and lamb. *Salmonella* serovars that are commonly isolated from food animals and the products thereof are also implicated in human salmonellosis. For example, serovar Enteritidis has been identified as one of the dominating serovars in the consumption of eggs, egg products, or foods containing raw eggs such as ice creams or cream pastry fillings (European Food Safety Authority, 2009b).

Salmonella enterica serovar Typhimurium is frequently isolated from pigs worldwide. In the United States, among the top 20 most common serovars associated with human *Salmonella* infections, four serovars (Typhimurium, Heidelberg, Agona, and Infantis) are commonly isolated from porcine products, three (Enteritidis, Typhimurium, and Heidelberg) from chickens, and three (Typhimurium, Newport, and Montevideo) from bovine products (Centers for Disease Control and Prevention, 2006).

Results of monitoring data on the prevalence of *Salmonella* in foodstuffs are often difficult to compare because of nonharmonized sampling plans, sample size, duration of sampling, and type of samples. In the European Union, the prevalence of *Salmonella* in fresh

Rapid Detection, Identification, and Quantification of Foodborne Pathogens, ed. by J. Hoorfar

eggs and in raw eggs at processing and at retail was reported to be below 5.8%, with an average of 0.8% (European Food Safety Authority, 2009b). The same report found *Salmonella* in fresh broiler meat at slaughter to be on average 5.5% and at processing up to 55.6% (Table 1).

In the United States, in 2009 the U.S. Department of Agriculture—Food Safety Inspection Service (USDA-FSIS) analyzed 29,116 verification samples across eight meat and poultry product classes with the following percent positive rate of *Salmonella* per product class: broilers (7.2%), market hog (2.3%), cow/bull (0.6%), steer/heifer (0.2%), ground beef (1.9%), ground chicken (18.2%), ground turkey (10. 7%), and turkey (3.8%) (U.S. Department of Agriculture—Food Safety Inspection Service, 2010). Worldwide prevalence data were previously summarized (European Commission, 2003; D'Aoust, 2000).

MEAT CHAIN FLOW, PRODUCTION AND HYGIENE MEASUREMENTS

The classical meat chain consists of the five steps primary production in herds, slaughtering, processing, retailing/storage, and preparation for consumption. However, each step can vary with regions, trade flows, and food customs. For example, the primary production of poultry is very diverse. The production system is divided into the production line poultry meat (carcasses and processed products), eggs for consumption (table eggs), and further processing (egg products). Various species are used in industrial poultry meat production: chickens, turkeys, ducks, and guinea fowl. Their importance is variable among regions and food customs. Alternative production systems, such as organic or free-range, are increasing.

Production of poultry meat or eggs is based on selection of pure lineages on precise genetic criteria, such as productivity, quality of products, and resistance against disease. The selection methods ensure a uniform quality of birds for further multiplication and production. Selection criteria differ according to the types of production. After the incubation time of eggs, stemming from this first crossing, the chicks are raised in breeding steps, giving rise to chicks intended for fattening for poultry carcasses, and pullets for laying of eggs for human consumption. The selected offspring from these are then multiplied in great-grandparent flocks and grandparent flocks, which are maintained at high health status. Chicks from grandparent flocks are used to populate parent flocks, e.g., broiler or layer breeder flocks, which are normally held by individual commercial companies. Eggs from these parent flocks are then hatched in commercial hatcheries to produce the commercial generation of birds. Different genetic lines of birds are used for meat- and egg-producing flocks of chickens. There are also different genetic lines of birds for conventional and free-range or organic production systems.

TABLE 1 Prevalence of *Salmonella* in table eggs, poultry meat, and pork

Food matrix	% Prevalence of *Salmonella*			
	European Union[a]		United States[b]	
	2007	2006	2009	2008
Table eggs	0.8	0.8	ND[c]	ND
Broiler meat	5.5	6.3	7.2	7.3
Turkey meat	6.8	6.4	3.8	6.2
Pork	1.1	1.0	ND	ND

[a] Data from European Food Safety Authority, 2009b.
[b] Data from U.S. Department of Agriculture—Food Safety Inspection Service, 2010.
[c] ND, not determined.

Pork and beef production shows a great variety of organizational structures and farm size patterns. The organizational pattern ranges from small independent farmers, who mainly produce for home consumption and local markets, to vertically integrated agribusiness companies that sell their products on national and international markets. The centers of pig production are in most cases closely linked to large pig farms. In addition, the availability of feed, either from domestic production or from import, plays an important role in the development of regions with intensive pig and beef production.

To make food safer, the industry is required to implement a control system that prevents the spread and multiplication of microbial risks associated with the food produced. Therefore, hazard analysis and critical control points (HACCP) has been implemented in the meat production chain for the last 2 decades (Sofos, 1993). HACCP is an internationally recognized way of managing food safety and protecting consumers. In meat plants, HACCP plans focus on control measures that can reduce the likelihood of contamination of meat from microbiological hazards, such as *Salmonella* during production. In 1996, the USDA-FSIS established the Pathogen Reduction and HACCP system program for meat- and poultry-processing plants and advocates its implementation (U.S. Department of Agriculture—Food Safety Inspection Service, 1998a).

In Europe, three new food hygiene regulations (published in 2004) have since 2006 become mandatory for the food industry (European Parliament, 2004a, 2004b, 2004c) and have replaced the carcass-testing requirements from the meat (HACCP) regulations of 2002. They include general rules for food business operators that are to be applied at all stages of production, processing, and distribution of food. The new regulations comprise a number of annexes to regulate the inspection of meat, sampling, and control measures including new criteria for *Salmonella* for red meat and poultry carcasses. In addition, food business operators are to comply with microbiological criteria established by Commission Regulation (EC) 2073/2005.

The two types of microbiological criteria considered are food safety and process hygiene. Food safety criteria define the acceptability of the batch for products placed on the market, whereas process hygiene criteria indicate the acceptable functioning of the process during the production process or at the end of it.

Microbiological criteria specify the number of sample units to be collected, the analytical method, and the number of analytical units that should conform to the limits. For meat, eggs, and the products thereof, usually *Salmonella* has to be absent in 10 to 25 g in five samples.

Special attention is given to the optimal storage of table eggs. However, two risk assessment studies concluded that if eggs are cooled after lay to an internal temperature of 7°C and if this temperature is maintained throughout egg processing and distribution, the risk of human salmonellosis can be minimized (U.S. Department of Agriculture—Food Safety Inspection Service, 1998b; World Health Organization/Food and Agriculture Organization, 2002). Disruption of the cold chain can increase the risk of condensation, leading to increased penetration of attached bacteria from the egg shell into the egg. However, a quantitative risk assessment approach is needed to assess the cost-benefit of egg cooling (European Food Safety Authority, 2009a).

RAPID DETECTION

The international accepted traditional culture-based method for detection and isolation of *Salmonella* in food is laid down in EN ISO standard 6579:2002 (International Organization for Standardization, 2002) and needs 4 to 5 working days for confirmation. The sensitivity can be 1 CFU per 25 g food analyzed. Thus, rapid detection methods should as a minimum fulfill this criterion. Nevertheless, foodstuffs can because of complexity and variety interfere with the rapid detection methods, especially when target bacteria are present at very low levels.

Thus, methods to sequester target pathogenic bacteria from interfering food components and to concentrate them in small volumes are needed to enable the efficient application of rapid detection and identification methods. For rapid detection of *Salmonella* in food including meat and eggs, three basic analytical principles are applied in practice: modified traditional culture methods, immunological methods, and nucleic acid-based methods.

Preanalytical Sample Preparation

The required detection limit in the culture-based standard method is 1 CFU per 25 g food including meat and eggs (International Organization for Standardization, 2002). Therefore, a sample preparation step is needed to accommodate samples to the small volumes used in molecular detection methods.

For nucleic acid amplification methods, a simple thermal cell lysis procedure in the presence of a chelating resin, e.g., 6% (wt/vol) Chelex-100 suspension, after nonselective enrichment for 12 to 16 h in buffered peptone water (BPW) at 37°C has worked for meat and meat products as well for egg (Malorny et al., 2004). A thermal cell lysis protocol is given below.

Protocol: Pre-PCR Thermal Cell Lysis Protocol

1. Weigh 25 g of raw meat or egg in a 750-ml Erlenmeyer flask, and add 225 ml BPW with loose lead.
2. Homogenize the sample manually by slewing the flask several times and incubate at 37°C for 12 to 16 h without shaking. If eggs are used, add optionally 50 ng/ml ferrioxamine E (Rijpens et al., 1999).
3. Centrifuge a 1-ml aliquot of the enriched sample for 5 min at 10,000 × *g*. The aliquot should not contain food debris. Discard the supernatant carefully with a pipette.
4. Resuspend the cell (pellet) in 300 μl 6% (wt/vol) Chelex-100 suspension or a similar product (Bio-Rad, Hercules, CA). Incubate the suspension at 56°C for 15 to 20 min. Mix the tube every 3 min briefly to keep the resin in suspension.
5. Place the tube immediately at 95 to 100°C and incubate for 8 min. Chill down the tube for 2 min on ice.
6. Centrifuge the tube for 5 min at 14,000 × *g* and 4°C.
7. Take 2 to 5 μl of the DNA-containing supernatant for PCR. The supernatant may not contain any particles of the resin. If inhibitors are expected, dilute the DNA 1:5 or 1:10 in Tris-EDTA buffer (10 mM Tris, 0.1 mM EDTA, pH 8.0).

Physical nondestructive purification methods were developed based on the principle of bacterial cell density properties, such as the aqueous two-phase systems (Lantz et al., 1997), buoyant density centrifugation (Wolffs et al., 2007), filtration (Wolffs et al., 2006), and flotation (Wolffs et al., 2004, 2007). Flotation is based on a density gradient centrifugation. This can separate biological particles and microorganisms that differ in buoyant density due to their density being lower than that of the medium, which allows the cells to float. The method separates *Salmonella* from the food matrix and can distinguish between living and dead cells (Wolffs et al., 2005).

Antibody-based sample preparation methods use *Salmonella*-specific antibodies to bind and separate cells from the food matrix. The magnetic beads are coated with the antibodies in order to separate them from the remaining debris. The technique is referred to as immunomagnetic separation (IMS) and was for some years used as a pre-PCR concentration step (Rijpens et al., 1999; Cudjoe and Krona, 1997). The disadvantage of IMS is the large variation in the target specificity depending on the antibody used.

Detection Technologies and Instruments

Modified Traditional Culture Methods

Both the USDA-FSIS (2008) and the FDA (U.S. Food and Drug Administration, 2007) methods have recently been modified for improved isolation of *Salmonella* from meat, poultry, and eggs. Although there are some minor differences in the use of enrichment media and the sample size, the overall procedures for the two methods are similar. In general, depending on the nature of the test sample, in the preenrichment step, lactose broth, tryptic soy broth with ferrous sulfate, nutrient broth, or BPW is used. The selective enrichment is done using Rappaport-Vassiliadis (RV), modified RV, selenite cystine broth, or Mueller-Kauffmann tetrathionate (TT) broth. Selective plating is done on Hektoen enteric (HE) agar, xylose lysine deoxycholate (XLD), xylose lysine Tergitol 4 (XLT4), double-modified lysine iron agar, bismuth sulfite (BS), triple sugar iron (TSI) and/or brilliant green sulfa (BGS) agar. For follow-up confirmation, isolated colonies from these plates are analyzed for their biochemical properties or carbon utilization patterns by biochemical assays including commercially available API test strips (Bio-Mérieux, Marcy l'Etoile, France), BBL Enterotube II (BD Diagnostics, Sparks, MD), or 96-well plate culture identification systems (BioLog, Hayward, CA). Isolated colonies are also confirmed by testing with polyvalent antisera to H and O antigens.

In the USDA-FSIS method, for analysis of ready-to-eat foods, 325 g of a composite sample is blended with 2,925 ml of BPW, while for raw meat and eggs, 25-g samples are enriched in 225 ml of BPW at 35°C for 20 to 24 h. A 0.5-ml or 0.1-ml volume of enriched sample is then transferred to 10 ml of TT or modified RV broth, respectively, and incubated at 42°C for 22 to 24 h. A loopful (10 μl) of enriched samples from each broth is streaked onto BGS agar and onto either XLT4 or double-modified lysine iron agar and then incubated at 35°C for 18 to 24 h. Suspected colonies are picked for further analysis.

In the FDA method, for testing of whole eggs, 25 ml of liquid egg from each of 15 eggs are combined to make 375 ml in an Erlenmeyer flask and held at room temperature for 96 h. Then, 3,375 ml of tryptic soy broth with ferrous sulfate is mixed with egg homogenate and held at room temperature for 1 h and then incubated at 35°C for 24 h. For meat or meat by-products, 25-g samples are blended with 225 ml of lactose broth and held at room temperature for 1 h. A 2.25-ml volume of steamed (15 min) Targitol Anionic 7 or Triton X-100 is added, and the sample is incubated at 35°C for 24 h. A 0.1-ml volume of enriched sample is transferred to 10 ml RV broth and another 1-ml sample to TT broth to achieve selective enrichment. For a heavily contaminated sample, RV is incubated at 42°C and TT at 43°C for 24 h; and for a sample with a low microbial load, RV is incubated at 42°C while TT is incubated at 35°C for 24 h. A loopful of cultures are then streaked onto BS, XLD, and HE agar plates for colony isolation. About two or more typical *Salmonella* colonies are picked from HE (blue-green to blue colonies with or without black centers), XLD (pink colonies with or without black centers), and BS (brown, gray, or black colonies).

Recently, BBL CHROMagar Salmonella (Webb and Ritter, 2009; Ritter and Dick, 2009) and RAPID'Salmonella (Lauer and Martinez, 2009) were evaluated and have been suggested as substitutes for selective media that could be used with the FDA procedure. In a separate study, the performances of three chromogenic plating media, i.e., AES *Salmonella* Agar Plate, Oxoid *Salmonella* Chromogen Media, and Miller-Mallinson agar, were compared with brilliant green (BG), XLD, and XLT4 agar for their ability to recover *Salmonella* colonies. Only Miller-Mallinson agar's performance was better than those of all the other selective media tested (Schönenbrücher et al., 2008). Furthermore, two culture-based methods, European gold standard (DIN EN 12824:1998) and ISO 6579:2002, were compared and showed similar levels of performance when large numbers of presumably naturally contaminated samples were tested (Schönenbrücher et al., 2008).

The disadvantages of the traditional culture-based method are the length of time required to complete a test and the fact that trained personnel are usually needed to perform the test. However, most importantly, it provides us with pure isolated colonies for identification. Isolated strains are also required when source tracking of cultures needs to be performed through genetic fingerprinting in the event the product was involved in an outbreak and for the regulatory agencies to initiate a product recall.

Nonisolating culture techniques (such as impedance microbiology) do not provide an isolate but can be faster and more convenient. Methods based on conductance or impedance rely on the changes arising from microbial growth in a medium (Smith et al., 1989; Yang and Bashir, 2008). The selectivity of the methods depends on the characteristics of the medium. The main advantage is the possibility for automation (Quinn et al., 1995).

Immunochemical Methods

Immunochemical methods have been used widely for detection or serotyping of microorganisms or toxins from food or clinical samples (Banada and Bhunia, 2008). These methods are relatively simple to perform and easy to interpret and are used in the development of rapid methods for pathogen testing. Often, the assay specificity depends on the avidity and specificity of the antibody, optimum antigen expression on the target organism, and the design of the assay configuration. For detection of whole cells of *Salmonella* or its extracellular antigens including cell surface antigen, lipopolysaccharides, flagella, or fimbriae, antibodies are generated against these targets to provide specificity. Optimal antigen expression on cells is critical to obtain positive results with immunoassays, since food environment or selective enrichment conditions can severely affect antibody-specific antigen expression (Banada and Bhunia, 2008; Hahm and Bhunia, 2006).

The major drawbacks of immunoassays are the need for large numbers of bacterial cells or toxins before a positive result can be obtained. Alternatively, to improve the sensitivity of an immunoassay, in recent years IMS coupled with enzyme immunoassay has been used for detection of *Salmonella* (Leon-Velarde et al., 2009), as have biosensor platforms (immunosensors). The sensitivity of immunosensors has been significantly improved, allowing detection in the range of 10^3 to 10^6 CFU (Bhunia, 2008; Byrne et al., 2009). The two most common biosensor platforms, i.e., fiber-optic sensors (Kramer and Lim, 2004; Leung et al., 2007; Valadez et al., 2009) and surface plasmon resonance sensors, have been used for detection of *Salmonella* from varieties of samples (Koubova et al., 2001; Mazumdar et al., 2010). Other sensors that are used for detection of *Salmonella* include time-resolved immunofluorescence assay (Tu et al., 2007), fluorescence resonance energy transfer (Ko and Grant, 2006), microflow cytometer (Kim et al., 2009), and electrochemical immunosensor (Salam and Tothill, 2009).

The more recent advances in immunological techniques include the use of protein microarray, where spots or parallel printed stripes containing antibodies against *Salmonella* are printed on glass or microtiter plates (Gehring et al., 2008; Blixt et al., 2008). This approach might in the future be a more rapid, precise, and low-cost alternative for screening of humans and livestock for *Salmonella* infections but needs further improvement and validation.

Nucleic-Acid-Based Methods

Already in 1992, Rahn et al. (1992) published a PCR method for the detection of *Salmonella* based on the invasion gene *invA*, and to date, more than 30 *Salmonella*-specific genes have been used for PCR analysis (Malorny et al., 2009; Levin, 2009). For diagnostic use, a real-time PCR with sequence-specific fluorescence-labeled probes should be selected due to its better specificity and sensitivity compared to double-stranded-DNA binding dyes. Another amplification technique is loop-mediated isothermal amplification or LAMP. The method relies on an autocycling strand displacement DNA synthesis performed by the *bst* polymerase large fragment (Ueda and Kuwabara, 2009). In comparison to PCR, this method was shown to be more sensitive in the *Salmonella* analysis of food and human materials (Ueda and Kuwabara, 2009).

Both conventional PCR and real-time PCR allow multiplex detection. By using the microarray technology, however, a large number of targets can be detected simultaneously, and this enables the detection of multiple pathogens or multiple (relevant) genes within one single assay. Wang et al. (2007) developed a microarray assay for the detection of 23 food microbial pathogens, including *Salmonella* spp. This assay was based on the application of universal PCR primers to amplify the in-between variable region of the 16S rRNA gene followed by microarray hybridization. A similar microarray-based method was reported by Suo et al. (2010) for the detection of multiple pathogens including *Salmonella enterica* from fresh meat samples.

Salmonella microarrays have been developed to reveal data on the gene repertoire and biological properties of a strain in order to estimate the risk for human consumption. These arrays were based on careful selection of various *Salmonella* genetic markers comprising the detection of flagellar and somatic antigens, important virulence genes, phage-associated elements, and antibiotic resistance determinants (Huehn et al., 2009; van Hoek and Aarts, 2008).

Open-Formula Real-Time PCR Protocol

The detection of *Salmonella* by real-time PCR precedes a preenrichment and sample preparation, e.g., according to the thermal cell lysis using Chelex (see "Preanalytical Sample Preparation," above). The analytical assay is based on a 5′ nuclease (TaqMan) real-time PCR (Malorny et al., 2004). Slightly modified assays were later published (Josefsen et al., 2007; Reynisson et al., 2006). The assay uses specifically designed primers and a probe located in the *ttrRSBCA* locus and contains an internal amplification control (IAC). It encodes proteins required for tetrathionate respiration in *Salmonella* (Hensel et al., 1999). The method has been validated and approved by the Nordic Organization for Validation of Alternative Methods (NordVal) for qualitative analysis of salmonellae in raw meat samples and carcass swabs (NordVal/NMKL, 2007a; Löfström et al., 2009) and has become an official method in Germany as well, according to the code of law of §64 LFGB (Foodstuffs, Consumer Good and Feedstuffs Law) (Malorny et al., 2007). Reagents and cycling conditions for real-time PCR are listed in Table 2. The real-time cycler should be in compliance with ISO/TS 20836 (International Organization for Standardization, 2005). Fluorescence measurements and threshold settings should be set according to the recommendations of the manufacturer. Data should be collected in the annealing step in the FAM (*Salmonella* target sequence) and HEX channel (IAC sequence) of the instrument. A sample is considered positive if a cycle threshold value for the FAM probe signal is below 40 and the controls are as expected. The IAC probe signal (cycle threshold value, ~30) has to be positive if the FAM signal is negative. In the case of strong positive probe FAM signals, the IAC probe signal can be negative.

Rapid Enumeration Methods

Usually, the generation of quantitative data is laborious and time-consuming. Currently, enumeration of bacteria is based on traditional bacteriological methods such as the most-probable-number (MPN) test or cell count using agar plates. Faster, easier, and cheaper enumeration methods could provide risk assessors with quantitative data for improving the assessment studies. Nevertheless, the *Salmonella* level during the production process could be an important hint to the source of contamination.

For enumeration of salmonellae, diagnostic laboratories need a highly sensitive and specific method because usually the number of cells in matrices from the primary production or ready-to-eat food is low and background flora might strongly influence the growth of the target bacteria. Currently, the MPN test is particularly useful for the determination of

TABLE 2 PCR reagents and cycle conditions for blockcyclers of typical *ttr* real-time PCR

Components	Vol (μl per reaction)	End concn in reaction
Deoxynucleoside triphosphates-dUTP mix (2 mM each dGTP, dATP, dCTP; 4 mM dUTP)	5.0	200 μM each dGTP, dATP, dCTP; 400 μM dUTP
10× PCR buffer (200 mM Tris-HCl [pH 8.4], 500 mM KCl) (Invitrogen, Karlsruhe, Germany)	5.0	1×
50 mM $MgCl_2$	4.5	4.5 mM
Platinum *Taq* polymerase (Invitrogen, Karlsruhe, Germany)	0.2	1 U
Primer ttr-6 (10 pmol/μl) AGCTCAGACCAAAAGTGACCATC	2.0	0.4 μM
Primer ttr-4 (10 pmol/μl) CTCACCAGGAGATTACAACATGG	2.0	0.4 μM
Probe ttr-5 (12.5 pmol/μl) FAM-CACCGACGGCGAGACCGACTTT-DQ	0.5	0.125 μM
Probe IAC-ttr (12.5 pmol/μl) HEX-CACACGGCGACGCGAACGCTTT-DQ	0.5	0.125 μM
Optional: uracil-*N*-glycosylase	1.0	0.25 U
Optional: 10 mg/ml fraction V bovine serum albumin (proteinase free)	5.0	1 μg/μl
Internal amplification control DNA	1.0	~100 copies
H_2O	18.3	
Sample DNA	5.0	
Total vol	50	
Cycle conditions		
Optional: uracil-*N*-glycosylase incubation	50°C	2 min
Initial denaturation	95°C	1 min
45 cycles:		
Denaturation	95°C	15 s
Annealing and synthesis	65°C	1 min

low concentrations of microorganisms and provides confirmed results within 4 to 5 days. Here, triplicates or 5-fold replicates are prepared from 10-fold serial dilutions. All samples are then tested by the horizontal culture method. A MPN/g value can be deduced from the ratio of positive to negative samples in relation to their concentrations.

The MPN method assumes that bacteria are distributed randomly within the sample and are separated (not clustered together). The growth medium and conditions of incubation have been chosen so that a single viable cell can be multiplied and detected. For *Salmonella*, a new ISO standard based on the so-called mini-MPN technique is under development (International Organization for Standardization, 2007). It describes a miniaturized version of the standard MPN method but with less manual handling and fewer reagents (Fravalo et al., 2003). Currently, the most robust and broadly applicable method to enumerate salmonellae in low numbers in meat and eggs is a combination of the mini-MPN technique and real-time PCR for qualitative detection of *Salmonella* from suspected samples. A protocol for enumeration of salmonellae is given below.

Protocol: Mini-MPN Protocol for Enumeration of Low Numbers of Salmonellae

1. Prepare a 1:10 resuspension in BPW (initial suspension) of the matrix to be determined.
2. Transfer 2.5 ml of the initial suspension to each well of the first empty row of three wells in a 12-well microtiter plate (e.g., Greiner bio-one, Solingen, Germany). Transfer a 0.5-ml aliquot of each well from the first row into 2 ml BPW in the successive wells in the second row, obtaining the first 1:5 dilution. Perform two more successive dilutions in the same way, mixing the suspensions each time before transferring the 0.5 ml into the next well. Seal the 12-well microtiter plate with a breathable sealing film (such as AeraSeal; Excel Scientific, Victorville, CA).
3. Incubate the plate for 18 h ± 2 h at 37°C ± 1°C.
4. Prepare a 12-well microtiter plate containing 2 ml of modified semisolid RV medium (MSRV) per well for selective enrichment. Transfer a 20-μl BPW aliquot of each well of the overnight culture into a corresponding well containing 2 ml of MSRV.
5. Incubate the inoculated MSRV 12-well microtiter plate for 24 to 48 h at 41.5°C ± 1°C.
6. After incubation, confirm a characteristic migration of salmonellae onto the MSRV agar by *ttr*-real-time PCR using DNA extracted from the MSRV wells by thermal cell lysis (see "Protocol: Pre-PCR Thermal Cell Lysis Protocol").
7. Calculate the MPN value by using the MPN calculator, for example, available at the website http://www.i2workout.com/mcuriale/mpn/index.html.

In comparison, higher levels of bacterial cells can be determined by the classical colony count method by using the direct isolation selective agar. One problem with the classical colony count is that high levels of background flora can disturb the growth of the target cells and lead to colony misidentification. In addition, selective media may inhibit the growth of stressed cells.

Quantitative real-time PCR for enumeration offers several advantages in terms of speed, detection limit, costs, and high throughput (Malorny et al., 2008a) and has been used to generate quantitative data on *Salmonella* in various matrices including pork and poultry meat (Wolffs et al., 2006; Guy et al., 2006; Seo et al., 2006, Wolffs et al., 2007). However, pre-PCR processing methods that do not include culture enrichment, such as flotation (Wolffs et al., 2004, 2007), paramagnetic beads (Warren et al., 2007), and filtration (Wolffs et al., 2006), have a limit of quantification of approximately 10^2 to 10^3 CFU/ml or gram of sample due to the loss of target material during the sample preparation and the small volumes analyzed. This limit of quantification is usually still too high, since most samples in the food production chain are contaminated with $<10^2$ salmonellae per gram (Fegan et al., 2004; Boughton et al., 2007). Recently, Löfström et al. (2010) developed a flotation protocol that allowed quantification of approximately 10^2 CFU/carcass swab sample, thereby getting closer to the needed quantification limit. An unconventional approach combines a short preenrichment (where the bacterial growth is, for most of the targets, in the log phase) and an analytical real-time PCR-based quantification assay (Krämer et al., in press). The procedure consists of an approximately 8-h preenrichment at 37°C or alternatively a 15-h preenrichment at 32°C of the sample diluted 1:10 in nonselective BPW followed by a short microbial DNA extraction, and finally the specific *Salmonella* detection and quantification by real-time PCR (Fig. 1). Enrichment of the cells into the log phase ensures the multiplication to a reasonable level for detection in the real-time assay and excludes the risk of detecting dead cells.

The enrichment time can slightly vary depending on the viability of the *Salmonella* cells present in the matrix. Stressed cells caused by

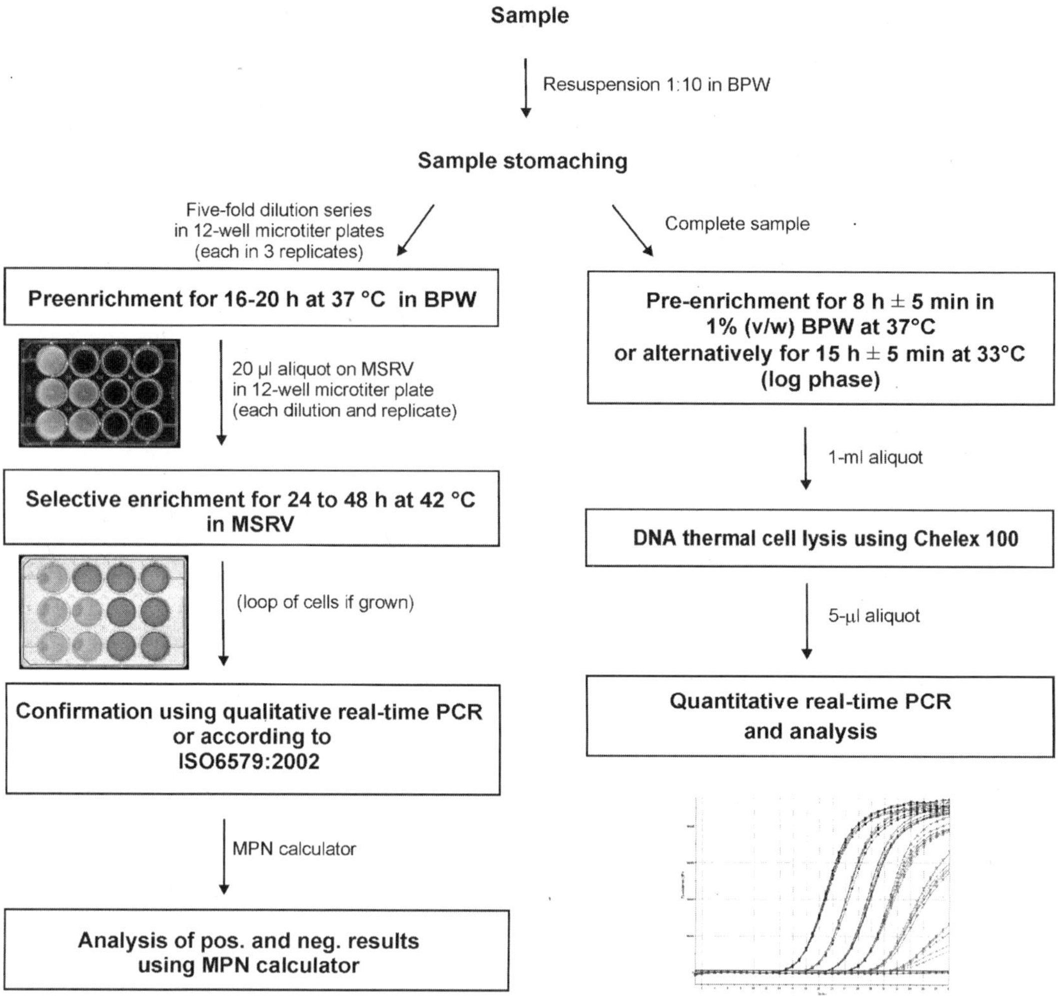

FIGURE 1 Comparison of quantitative real-time PCR and mini-MPN technique. Steps of procedures are displayed for real-time PCR (right) and mini-MPN (left).

refrigeration or a low-pH environment will need more time for reconstitution. The quantification by real-time PCR is not based on the end point signal but rather on the exponential increase of the initial DNA amount with regard to the performed PCR cycles (Mackay, 2004). A serial dilution of a known number of target copies can be used to set up a standard curve, which is applied to determine an unknown amount of DNA in a sample (absolute quantification). The automation of DNA sample preparation methods and the real-time PCR setup itself are undoubtedly advantageous to generate a great amount of quantitative data at a lower cost than culture methods (Malorny et al., 2008a).

Rapid Characterization and Subtyping

Subtyping of *Salmonella* is an approach often applied to facilitate the epidemiological investigation of outbreaks and to identify the source of entry into the food chain. Newer methods for *Salmonella* typing and subtyping include genome-based methods such as pulsed-field

TABLE 3 Reagents, cycling conditions, and fragment analyzer run parameters used in the MLVA-PCR for serovar Typhimurium or serovar Enteritidis

Sample preparation	Resuspend three to four *Salmonella* colonies of the test strain grown overnight on Mueller-Hinton agar (Oxoid GmbH, Wesel, Germany) at 37°C in 150 μl double-distilled water		
PCR			
Reagent	Vol (μl)	Concn of stock solution	Sequence or other relevant information
Serovar Typhimurium			
Multiplex PCR Mix	6.25	2×	Qiagen, Hilden, Germany, no. 206143
Q-solution (Qiagen)	1.25	2×	Included in Qiagen multiplex PCR mix
Primers	0.25	5 pmol/μl	STTR3-F: *HEX*-CCC-CCT-AAG-CCC-GAT-AAT-GG
	0.25	5 pmol/μl	STTR3-R: TGA-CGC-CGT-TGC-TGA-AGG-TAA-TAA
	0.125	5 pmol/μl	STTR5-F: *HEX*-ATG-GCG-AGG-CGA-GCA-GCA-GT
	0.125	5 pmol/μl	STTR5-R: GGT-CAG-GCC-GAA-TAG-CAG-GAT
	0.25	5 pmol/μl	STTR6-F: *FAM*-TCG-GGC-ATG-CGT-TGA-AA
	0.25	5 pmol/μl	STTR6-R: CT-GGT-GGG-GAG-AAT-GAC-TGG
	0.125	5 pmol/μl	STTR9-F: *FAM*-AGA-GGC-GCT-GCG-ATT-GAC-GAT-A
	0.125	5 pmol/μl	STTR9-R: CAT-TTT-CCA-CAG-CGG-CAG-TTT-TTC
	0.125	5 pmol/μl	STTR10pl-F: *TET*-CGG-GCG-CGG-CTG-GAG-TAT-TTG
	0.125	5 pmol/μl	STTR10pl-R: G-AAG-GGG-CCG-GGC-AGA-GAC-AGC
Double-distilled water	2.25		
DNA template (cells)	1.0		Three to four colonies grown on Mueller-Hinton agar resuspended in150 μl double-distilled water
Total vol/sample	12.5		
Cycling conditions	95°C for 15 min, (94°C for 30 s, 60°C for 90 s, 72°C for 90 s) × 25, 72°C for 10 min		
Serovar Enteritidis			
Multiplex PCR mix	6.25	2×	Qiagen, Hilden, Germany, No. 206143
Q-solution (Qiagen)	1.25	2×	Included in Qiagen multiplex PCR mix
Primers	0.25	20 pmol/μl	SENTR1-F: *HEX*-GCA-ACA-GCA-GCA-GCA-ACA-G
	0.25	20 pmol/μl	SENTR1-R: CCG-AGC-TGA-GAT-CGC-CAA-G
	0.25	20 pmol/μl	SENTR2-F: *FAM*-CAC-TGG-ACG-ATC-TGG-ATT-TCT-C
	0.25	20 pmol/μl	SENTR2-R: GTC-GCC-GTT-ACG-CAT-CAA-C
	0.25	20 pmol/μl	SENTR3-F: *TET*-CTA-AAC-AAG-CCG-CTC-ATC-CG
	0.25	20 pmol/μl	SENTR3-R: ACA-ACC-TGC-TGC-TGT-GCT-G
	0.25	5 pmol/μl	SENTR4-F: *HEX*-GAC-CAA-CAC-TCT-ATG-AAC-CAA-TG
	0.25	5 pmol/μl	SENTR4-R: ACC-AGG-CAA-CTA-TTC-GCT-ATC
	0.25	5 pmol/μl	SENTR5-F: *FAM*-CAC-CGC-ACA-ATC-AGT-GGA-AC
	0.25	5 pmol/μl	SENTR5-R: GCG-TTG-AAT-ATC-GGC-AGC-ATG
	0.25	5 pmol/μl	SENTR6-F: *TET*-ATG-GAC-GGA-GGC-GAT-AGA-C
	0.25	5 pmol/μl	SENTR6-R: AGC-TTC-ACA-ATT-TGC-GTA-TTC-G
	0.25	5 pmol/μl	SENTR7-F: *FAM*-ACG-ATC-ACC-ACG-GTC-ACT-TC
	0.25	5 pmol/μl	SENTR7-R: CGG-ATA-ACA-ACA-GGA-CGC-TTC
	0.25	5 pmol/μl	SE-3F: *HEX*-CAA-CAA-AAC-AAC-AGC-AGC-AT
	0.25	5 pmol/μl	SE-3R: GGG-AAA-CGG-TAA-TCA-GAA-AGT
	0.25	5 pmol/μl	SE-7F: *HEX*-GAT-AAT-GCT-GCC-GTT-GGT-AA
	0.25	5 pmol/μl	SE-7R: ACT-GCG-TTT-GGT-TTC-TTT-TCT
DNA template (cells)	1.0		Three to four colonies grown on Mueller-Hinton agar resuspended in 150 μl double-distilled water
Total vol/sample	12.5		
Cycling conditions	95°C for 15 min, (94°C for 30 s, 55°C for 90 s, 72°C for 90 s) × 25, 72°C for 10 min		

(continued on next page)

TABLE 3 Reagents, cycling conditions, and fragment analyzer run parameters used in the MLVA-PCR for serovar Typhimurium or serovar Enteritidis (*Continued*)

Preparation of sample for fragment separation	After cycling, dilute a 5-μl aliquot of the PCR product 1:10 in double-distilled water. Mix a 1-μl aliquot of this dilution with 1.25 μl of the Geneflo-625-TAMRA (CHIMERx, Madison, WI) internal size marker for serovar Typhimurium strains or MapMarker 1000-TAMRA (Eurogentec, Seraing, Belgium) size marker for serovar Enteritidis strains and 12 μl TS reagent containing formamide (Applied Biosystems). Denature the samples for 2 min at 94°C.
Fragment separation conditions	The injection voltage is 15 kV for 15 s. The electrophoresis is run at 60°C for 50 min using POP4 polymer (Applied Biosystems) with a run voltage of 15 kV.
Analyzing	Identify each peak according to color and size. Assign each repeat size within each locus to a distinct allele number. Further analysis can be performed using, e.g., BioNumerics v5.1 (Applied Maths).

gel electrophoresis, multilocus variable-number tandem-repeat analysis (MLVA), multilocus sequence typing, and (multiplex-) PCR-based methods. Pulsed-field gel electrophoresis is currently the method of choice for molecular subtyping *Salmonella* serovars. It has been proven to be a useful discriminatory method, which was standardized by the PulseNet Consortium (Gerner-Smidt et al., 2006). However, a new high-resolution approach is the MLVA, which is based on the genetic analysis of various loci within a strain that consists of repeating units of approximately 1 to 100 bp in length. MLVA has great potential for tracing purposes; it is fast, cheap, and easy to perform. However, MLVA protocols are only applicable for a few *Salmonella* serovars, especially for serovars Typhimurium (Lindstedt et al., 2004), Enteritidis (Malorny et al., 2008b; Boxrud et al., 2007), Infantis (Ross and Heuzenroeder, 2008), Typhi (Ramisse et al., 2004), and Newport (Davis et al., 2009). Since mostly serovar Typhimurium is found in pork and beef and mostly serovar Enteritidis is found in eggs and poultry meat, MLVAs for those serovars become more and more important for tracing and outbreak studies. Protocols for both serovars are given in Table 3.

Another approach is the revealing of genomic information of a strain useful to estimate the risk of the particular type for humans in terms of biological significance. Various *Salmonella*-associated DNA microarrays have successfully been developed, specifically targeting genes involved in antibiotic resistance profiles (van Hoek et al., 2008), phage types (Pelludat et al., 2005), or serovars (Wattiau et al., 2007; Yoshida et al., 2007) or a combination of condensed *Salmonella* targets including virulence gene markers for the discriminative characterization of *Salmonella* serovars (Huehn et al., 2009). The disadvantage of microarrays is that the analysis process per sample is laborious and currently not suitable for routine tracing properties.

CONCLUSION AND FUTURE TRENDS

Although a large number of various rapid detection methods for *Salmonella* are reported, validated protocols according to internationally accepted criteria are still rare. This hampers the application of open-formula methods, despite their many advantages in comparison to traditional culture-base methods, and delays their broad acceptance, e.g., by legal authorities. Furthermore, legislators can be hesitant to adopt alternative methods, i.e., other than currently accepted reference methods. While real-time PCR is being increasingly implemented for screening of (especially) negative samples, culture-based techniques are maintained in order to isolate strains for further characterization and source attribution studies. Enumeration of salmonellae comes increasingly into focus as a tool in risk assessment studies as well as in diagnostic settings. A better estimation is needed on how many salmonellae were present in the initial sample, how many were lost due to sample preparation, and what is being detected in the final assay. However, it will still require some years to develop and validate simple, convenient, and robust molecular quantification methods.

ACKNOWLEDGMENTS

This work was funded in part by the European Union project BIOTRACER (FOOD-2006-CT-036272) and by the U.S. Department of Agriculture project number 1935-42000-035 (AB).

REFERENCES

Banada, P. P., and A. K. Bhunia. 2008. Antibodies and immunoassays for detection of bacterial pathogens, p. 567–602. *In* M. Zourob, S. Elwary, and A. Turner (ed.), *Principles of Bacterial Detection: Biosensors, Recognition Receptors and Microsystems.* Cambridge University, Manchester, United Kingdom.

Berrang, M. E., N. A. Cox, J. F. Frank, R. J. Buhra, and J. S. Bailey. 2000. Hatching egg sanitization for prevention or reduction of human enteropathogens: a review. *J. Appl. Poult. Res.* **9:**279–284.

Bhunia, A. K. 2008. Biosensors and bio-based methods for the separation and detection of foodborne pathogens. *Adv. Food Nutr. Res.* **54:**1–44.

Blixt, O., J. Hoffmann, S. Svenson, and T. Norberg. 2008. Pathogen specific carbohydrate antigen microarrays: a chip for detection of *Salmonella* O-antigen specific antibodies. *Glycoconj. J.* **25:**27–36.

Boughton, C., J. Egan, G. Kelly, B. Markey, and N. Leonard. 2007. Quantitative examination of *Salmonella* spp. in the lairage environment of a pig abattoir. *Foodborne Pathog. Dis.* **4:**26–32.

Boxrud, D., K. Pederson-Gulrud, J. Wotton, C. Medus, E. Lyszkowicz, J. Besser, and J. M. Bartkus. 2007. Comparison of multiple-locus variable-number tandem repeat analysis, pulsed-field gel electrophoresis, and phage typing for subtype analysis of *Salmonella enterica* serotype Enteritidis. *J. Clin. Microbiol.* **45:**536–543.

Byrne, B., E. Stack, N. Gilmartin, and R. O. Kennedy. 2009. Antibody-based sensors: principles, problems and potential for detection of pathogens and associated toxins. *Sensors* **9:**4407–4445.

Centers for Disease Control and Prevention. 2006. Salmonella *Surveillance: Annual Summary Report, 2006.* Centers for Disease Control and Prevention, Atlanta, GA. http://www.cdc.gov/ncidod/dbmd/phlisdata/Salmonella.htm#2006.

Cudjoe, K. S., and R. Krona. 1997. Detection of *Salmonella* from raw food samples using dynabeads anti-*Salmonella* and a conventional reference method. *Int. J. Food Microbiol.* **37:**55–62.

D'Aoust, J. Y. 2000. *Salmonella*, p. 1233–1299. *In* B. M. Lund, A. C. Baird-Parker, and G. W. Gould (ed.), *The Microbiological Safety and Quality of Food,* vol. II . Aspen Publishers Inc., Gaithersburg, MD.

Davis, M. A., K. N. Baker, D. R. Call, L. D. Warnick, Y. Soyer, M. Wiedmann, Y. Grohn, P. L. McDonough, D. D. Hancock, and T. E. Besser. 2009. Multilocus variable-number tandem-repeat method for typing *Salmonella enterica* serovar Newport. *J. Clin. Microbiol.* **47:**1934–1938.

European Commission. 2003. *Scientific Opinion on Salmonellae in Foodstuffs, Adopted 14–15 April 2003.* European Commission, Scientific Committee on Veterinary Measures Related to Public Health. http://ec.europa.eu/food/fs/sc/scv/outcome_en.html.

European Food Safety Authority. 2009a. Scientific opinion of the panel on biological hazards on a request from the European Commission on special measures to reduce the risk for consumers through *Salmonella* in table eggs—e.g. cooling of table eggs. *EFSA J.* **957:**1–29.

European Food Safety Authority. 2009b. *Salmonella* in food. The Community summary report on trends and sources of zoonoses and zoonotic agents in the European Union in 2007. *The EFSA J.* **223:** 30–64.

European Parliament. 2004a. Regulation (EC) No 852/2004 of the European Parliament and of the council of 29 April 2004 on the hygiene of foodstuffs. *Off. J. Eur. Union* L **139:**3–21.

European Parliament. 2004b. Regulation (EC) No 853/2004 of the European Parliament and of the council of 25 June 2004 laying down specific hygiene rules for food of animal origin. *Off. J. Eur. Union* L **139:**22–82.

European Parliament. 2004c. Regulation (EC) No 854/2004 of the European Parliament and of the council of 25 June 2004 laying down specific rules for the organisation of official controls on products of animal origin intended for human consumption. *Off. J. Eur. Union* L **139:**83–127.

Fegan, N., P. Vanderlinde, G. Higgs, and P. Desmarchelier. 2004. Quantification and prevalence of *Salmonella* in beef cattle presenting at slaughter. *J. Appl. Microbiol.* **97:**892–898.

Fravalo, P., Y. Hascoet, M. Le Fellic, S. Queguiner, J. Petton, and G. Salvat. 2003. Convenient method for rapid and quantitative assessment of *Salmonella enterica* contamination: the mini-MSRV MPN technique. *J. Rapid Methods Automat. Microbiol.* **11:**81–88.

Gehring, A. G., D. M. Albin, S. A. Reed, S. I. Tu, and J. D. Brewster. 2008. An antibody microarray, in multiwell plate format, for multiplex screening of foodborne pathogenic bacteria and biomolecules. *Anal. Bioanal. Chem.* **391:**497–506.

Gerner-Smidt, P., K. Hise, J. Kincaid, S. Hunter, S. Rolando, E. Hyytiä-Trees, E. M. Ribot, and B. Swaminathan. 2006. PulseNet USA: a five-year update. *Foodborne Pathog. Dis.* **3:**9–19.

Guy, R. A., A. Kapoor, J. Holicka, D. Shepherd, and P. A. Horgen. 2006. A rapid molecular-based assay for direct quantification of viable bacteria in slaughterhouses. *J. Food Prot.* **69:** 1265–1272.

Hahm, B. K., and A. K. Bhunia. 2006. Effect of environmental stresses on antibody-based detection of *Escherichia coli* O157:H7, *Salmonella enterica* serotype Enteritidis and *Listeria monocytogenes*. *J. Appl. Microbiol.* **100:**1017–1027.

Hald, T., D. Vose, H. C. Wegener, and T. Koupeev. 2004. A Bayesian approach to quantify the contribution of animal-food sources to human salmonellosis. *Risk Anal.* **24:**255–269.

Hensel, M., A. P. Hinsley, T. Nikolaus, G. Sawers, and B. C. Berks. 1999. The genetic basis of tetrathionate respiration in *Salmonella typhimurium*. *Mol. Microbiol.* **32:**275–287.

Huehn, S., C. Bunge, E. Junker, R. Helmuth, and B. Malorny. 2009. Poultry-associated *Salmonella enterica* subsp. *enterica* serovar 4,12:d:- reveals high clonality and a distinct pathogenicity gene repertoire. *Appl. Environ. Microbiol.* **75:**1011–1020.

International Organization for Standardization. 2002. *Microbiology of Food and Animal Feeding Stuffs—Horizontal Method for the Detection of* Salmonella *spp., ISO 6579:2002.* International Organization for Standardization, Geneva, Switzerland.

International Organization for Standardization. 2005. *Microbiology of Food and Animal Feeding Stuffs—Polymerase Chain Reaction (PCR) for the Detection of Food-Borne Pathogens—Performance Testing for Thermal Cyclers, ISO/TS 20836:2005.* International Organization for Standardization, Geneva, Switzerland.

International Organization for Standardization. 2007. *Microbiology of Food, Animal Feeding Stuffs, Animal Faeces and Environmental Samples from the Primary Production Stage—Horizontal Method for the Enumeration of Salmonella by a Miniaturized MPN Technique. ISO/WD TS 13305, DRAFT 011007.* International Organization for Standardization, Geneva, Switzerland.

Josefsen, M. H., M. Krause, F. Hansen, and J. Hoorfar. 2007. Optimization of a 12-hour TaqMan PCR-based method for detection of *Salmonella* bacteria in meat. *Appl. Environ. Microbiol.* **73:**3040–3048.

Kim, J. S., G. P. Anderson, J. S. Erickson, J. P. Golden, M. Nasir, and F. S. Ligler. 2009. Multiplexed detection of bacteria and toxins using a microflow cytometer. *Anal. Chem.* **81:**5426–5432.

Ko, S., and S. A. Grant. 2006. A novel FRET-based optical fiber biosensor for rapid detection of *Salmonella typhimurium*. *Biosens. Bioelectron.* **21:** 1283–1290.

Koubova, V., E. Brynda, L. Karasova, J. Skvor, J. Homola, J. Dostalek, P. Tobiska, and J. Rosicky. 2001. Detection of foodborne pathogens using surface plasmon resonance biosensors. *Sens. Actuators B* **74:**100–105.

Kramer, M. F., and D. V. Lim. 2004. A rapid and automated fiber optic-based biosensor assay for the detection of *Salmonella* in spent irrigation water used in the sprouting of sprout seeds. *J. Food Prot.* **67:**46–52.

Krämer, N., C. Löfström, H. Vigre, J. Hoorfar, C. Bunge, and B. Malorny. A novel strategy to obtain quantitative data for modelling: combined enrichment and real-time PCR for enumeration of salmonellae from pig carcasses. *Int. J. Food Microbiol.*, in press.

Lantz, P.-G., M. Matsson, T. Wadström, and P. Rådström. 1997. Removal of PCR inhibitors from human faecal samples through the use of an aqueous two-phase system for sample preparation prior to PCR. *J. Microbiol. Methods* **28:**159–167.

Lauer, W., and F. Martinez. 2009. RAPID'Salmonella chromogenic medium. Performance tested method 050701. *J. AOAC Int.* **92:** 1871–1875.

Leon-Velarde, C. G., L. Zosherafatein, and J. A. Odumeru. 2009. Application of an automated immunomagnetic separation-enzyme immunoassay for the detection of *Salmonella enterica* subspecies enterica from poultry environmental swabs. *J. Microbiol. Methods* **79:**13–17.

Leung, A., P. M. Shankar, and R. Mutharasan. 2007. A review of fiber-optic biosensors. *Sens. Actuators B* **125:**688–703.

Levin, R. E. 2009. The use of molecular methods for detecting and discriminating *Salmonella* associated with foods—a review. *Food Biotechnol.* **23:** 313–367.

Lindstedt, B. A., T. Vardund, L. Aas, and G. Kapperud. 2004. Multiple-locus variable-number tandem-repeats analysis of *Salmonella enterica* subsp. *enterica* serovar Typhimurium using PCR multiplexing and multicolor capillary electrophoresis. *J. Microbiol. Methods* **59:**163–172.

Löfström, C., M. Krause, M. H. Josefsen, F. Hansen, and J. Hoorfar. 2009. Validation of a same-day real-time PCR method for screening of meat and carcass swabs for *Salmonella*. *BMC Microbiol.* **9:**85.

Löfström, C., J. Schelin, B. Norling, H. Vigre, J. Hoorfar, and P. Rådström. 2010. Culture-independent quantification of *Salmonella enterica* in carcass gauze swabs by flotation prior to real-time PCR. *Int. J. Food Microbiol.* Epub ahead of print. doi:10.1016/j.ijfoodmicro.2010.03.042.

Mackay, I. M. 2004. Real-time PCR in the microbiology laboratory. *Clin. Microbiol. Infect.* **10:**190–212.

Malorny, B., C. Löfström, M. Wagner, N. Krämer, and J. Hoorfar. 2008a. Enumeration of *Salmonella* in food and feed samples by real-time PCR for quantitative microbial risk assessments. *Appl. Environ. Microbiol.* **74:**1299–1304.

Malorny, B., S. Huehn, H. Dieckmann, N. Krämer, and R. Helmuth. 2009. Polymerase chain reaction for the rapid detection and serovar-identification of *Salmonella* in food and feeding stuff. *Food Anal. Methods* **2:**81–95.

Malorny, B., E. Junker, and R. Helmuth. 2008b. Multi-locus variable-number tandem repeat analysis for outbreak studies of *Salmonella enterica* serotype Enteritidis. *BMC Microbiol.* **8:**84.

Malorny, B., D. Mäde, P. Teufel, C. Berghof-Jäger, I. Huber, A. Anderson, and R. Helmuth. 2007. Multicenter validation study of two blockcycler- and one capillary-based real-time PCR methods for the detection of *Salmonella* in milk powder. *Int. J. Food Microbiol.* **117:**211–218.

Malorny, B., E. Paccassoni, P. Fach, C. Bunge, A. Martin, and R. Helmuth. 2004. Diagnostic real-time PCR for the detection of *Salmonella* in food. *Appl. Environ. Microbiol.* **70:**7046–7052.

Mazumdar, S. D., B. Barlen, P. Kämpfer, and M. Keusgen. 2010. Surface plasmon resonance (SPR) as a rapid tool for serotyping of *Salmonella*. *Biosens. Bioelectron.* **25:**967–971.

NordVal/NMKL. 2007. *NordVal Certificate no 031.* NordVal/NMKL, Søborg, Denmark.

Pelludat, C., R. Prager, H. Tschäpe, W. Rabsch, J. Schuchhardt, and W. D. Hardt. 2005. Pilot study to evaluate microarray hybridization as a tool for *Salmonella enterica* serovar Typhimurium strain differentiation. *J. Clin. Microbiol.* **43:**4092–4106.

Quinn, C., J. Ward, M. Griffin, D. Yearsley, and J. Egan. 1995. A comparison of conventional culture and three rapid methods for the detection of *Salmonella* in poultry feeds and environmental samples. *Lett. Appl. Microbiol.* **20:**89–91.

Rahn, K., S. A. De Grandis, R. C. Clarke, S. A. McEwen, J. E. Galán, J. E. C. Ginocchio, R. Curtiss III, and C. L. Gyles. 1992. Amplification of an *invA* gene sequence of *Salmonella typhimurium* by polymerase chain reaction as a specific method of detection of *Salmonella*. *Mol. Cell. Probes* **6:**271–279.

Ramisse, V., P. Houssu, E. Hernandez, F. Denoeud, V. Hilaire, O. Lisanti, F. Ramisse, J. D. Cavallo, and G. Vergnaud. 2004. Variable number of tandem repeats in *Salmonella enterica* subsp. *enterica* for typing purposes. *J. Clin. Microbiol.* **42:**5722–5730.

Reynisson, E., M. H. Josefsen, M. Krause, and J. Hoorfar. 2006. Evaluation of probe chemistries and platforms to improve the detection limit of real-time PCR. *J. Microbiol. Methods* **66:**206–216.

Rijpens, N., L. Herman, F. Vereecken, G. Jannes, J. De Smedt, and L. De Zutter. 1999. Rapid detection of stressed *Salmonella* spp. in dairy and egg products using immunomagnetic separation and PCR. *Int. J. Food Microbiol.* **46:**37–44.

Ritter, V., and N. Dick. 2009. USDA FSIS and FDA BAM culture methods BBL CHROMagar *Salmonella* prepared plated and Difco dehydrated culture media. *J. AOAC Int.* **92:**459–470.

Ross, I. L., and M. W. Heuzenroeder. 2008. A comparison of three molecular typing methods for the discrimination of *Salmonella enterica* serovar Infantis. *FEMS Immunol. Med. Microbiol.* **53:**375–384.

Salam, F., and I. E. Tothill. 2009. Detection of *Salmonella* typhimurium using an electrochemical immunosensor. *Biosens. Bioelectron.* **24:**2630–2636.

Schönenbrücher, V., E. T. Mallinson, and M. Bülte. 2008. A comparison of standard cultural methods for the detection of foodborne *Salmonella* species including three new chromogenic plating media. *Int. J. Food Microbiol.* **123:**61–66.

Seo, K. H., I. E. Valentin-Bon, and R. E. Brackett. 2006. Detection and enumeration of *Salmonella* Enteritidis in homemade ice cream associated with an outbreak: comparison of conventional and real-time PCR methods. *J. Food Prot.* **69:**639–643.

Smith, P. J., A. Boardman, and P. C. Shutt. 1989. Detection of salmonellas in animal feeds by electrical conductance. *J. Appl. Bacteriol.* **67:**575–588.

Sofos, J. N. 1993. Current microbiological considerations in food preservation. *Int. J. Food Microbiol.* **19:**87–108.

Suo, B., Y. He, G. Paoli, A. Gehring, S.-I. Tu, and X. Shi. 2010. Development of an oligonucleotide-based microarray to detect multiple foodborne pathogens. *Mol. Cell. Probes* **24:**77–86.

Tu, S., A. Gehring, and G. Paoli. 2007. Detection of *Salmonella* Enteritidis in shell egg contents by immunomagnetic capture and time-resolved fluorescence. *J. Rapid Methods Automat. Microbiol.* **15:**107–119.

Ueda, S., and Y. Kuwabara. 2009. The rapid detection of *Salmonella* from food samples by loop-mediated isothermal amplification (LAMP). *Biocontrol Sci.* **14:**73–76.

U.S. Department of Agriculture—Food Safety and Inspection Service. 1998a. *Supervisory Guideline for the Pathogen Reduction/HACCP Regulatory*

Requirements. http://smas.chemeng.ntua.gr/miram/files/publ_275_11_2_2005.pdf.

U.S. Department of Agriculture—Food Safety and Inspection Service. 1998b. Salmonella *Enteritidis Risk Assessment—Shell Eggs and Egg Products. Final Report Prepared for the Food Safety and Inspection Service by the Salmonella Enteritidis Risk Assessment Team*. http://www.fsis.usda.gov/ophs/risk/contents.htm. Accessed 26 January 2009.

U.S. Department of Agriculture—Food Safety and Inspection Service. 2008. *USDA-FSIS Method for Isolation and Identification of* Salmonella *from Meat, Poultry, and Egg Products, QD-F-Micro-0004.03*. http://www.fsis.usda.gov/PDF/MLG_4_04.pdf. Accessed 8 April 2010.

U.S. Department of Agriculture—Food Safety and Inspection Service. 2010. *Progress Report on* Salmonella *Testing of Raw Meat and Poultry Products, 1998–2009*. http://www.fsis.usda.gov/science/progress_report_salmonella_testing/index.asp. Last modified 8 June 2009.

U.S. Food and Drug Administration. 2007. *Salmonella. Bacteriological Analytical Manual*, chapt. 5. http://www.fda.gov/Food/ScienceResearch/LaboratoryMethods/BacteriologicalAnalyticalManualBAM/ucm070149.htm#Isol. Accessed 8 April 2010.

Valadez, A., C. Lana, S.-I. Tu, M. Morgan, and A. Bhunia. 2009. Evanescent wave fiber optic biosensor for *Salmonella* detection in food. *Sensors* **9:**5810–5824.

van Hoek, A. H., and H. J. M. Aarts. 2008. Microarray-based detection of antibiotic resistance genes in *Salmonella*. *Food Anal. Methods* **1:**95–108.

Wang, X. W., L. Zhang, L. Q. Jin, M. Jin, Z. Q. Shen, S. An, F. H. Chao, and J. W. Li. 2007. Development and application of an oligonucleotide microarray for the detection of food-borne bacterial pathogens. *Appl. Environ. Microbiol.* **76:**225–233.

Warren, B. R., H. G Yuk, and K. R. Schneider. 2007. Detection of *Salmonella* by flow-through immunocapture real-time PCR in selected foods within 8 hours. *J. Food Prot.* **70:**1002–1006.

Wattiau, P., M. Van Hessche, C. Schlicker, V. H. Vander, and H. Imberechts. 2008. Comparison of classical serotyping and PremiTest assay for routine identification of common *Salmonella enterica* serovars. *J. Clin. Microbiol.* **46:**4037–4040.

Webb, K., and V. Ritter. 2009. CHROMagar *Salmonella* detection test kit. Performance tested method 020502. *J. AOAC Int.* **92:**1906–1909.

Wolffs, P., R. Knutsson, B. Norling, and P. Rådström. 2004. Rapid quantification of *Yersinia enterocolitica* in pork samples by a novel sample preparation method, flotation, prior to real-time PCR. *J. Clin. Microbiol.* **42:**1042–1047.

Wolffs, P., B. Norling, J. Hoorfar, M. Griffiths, and P. Rådström. 2005. Quantification of *Campylobacter* spp. in chicken rinse samples by using flotation prior to real-time PCR. *Appl. Environ. Microbiol.* **71:**5759–5764.

Wolffs, P. F., K. Glencross, R. Thibaudeau, and M. W. Griffiths. 2006. Direct quantitation and detection of salmonellae in biological samples without enrichment, using two-step filtration and real-time PCR. *Appl. Environ. Microbiol.* **72:**3896–3900.

Wolffs, P. F., K. Glencross, B. Norling, and M. W. Griffiths. 2007. Simultaneous quantification of pathogenic *Campylobacter* and *Salmonella* in chicken rinse fluid by a flotation and real-time multiplex PCR procedure. *Int. J. Food Microbiol.* **117:**50–54.

World Health Organization/Food and Agriculture Organization of the United Nations. 2002. Risk assessment for *Salmonella* in eggs and broiler chickens. Microbiological Risk Assessment Series No 2. World Health Organization, Food and Agriculture Organization of the United Nations. http://www.fao.org/DOCREP/005/Y4392E/Y4392E00.HTM.

Yang, L., and R. Bashir. 2008. Electrical/electrochemical impedance for rapid detection of foodborne pathogenic bacteria. *Biotechnol. Adv.* **26:**135–150.

Yoshida, C., K. Franklin, P. Konczy, J. R. McQuiston, P. I. Fields, J. H. Nash, E. N. Taboada, and K. Rahn. 2007. Methodologies towards the development of an oligonucleotide microarray for determination of *Salmonella* serotypes. *J. Microbiol. Methods* **70:**261–271.

YERSINIA ENTEROCOLITICA IN PORK

M. Fredriksson-Ahomaa, T. Nesbakken, M. Skurnik, S. Thisted Lambertz, J. S. Dickson, J. Hoorfar, and H. Korkeala

14

Foodborne bacteria transmitted to humans via contaminated pork have a major impact, especially in developed countries. Among these bacteria, *Yersinia enterocolitica* has been characterized by one of the highest scores of risk for pork consumers (Fosse et al., 2009). Pigs are considered to be the most important reservoir for human pathogenic *Y. enterocolitica* strains. The most common bioserotype in pigs in Europe is 4/O:3, whereas bioserotypes 2/O:5 and 2/O:9 are only sporadically found. However, in England, bioserotypes 2/O:5 and 2/O:9 are commonly found in pigs at slaughter (Martínez et al., 2010), while in the United States, serotype O:5 (Bhaduri and Wesley, 2006) is common. By use of both culture and serology, the prevalence of *Y. enterocolitica* in pigs has been shown to vary among pig farms (Skjerve et al., 1998; Laukkanen et al., 2009). This indicates that there are specific farm factors affecting *Y. enterocolitica* prevalence in pigs. There is evidence that *Y. enterocolitica* is more common in specialized slaughter pig production than in farrow-to-finish production (Skjerve et al., 1998).

Pigs are mostly asymptomatic carriers of pathogenic *Y. enterocolitica*, and *Y. enterocolitica*-positive pigs cannot be identified during slaughtering. Pathogenic *Y. enterocolitica* strains have been isolated from the tonsils, feces, and submaxillary lymph nodes of asymptomatic pigs at slaughter. The prevalence rates have been shown to be higher among fattening pigs than among sows and higher in tonsils than in feces and lymph nodes, with both culture and PCR (Fredriksson-Ahomaa and Korkeala, 2003; Nesbakken et al., 2003). During the slaughter process, this pathogen can easily be transmitted from tonsils and intestinal content to other parts of the carcass and offal and to the environment (Fredriksson-Ahomaa et al., 2006; Laukkanen et al., 2009). Pathogenic *Y. enterocolitica* strains have only occasionally been isolated from the environment in slaughterhouses and butchers' shops, but by use of PCR this pathogen has been demonstrated to

M. Fredriksson-Ahomaa and H. Korkeala, Department of Food Hygiene and Environmental Health, Faculty of Veterinary Medicine, P.O. Box 66, University of Helsinki, FI-00014 Helsinki, Finland. *T. Nesbakken*, Norwegian School of Veterinary Science, Dept. of Food Safety and Infection Biology, P.O. Box 8146 Dep., 0033 Oslo, Norway. *M. Skurnik*, Haartman Institute, Department of Bacteriology and Immunology, P.O. Box 21, University of Helsinki, FI-00014 Helsinki, Finland. *S. Thisted Lambertz*, National Food Administration, P.O. Box 622, SE-75126 Uppsala, Sweden. *J. S. Dickson*, Department of Animal Science, Inter-Department Program in Microbiology, 215F Meat Laboratory, Iowa State University, Ames, IA 50011. *J. Hoorfar*, National Food Institute, Technical University of Denmark, Mørkhøj Bygade 19, DK-2860 Søborg, Denmark.

Rapid Detection, Identification, and Quantification of Foodborne Pathogens, ed. by J. Hoorfar

be widely spread in the slaughterhouse environment. There are several slaughter and meat inspection procedures, such as evisceration and incision of the submaxillary lymph nodes, whereby contamination can occur; therefore, strict slaughter hygiene remains important in reducing contamination in slaughterhouses.

Consumption of pork has been associated with *Y. enterocolitica* infections in epidemiological studies (Boqvist et al., 2009). Raw pork products have been widely investigated due to the link between *Y. enterocolitica* and pigs. However, the isolation rates of pathogenic *Y. enterocolitica* have been low, probably due to the low sensitivity of the culture methods (Fredriksson-Ahomaa and Korkeala, 2003). By use of PCR, pathogenic *Y. enterocolitica* has been detected frequently, especially on pig tongues but also on the surface of freshly slaughtered pig carcasses, in minced pork, and in fermented sausages (Lindblad et al., 2007; Fredriksson-Ahomaa et al., 2010). *Y. enterocolitica* can easily be killed by heat; thus, cross-contamination from raw products to heat-treated products should be avoided.

CHAIN FLOW: SAMPLE TYPES, SIZE, TREATMENT, AND STORAGE

Meat Chain

There are several points in the meat chain, from farm to table, at which control measures can be implemented to prevent or minimize the spread of *Y. enterocolitica* from slaughter

TABLE 1 Sampling and testing based on the most significant preventive measures to minimize spread of *Y. enterocolitica* from slaughter pigs, via meat and meat products, to humans

Process step	Significance	Preventive action	Sampling site	Methods for different sample types	
				PCR/culture	Serology
Herd level	Contamination from *Y. enterocolitica*-positive pigs	Purchase of *Y. enterocolitica*-negative pigs (CP)	Herd	Feces	Blood
Herd level	Prevalence studies	Avoidance of cross-contamination (CP)	Abattoir[a]	Tonsils, feces	Blood, meat juice
Abattoir: evisceration	Documentation of slaughter hygiene	Avoidance of cross-contamination (CP)	At the end of the slaughter line	Carcass surfaces	
Abattoir: decontamination[b]	Documentation of successful process	Time and temperature (CCP)	After decontamination	Carcass surface	
Processing	End-point control	Avoidance of recontamination (CP)	After heat treatment and possible recontamination	Meat samples	
Packaging[c]	Documentation of food safety	Packaging (CP)	End of maximum storage period	Meat samples	

[a] The abattoir is the most practical sampling site because tonsil and meat samples can easily be collected after evisceration and blood samples during bleeding.

[b] In Europe, surface pasteurization of carcasses is allowed only with hot water or steam, but in the United States, organic acids are also allowed.

[c] A CO_2 level of about 60% inhibits the growth of *Y. enterocolitica* (Nissen et al., 2000).

pigs, via meat and meat products, to humans (Table 1). Sampling might be performed at different stages and might represent documentation of critical points (CPs) connected to good hygiene practice and critical control points (CCPs) connected to hazard analysis and critical control points (Table 1). Interventions based on risk factors and procedures that might be controlled and/or documented by testing are (i) pigs and piglets purchased from other herds with an unknown carrier state for human pathogenic *Y. enterocolitica*, (ii) evisceration procedures, (iii) decontamination, (iv) heat treatment (lethality), and (v) packaging procedures.

Regarding the purchase of pigs and piglets from other herds, Skjerve et al. (1998) have shown that occurrence of antibodies against *Y. enterocolitica* O:3 was less frequent in mixed breeding-finishing herds than in fattening herds in which piglets are purchased from other herds. In regard to the evisceration procedures, the sealing or tying off of the rectum and covering with a plastic bag immediately after it has been freed can significantly reduce the spread of human pathogenic *Y. enterocolitica* (Nesbakken, 2009). It is also important that the operator who is responsible for the offal (plucks, i.e., tongue, heart, lungs, diaphragm, and liver) evisceration remove the tonsils together as a whole with the tongue and the gullet. During the circumanal incision and the removal of the plucks, the two-knife method is compulsory. In this method, one knife is in use and the other is immersed in water at around 80°C. After every unclean operation, the knife is rinsed and placed in the water decontaminator (temperature, ~80°C).

The use of surface pasteurization of pig carcasses represents an efficient approach to achieve a general reduction in the number of enteric pathogens, such as *Y. enterocolitica*, on carcasses. Treatment of carcasses with hot water (>70°C) or stream techniques to remove surface contamination is an option for post-harvest interventions (Nesbakken, 2009). In the countries in which organic acid rinses are allowed, lactic or acetic acid in combination with water washing has the potential to lower microbial contamination on the carcass surfaces (Eggenberger-Solorzano et al., 2002). These decontamination techniques might significantly reduce or even eliminate the occurrence of *Y. enterocolitica* on pig carcasses. However, these techniques are so far not widely used in pig slaughterhouses in the European Union. In North America, most slaughter establishments use at least one form of decontamination intervention. Regarding heat treatment, the fact that recontamination after that procedure may result in relatively unhindered growth of the pathogenic *Y. enterocolitica* during refrigerated storage emphasizes the need for appropriate measures and documentation targeted at prevention of recontamination after heat treatment (Nesbakken, 2009). Regarding packaging, Nissen et al. (2000) have shown that the growth of *Y. enterocolitica* 4/O:3 was nearly totally inhibited both at 4°C and at 10°C in modified atmosphere containing >60% CO_2.

Sampling

Blood Samples

Blood is collected in vacutainers from the *vena jugularis externa*, or blood samples are taken during bleeding of the animals after stunning. Blood samples are kept at 4°C and brought to the laboratory. Samples are centrifuged (for 10 min at 3,500 × *g*), and the serum is collected and frozen at −20°C until further transport and analysis. Sera are analyzed for antibodies against *Y. enterocolitica* O:3 and O:9 (European Food Safety Authority, 2009), for example, by an indirect pig immunoglobulin lipopolysaccharide enzyme-linked immunosorbent assay (ELISA) (Nielsen et al., 1996). Sera can also be analyzed for antibodies against *Yersinia* outer proteins (Yops) of pathogenic *Yersinia* spp. by using an ELISA (Pigtype Yopscreen; LDL, Leipzig, Germany). A basic cutoff of optical density is used to maximize the specificity of the ELISA.

Meat Samples (Muscle Fluid) for Serology

Muscle tissue juice or meat "juice" from diaphragm, sampled during *Salmonella* monitoring, is also suitable for detection of pathogenic

Yersinia. A meat sample consisting of at least 10 g is collected from the diaphragm pillar. The sample is tagged for identification and frozen for later shipment to the laboratory. The frozen meat samples can be stored for several months at −20°C before testing. The muscle fluid is released from the thawed meat (for a description of the meat, see Hoorfar et al., 1997).

Feces

Fecal samples are aseptically collected from the rectum of the pigs by use of a clean plastic glove, put into sterile plastic bags intended for use in a peristaltic homogenizer ("stomacher"), placed into a box, and held at 4°C. The samples are transported to the laboratory where analyses are initiated within 24 h after collection.

Tonsils

According to the European Food Safety Authority (2009), "On the slaughter line, tonsil samples are collected from plucks with intact tonsils." The plucks may be conveniently transferred to a tray or hook during sampling. By use of disposable plastic gloves, the tonsils are removed with sterile utensils, i.e., scissors, and placed in a container for transport to the laboratory. In case the slaughter procedure leaves the tonsils in the head region of the carcass, samples must be taken from the carcass. In this case, sampling should also be made from intact tonsils. The removed tonsils are then placed in a container for transport to the laboratory.

Samples may conveniently be collected at the platform used by the meat inspection. Sinks with soap and hot water should be within easy reach, as well as a sterilizer with running hot water (82°C). Gloves should be changed after each sample. The gloves may be ordinary disposable gloves, used routinely for laboratory work, as they may be regarded as commercially sterile. The utensils used for collecting the tonsils should be sterile. This may be achieved by using disposable utensils or by sterilizing the utensils between operations. Sterilization may be done using ethanol or hot water. The utensils should be cleaned by washing prior to sterilization. The samples are transported to the laboratory, where analyses are initiated within 24 h after collection.

Carcass Surface Samples

Carcass surface samples are often collected just subsequent to meat inspection, before the carcasses are chilled. Using a nondestructive method, sample sites are rubbed thoroughly with sterile swabs dipped into sterile peptone water, with sterile sponges moistened with 10 ml of peptone water, or with a sterile gauze pad moistened with 10 ml of peptone water, which are put into sterile plastic bags. According to the decision of the Commission of the European Communities (2001), the sampling area (cheek, belly, back, and ham) should cover at least 100 cm^2 per sampling site (total area, 400 cm^2). The samples should be stored refrigerated at 4°C until examination. The samples are transported to the laboratory, where analyses are initiated within 24 h after collection.

Meat Samples at Retail

Either the entire product (raw pork, sausages, minced meat, etc.) or a representative portion of 100 to 150 g should be taken initially. The food items are then transported and kept chilled or refrigerated until immediately before analysis, which should take place either the same day or the day after arrival at the laboratory. Using a sterile instrument and an aseptic technique, the test portion of between 10 and 25 g (preferably 25 g) should be sampled. Cuts of raw pork must include small pieces taken in a manner to represent different places of the surface of the cut. When a packaged product is sliced or cut, the respective sample is taken from more than one of the slices or cuts. The minced meat should be sampled to include both the surface and the inside of the initial test portion. To the test portion (25 g), 9 volumes (225 ml) of the diluent is added, and subsequently the mixture is homogenized in a stomacher or a pulsifier for

30 to 60 s before direct plating and enrichment. For quantification, direct plating can be used.

Water

Water samples should be collected in sterile flasks, and if necessary, a chemical should be added to neutralize any chlorine, and then samples should be placed into a box and held at 4°C. The samples are transported to the laboratory, where analyses are initiated as soon as possible and within 24 h after collection. From each sample, at least 100 ml (1,000 ml might be a more desirable volume) is passed through a 0.45-μm-pore-size filter, which might be placed face up on a selective agar or in a selective medium for enumeration or following a protocol for PCR.

RAPID DETECTION

Isolation of pathogenic *Y. enterocolitica* by conventional bacteriological analysis is laborious and may take up to 4 weeks. Thus, investigations have been focused on the development of rapid and reliable methods from animal carriers and food and environmental samples. PCR is so far the most frequently used rapid method (Skurnik et al., 2009). The latest published PCR methods are based on real-time PCR. Recently, microarray technology has also been developed to detect different genes directly in foods without bacterial isolation. A DNA microarray based on multiplex PCR and hybridization of four genes, *virF, ail, yst,* and *blaA*, for rapid detection of pathogenic *Y. enterocolitica* in fresh produce has been described (Siddique et al., 2009).

Preanalytical Sample Preparation

The sampling and sample preparation have a great influence on the sensitivity of the PCR. Since food surfaces are the primary site of contamination, a nondestructive swabbing procedure for enrichment prior to PCR reduces the amount of nonspecific food-derived DNA and possible inhibitory factors in the food matrix. Several sample preparation methods prior to PCR have been reported for detection in naturally contaminated samples (Fredriksson-Ahomaa and Korkeala, 2003; Skurnik et al., 2009). An enrichment step prior to PCR, which increases sensitivity and ensures the detection of viable cells, has been applied in most procedures. To inhibit the growth of competing microflora, selective enrichment broth can be used. Knutsson et al. (2002) have constructed a selective method in order to improve the sensitivity of the PCR. This medium, named Yersinia-PCR-compatible-enrichment medium, used prior to PCR, allows detection of low concentrations of target in the presence of high levels of background flora. For quantification, the DNA has to be extracted directly from the sample matrix; however, the detection limit is poor due to small volumes and high amount of nontemplate DNA (Skurnik et al., 2009). Various protocols are available for extraction of DNA from meat, and several companies deliver well-functioning, easy-to-use DNA extraction kits for different matrices. In most of the PCR methods designed for the detection of pathogenic *Y. enterocolitica* in pork, commercial DNA extraction kits have been used (Table 2).

Detection Technologies

The presence of *Y. enterocolitica* can be determined in several ways. Traditional culture methods or PCR may be used for pig tonsils, fecal material, and carcasses sampled at the slaughterhouse and for pork samples collected at retail level. When large numbers of animals are to be tested, prescreening with serological testing (for example, by ELISA) could be used to identify infected herds from which pathogenic *Y. enterocolitica* thereafter can be detected by PCR and culture methods.

Serological Methods

The antibody response measured in the ELISA is most likely stimulated by an invasion of the tonsils, as described by Shiozawa et al. (1991) and colonization of the intestines. A persistent serological response may similarly be caused by a chronic infection of the tonsils according

TABLE 2 PCR methods for detection of pathogenic *Y. enterocolitica* in pork samples

Target	Detection	Enrichment/DNA extraction	Reference
yadA	AGE[a]	Nonselective enrichment/NaOH and heating	Anonymous, 1998 (method 1)
ail, virF	AGE	Directly or with nonselective enrichment/ proteinase K and heating	Anonymous, 1998 (method 2)
ail	Real-time[b]	Selective enrichment/DNA extraction kit	Jourdan et al., 2000
ail	Real-time	Selective enrichment/DNA extraction kit	Boyapalle et al., 2001
ail, yadA	AGE		
yst	Real-time	Selective enrichment/DNA extraction kit	Visnubhatla et al., 2001
ail, virF	AGE	Selective enrichment/phenol and chloroform	Hudson et al., 2008
ail	Real-time	Nonselective enrichment/DNA extraction kit	Lambertz et al., 2008

[a] AGE, agarose gel electrophoresis.
[b] Real-time monitoring.

to Nielsen et al. (1996), who presented an indirect ELISA based on purified lipopolysaccharide from pathogenic *Y. enterocolitica* for screening herds, as demonstrated for serotype O:3. The commercial serological tests only cover serotypes O:3 and O:9, which may be a limitation, although the vast majority of European pathogenic *Y. enterocolitica* strains belong to the serotype O:3. To circumvent this drawback, a commercial kit detecting antibodies against Yops of pathogenic *Yersinia* spp. (Pigtype Yopscreen) can be used. However, by detecting Yops, it is not possible to differentiate between pathogenic *Y. enterocolitica* and *Yersinia pseudotuberculosis*. Serological testing may be performed on serum or muscle fluid, which may conveniently be collected at the time of slaughter.

Serological analysis could be used as an alternative method for the estimation of the prevalence of *Y. enterocolitica* in pig herds (Nielsen et al., 1996), since it is not as expensive and time-consuming as traditional culture methods. However, serological diagnosis is not equivalent to classical microbiological detection of the organism, as the serological response is delayed in comparison to the time of infection. Nonetheless, this could also be an advantage, as it allows identification of herds that have been or currently are infected with *Y. enterocolitica*. Furthermore, infection that has occurred during transport or lairage will not interfere with the results, nor will contamination that occurs during slaughtering. However, as the result is not equivalent to that obtained by classical detection, serological analysis should be performed as an additional method and should not stand alone, especially if collection of *Y. enterocolitica* strains for verification and typing may be of interest. Serological evaluation of herds may be a predictor of future contamination of meat from herds that are infected to a greater degree than others.

The serological diagnosis of *Y. enterocolitica* infection may be complicated by the existence of cross-reactions between *Y. enterocolitica*, most notably serotype O:9, and such organisms as *Y. pseudotuberculosis, Brucella, Vibrio, Salmonella, Proteus*, and *Escherichia coli* (Nielsen et al., 1996; Weynants et al., 1996). The correlation between the occurrence of *Y. enterocolitica* bioserotype 4/O:3 in tonsils and feces and the presence of antibodies in blood is presented in Fig. 1 and could be summarized as follows (Nesbakken et al., 2006):

1. Serological testing of pigs could be performed at all ages from about 100 days, including at slaughter when the pigs are 150 to 180 days old.

2. Bacteriological examination of feces could be used at all ages from 85 days, until about 135 days.

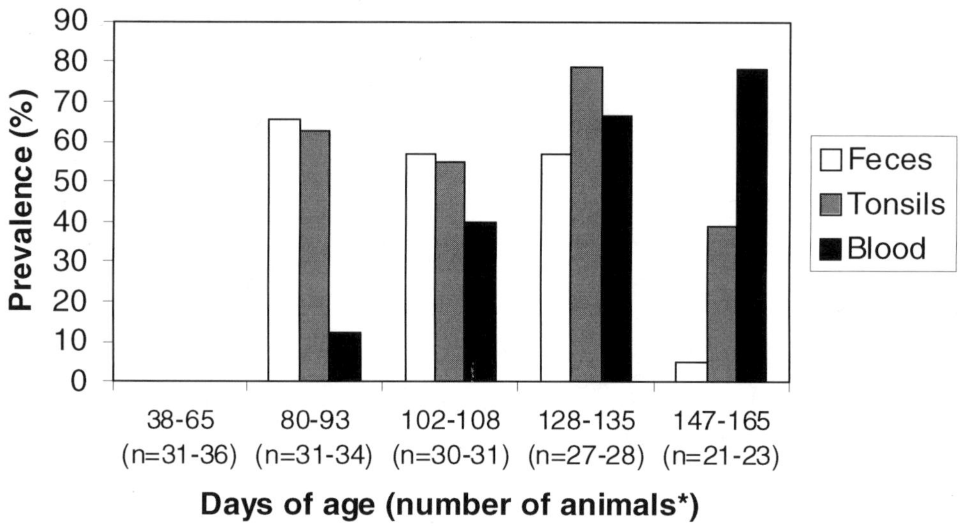

FIGURE 1 Occurrence of *Y. enterocolitica* bioserotype 4 / O:3 in samples of feces and tonsils, and occurrence of antibodies against *Y. enterocolitica* O:3 in blood samples, from different age groups of pigs from the two herds investigated by Nesbakken et al. (2009). Adapted from Lium et al. (2006) and Nesbakken (2008). *, numbers of animals differ since feces, tonsils, or blood samples were not tested from all animals due to unsuccessful sampling from some animals. Reprinted from the *Proceedings of the 19th International Pig Veterinary Society Congress, Copenhagen, Denmark*, p. 399 (Lium et al., 2006), with permission from the publisher.

3. Bacteriological examination of tonsils could be used at all ages from 85 days, including at slaughter when the pigs are 150 to 180 days old.

PCR Methods

PCR methods are designed to target either chromosomal or virulence plasmid-borne genes or both (Table 2). The *ail* gene located in the chromosome of pathogenic *Y. enterocolitica* strains is the most frequently used target. However, there is recent evidence that the *ail* gene can also be present in some biotype 1A strains (L. Sihvonen, S. Hallanvuo, K. Haukka, M. Skurnik, and A. Siitonen, submitted for publication). To be fully virulent in human, pathogenic *Y. enterocolitica* strains must harbor a virulence plasmid. Therefore, several methods use primers targeting the *virF* or *yadA* gene located on the plasmid. Since the virulence plasmid may be lost during subculturing, the plasmid alone is not considered a suitable target for the identification of this pathogen; thus, detection of both chromosomal and plasmid-borne genes is recommended. Two standardized conventional PCR methods for detection of pathogenic *Y. enterocolitica* in food have been reported by the Nordic Committee on Food Analysis (Anonymous, 1998). The increased interest in implementation of quality assurance and control measures in laboratories using PCR will ensure reliable PCR results in the future. According to the ISO standard 22174:2005, inclusion of controls to monitor the performance of the analysis is considered mandatory and must be implemented in each PCR. The inclusion of an internal amplification control will serve as an aid to indicate PCR inhibition and thus monitor for the occurrence of false-negative PCR results.

Culture

Validated methods like ISO 10273:2003 in Europe should be used for isolation of pathogenic *Y. enterocolitica*. The workload can be reduced considerably by examining the morphology of the colony grown on cefsulodin-irgasan-novobiocin and/or

Salmonella-Shigella agar with sodium deoxycholate and calcium chloride agar, using a stereomicroscope, when picking presumptive colonies for confirmation (Wauters et al., 1988; Hallanvuo et al., 2006). For confirmation, a reduced biotyping scheme has been presented (Hallanvuo et al., 2006). This scheme includes some simple pathogenicity-associated tests like esculin, salicin, and pyrazinamidase and use of a Congo red-magnesium oxalate agar plate to detect the presence of the virulence plasmid. Instead of phenotypic virulence tests, PCR can be used to detect pathogenic *Y. enterocolitica* isolates, as long as the PCR detects both chromosomal and plasmid virulence genes. If there are numerous suspected *Y. enterocolitica* colonies, 5 to 10 colonies can be pooled and tested in the same PCR run. Here it is recommended to use a PCR assay targeting a chromosomally located gene sequence, for example, the *ail* sequence. Presence of the virulence plasmid can be confirmed by using PCR targeting, for example, the *virF* and *yadA* genes. DNA can be extracted from colonies simply by boiling *Yersinia* bacteria in 100 μl of water for 10 min.

Example of a Real-Time PCR Method

The following probe-based real-time PCR method has recently been developed by Lambertz et al. (2008) for rapid detection of pathogenic *Y. enterocolitica* in meat. The PCR assay targets a sequence of the chromosomally located *ail* gene, which is present in all human pathogenic bioserotypes of *Y. enterocolitica.* This *ail*-specific real-time PCR is described in detail below.

Protocol: Enrichment of Pork Samples

1. Dilute 10 to 25 g in 90 to 225 ml (= 1: 10) of tryptone soya broth (30 g/liter) supplemented with yeast (6.0 g/liter) (= TSBY, pH 7.3).
2. Homogenize for 30 to 60 s.
3. Enrichment for 18 to 20 h at 25°C.
4. Mix the enrichment, leave to sediment 15 to 30 min at 4 to 8°C.
5. Transfer 1 ml of the homogenate for DNA extraction.

Protocol: DNA Extraction from the Enriched Homogenate

1. Centrifuge 1 ml (usually) of the enriched homogenate (1:10) for 10 min at approximately 13,000 × *g* to sediment the cells.
2. Discard the supernatant and extract the DNA from the precipitate using a ready-to-use tissue kit following the manufacturer's instruction. Use 5 μl of the extraction for PCR analysis. Store the remaining volume at −20°C for possible detection of virulence plasmid-borne genes like *virF*.

Real-Time PCR. The primer and probe sequences, the pipetting scheme, and the real-time conditions are given in Tables 3 to 5. The primers and the TaqMan-based probe are used for amplification and for detection of a 163-bp fragment from the chromosomally located *ail* gene (Table 3).

The PCR components have been optimized for the 7500 ABI system or a similar platform (Table 4). These components can be replaced by the 10× TaqMan Universal PCR Master mix (Applied Biosystems, Foster City,

TABLE 3 Primers and probe used for detection of *ail*-positive *Y. enterocolitica* by real-time PCR

Primer or probe	Sequence (5′–3′)	Position	Temp (°C)
F-real 10A	ATGATAACTGGGGAGTAATAGGTTCG	2379–2405	63.0
R-real 9A	CCCAGTAATCCATAAAGGCTAACATAT	2516–2542	61.5
YE probe	FAM-TCTATGGCAGTAATAAGTTTGGTCACGGTGATCT-TAMRA	2437–2470	68.1

TABLE 4 Pipetting scheme for the TaqMan real-time PCR assay

Component (stock concn)	Vol per reaction (final concn)
10× TaqMan buffer A	2.5 μl (1×)
$MgCl_2$ (25 mM)	3.5 μl (3.5 mM)
dNTPs (50 mM)	0.1 μl (200 μM)
AmpliTaqGold® (5 U/μl)	0.25 μl (0.02 U/μl)
Primer 1 (10 μM)	0.75 μl (300 nM)
Primer 2 (10 μM)	0.75 μl (300 nM)
Probe (20 μM)	0.25 μl (200 nM)
IAC	50–100 copies or about 1 fg of pUC 19 DNA
IAC forward primer	1.25 μl (500 nM)
IAC reverse primer	1.25 μl (500 nM)
IAC probe	0.25 μl (200 nM)
Test sample	5 μl
Adjust the volume to 25 μl using sterile distilled water	

CA) or by similar buffers from other suppliers. When naturally contaminated samples are tested, it is strongly recommended to use buffers containing dUTP instead of dTTP to prevent carryover. An open-formula pUC 19-based internal amplification control (IAC) developed by Fricker et al. (2007) or a commercially available TaqMan exogenous internal positive control (Applied Biosystems) can be used. The final volume for PCR is 25 μl including 5 μl of test sample (Table 4).

The PCR cycling parameters were optimized for the 7500 ABI system (Table 5). The same protocol was tested using the Biorad CFX96 real-time instrument. If dTTP is replaced by dUTP, uracil-*N*-glycosylase is added to the PCR mixture to inactivate the old amplified PCR products by adding an incubation step of 50°C for 2 min to the cycle parameters.

TABLE 5 PCR cycling parameters for the TaqMan real-time PCR assay

Cycles and steps	Temp-time combinations
Cycle 1 (× 1)	Denaturation
Step 1	95°C, 10 min
Cycle 2 (× 45)	Amplification
Step 1	95°C, 15 s
Step 2	60°C, 1 min

A PCR-negative result obtained both by the amplification control and by the target DNA is an indication of a PCR inhibition. These samples should be rerun after dilution of the sample by at least 1:10. Confirmation of PCR-positive results by culturing and characterization of the isolates (bio- and serotyping or DNA fingerprinting) is encouraged but is usually difficult to achieve due to the low detection efficiency of culturing methods. However, to increase the possibilities to isolate colonies, it is recommended to treat a volume of the enriched homogenate with a potassium hydroxide (KOH) solution before spreading onto cefsulodin-irgasan-novobiocin agar, proceeding as described in ISO 10273:2003. If characteristic colonies are not present after 24 h of enrichment, further incubation for another 24 h can be favorable and is highly recommended.

Rapid Enumeration Methods

Enumeration of pathogenic *Y. enterocolitica*, especially in food, is problematic due to the insensitive culture methods available. Hudson et al. (2008) have recently described a method for enumeration of pathogenic *Y. enterocolitica* in samples from pork surfaces. Pathogenic *Y. enterocolitica* was detected by a conventional multiplex PCR system targeting *ail* and *virF* genes, and the enumeration was achieved by adapting the enrichment to a most-probable-number method format. However, this method is quite laborious and expensive. The real-time PCR technique simplifies the quantification, but so far no method is available for quantification of *Y. enterocolitica* in naturally contaminated samples. This topic needs more research.

Rapid Characterization and Subtyping Methods

Rapid and specific identification of pathogenic *Y. enterocolitica* strains is necessary to eliminate confusion with nonpathogenic *Y. enterocolitica* and other *Yersinia* strains. Phenotypic identification of *Y. enterocolitica* strains is difficult due to the biochemical heterogeneity of this species. Furthermore, biochemical identification is laborious and time-consuming. This can be overcome by using PCR targeting chromosomal and/or virulence plasmid-borne genes, which are *Y. enterocolitica* specific (Fredriksson-Ahomaa et al., 2006). By using multiplex PCR, several virulence genes (*ail, inv, yst, yadA, virF, yopT*, etc.) can be detected in parallel (Skurnik et al., 2009). Oligonucleotide arrays based on a PCR method containing a combination of probes for taxonomic markers and species-specific virulence genes have been developed for several foodborne bacteria but so far not for characterization of *Y. enterocolitica.*

Subtyping has previously mostly relied on phenotypic characteristics such as biochemical properties (biotyping), O and H antigens (serotyping), antimicrobial susceptibility (antibiogram typing), and bacteriophage lysis patterns (phage typing). Biotyping is widely used and recommended by the European Food Safety Authority (2009) to identify strains belonging to biotypes 1B and 2 through 5, which are associated with human disease. However, this biotyping is laborious and time-consuming and no commercial kits are currently available. Serotyping can be rapidly done with commercially available antisera for serotypes O:1, O:3, O:5, O:8, O:9, and O:27, which are associated with human disease, but the same serotypes can also be found among nonpathogenic *Y. enterocolitica* and other *Yersinia* strains.

DNA-based molecular methods with high discriminatory power are needed for subtyping of *Y. enterocolitica* strains belonging to the same bioserotype to study transmission routes of this pathogen through the whole food chain. Several DNA-based methods have been used in molecular typing of *Y. enterocolitica* (Fredriksson-Ahomaa et al., 2006). Pulsed-field gel electrophoresis, which is so far the gold standard for typing, has proved to be highly discriminatory for strains belonging to the same bioserotype. However, this method is very laborious and time-consuming and does not allow automation. Automated ribotyping has widely been used to identify and characterize foodborne pathogenic bacteria. It allows to some extent rapid subtyping of strains within a given bioserotype. Amplified fragment length polymorphism has been applied successfully to a variety of foodborne pathogenic bacteria including *Y. enterocolitica.* Amplified fragment length polymorphism in combination with automated laser fluorescence analyzer allows relatively rapid subtyping of *Yersinia* strains. Multilocus variable number tandem repeat regions (VNTR) analysis is a PCR-based method that has recently been adapted for characterization of *Y. enterocolitica* strains. It is a promising method for the identification of different *Y. enterocolitica* strains belonging to the same bioserotype. This method is faster and less laborious than pulsed-field gel electrophoresis, and it allows high throughput and automation.

Gierczynski et al. (2007) have recently developed a multilocus VNTR-based subtyping method for strains belonging to bioserotype 4/O:3, the most common human pathogenic type found in pigs and pork. In this method, DNA is extracted from overnight culture incubated at 27°C by using a commercial kit.

Eight VNTR markers are amplified by using the primers at a concentration of 0.2 μM and a commercial master mix in a volume of 20 μl. The following PCR cycling parameters are used for all VNTR markers: initial denaturation at 94°C for 10 min; followed by 35 cycles each of denaturation at 94°C for 30 s, annealing at 58°C for 30 s, and elongation at 72°C for 30 s; and final extension at 72°C for 3 min. Electrophoresis analysis can be performed by using a 3% agarose gel in 1× Tris-borate-EDTA buffer at a constant voltage of 100 V for 7 h and visualized after staining with ethidium bromide (2 μg/ml). In order to obtain better resolution, VNTR markers can be separated by denaturing gel electrophoresis according to Gierczynski et al. (2007). A 20- or 50-bp DNA ladder is used as the size standard marker.

CONCLUSION AND FUTURE TRENDS

The serological analysis method can be used for the estimation of the prevalence of *Y. enterocolitica* in pig herds. The method has an advantage further down in the meat chain, since infection of pigs during transport and in the lairage will not interfere with the results. Even cross-contamination during slaughtering and dressing will not affect the result. Not only blood but also muscle fluid can be used as the basis for the analysis.

Real-time PCR is a powerful advancement of the basic PCR technique. A future challenge for sample preparation is to design PCR protocols that integrate DNA extraction and amplification in an automated manner. Additionally, high-throughput technologies such as DNA-microarray, mass spectrometry-PCR, and whole-genome sequencing platforms will strengthen the portfolio of molecular diagnostic tools for pathogenic *Yersinia* spp. The number of PCR and microarray assays based on novel biomarkers will increase in the future due to the growing number of sequenced *Yersinia* strains. This, along with the development of novel detection devices, will most likely allow the design of high-throughput systems that can cost-efficiently and simultaneously detect different pathogenic *Yersinia* strains from different types of samples.

Quantitative data are needed for quantitative microbial risk assessments. However, enumeration of *Y. enterocolitica* in foods is challenging. The methods available are time-consuming and laborious. There is thus an urgent need for rapid and specific enumeration methods. One alternative could be direct quantitation of *Y. enterocolitica* in food by real-time PCR. For that, sample preparation methods that separate pathogenic *Y. enterocolitica* from an appropriate amount of the food sample and concentrate the DNA in a small volume have to be further developed.

ACKNOWLEDGMENT

The Academy of Finland has partly supported the works of Hannu Korkeala and Maria Fredriksson-Ahomaa (project no. 118602) and the work of Mikael Skurnik (project no. 114075).

REFERENCES

Anonymous. 1998. *Pathogenic Yersinia enterocolitica. PCR Methods for Detection in Foods. NMKL Method NO. 163.* Nordic Committee on Food Analysis, Oslo, Norway.

Bhaduri, S., and I. Wesley. 2006. Isolation and characterization of *Yersinia enterocolitica* from swine feces recovered during the National Animal Health Monitoring System Swine 2000 Study. *J. Food Prot.* **69:**2107–2112.

Boqvist, S., H. Pettersson, A. Svensson, and Y. Andersson. 2009. Sources of sporadic *Yersinia enterocolitica* infection in children in Sweden, 2004: a case-control study. *Epidemiol. Infect.* **137:**897–905.

Boyapalle, S., I. V. Wesley, H. S. Hurd, and P. G. Reddy. 2001. Comparison of culture, multiplex, and 5′ nuclease polymerase chain reaction assay for the rapid detection of *Yersinia enterocolitica* in swine and pork products. *J. Food Prot.* **64:**1352–1361.

Commission of the European Community. 2001. Commission decision of 8 June 2001. *Off. J. Eur. Commun. L* **165:**8–53.

Eggenberger-Solorzano, L., S. E. Niebuhr, G. R. Acuff, and J. S. Dickson. 2002. Hot water and organic acid interventions to control microbiological contamination on hog carcasses during processing. *J. Food Prot.* **65:**1248–1252.

European Food Safety Authority. 2009. Technical specification for harmonized national surveys on

Yersinia enterocolitica in slaughter pigs. *EFSA J.* **7:** 1374–1397.

Fosse, J., H. Seegers, and C. Magars. 2009. Prevalence and risk factors for bacterial food-borne zoonotic hazards in slaughter pigs: a review. *Zoonosis Public Health* **56:**429–454.

Fredriksson-Ahomaa, M., and H. Korkeala. 2003. Low occurrence of pathogenic *Yersinia enterocolitica* in clinical, food, and environmental samples: a methodological problem. *Clin. Microbiol. Rev.* **16:** 220–229.

Fredriksson-Ahomaa, M., M. Lindström, and H. Korkeala. 2010. *Yersinia enterocolitica* and *Yersinia pseudotuberculosis*, p. 164–180. *In* V. K. Juneja and N. J. Sofos (ed.), *Pathogens and Toxins in Foods: Challenges and Interventions*. ASM Press, Washington, DC.

Fredriksson-Ahomaa, M., A. Stolle, and H. Korkeala. 2006. Molecular epidemiology of *Yersinia enterocolitica* infections. *FEMS Immunol. Med. Microbiol.* **47:**315–329.

Fricker, M., U. Messelhausser, U. Busch, S. Scherer, and M. Ehling-Schulz. 2007. Diagnostic real-time PCR assays for the detection of emetic *Bacillus cereus* strains in foods and recent food-borne outbreaks. *Appl. Environ. Microbiol.* **73:** 1892–1898.

Gierczynski, R., A. Golubov, H. Neubauer, J. N. Pham, and A. Rakin. 2007. Development of multiple-locus variable-number tandem-repeat analysis for *Yersinia enterocolitica* subsp. *palearctica* and its application to bioserogroup 4/O:3 subtyping. *J. Clin. Microbiol.* **45:**2508–2515.

Hallanvuo, S., J. Peltola, T. Heiskanen, and A. Siitonen. 2006. Simplified phenotypic scheme evaluated by 16S rRNA sequencing for differentiation between *Yersinia enterocolitica* and *Y. enterocolitica*-like species. *J. Clin. Microbiol.* **44:**1077–1080.

Hoorfar, J., A. Wedderkopp, and P. Lind. 1997. Detection of antibodies to *Salmonella* lipopolysaccharides in muscle fluid from cattle. *Am. J. Vet. Res.* **58:**334–337.

Hudson, J. A., N. J. King, A. J. Cornelius, T. Bigwood, K. Thom, and S. Monson. 2008. Detection, isolation and enumeration of *Yersinia enterocolitica* from raw pork. *Int. J. Food Microbiol.* **123:** 25–31.

Jourdan, A. D., S. C. Johnson, and I. V. Wesley. 2000. Development of a fluorogenic 5′ nuclease PCR assay for detection of the *ail* gene of pathogenic *Yersinia enterocolitica*. *Appl. Environ. Microbiol.* **66:**3750–3755.

Knutsson, R., M. Fontanesi, H. Grage, and P. Rådström. 2002. Development of a PCR-compatible enrichment medium for *Yersinia enterocolitica*: amplification precision and dynamic detection range during cultivation. *Int. J. Food Microbiol.* **72:**185–201.

Lambertz, S. T., C. Nilsson, S. Hallanvuo, and M. Lindblad. 2008. Real-time PCR method for detection of pathogenic *Yersinia enterocolitica* in food. *Appl. Environ. Microbiol.* **74:**6060–6067.

Laukkanen, R., P. O. Martínez, K.-M. Siekkinen, J. Ranta, R. Maijala, and H. Korkeala. 2009. Contamination of carcasse with human pathogenic *Yersinia enterocolitica* 4/O:3 originates from pigs infected on farms. *Foodborne Pathog. Dis.* **6:**681–688.

Lindblad, M., H. Lindmark, S. T. Lambertz, and R. Lindqvist. 2007. Microbiological baseline study of swine carcasses at Swedish slaughterhouses. *J. Food Prot.* **70:**1790–1797.

Lium, B., T. Iversen, K. Eckner, and T. Nesbakken. 2006. Prevalence of *Yersinia enterocolitica* O:3 in different age groups of pigs in naturally infected integrated herds, p. 388. *Proceedings of the 19th International Pig Veterinary Society Congress, 2006, Copenhagen.*

Martínez, P. O., S. Mylona, I. Drake, M. Fredriksson-Ahomaa, H. Korkeala, and J. Corry. 2010. Wide variety of bioserotypes of enteropathogenic *Yersinia* in tonsils of English pigs at slaughter. *Int. J. Food Microbiol.* **139:**64–69.

Nesbakken, T. 2009. Control of human pathogenic *Yersinia enterocolitica* in the meat chain. Thesis. Norwegian School of Veterinary Science, Oslo, Norway.

Nesbakken, T., K. Eckner, H. K. Høidal, and O. J. Røtterud. 2003. Occurrence of *Yersinia enterocolitica* and *Campylobacter* spp. in slaughter pigs and consequences for meat inspection, slaughtering and dressing procedures. *Int. J. Food Microbiol.* **80:** 231–240.

Nesbakken, T., T. Iversen, K. Eckner, and B. Lium. 2006. Testing of pathogenic *Yersinia enterocolitica* in pig herds based on the natural dynamic of infection. *Int. J. Food Microbiol.* **111:**99–104.

Nielsen, B., C. Heisel, and A. Wingstrand. 1996. Time course of the serological response to *Yersinia enterocolitica* O:3 in experimentally infected pigs. *Vet. Microbiol.* **48:**293–303.

Nissen, H., O. Alvseike, S. Bredholt, A. Hock, and T. Nesbakken. 2000. Comparison between the growth of *Yersinia enterocolitica, Listeria monocytogenes* and *Escherichia coli* O157:H7 and *Salmonella* spp. in ground beef packed by three commercially used packaging techniques. *Int. J. Food Microbiol.* **59:**211–220.

Shiozawa, K., T. Nishina, Y. Miwa, T. Mori, S. Akahane, and K. Ito. 1991. Colonization in the tonsils of swine by *Yersinia enterocolitica*. *Contrib. Microbiol. Immunol.* **12:**63–67.

Siddique, N., D. Sharma, and S. F. Khaldi. 2009. Detection of *Yersinia enterocolitica* in alfalfa, mung bean, cilantro, and mamey sapota food matrices using DNA microarray chip hybridization. *Curr. Microbiol.* **59:**233–239.

Skjerve, E., B. Lium, B. Nielsen, and T. Nesbakken. 1998. Control of *Yersinia enterocolitica* in pigs at herd level. *Int. J. Food Microbiol.* **45:**195–203.

Skurnik, M., P. Rådström, R. Knutsson, B. Segerman, S. Hallanvuo, S. Thisted Lambertz, H. Korkeala, and M. Fredriksson-Ahomaa. 2009. *Yersinia*, p. 501–518. *In* D. Liu (ed.), *Molecular Detection of Foodborne Pathogens*. CRC Press, Boca Raton, FL.

Vishnubhatla, A., R. D. Oberst, D. Y. C. Fung, W. Wonglumson, M. P. Hays, and T. G. Nagaraja. 2001. Evaluation of a 5′-nuclease (TaqMan) assay for the detection of virulent strains of *Yersinia enterocolitica* in raw meat and tofu samples. *J. Food Prot.* **64:**355–360.

Wauters, G., V. Goossens, M. Janssens, and J. Vandepitte. 1988. New enrichment method for isolation of pathogenic *Yersinia enterocolitica* serogroup O:3 from pork. *Appl. Environ. Microbiol.* **54:** 851–854.

Weynants, V., A. Tibor, P. A. Denoel, C. Saegerman, J. Godfroid, P. Thiange, and J.-J. Letesson. 1996. Infection of cattle with *Yersinia enterocolitica* O:9 a cause of the false positive serological reactions in bovine brucellosis diagnostic tests. *Vet. Microbiol.* **48:**101–112.

CAMPYLOBACTER IN POULTRY, PORK, AND BEEF

M. H. Josefsen, C. Carroll, K. Rudi, E. Olsson Engvall, and J. Hoorfar

15

INTRODUCTION TO THE PATHOGEN AND THE COMMODITY

Campylobacter infection has become one of the most important zoonoses worldwide. Close to 200,000 confirmed cases of human *Campylobacter* infection were reported to the European Food Safety Authority in 2008, corresponding to 40.7 cases per 100,000 population (European Food Safety Authority, 2010a). Finding similar numbers, the disease surveillance system of the United States, FoodNet, reported an incidence of human infection of 12.7 per 100,000; however, the actual number has been estimated to be 432 per 100,000 (Olson et al., 2008; Samuel et al., 2004). *Campylobacter* is also associated with a high number of human infections in developing countries, especially in the early years of life.

Campylobacters are commensals of a wide range of wild, domesticated, and food production animals. As they are able to persist in the environment for prolonged periods of time, even under unfavorable conditions (Rollins and Colwell, 1986), the routes of transmission to food production animals can be numerous. In most cases, they quickly establish themselves in the intestinal environment of the entire flock of food production animals, through fecal shedding (>10^6 CFU/g feces), coprophagy, and possibly other vectors, without causing any form of clinical symptoms (Dhillon et al., 2006; Kastma et al., 2007).

Transmission to human's is regarded to be mainly foodborne and related to food items of animal origin. Human infection is most often sporadic, including only a single family or community member at any one time, and the source of infection is seldom identified. This renders the source attribution (i.e., identifying the source of food causing the infection) of this pathogen difficult, but case-control studies indicate poultry and cross-contamination to other food items from poultry as the major sources of infection (Doorduyn et al., 2010; Eberhart-Phillips et al., 1997; Friedmann et al., 2004; Harris et al., 1986). Currently, a minor number of source attribution approaches have supported these findings (European Food Safety Authority, 2010b). The importance of poultry as a major source of human infection

M. H. Josefsen and J. Hoorfar, National Food Institute, Technical University of Denmark, Mørkhøj Bygade 19, DK-2860 Søborg, Denmark. *C. Carroll*, School of Natural Sciences, National University of Ireland, Galway, University Rd., Galway City, Ireland. *K. Rudi*, Nofima Mat, Norwegian Food Research Institute, Osloveien 1, 1430 Ås, Norway. *E. Olsson Engvall*, National Veterinary Institute, SE- 751 89 Uppsala, Sweden.

Rapid Detection, Identification, and Quantification of Foodborne Pathogens, ed. by J. Hoorfar

was emphasized during the Belgian dioxin crisis in 1999, when all poultry products were withdrawn from the market and the number of human infections declined accordingly (Vellinga and van Loock, 2002). The number of human infections also declined markedly in Iceland in 2000, as a consequence of a broad intervention scheme for poultry including freezing of *Campylobacter*-positive broiler flocks (Stern et al., 2003).

Of the thermotolerant campylobacters, *Campylobacter coli* is by far most often isolated from pig herds, presumably due to adaptation to this niche. The herd prevalence of *Campylobacter* in finishing pigs is close to 100%, and the individual shedding has been shown to be ~66% (Fosse et al., 2009). The prevalence in cattle herds has likewise been shown to be ~54% (Minihan et al., 2004; Pezzotti et al., 2003). In a 2008 European baseline survey, 71% of the broiler batches were shown to be colonized by *Campylobacter* at slaughter (European Food Safety Authority, 2010c).

Carcass contamination with intestinal contents often occurs during slaughter. Especially poultry can become extensively contaminated due to the nature of the slaughter process and the high levels of *Campylobacter* in feces. However, the intestinal content of *Campylobacter* in cattle and pigs has been reported to be much lower than the levels found in poultry (Stanley et al., 1998). Furthermore, postslaughter air chilling of red meat carcasses contributes to the reduction of contamination on beef and pork.

A low prevalence of *Campylobacter* is generally found in beef and pork at retail (European Food Safety Authority, 2010a; Ghafir et al., 2007; Wong et al., 2007), although they may still be sources of infection (Little et al., 2008). A high frequency of contamination has been reported for various kinds of red meat offal (Kramer et al., 2000; Little et al., 2008). In 2008, an average prevalence of 30% in European broiler meat sampled at slaughter, processing, and retail was reported to the European Food Safety Authority. However, a worldwide literature survey by Suzuki and Yamamoto (2009) shows an average prevalence in retail poultry meat of 60% (based on studies including analysis of more than 25 samples).

Based on the high prevalence of poultry-associated infections, this chapter will mainly focus on rapid methods for detection of *Campylobacter* in this particular production chain, also describing the routes of transmission and sampling in the different levels as well as intervention strategies.

Campylobacter jejuni subsp. *jejuni* and *C. coli* are by far the most important foodborne human pathogens in the genus and account for more than 95% of all human clinical isolates of *Campylobacter* worldwide (Lastovica and Allos, 2008). This number might be biased, since traditional detection methods favor isolation of these species compared to other potentially pathogenic but more sensitive *Campylobacter* species. The above-mentioned two thermotolerant species will albeit be the focus of this chapter, and in the following the term *Campylobacter* will refer to them unless otherwise stated.

CAMPYLOBACTER IN THE POULTRY PRODUCTION CHAIN

This section focuses on the introduction, infection dynamics, and sampling of *Campylobacter* throughout the poultry production chain, from farm to consumer level.

Farm Level

Commercially reared broiler flocks are easily colonized by *Campylobacter* spp. The reported prevalence of positive flocks varies between countries (European Food Safety Authority, 2010c). At the European level, 71.2% of broiler batches brought to slaughter were positive for *Campylobacter*, with a span in prevalence ranging from 2% to 100%. In countries with a colder climate, there is typically a seasonal variation in flock prevalence with low numbers of positive flocks during the winter and a peak during the summer (Hańsson et al., 2007a; Wedderkopp et al., 2000).

Age has been identified as a risk factor for *Campylobacter* colonization of broilers, partly ascribed to maternally derived immunity

(Bouwknegt et al., 2004). Chickens seldom become colonized before 2 to 3 weeks of age, and the risk for colonization increases with age (Corry and Atabay, 2001; Hansson et al., 2007a). Once the bacteria have been introduced into a flock, all birds are rapidly colonized, usually within a week (Jacobs-Reitsma et al., 1995). However, low within-flock prevalence at the time of slaughter has been reported, probably as a consequence of late introduction of the bacteria into the flock (Hansson et al., 2004).

Many routes for transmission of *Campylobacter* into a broiler flock have been identified, but the significance of different sources for flock colonization could vary between farms and regions. Campylobacters are mainly transmitted horizontally, from the environment or from a previously positive flock into a new flock. Vertical transmission, from the hen via the egg to the chicken, may occur but is generally considered of low significance (Newell and Fearnley, 2003). The sources in the surroundings could be domestic or wild animals, water puddles, or contaminated soil and litter. The bacteria are brought into the broiler house by the staff, catching teams, contaminated equipment, insects, or by air (Bull et al., 2006; Hald et al., 2004; Johnsen et al., 2006). In order to prevent the transmission, strict biosecurity measures with hygiene barriers should be applied (Hansson et al., 2007b). However, the high level of hygiene barriers is difficult to maintain in some countries and for some types of production (e.g., organic), and therefore, other on-farm intervention strategies are needed. The efficacy of immunization and treatment with bacteriophages or bacteriocins is currently investigated (Cox and Pavic, 2009).

In studying the epidemiology of *Campylobacter* in the primary production, samples are taken from broilers and from items in the environment that may be contaminated (Bull et al., 2006; Hansson et al., 2007b; Johnsen et al., 2006; Rosenquist et al., 2006). On-farm samples from broilers could consist of swab samples from fresh cecal droppings or sock samples (gauze socks pulled over the shoes). Cloaca swab samples could also be taken when birds are collected for transport to slaughter. Multiple swabs (5 to 10) are usually pooled into one composite sample.

For sampling the equipment and environment (ventilation shafts, floors, etc.), 10- by 10-cm gauze swabs, moistened with, e.g., buffered peptone water or a transport medium, could be used. Flying insects are collected and pooled and added to a diluent. Water of varying volumes (400 ml to 3 liters) is taken directly from the tap in the broiler house or outdoor water supplies and collected in sterile containers. Air can be sampled by special air samplers (Johnsen et al., 2006; Olsen et al., 2009).

Transportation to Slaughter

During the transportation of colonized birds to slaughter, excreted feces contaminate the exterior of the birds, the crates, and other equipment (Allen et al., 2008). To reduce the risk for cross-contamination of on-farm negative chickens, the crates used for transportation should be properly cleaned, disinfected, and dried before being dispatched to a new farm. However, inadequate cleaning of crates is a well-recognized problem (Hansson et al., 2005; Slader et al., 2002). To investigate the *Campylobacter*-free status after cleaning, samples of the crates could be taken by sterile gauze swabs, moistened with a diluent or transport medium.

Slaughter and Processing

The number of *Campylobacter* in ceca of colonized birds is usually very high, up to 10^{10} CFU/g cecum content or more (Stern and Robach, 2003). During processing, bacteria are transmitted to the processing water, equipment, and hands of the slaughterhouse workers (Corry and Atabay, 2001). The carcasses of colonized flocks easily become contaminated either directly via intestinal contents or indirectly via the water or equipment, but also noncolonized flocks may end up contaminated at the end of the slaughter line. In studies of prevalence of intestinally colonized versus carcass-contaminated flocks, the prevalence of flocks with carcass contamination usually exceeds the prevalence of colonized flocks. This was also found in the 2008 European Union baseline survey, which reported that the prevalence of *Campylobacter*-colonized broiler batches was 71.2% and that of contaminated broiler carcasses was 75.8% (European Food Safety Authority, 2010c).

Campylobacter bacteria will not multiply outside the gut of the host and are sensitive to dehydration but survive well in the wet environment during processing. The concentration of contaminating *Campylobacter* on the carcasses therefore depends on how much fecal contamination they are exposed to and to what extent different processing steps may remove the contamination. The mere presence of *Campylobacter* on carcasses at the beginning of the processing does not necessarily mean that the final product is contaminated. However, there is a positive correlation between the level of *Campylobacter* entering the slaughterhouse and the risk of carcass contamination (Lindblad et al., 2006; Rosenquist et al., 2003).

The poultry slaughter procedures are almost identical worldwide and include stunning, bleeding, scalding, defeathering, evisceration, washing, and chilling, but the processing machinery and application of automated equipment vary. This could have an impact on the potential to control the contamination of carcasses. During the slaughter process, bacterial levels vary at the different steps (Rosenquist et al., 2006). A decrease has been reported after scalding, but cross-contamination can occur in the scald tank; and after evisceration and defeathering, increasing numbers of *Campylobacter* are seen (Izat et al., 1988; Oosterom et al., 1983). The chilling process has been found to reduce the level of carcass contamination in several studies. The reduction has been reported to vary between chilling methods, but there is no consensus about the optimal method (Lindblad et al., 2006; Rosenquist et al., 2006).

Many actions/intervention strategies have been applied to decrease the prevalence of contaminated carcasses and reduce the numbers of *Campylobacter* on contaminated carcasses/poultry meat. The strategies include scheduled slaughter, improved hygiene measures, and decontamination. Scheduled slaughter is possible if the *Campylobacter* status of the flocks is known prior to transportation to the slaughterhouse. Positive flocks can be sent to dedicated slaughterhouses or be slaughtered at the end of the day to avoid cross-contamination of successive flocks (Havelaar et al., 2007; Hofshagen and Kruse, 2005). On-farm sampling can be performed by using fresh cecal droppings or sock samples. The samples must to be taken as close to the time of slaughter as possible, and rapid and reliable methods for analysis are needed in order to enable correct sorting of flocks prior to slaughter. The need for more rapid methods here is emphasized by the fact that almost 25% of negative flocks are contaminated between the time of the last sampling at the farm and the time of arrival at slaughterhouse (Rosenquist et al. 2006). Using semicontinuous air sampling in chicken houses might be a cost-effective solution.

Samples are collected at the slaughterhouse level, both for estimating the prevalence of colonized broiler flocks and to control the hygiene and identify targets for intervention measures. The types of samples include chicken ceca, neck skins, and carcasses and samples from different slaughter stages, i.e., scalding tank and equipment.

Retail and Consumer Levels

A significant portion of poultry meat at retail is reported to be contaminated with *Campylobacter* (European Food Safety Authority, 2009, 2010a). Consumption of undercooked contaminated meat or of ready-to-eat food items that have been cross-contaminated constitutes a risk for human health. In general, the risk of becoming ill depends on the numbers of bacteria and is more likely when the contamination level is higher (Nauta et al., 2005).

Different strategies for interventions could be applied at this level, such as freezing of poultry meat, which has been shown to reduce the numbers of *Campylobacter* by 2 $\log_{10}$ units (Oosterom et al., 1983), heat treatment, minimizing the handling of raw meat, improved kitchen hygiene, and consumer education.

Samples taken from retail products are analyzed for the presence and concentration of *Campylobacter* in order to identify positive samples and provide quantitative data for risk assessment studies.

RAPID DETECTION

Standardized reference culture methods, issued by national and international standardization organizations, are applied in control laboratories throughout the world. This culture-based approach is, however, not optimal, due to poor specificity, lack of detection of viable but nonculturable (VBNC) cells, and labor-intensive and time-consuming protocols (up to 6 days). The current International Organization for Standardization method (ISO 10272-1:2006) for the detection of *Campylobacter* spp. in food and animal feeding stuffs (International Organization for Standardization, 2006) is based on sample enrichment in Bolton broth (test portion/enrichment media ratio, 1:10) at 37°C for 4 h to 6 h, followed by 41.5°C for 44 ± 4 h in a microaerobic atmosphere. The sample is streaked onto selective media, and confirmation tests are performed with isolated colonies. In total, this method requires 6 days to produce a confirmed positive result. From a risk management perspective, such a slow method does not ensure the microbiological safety of products being released to the consumer.

Culture-based, immunological, and molecular methods for rapid detection, characterization, and enumeration have been described for *Campylobacter*. Rapid methods can generally be also more sensitive and specific than culture-based methods, and other advantages can be a high possibility of automation and detection of VBNC cells. The strength of rapid methods lies in their ability to screen large numbers of samples, identify the negative ones, allowing resources to be focused on confirming and culturing of presumptive positive samples to produce isolates for further characterization (see chapter 1).

The choice of a rapid method will always depend on the requested information and be influenced by the relevant matrix and the expected level of contamination. In the following sections, the most widely used rapid methods for *Campylobacter* will be described.

Preanalytical Sample Preparation

Sample preparation is the first and most important step in applying any rapid method for detection of foodborne pathogens. No matter what the downstream analysis (detection/identification/enumeration), the target bacteria or their DNA has to be separated from the inhibitory compounds of the food matrix and background flora, and a sample size reduction concentration is required for detection. Regarding *Campylobacter*, monitoring of poultry flocks *ante mortem* and throughout the production chain, as described in "Campylobacter in the Poultry Production Chain" above, means that the fecal sample matrix is also relevant, posing even higher demands on preanalytical sample preparation.

For nucleic acid-based rapid methods such as PCR, the exact components of the food/fecal/environmental sample directly affect the sensitivity and amplification efficiency of the assay, and therefore, no universal DNA extraction procedure exists. Each matrix and even varying samples within a matrix present a different challenge whereby the DNA extraction must be adjusted on a case-by-case basis.

For qualitative detection of *Campylobacter* in food with relatively low numbers of this pathogen, a culture-based enrichment step is often applied. Though extending the time of analysis, several advantages are obtained: the inhibitory constituents are diluted, the level of *Campylobacter* is increased, resuscitation of stressed and injured cells can take place, and finally the ratio between viable and dead cells is increased substantially. Following the enrichment step, a DNA extraction or a simple dilution step can be necessary due to the possible inhibition caused by the enrichment medium itself.

The need for quantitative data on *Campylobacter* contamination in the food production chains is, however, increasing. Rapid methods able to enumerate these bacteria are needed for modeling, risk assessment, evaluation of the effect of different intervention/reduction strategies, and knowledge on dynamics, distribution, and survival throughout the production chain.

In order to quantify a pathogen present in low levels (without preceding culture-based enrichment), other means of concentration or amplification have to be employed. Separation and concentration methods have been applied widely in food microbiology to separate the target organism from the food matrix and accompanying flora in very reduced volumes. Extracting DNA after removing the target cells from the food sample improves the yield and reduces the amount of inhibitory substances. Immunocapture is an example of a widely used separation strategy, whereby paramagnetic beads coated with, for example, antibodies to form a complex with the target organism are added and circulated with the sample before they are removed magnetically.

Using immunocapture of *Campylobacter* prior to PCR has been attempted with varying success. Liu et al. (2006) reported an improved limit of detection of 10 CFU/ml of food and water sample by using direct immunocapture followed by PCR, whereas Morales-Rayas et al. (2008) found poor recovery rates of *Campylobacter* cells in the complex food matrices. Waller and Ogata (2000) described a quantitative immunocapture PCR with a total time of analysis of 8 hours, which was able to detect 1 CFU/ml artificially inoculated milk or chicken skin wash. However, this needs further validation and reproduction.

Wolffs et al. (2005) described the use of flotation to single out viable and VBNC *Campylobacter* prior to quantitative PCR. The limit of detection of this method was determined to be 8.6×10^2 CFU/ml, and quantification was possible in the range of 2.6×10^3 to 2.6×10^7 CFU/ml in chicken carcass rinse. Also, buoyant density gradient centrifugation, whose routine application can be limited due to technical difficulty, has been demonstrated to concentrate *Campylobacter* 250-fold, allowing detection and quantification of 1×10^1 to 1×10^3 CFU/g (Fukushima et al., 2007).

An integrated cell concentration and DNA extraction approach followed by real-time PCR was shown to quantify *Campylobacter* in fecal samples from the poultry production with a limit of detection of 2 to 25 CFU/PCR. In this method, the same paramagnetic beads were added to the sample for capture of cells, separated from the fecal matrix, and transferred to DNA extraction (Rudi et al., 2004).

An integrated lab-on-a-chip technology, combining sampling of airborne particles, lysis of microorganisms, amplification of nucleic acids, signal generation, and wireless transmission in a microchip-based device (Jensen et al., 2005), has been described for bioterror agents. This technology is being developed with reference to semiautomatic monitoring of food production animals in the stables, and the feasibility of the approach was demonstrated by detection of *Campylobacter* in the air of broiler houses (Olsen et al., 2009).

Rapid Methods for Detection and Enumeration

Culture-Based Methods

Modifications to conventional culture-based detection constitute a significant part of rapid methods for detection of foodborne pathogens, including *Campylobacter*. Chromogenic culture media are increasingly used. Specific chromogenic enzyme substrates indicate the presence of campylobacters as specific colored colonies, and as such, they facilitate the direct selection and presumptive identification of campylobacters (in particular *C. jejuni* and *C. coli*) without the need for subculture and confirmatory tests. Their use incurs a number of associated disadvantages including a greater expense compared to conventional media, they confine detection to usually one or two targeted microorganisms, and they may be less sensitive than conventional media due to their high selectivity.

Rapid methods based on a combination of filtration and enrichment have also been developed for early detection of *Campylobacter*. In most cases, the enrichment time can be reduced by 1 day, and increased sensitivity compared with standard culture methods has been reported (Wisessombat et al., 2009).

Another culture-based rapid method employs a specially designed broth medium that allows for detection of low numbers of stressed *Campylobacter* cells by using capacitance measurement (Line and Pearson, 2003). In general, conductance and impedance techniques require specialized instrumentation with significant up-front costs and necessitate the establishment of stringently reproducible growth curve profiles of isolated strains under each temperature or culture condition of interest. Methods in this category have not been applied for detection of *Campylobacter* with much success.

Molecular Methods

The majority of molecular assays for *Campylobacter* in food are currently used with enriched samples for detection or identification due to the low level of contamination, and as such, they are qualitative or semiquantitative at best, depending on the limit of detection.

Immunochemical Assays. Several immunochemical rapid methods for detection of *Campylobacter* have been developed and published. Compared to standard culture methods, most of them perform with equal sensitivity and specificity after the enrichment step, while the total time of analysis is reduced substantially. However, the limit of detection of these immunological methods is generally around 10^4 to 10^5 CFU/ml, rendering them unable to detect *Campylobacter* directly unless analyzing heavily contaminated samples.

Enzyme-linked immunosorbent assay (ELISA) methods for detection of *Campylobacter* in food have been found to produce results comparable to culture, PCR, and lateral flow devices following enrichment (Bohaychuk et al., 2005; Hochel et al., 2007).

A combination of immunochemistry and PCR (PCR-ELISA) has, however, been reported to increase the sensitivity of the conventional PCR method 100- to 1,000-fold with detection limits as low as 40 *Campylobacter* cells/ml chicken carcass rinse (Hong et al., 2003).

PCR-Based Assays. PCR-based methods for *Campylobacter* offer the advantages of short assay times and the ability to identify different *Campylobacter* spp. and to target different genes. Numerous conventional, gel-based PCR methods for detection of *Campylobacter* have been published, targeting a variety of genes and species and developed for many different sample types. A majority of methods have been developed for detection in poultry and poultry products, reflecting the importance of these foods as a source of human *Campylobacter* infections. A majority of PCR primers applied for detection of *Campylobacter* target the 16S rRNA gene, while species-specific assays are often designed to recognize the *mapA* or *hipO* gene for *C. jejuni* and the *ceuE* gene for *C. coli* (Abubakar et al., 2007).

For PCR detection of the bacteria in food in which a relatively low amount of *Campylobacter* is suspected, an enrichment step prior to PCR is often applied. Although adding to the total time of analysis, several advantages are obtained by this: the PCR inhibitory sample constituents are diluted, the level of *Campylobacter* is increased to allow detection, and the ratio between viable and dead cells is increased substantially, making the contribution to PCR response from dead cells negligible.

Quantitative PCR. Several real-time PCR methods for *Campylobacter* have been described in the literature, albeit most of them concern detection and quantification *after* enrichment, which is a contradiction in terms. It has, however, been shown that a linear relationship exists between the threshold cycle values obtained in real-time PCR and the CFU in enriched samples (Josefsen et al., 2004), although not reflecting the exact number of *Campylobacter* cells present in the initial sample.

Direct quantitative real-time PCR methods for *C. jejuni* and *C. coli* have been published, mostly developed for the poultry matrix (Botteldoorn et al., 2008; Debretsion et al., 2007; Hong et al., 2007; Rönner and Lindmark, 2007). The total time of analysis applying these methods is usually approximately 3 hours, and the limit of quantification ranges from 1 to 100 CFU/ml of sample. Bearing in mind that several of the current control strategies are focused on postslaughter reduction of the number of bacterial cells on the chicken carcass, the usefulness of real-time PCR assays for quantification could be limited, since they detect all *Campylobacter* organisms present in a sample, including the dead cells.

PCR-Based Quantification of Viable Cells. Separate detection of viable, VBNC, and dead bacteria indeed poses a particular diagnostic challenge (Barer and Harwood, 1999; Chaveerach et al., 2003). On the one hand, traditional culture-based techniques will not detect the VBNC state, while PCR-based techniques cannot distinguish dead bacteria from the viable and the VBNC state. For *C. jejuni*, there has been a long-standing debate whether this bacterium can enter a VBNC state or not (Chaveerach et al., 2003). Whether or not a VBNC state exists, *C. jejuni* can be difficult to diagnose both by PCR and by traditional means. Therefore, alternative strategies are needed in the diagnostics of this bacterium.

Strategies that are currently being evaluated for separate detection of viable and dead bacteria are based on diagnostics of RNA rather than DNA (Sheridan et al., 1998; Uyttendaele et al., 1997). RNA is unstable in environmental matrices and is rapidly degraded when a bacterium dies, rendering RNA-based assays difficult to standardize.

Physical separation of viable and dead *C. jejuni* has successfully been achieved by differences in densities (Wolffs et al., 2005). This promising approach combines cell concentration with viable/dead diagnostics. A challenge here is different densities among viable bacteria due to differences in their energy state.

The most widely used strategy for viable/dead differentiation of bacteria is based on membrane permeability properties. A range of such approaches has been developed for microscopy and flow cytometry applications. Protocols based on the BacLight kit are by far the most widely applied (Boulos et al., 1999). The difficulty here is the lack of differentiation among bacterial species. Combinations of cytometry with cell sorting have been used to solve this problem (Caron et al., 1998). Single cell sorting, however, both requires expensive equipment and is difficult to standardize.

The first application of combining cell integrity and DNA diagnostic as a means for viable/dead detection was developed more than 10 years ago by the use of DNase exposure of DNA in dead cells (Nogva et al., 2000). More recent developments included the use of molecules that do not enter cells with an intact membrane and that can be covalently attached to DNA and inhibit PCR amplification (Nogva et al., 2003). Ethidium bromide monoazide (EMA) was the first molecule to be explored (Rudi et al., 2005). Later, propidium bromide monoazide (PMA) was successfully synthesized and applied in a manner similar to that of EMA (Nocker et al., 2006). The main difference between EMA and PMA is the molecular charge, with EMA being more prone to enter bacterial cells (Nocker et al., 2006). The use of EMA and PMA involves the separation of the bacteria from the food matrix as the first step prior to the treatment. The reasons for this are dye-scavenging and light-scattering effects. Light is used for cross-linking the dye to DNA. The next step is adding the dye in the dark, since the molecules are light sensitive. Then, the samples are exposed to very strong light from a halogen lamp. The light cross-links the dye to DNA and inactivates the remainder of free molecules. After approximately 1 minute of light exposure, the DNA can be purified by the protocol of choice.

A protocol for culture-independent enumeration of viable thermotolerant *Campylobacter* in chicken carcass rinse by quantitative PMA-PCR was described by Josefsen et al. (2010) as shown in Fig. 1.

Since EMA-PCR and PMA-PCR are direct detection techniques, sensitivity is a critical issue. For sample matrices containing low levels of solid debris, bacterial cells can be concentrated either by filtration or centrifugation. If a high level of solids is present, then some kind of density gradient centrifugation can be used for cell separation. The conditions for the cell separation must not kill target bacteria—leading to an overestimation of the dead cell population. This is a special consideration for *C. jejuni* due to its sensitivity to oxygen exposure.

The main challenge with the use of membrane permeability dyes in viable/dead diagnostics is that there are cell states between the viable metabolizing state and the completely dead one that will not be recovered. In future, combinations of dyes with different membrane permeability properties or activity-labile compounds for metabolic and respiratory active cells could be used to differentiate more accurately between cell states (Nocker and Camper, 2009).

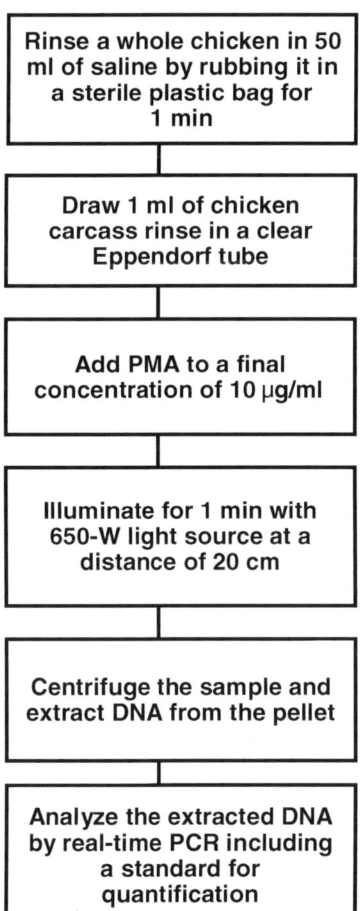

FIGURE 1 Schematic outline of the protocol for culture-independent enumeration of viable thermotolerant *Campylobacter* in chicken carcass rinse by quantitative PMA-PCR (Josefsen et al., 2010).

LAMP. Loop-mediated isothermal amplification (LAMP) enables the synthesis of larger amounts of DNA at a constant temperature between 60°C and 65°C, not requiring a thermal cycler. The method involves autocycling and strand displacement DNA synthesis involving the use of the large fragment of *Bst* DNA polymerase and a set of four primers recognizing six distinct regions on the target nucleotides, ensuring an extremely high specificity. Detection is performed by visual inspection of turbidity or by a colorimetric detection without electrophoresis. Direct analysis of *Campylobacter* in complex matrices by LAMP has shown slightly lower sensitivity and specificity than standard culture methods (Yamazaki et al., 2008), and this technique may be a solution for point-of-care analysis.

Biosensors. A surface plasmon resonance sensor based on direct recognition of unlabeled antibodies has been shown to quantify *C. jejuni*, simultaneously with three additional foodborne pathogens, in fluid matrices, with a limit of detection ranging from 4.4×10^4

to 1.2×10^5 CFU/ml (Taylor et al., 2006). This surface plasmon resonance biosensor approach was also reported to detect down to 1×10^3 CFU/ml *C. jejuni* in chicken carcass rinse, despite a high level of background from the carcass rinse matrix (Wei et al., 2007). An array biosensor using a sandwich assay with labeled antibodies has been shown to detect 9.7×10^2 CFU/ml *C. jejuni* in complex food matrices (Sapsford et al., 2004).

Other potential platforms in this category include an ion-channel biosensor based on antibodies against *Campylobacter* embedded into lipid membranes (Ivnitski et al., 2000), and an impedimetric immunosensor based on nanoparticles with immobilized anti-*flaA* antibodies for rapid detection of *C. jejuni* in fecal samples by electrochemical impedance spectroscopy (Huang et al., 2010).

Rapid Characterization and Subtyping

While *Campylobacter* was earlier subtyped by phenotypic methods like serotyping, biotyping, phage typing, and antibiotic sensitivity, the majority of subtyping today is done using genotypic methods (Wassenaar and Newell, 2000).

A variety of genotypic tests may be used for identification to the species level and subtyping of *Campylobacter* species. Each method has advantages and disadvantages, and often the best practice is to use a combination of different methods in a synergistic fashion. It is important, however, to keep in mind that the standard culture methods for *Campylobacter* isolation carry inherent limitations that may bias the results. The antimicrobial agents used in selective media, the high temperature, and the microaerobic atmosphere applied during enrichment can inhibit some of the more sensitive *Campylobacter* strains.

Species Identification

For identification of *Campylobacter* to the species level, traditional assays have combined PCR amplification of specific genomic targets with post-PCR restriction enzyme digestion of the PCR product to discriminate in particular *C. jejuni* and *C. coli*. Other assays have combined PCR amplification with a DNA probe hybridization step using nonradiolabeled oligonucleotide probes to detect and discriminate *Campylobacter* spp. based on the *ceuE* gene (Gonzalez et al., 1997), the *glyA* gene, and 16S/23S rRNA intergenic spacer (O'Sullivan et al., 2000). PCR/ELISAs have also been developed based on the 16S/23S rRNA intergenic spacer (Grennan et al., 2001). Species-specific microarrays have been described for the rapid identification of *Campylobacter* spp. targeting genes including *hipO, glyA, fur, cdtA* through *cdtC, ceuB* through *ceuE*, and *fliY*. Microarrays offer great potential for epidemiological studies, but their high associated costs make them impractical for routine use.

Novel multiplex PCR assays offer a rapid, easy, and cost-efficient way to distinguish between *Campylobacter* species in a test sample by using more than one pair of primers to simultaneously amplify many targets of interest in one reaction. Yamazaki-Matsune et al. (2007) describe a multiplex PCR assay for the identification of the six common *Campylobacter* species associated with human gastroenteritis, namely *C. coli, C. fetus, C. hyointestinalis* subsp. *hyointestinalis, C. jejuni, C. lari*, and *C. upsaliensis*. A real-time PCR assay to distinguish between the most important foodborne species (*C. jejuni, C. coli*, and *C. lari*) was published by Mayr et al. (2010), based on four different fluorescently labeled TaqMan probes.

Molecular Typing Methods

PFGE. Pulsed-field gel electrophoresis (PFGE) has been considered the "gold standard" in *Campylobacter* genotyping studies. PFGE allows analysis of the whole genome of *Campylobacter* species. *Campylobacter* species have a low G+C content (Vandamme, 2000), and an enzyme like SmaI with a G+C-rich recognition site will cut *Campylobacter* DNA infrequently, giving a limited number of bands that are easily discriminated. As a result of the infrequency of the cutting of the enzymes, the

DNA fragments obtained may be large, varying from 20 kb to 1 Mb. The banding patterns of strains can be compared to determine the genetic relatedness of the strains according to criteria developed by Tenover et al. (1995). Interpretation of PFGE results has been greatly enhanced through the development software packages specifically designed for the analysis of DNA fingerprint patterns. Analyses of band patterns and subsequent clustering of these patterns are performed using a number of statistical parameters such as the Dice coefficient, the Jaccard coefficient, and the unweighted-pair group method using average linkages.

AFLP. Amplified fragment length polymorphism (AFLP) analysis involves the digestion of chromosomal DNA with two restriction enzymes (a frequent cutter and a rare cutter). A number of restriction enzymes have been used in AFLP analysis of *Campylobacter*, with the combination of HindIII and HhaI regarded as optimal (Duim et al., 1999; Lindstedt et al., 2000). The method generates PCR products of 50 to 500 bp in size, which can be separated by polyacrylamide gel electrophoresis, generating a profile that can be used to determine epidemiological relationships between strains and to identify different *Campylobacter* species. This technology involves DNA sequencing of certain portions of the genome and is heavily reliant on specialized technology; and while it is expensive, it can be automated and generates data that are representative of the whole genome. This technique has been successfully applied for qualitative and quantitative differentiation of *Campylobacter* load on poultry carcasses during different stages of slaughter (Wieler and Laturnus, 2003).

RFLP. Restriction fragment length polymorphism (RFLP) of the flagellin gene (*fla* typing) is a PCR-based typing method. The flagellin gene locus of *C. jejuni* and *C. coli* contains two flagellin genes, *flaA* and *flaB*, of approximately 1.7 kb, arranged in tandem and separated by approximately 170 nucleotides. These genes are highly conserved, with 92% homology between the *flaA* and *flaB* genes in individual isolates. However, polymorphic regions exist in the flagellin genes, which provide the basis of a typing scheme. Such a conserved genetic structure serves as the basis for the design of primers, and thus, a product is obtained by PCR incorporating both the conserved and variable (polymorphic) regions of the *flaA* gene. The PCR product is digested with an appropriate restriction enzyme and the DNA fragment separated by gel electrophoresis. Many different *fla* typing methods exist, with considerable variation in the PCR-RFLP procedures, and so do not allow for direct comparison of profiles generated in different laboratories; there is currently no generally accepted protocol for *fla* typing of *Campylobacter*. The *fla* typing method developed by Nachamkin et al. (1993, 1996), which uses DdeI for digestion of PCR products, is widely used.

One of the disadvantages of using *flaA* typing is that the resultant profiles are not species specific, with some *C. jejuni* and *C. coli* isolates reported to share the same flagellin type (DeBoer et al., 2000). In addition, recombination within the *fla* genes can occur, making further testing necessary to confirm the PCR-RFLP results of *fla* typing for definitive epidemiological analysis.

MLST. Multilocus sequence typing (MLST) determines epidemiological relationships based on sequence polymorphisms in selected housekeeping genes. These genes evolve slowly, as they are under stabilizing selection for the conservation of metabolic function. The technique involves nucleotide sequencing of internal fragments (approximately 400 to 500 bp) of housekeeping genes (usually seven), so that the allelic profile or sequence type at each locus is defined directly for each isolate. Using genes that are not as susceptible to genetic variation as other loci such as the *Campylobacter* flagellin gene (*fla*) is a more phylogenetic approach providing less diversity in terms of types than many other

DNA-based techniques. The MLST technology employed can be reproducible (Maiden, 2000). Housekeeping genes used in MLST schemes for *Campylobacter jejuni* include *aspA, glnA, gltA, glyA, pgm, tkt*, and *uncA*. The University of Oxford, Oxford, United Kingdom, has established a publicly accessible MLST database and software website (http://pubmlst.org/campylobacter) for exchange of data and protocols. The use of MLST for short-term epidemiology is a subject of discussion. A review by Wassenaar and Newell (2000) recommends that for short-term epidemiological studies and source tracing, PFGE or flagellin PCR-RFLP could be applied, while AFLP and/or MLST are likely to give the most information regarding strain phylogeny for longer-term studies. MLST is expensive to set up for small routine laboratories.

Korczak et al. (2009) published a multiplex strategy for MLST, *fla* typing, and genetic determination of antimicrobial resistance of *C. jejuni* and *C. coli*. In this approach, 13 different primer sets, targeting the seven housekeeping genes, the two flagellin genes, the 16S rRNA and *rpoB* genes for identification and targeting the 23S rRNA and *gyrA* genes for macrolide and quinolone resistance, were amplified and divided into four groups as shown in Fig. 2.

Other genotyping methods include enterobacterial repetitive intergenic consensus sequencing (ERIC), repetitive PCR, and random amplified polymorphic DNA (RAPD). These are relatively simple PCR-based typing methods that have been applied with varying degrees of success to type *Campylobacter*. ERIC-PCR targets enterobacterial repetitive intergenic consensus sequences, while repetitive-PCR targets repetitive DNA elements distributed randomly over the genome. RAPD based on different primers and reaction conditions is also discriminatory in establishing genetic diversity among *C. jejuni* and *C. coli* isolates from different sources; 10-mer primers have been frequently used for RAPD analysis of *C. jejuni* and *C. coli* (Madden et al., 1996; Misawa et al., 2000; Payne et al., 1999). Ribotyping (RFLP of rRNA genes) that targets the 16S/23S rRNA intergenic spacer region conserved among *Campylobacter* spp. has been shown to be of value in typing a wide range of bacterial pathogens including *Campylobacter* (Nielsen et al., 2001); there are, however, conflicting opinions regarding the discriminatory power of ribotyping of *Campylobacter* (Pavlic and Griffiths, 2009).

PCR Binary Typing. PCR binary typing uses multiplex PCR to generate a code of positive or negative results of PCR analyses for a wide range of genes (18 core gene targets) widely distributed in *C. jejuni* genomes and extrachromosomal elements, which the authors conclude to be more discriminatory than MLST and SmaI-based PFGE typing (Cornelius et al., 2010).

Infrared (IR) spectroscopy has been successfully applied to the identification and classification of microorganisms at the species and subspecies levels. Mouwen et al. (2005, 2006) used Fourier transform infrared spectroscopy for the identification and typing of *C. jejuni* and *C. coli*. The technology exploits the fact that when intact microbial cells are radiated with a continuous spectrum of IR light, a characteristic resonance absorption band is obtained. The IR spectrum of a cluster of cells exhibits particular traits that make it a unique Fourier transform infrared fingerprint. Analysis is carried out using hierarchical cluster analysis to group elements on the basis of their perceived closeness.

CONCLUSION AND FUTURE TRENDS

The ultimate goal of rapid detection of *Campylobacter* is to prevent contaminated meat from entering the market, by a cost-effective and reliable approach. Cost-effective analysis can be performed if the methods are amenable to automation and easy to perform by untrained personnel. Currently, real-time PCR is fulfilling these criteria, but only to a certain extent. More-cost-effective methods seem to be on-farm and at-line monitoring at slaughterhouses, where the same detection platform

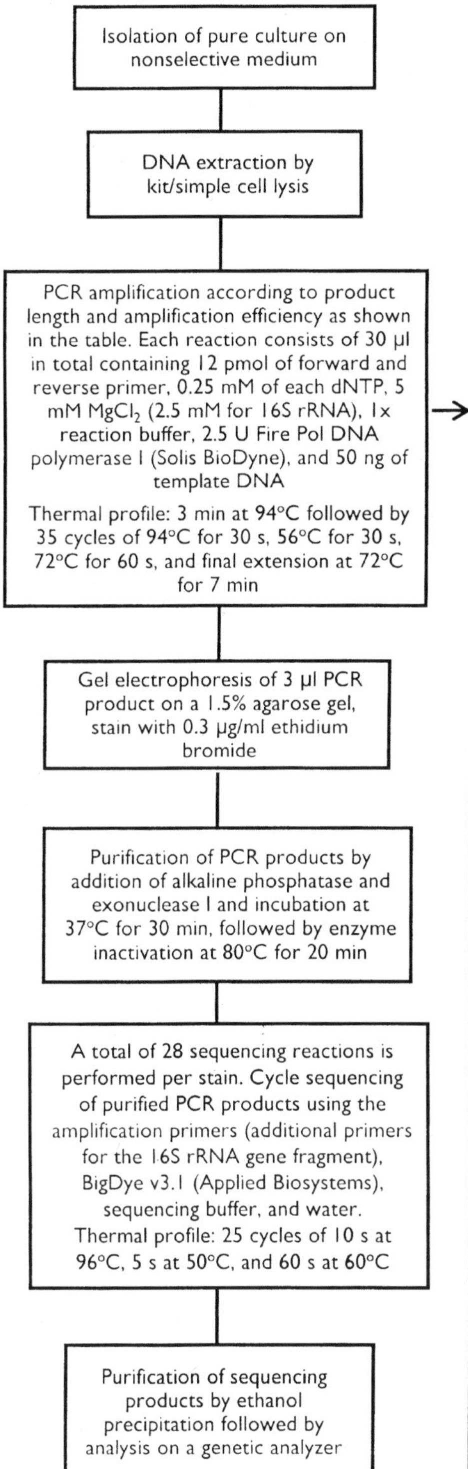

Amplification group	Gene product	Primer name	PCR fragment size (bp)
1	Flagellin	flaA_Cjc-L	471
		flaAB_Cjc-R	
		flaA_Cjc-L1	614
	23S rRNA	F1-campy-23S	508
		R1-campy-23S	
	Aspartase	aspA_Cjc-L	594
		aspA_Cjc-R	
	Phosphoglucosamine mutase	glmM_Cjc-L	685
		glmM_Cjc-R	
2	Gyrase	gyrA_Cjc-L	294
		gyrA_Cjc-R	
	Flagellin	flaB_Cjc-L	488
		flaAB_Cjc-R	
		flaB_Cjc-L1	602
	Transketolase	tkt_Cjc-L	606
		tkt_Cjc-R	
	Glutamine synthetase	glnA_Cjc-L	712
		glnA_Cjc-R	
3	RNA polymerase beta subunit	CamrpoB-L	524
		RpoB-R	
	Citrate synthase	gltA_Cjc-L	567
		gltA_Cjc-R	
	ATP synthase alpha subunit	atpA_Cjc-L	623
		atpA_Cjc-R	
	Serine hydroxylmethyltransferase	glyA_Cjc-L	701
		glyA_Cjc-R	
4	16S rRNA	16S UNI-L	1381
		16S UNI-R	

FIGURE 2 The workflow of multiplex PCR followed by MLST, *fla* typing, and genetic determination of antimicrobial resistance of *Campylobacter jejuni* and *C. coli* according to Korczak et al. (2009).

can also be used to detect *Salmonella* and even other pathogens of significance to animal health.

Another future trend is at the research level, where the use of metagenomic techniques is emerging, mainly as tools for quantitative risk assessments or epidemiological studies. The metagenomic approach aims at simultaneous detection and identification in one step, performed directly (without any culture enrichment) on the total DNA content of the sample. Developing these techniques for quantitative enumeration will be the next step. By the increasing use of high-throughput DNA sequencing at a low per-sample price, this could become a widespread technique to ensure future meat safety, whereby large conglomerates centralize their specialized testing capacities.

ACKNOWLEDGMENT

The work was supported in part by the European Union-funded Integrated Project BIOTRACER (Contract 036272) under the 6th RTD Framework.

REFERENCES

Abubakar, I., L. Irvine, C. F. Aldus, G. M. Wyatt, R. Fordham, S. Schelenz, L. Shepstone, A. Howe, M. Peck, and P. R. Hunter. 2007. A systematic review of the clinical, public health and cost-effectiveness of rapid diagnostic tests for the detection and identification of bacterial intestinal pathogens in faeces and food. *Health Technol. Assess.* **11:**1–216.

Allen, V. M., C. H. Burton, D. J. Wilkinson, R. T. Whyte, J. A. Harris, M. Howell, and D. B. Tinker. 2008. Evaluation of the performance of different cleaning treatments in reducing microbial contamination of transport crates. *Br. Poult. Sci.* **49:**233–240.

Barer, M. R., and C. R. Harwood. 1999. Bacterial viability and culturability. *Adv. Microb. Physiol.* **41:** 93–137.

Bohaychuk, V. M., G. E. Gensler, R. K. King, J. T. Wu, and L. M. McMullen. 2005. Evaluation of detection methods for screening meat and poultry products for the presence of foodborne pathogens. *J. Food Prot.* **68:**2637–2647.

Botteldoorn, N., E. Van Coillie, V. Piessens, G. Rasschaert, L. Debruyne, M. Heyndrickx, L. Herman, and W. Messens. 2008. Quantification of *Campylobacter* spp. in chicken carcass rinse by real-time PCR. *J. Appl. Microbiol.* **105:**1909–1918.

Boulos, L., M. Prévost, B. Barbeau, J. Coallier, and R. Desjardins. 1999. LIVE/DEAD BacLight™: application of a new rapid staining method for direct enumeration of viable and total bacteria in drinking water. *J. Microbiol. Methods* **37:**77–86.

Bouwknegt, M., A. W. van de Giessen, W. D. C. Dam-Deisz, A. H. Havelaar, N. J. D. Nagelkerke, and A. M. Henken. 2004. Risk factors for the presence of *Campylobacter* spp in Dutch broiler flocks. *Prev. Vet. Med.* **62:**35–49.

Bull, S. A., V. M. Allen, G. Domingue, F. Jorgensen, J. A. Frost, R. Ure, R. Whyte, D. Tinker, J. E. L. Corry, J. Gillard-King, and T. J. Humphrey. 2006. Sources of *Campylobacter* spp. colonizing housed broiler flocks during rearing. *Appl. Environ. Microbiol.* **72:**645–652.

Caron, G. N., P. Stephens, and R. A. Badley. 1998. Assessment of bacterial viability status by flow cytometry and single cell sorting. *J. Appl. Microbiol.* **84:**988–998.

Chaveerach, P., A. A. ter Huurne, L. J. Lipman, and F. van Knapen. 2003. Survival and resuscitation of ten strains of *Campylobacter jejuni* and *Campylobacter coli* under acid conditions. *Appl. Environ. Microbiol.* **69:**711–714.

Cornelius, A. J., B. Gilpin, P. Carter, C. Nicol, and S. L. W. On. 2010. Comparison of PCR binary typing (P-BIT), a new approach to epidemiological subtyping of *Campylobacter jejuni*, with serotyping, pulsed-field gel electrophoresis, and multilocus sequence typing methods. *Appl. Environ. Microbiol.* **76:**1533–1544.

Corry, J. E., and H. I. Atabay. 2001. Poultry as a source of *Campylobacter* and related organisms. *Symp. Ser. Soc. Appl. Microbiol.* **2001:**96S-114S.

Cox, J. M., and A. Pavic. 2009. Advances in enteropathogen control in poultry production. *J. Appl. Microbiol.* **108:**745–755.

DeBoer, P., B. Duim, A. Rigter, J. van der Plas, W. F. Jacobs-Reitsma, and J. A. Wagenaar. 2000. Computer-assisted analysis and epidemiological value of genotyping methods for *Campylobacter jejuni* and *Campylobacter coli*. *J. Clin. Microbiol.* **38:** 1940–1946.

Debretsion, A., T. Habtemariam, S. Wilson, D. Nganwa, and T. Yehualaeshet. 2007. Real-time PCR assay for rapid detection and quantification of *Campylobacter jejuni* on chicken rinses from poultry processing plant. *Mol. Cell. Probes* **21:**177–181.

Dhillon, A. S., H. L. Shivaprasad, D. Schaberg, F. Wier, S. Weber, and D. Bandli. 2006. *Campylobacter jejuni* infection in broiler chickens. *Avian Dis.* **50:**55–58.

Doorduyn, Y., W. E. van Brandhof, Y. T. van Duynhoven, B. J. Breukink, J. A. Wagenaar,

and W. van Pelt. 2010. Risk factors for indigenous *Campylobacter jejuni* and *Campylobacter coli* infections in The Netherlands: a case-control study. *Epidemiol. Infect.* **12:**1–14.

Duim, B., T. M. Wassenaar, A. Rigter, and J. Wagenaar. 1999. High-resolution genotyping of *Campylobacter* strains isolated from poultry and humans with amplified fragment length polymorphism fingerprinting. *Appl. Environ. Microbiol.* **65:** 2369–2375.

Eberhart-Phillips, J., N. Walker, N. Garret, D. Bell, D. Sinclair, W. Rainger, and M. Bates. 1997. Campylobacteriosis in New Zealand: results of a case-control study. *J. Epidemiol. Commun. Health* **51:**686–691.

European Food Safety Authority. 2009. The community summary report on trends and sources of zoonoses and zoonotic agents in the European Union in 2007. European Food Safety Authority, Parma, Italy. http://www.efsa.europa.eu/en/efsajournal/doc/223r.pdf.

European Food Safety Authority. 2010a. The community summary report on trends and sources of zoonoses, zoonotic agents and food-borne outbreaks in the European Union in 2008. *EFSA J.* **8:** 1496.

European Food Safety Authority. 2010b. EFSA panel on biological hazards (BIOHAZ). Scientific opinion on quantification of the risk posed by broiler meat to human campylobacteriosis in the EU. *EFSA J.* **8:**1437. doi:10.2903/j.efsa.2010.1437.

European Food Safety Authority. 2010c. EFSA analysis of the baseline survey on the prevalence of *Campylobacter* in broiler batches and of *Campylobacter* and *Salmonella* on broiler carcasses in the EU, 2008. Part A. *Campylobacter* and *Salmonella* prevalence estimates. *EFSA J.* **8:**1503. doi:10.2903/j.efsa.2010.1503.

Fosse, J., H. Seegers, and C. Magras. 2009. Prevalence and risk factors for bacterial food-borne zoonotic hazards in slaughter pigs: a review. *Zoonoses Public Health* **56:**429–454. doi:10.1111/j.1863-2378.2008.01185.x.

Friedmann, C., R. Hoekstra, M. Samuel, R. Marcus, J. Bender, B. Shiferaw, V. Reddy, S. Ahuja, D. Helfrick, F. Hardnett, M. Carter, B. Andersson, and R. Tauxe. 2004. Risk factors for sporadic *Campylobacter* infection in the United States: a case-control study in FoodNet sites. *Clin. Infect. Dis.* **38**(Suppl. 3)**:**S285–S296.

Fukushima, H., K. Katsube, Y. Hata, R. Kishi, and S. Fujiwara. 2007. Rapid separation and concentration of food-borne pathogens in food samples prior to quantification by viable-cell counting and real-time PCR. *Appl. Environ. Microbiol.* **73:**92–100. doi:10.1128/AEM.01772-06.

Ghafir, Y., B. China, K. Dierick, L. De Zutter, and G. Daube. 2007. A seven-year survey of *Campylobacter* contamination in meat at different production stages in Belgium. *Int. J. Food Microbiol.* **116:**111–120. doi:10.1016/j.ijfoodmicro.2006.12.012.

Gonzalez, I., K. A. Grant, P. T. Richardson, S. F. Park, and M. D. Collins. 1997. Specific identification of the enteropathogens *Campylobacter jejuni* and *Campylobacter coli* by using a PCR test based on the *ceuE* gene encoding a putative virulence determinant. *J. Clin. Microbiol.* **35:**759–763.

Grennan, B., N. A. O'Sullivan, R. Fallon, C. Carroll, T. Smith, M. Glennon, and M. Maher. 2001. PCR-ELISAs for the detection of *Campylobacter jejuni* and *Campylobacter coli* in poultry samples. *BioTechniques* **30:**602–606, 608–610.

Hald, B., H. Skovgård, D. D. Bang, K. Pedersen, J. Dybdahl, J. B. Jespersen, and M. Madsen. 2004. Flies and *Campylobacter* infection of broiler flocks. *Emerg. Infect. Dis.* **10:**1490–1492.

Hansson, I., E. O. Engvall, J. Lindblad, A. Gunnarsson, and I. Vågsholm. 2004. The *Campylobacter* surveillance program for broilers in Sweden July 2001–June 2002. *Vet. Rec.* **155:**193–196.

Hansson, I., I. Vågsholm, L. Svensson, and E. Olsson Engvall. 2007b. Correlations between *Campylobacter* spp. prevalence in the environment and broiler flocks. *J. Appl. Microbiol.* **103:**640–649.

Hansson, I., L. Plym Forshell, P. Gustafsson, S. Boqvist, J. Lindblad, E. Olsson Engvall, Y. Andersson, and I. Vågsholm. 2007a. Summary of the Swedish *Campylobacter* program in broilers 2001–2005. *J. Food Prot.* **70:**2008–2014.

Hansson, I., M. Ederoth, L. Andersson, I. Vågsholm, and E. Olsson Engvall. 2005. Transmission of *Campylobacter* spp. to chickens during transport to slaughter. *J. Appl. Microbiol.* **99:**1149–1157.

Harris, N., N. Weiss, and C. Nolan. 1986. The role of poultry and meats in the etiology of *Campylobacter jejuni/coli* enteritis. *Am. J. Public Health* **76:** 407–411.

Havelaar, A. H., M. J. Mangen, A. A. de Koeijer, M. J. Bogaardt, E. G. Evers, W. F. Jacobs-Reitsma, W. van Pelt, J. A. Wagenaar, G. A. de Wit, H. van der Zee, and M. J. Nauta. 2007. Effectiveness and efficiency of controlling *Campylobacter* on broiler chicken meat. *Risk Anal.* **27:**831–844.

Hochel, I., D. Slavickova, D. Viochna, J. Skvor, and I. Steinhauserova. 2007. Detection of *Campylobacter* species in foods by indirect competitive ELISA using hen and rabbit antibodies. *Food Agric. Immunol.* **18:**151–167.

Hofshagen, M., and H. Kruse. 2005. Reduction in flock prevalence of *Campylobacter* spp. in broilers in Norway after implementation of an action plan. *J. Food Prot.* **68:**2220–2223.

Hong, J., W. K. Jung, J. M. Kim, S. H. Kim, H. C. Koo, J. Ser, and Y. H. Park. 2007. Quantification and differentiation of *Campylobacter jejuni* and *Campylobacter coli* in raw chicken meats using a real-time PCR method. *J. Food Prot.* **70:**2015–2022.

Hong, Y., M. E. Berrang, T. Liu, C. L. Hofacre, S. Sanchez, L. Wang, and J. J. Maurer. 2003. Rapid detection of *Campylobacter coli, C. jejuni,* and *Salmonella* enterica on poultry carcasses by using PCR-enzyme-linked immunosorbent assay. *Appl. Environ. Microbiol.* **69:**3492–3499.

Huang, J., G. Yang, W. Meng, L. Wu, A. Zhu, and X. Jiao. 2010. An electrochemical impedimetric immunosensor for label-free detection of *Campylobacter jejuni* in diarrhea patients' stool based on O-carboxymethylchitosan surface modified Fe3O4 nanoparticles. *Biosens. Bioelectron.* **25:**1204–1211. doi:10.1016/j.bios.2009.10.036.

International Organization for Standardization. 2006. *BS EN ISO 10272-1:2006 Microbiology of Food and Animal Feeding Stuffs—Horizontal Method for Detection and Enumeration of* Campylobacter *spp.* BSI, London, United Kingdom.

Ivnitski, D., E. Wilkins, H. T. Tien, and A. Ottova. 2000. Electrochemical biosensor based on supported planar lipid bilayers for fast detection of pathogenic bacteria. *Electrochem. Commun.* **2:**457–460.

Izat, A. L., F. A. Gardner, J. H. Denton, and F. A. Golan. 1988. Incidence and level of *Campylobacter jejuni* in broiler processing. *Poult. Sci.* **67:** 1568–1572.

Jacobs-Reitsma, W. F., A. W. van de Giessen, N. M. Bolder, and R. W. A. W. Mulder. 1995. Epidemiology of *Campylobacter* spp. at two Dutch broiler farms. *Epidemiol. Infect.* **114:**413–421.

Jensen, G. B., L. Thomsen, and O. R. Veltman. September 2005. Method, chip, device and integrated system for detection of biological materials. International patent WO2005083426.

Johnsen, G., H. Kruse, and M. Hofshagen. 2006. Genetic diversity and description of transmission routes for *Campylobacter* on broiler farms by amplified-fragment length polymorphism. *J. Appl. Microbiol.* **101:**1130–1139.

Josefsen, M. H., C. Lofstrom, T. B. Hansen, L. S. Christensen, J. E. Olsen, and J. Hoorfar. 2010. Rapid quantification of viable *Campylobacter* on chicken carcasses by real-time PCR and propidium monoazide as a tool for quantitative risk assessment. *Appl. Environ. Microbiol.* **76:**5097–5104. doi:10.1128/AEM.00411-10.

Josefsen, M. H., N. R. Jacobsen, and J. Hoorfar. 2004. Enrichment followed by quantitative PCR both for rapid detection and as a tool for quantitative risk assessment of food-borne thermotolerant campylobacters. *Appl. Environ. Microbiol.* **70:**3588–3592. doi:10.1128/AEM.70.6.3588-3592.2004.

Kastma, E., A. A. De Koeijer, W. F. Jacobs-Reitsma, M. J. Mangen, and J. A. Wagenaar. 2007. Assessing interventions to reduce the risk of *Campylobacter* prevalence in broilers. *Risk Anal.* **27:** 863–876.

Korczak, B. M., M. Zurfluh, S. Emler, J. Kuhn-Oertli, and P. Kuhnert. 2009. Multiplex strategy for multilocus sequence typing, *fla* typing, and genetic determination of antimicrobial resistance of *Campylobacter jejuni* and *Campylobacter coli* isolates collected in Switzerland. *J. Clin. Microbiol.* **47:** 1996–2007. doi:10.1128/JCM.00237-09.

Kramer, J. M., J. A. Frost, F. J. Bolton, and D. R. Wareing. 2000. *Campylobacter* contamination of raw meat and poultry at retail sale: identification of multiple types and comparison with isolates from human infection. *J. Food Prot.* **63:** 1654–1659.

Lastovica, A. J., and B. M. Allos. 2008. Clinical significance of *Campylobacter* and related species other than *Campylobacter jejuni* and *Campylobacter coli*, p. 123–149. *In* I. Nachamkin, C. M. Szymanski and M. J. Blaser (ed.), *Campylobacter.* ASM Press, Washington, DC.

Lindblad, M., I. Hansson, I. Vågsholm, and R. Lindqvist. 2006. Postchill *Campylobacter* prevalence on broiler carcasses in relation to slaughter group colonization level and chilling system. *J. Food Prot.* **69:**495–499.

Lindstedt, B. A., E. Heir, T. Vardund, K. K. Melby, and G. Kapperud. 2000. Comparative fingerprinting analysis of *Campylobacter jejuni* subsp. *jejuni* strains by amplified-fragment length polymorphism genotyping. *J. Clin. Microbiol.* **38:**3379–3387.

Line, J. E., and K. G. Pearson. 2003. Development of a selective broth for the detection of injured *Campylobacter jejuni* by capacitance monitoring. *J. Food Prot.* **66:**1752–1755.

Little, C. L., J. F. Richardson, R. J. Owen, E. de Pinna, and E. J. Threlfall. 2008. *Campylobacter* and *Salmonella* in raw red meats in the United Kingdom: prevalence, characterization and antimicrobial resistance pattern, 2003–2005. *Food Microbiol.* **25:**538–543.

Liu, G., Y. Han, X. Li, and S. Song. 2006. Applicability of a rapid method based on immunomagnetic capture-fluorescent PCR assay for *Campylobacter jejuni*. *Food Control* **17:**527–532.

Madden, R. H., L. Moran, and P. Scates. 1996. Sub-typing of animal and human *Campylobacter* spp. using RAPD. *Lett. Appl. Microbiol.* **23:**167–170.

Maiden, M. C. 2000. High-throughput sequencing in the population analysis of bacterial pathogens of humans. *Int. J. Med. Microbiol.* **290:**183–190.

Mayr, A. M., S. Lick, J. Bauer, D. Tharigen, U. Busch, and I. Huber. 2010. Rapid detection and differentiation of *Campylobacter jejuni, Campylobacter coli*, and *Campylobacter lari* in food, using multiplex real-time PCR. *J. Food Prot.* **73:**241–250.

Minihan, D., P. Whyte, M. O'Mahony, S. Fanning, K. McGill, and J. D. Collins. 2004. *Campylobacter* spp. in Irish feedlot cattle: a longitudinal study involving pre-harvest and harvest phases of the food chain. *J. Vet. Med. B* **51:**28–33.

Misawa, N., S. Shinohara, H. Satoh, H. Itoh, K. Shinohara, K. Shimomura, F. Kondo, and K. Itoh. 2000. Isolation of *Campylobacter* species from zoo animals and polymerase chain reaction-based random amplified polymorphism DNA analysis. *Vet. Microbiol.* **71:**59–68.

Morales-Rayas, R., P. F. Wolffs, and M. W. Griffiths. 2008. Immunocapture and real-time PCR to detect *Campylobacter* spp. *J. Food Prot.* **71:** 2543–2547.

Mouwen, D. J. M., M. J. B. M. Weijtens, R. Capita, C. Alonso-Calleja, and M. Prieto. 2005. Discrimination of enterobacterial repetitive intergenic consensus PCR types of *Campylobacter coli* and *Campylobacter jejuni* by Fourier transform infrared spectroscopy. *Appl. Environ. Microbiol.* **71:** 4318–4324.

Mouwen, D. J. M., R. Capita, C. Alonso-Calleja, J. Prieto-Gómez, and M. Prieto. 2006. Artificial neural network based identification of *Campylobacter* species by Fourier transform infrared spectroscopy. *J. Microbiol. Methods* **67:**131–140.

Nachamkin, I., K. Bohachick, and C. M. Patton. 1993. Flagellin gene typing of *Campylobacter jejuni* by restriction fragment length polymorphism analysis. *J. Clin. Microbiol.* **31:**1531–1536.

Nachamkin, I., K. Bohachick, and C. M. Patton. 1996. Analysis of HL and O serotypes of *Campylobacter* strains by the flagellin gene typing system. *J. Clin. Microbiol.* **34:**277–281.

Nauta, M. J., W. F. Jacobs-Reitsma, E. G. Evers, W. van Pelt, and A. H. Havelaar. 2005. *Risk Assessment of Campylobacter in The Netherlands via Broiler Meat and Other Routes. RIVM Report 250911006/2005.* RIVM, Bilthoven, The Netherlands.

Newell, D. G., and C. Fearnley. 2003. Sources of *Campylobacter* colonization in broiler chickens. *Appl. Environ. Microbiol.* **69:**4343–4351.

Nielsen, E. M., J. Engberg, and V. Fussing. 2001. Genotypic and serotypic stability of *Campylobacter jejuni* strains during in vitro and in vivo passage. *Int. J. Med. Microbiol.* **291:**379–385.

Nocker, A., and A. K. Camper. 2009. Novel approaches toward preferential detection of viable cells using nucleic acid amplification techniques. *FEMS Microbiol. Lett.* **291:**137–142. doi:10.1111/j.1574-6968.2008.01429.x.

Nocker, A., C. Y. Cheung, and A. K. Camper. 2006. Comparison of propidium monoazide with ethidium monoazide for differentiation of live vs. dead bacteria by selective removal of DNA from dead cells. *J. Microbiol. Methods* **67:**310–320.

Nogva, H., A. Bergh, A. Holck, and K. Rudi. 2000. Application of the 5′-nuclease PCR assay in the evaluation and development of methods for quantitative detection of *Campylobacter jejuni. Appl. Environ. Microbiol.* **66:**4029-4036.

Nogva, H. K., S. M. Dromtorp, H. Nissen, and K. Rudi. 2003. Ethidium monoazide for DNA-based differentiation of viable and dead bacteria by 5′-nuclease PCR. *BioTechniques* **34:**804–808, 810, 812–813.

Olsen, K. N., M. Lund, J. Skov, L. S. Christensen, and J. Hoorfar. 2009. Detection of *Campylobacter* bacteria in air samples for continuous real-time monitoring of *Campylobacter* colonization in broiler flocks. *Appl. Environ. Microbiol.* **75:**2074–2078.

Olson, C. K., S. Ethelberg, W. van Pelt, and R. V. Tauxe. 2008. Epidemiology of *Campylobacter jejuni* infections in industrialized nations, p. 163–189. *In* I. Nachamkin, C. M. Szymanski, and J. Blaser (ed.), *Campylobacter*, 3rd ed. ASM Press, Washington, DC.

Oosterom, J., G. J. A. de Wilde, E. de Boer, L. H. de Blaauw, and H. Karman. 1983. Survival of *Campylobacter jejuni* during poultry processing and pig slaughtering. *J. Food Prot.* **46:**702–706.

O'Sullivan, N. A., R. Fallon, C. Carroll, T. Smith, and M. Maher. 2000. Detection and differentiation of *Campylobacter jejuni* and *Campylobacter coli* in broiler chicken samples using a PCR/DNA probe membrane based colorimetric detection assay. *Mol. Cell. Probes* **14:**7–16.

Pavlic, M., and M. W. Griffiths. 2009. Principles, applications, and limitations of automated ribotyping as a rapid method in food safety. *Foodborne Pathog. Dis.* **6:**1047–1055.

Payne, R. E., M. D. Lee, D. W. Dreesen, and H. M. Barnhart. 1999. Molecular epidemiology of *Campylobacter jejuni* in broiler flocks using randomly amplified polymorphic DNA-PCR and 23S rRNA-PCR and role of litter in its transmission. *Appl. Environ. Microbiol.* **65:**260–263.

Pezzotti, G., A. Serafin, I. Luzzi, R. Mioni, M. Milan, and R. Perin. 2003. Occurrence and resistance to antibiotics of *Campylobacter jejuni* and *Campylobacter coli* in animals and meat in northeastern Italy. *Int. J. Food Microbiol.* **82:**281–287.

Rollins, D. M., and R. R. Colwell. 1986. Viable but nonculturable stage of *Campylobacter jejuni* and its role in survival in the natural aquatic environment. *Appl. Environ. Microbiol.* **52:**531–538.

Rönner, A. C., and H. Lindmark. 2007. Quantitative detection of *Campylobacter jejuni* on fresh chicken carcasses by real-time PCR. *J. Food Prot.* **70:**1373–1378.

Rosenquist, H., H. M. Sommer, N. L. Nielsen, and B. B. Christensen. 2006. The effect of slaughter operations on the contamination of chicken carcasses with thermotolerant *Campylobacter*. *Int. J. Food Microbiol.* **108:**226–232.

Rosenquist, H., N. L. Nielsen, H. M. Sommer, B. Nørrung, and B. B. Christensen. 2003. Quantitative risk assessment of human campylobacteriosis associated with thermophilic *Campylobacter* species in chickens. *Int. J. Food Microbiol.* **83:**87–103.

Rudi, K., H. K. Hoidal, T. Katla, B. K. Johansen, J. Nordal, and K. S. Jakobsen. 2004. Direct real-time PCR quantification of *Campylobacter jejuni* in chicken fecal and cecal samples by integrated cell concentration and DNA purification. *Appl. Environ. Microbiol.* **70:**790–797.

Rudi, K., B. Moen, S. M. Dromtorp, and A. L. Holck. 2005. Use of ethidium monoazide and PCR in combination for quantification of viable and dead cells in complex samples. *Appl. Environ. Microbiol.* **71:**1018–1024.

Samuel, M., D. Vugia, S. Shallow, R. Marcus, S. Segler, T. McGivern, H. Kassenborg, K. Reilly, M. Kennedy, F. Angulo, and R. Tauxe. 2004. Epidemiology of sporadic *Campylobacter* infection in the United States and the declining trend in incidence. *Clin. Infect. Dis.* **38**(Suppl. 3)**:**S165–S174.

Sapsford, K. E., A. Rasooly, C. R. Taitt, and F. S. Ligler. 2004. Detection of *Campylobacter* and *Shigella* species in food samples using an array biosensor. *Anal. Chem.* **76:**433–440. doi:10.1021/ac035122z.

Sheridan, G. E., C. I. Masters, J. A. Shallcross, and B. M. MacKey. 1998. Detection of mRNA by reverse transcription-PCR as an indicator of viability in *Escherichia coli* cells. *Appl. Environ. Microbiol.* **64:**1313–1318.

Slader, J., G. Dominue, F. Jorgensen, K. McAlpine, R. J. Owen, F. J. Bolton, and T. J. Humphrey. 2002. Impact of transport crate reuse and of catching and processing on *Campylobacter* and *Salmonella* contamination of broiler chickens. *Appl. Environ. Microbiol.* **68:**713–719.

Stanley, K. N., J. S. Wallace, J. E. Currie, P. J. Diggle, and K. Jones. 1998. The seasonal variation of thermophilic campylobacters in beef cattle, dairy cattle and calves. *J. Appl. Microbiol.* **85:**472–480.

Stern, N. J., and M. C. Robach. 2003. Enumeration of *Campylobacter* spp. in broiler feces and in corresponding processed carcasses. *J. Food. Prot.* **66:** 1557–1563.

Stern, N. J., K. L. Hiett, G. A. Alfredsson, K. G. Kristinsson, J. Reiersen, H. Hardardottir, H. Briem, E. Gunnarsson, F. Georgsson, R. Lowman, E. Berndtson, A. M. Lammerding, G. M. Paoli, and M. T. Musgrove. 2003. *Campylobacter* spp. in Icelandic poultry operations and human disease. *Epidemiol. Infect.* **130:**23–32.

Suzuki, H., and S. Yamamoto. 2009. *Campylobacter* contamination in retail poultry meats and by-products in the world: a literature survey. *J. Vet. Med. Sci.* **71:**255–261.

Taylor, A. D., J. Ladd, Q. Yu, S. Chen, J. Homola, and S. Jiang. 2006. Quantitative and simultaneous detection of four foodborne bacterial pathogens with a multi-channel SPR sensor. *Biosens. Bioelectron.* **22:**752–758. doi:10.1016/j.bios.2006.03.012.

Tenover, F. C., R. D. Arbeit, R. V. Goering, P. A. Mickelsen, B. E. Murray, D. H. Persing, and B. Swaminathan. 1995. Interpreting chromosomal DNA restriction patterns produced by pulsed-field gel electrophoresis: criteria for bacterial strain typing. *J. Clin. Microbiol.* **33:**2233–2239.

Uyttendaele, M., A. Bastiaansen, and J. Debevere. 1997. Evaluation of the NASBA nucleic acid amplification system for assessment of the viability of *Campylobacter jejuni*. *Int. J. Food Microbiol.* **37:**13–20.

Vandamme, P. 2000. Taxonomy of the family *Campylobacteraceae*. *In* I. Nachamkin and M. J. Blaser (ed.), *Campylobacter*, 2nd ed. ASM Press, Washington, DC.

Vellinga, A., and F. van Loock. 2002. The dioxin crisis as experiment to determine poultry-related *Campylobacter* enteritis. *Emerg. Infect. Dis.* **8:**19–22.

Waller, D. F., and S. A. Ogata. 2000. Quantitative immunocapture PCR assay for detection of *Campylobacter jejuni* in foods. *Appl. Environ. Microbiol.* **66:**4115–4118.

Wassenaar, T. M., and D. G. Newell. 2000. Genotyping of *Campylobacter* spp. *Appl. Environ. Microbiol.* **66:**1–9.

Wedderkopp, A., E. Rattenberg, and M. Madsen. 2000. National surveillance of *Campylobacter* in broilers at slaughter in Denmark 1998. *Avian Dis.* **44:**993–999.

Wei, D., O. A. Oyarzabal, T. S. Huang, S. Balasubramanian, S. Sista, and A. L. Simonian. 2007. Development of a surface plasmon resonance biosensor for the identification of *Campylobacter je-*

juni. J. Microbiol. Methods **69:**78–85. doi:10.1016/j.mimet.2006.12.002.

Wieler, L. H., and C. Laturnus. 2003. Classification and typing of *Campylobacter*: application of genotypical methods in veterinary medicine. *Genome Lett.* **2:**53–61.

Wisessombat, S., K. Kittiniyom, P. Srimanote, W. Wonglumsom, and S. P. Voravuthikunchai. 2009. A novel method and simple apparatus for the detection of thermophilic *Campylobacter* spp. in chicken meat products. *J. Microbiol. Methods* **76:** 169–173. doi:10.1016/j.mimet.2008.10.007.

Wolffs, P., B. Norling, J. Hoorfar, M. Griffiths, and P. Radstrom. 2005. Quantification of *Campylobacter* spp. in chicken rinse samples by using flotation prior to real-time PCR. *Appl. Environ. Microbiol.* **71:**5759–5764. doi:10.1128/AEM.71.10.5759-5764.2005.

Wong, T. L., L. Hollis, A. Cornelius, C. Nicol, R. Cook, and J. A. Hudson. 2007. Prevalence, numbers, and subtypes of *Campylobacter jejuni* and *Campylobacter coli* in uncooked retail meat samples. *J. Food Prot.* **70:**566–573.

Yamazaki, W., M. Taguchi, M. Ishibashi, M. Kitazato, M. Nukina, N. Misawa, and K. Inoue. 2008. Development and evaluation of a loop-mediated isothermal amplification assay for rapid and simple detection of *Campylobacter jejuni* and *Campylobacter coli. J. Med. Microbiol.* **57:**444–451. doi:10.1099/jmm.0.47688-0.

Yamazaki-Matsune, W., M. Taguchi, K. Seto, R. Kawahara, K. Kawatsu, Y. Kumeda, M. Kitazato, M. Nukina, N. Misawa, and T. Tsukamoto. 2007. Development of a multiplex PCR assay for identification of *Campylobacter coli, Campylobacter fetus, Campylobacter hyointestinalis subsp. hyointestinalis, Campylobacter jejuni, Campylobacter lari* and *Campylobacter upsaliensis. J. Med. Microbiol.* **56:** 1467–1473.

SHIGA TOXIN-PRODUCING *ESCHERICHIA COLI* IN FOOD

R. Stephan, C. Zweifel, P. Fach, S. Morabito, and L. Beutin

16

CHARACTERISTICS OF STEC

Shiga toxin-producing *Escherichia coli* (STEC) strains are important foodborne pathogens responsible for a number of human gastrointestinal diseases, including watery or bloody diarrhea and hemorrhagic colitis (HC). In a proportion of individuals, mainly children and the elderly, these symptoms may be complicated by neurological and renal sequelae, including hemolytic-uremic syndrome (HUS) (Tarr et al., 2005).

Most outbreaks and sporadic cases of HC and HUS have been attributed to STEC strains of serotype O157:H7. However, especially in continental Europe, STEC strains belonging to serotypes O26:H11/H-, O91:H21/H-, O103:H2, O111:H-, O113:H21, O121:H19, O128:H2/H, and O145:H28/H- are increasingly reported as causes of HC and HUS (Johnson et al., 2006). In general, these STEC strains are termed non-O157.

Genetic Determinants Useful for Diagnostics

Clinical laboratories historically screened only for serogroup O157, leading to possible underreporting of non-O157 STEC-associated diseases. The main virulence factor of STEC is the production of Shiga toxin 1 (Stx1) and/or Stx2 or its variants. STEC strains causing severe disease in humans mainly possess the Stx2 (sub-) type and carry genes encoding additional virulence factors such as the outer membrane protein intimin, essential for the intimate attachment and the formation of attaching and effacing lesions on gastrointestinal epithelial cells. The genetic determinant of these cells is harbored by the locus of enterocyte effacement (LEE) pathogenicity island (PAI) and/or the plasmid-encoded enterohemolysin (Paton and Paton, 1998b; Friedrich et al., 2002).

In addition to LEE, several other putative PAIs have been identified in STEC strains. Both O157:H7 strains (Kim et al., 1999; Wick et al., 2005) and non-O157 strains (Karmali et al., 2003; Wickham et al., 2006) present a variable repertoire of effector genes that en-

R. Stephan and C. Zweifel, Institute for Food Safety and Hygiene, University of Zurich, Winterthurerstrasse 272, CH-8057 Zurich, Switzerland. *P. Fach*, Laboratory for Study and Research on Food Quality and Processes (LERQAP), AFSSA (French Food Safety Agency), 23 Av. du Général De Gaulle, Fr-94706 Maisons-Alfort, France. *S. Morabito*, Food-Borne Zoonoses Unit, Department of Food Safety and Veterinary Public Health, Istituto Superiore di Sanità, 00161 Rome, Italy. *L. Beutin*, National Reference Laboratory for Escherichia coli (NRL-E.coli), Federal Institute for Risk Assessment (BfR), Diedersdorfer Weg 1, D-12277 Berlin, Germany.

Rapid Detection, Identification, and Quantification of Foodborne Pathogens, ed. by J. Hoorfar

code potential virulence determinants (*nle* [acronym for non-LEE-encoded effector] genes) (Coombes et al., 2008; Konczy et al., 2008). Among the *nle* genes, those encoded in the putative PAIs termed O-island 122 (OI-122) and O-island 71 (OI-71) in the STEC O157 reference strain EDL 933 have been proposed to distinguish STEC strains constituting a severe risk for human health from STEC that are not associated with severe and epidemic disease (Coombes et al., 2008; Konczy et al., 2008).

It has been postulated that the presence of these genes may be used to make a "molecular risk assessment" to predict the virulence potential of STEC strains (Coombes et al., 2008). A recent report showed that the *nle* genes of the OI-122 (*ent/EspL2, nleB*, and *nleE*) were detected in all typical and new emerging pathogenic STEC (Bugarel et al., 2010), while those carried by the OI-71 (*nleF, nleH1-2*, and *nleA*) were less common in these strains. Among the other *E. coli* pathogroups, the *nle* genes were detected only in some enteropathogenic *E. coli*. This finding, together with the uneven distribution of these genetic determinants among STEC, suggests that *nle* genes may be strain and/or serotype specific, probably reflecting adaptation of the strains to different host or environmental niches.

Despite the wealth of data gathered in the recent years by molecular microbiologists, the definite combination of virulence attributes of STEC has not been settled conclusively (Beutin et al., 2004). Karmali et al. grouped STEC strains into five seropathotypes (SPTs), from A through E, according to their reported frequencies in human illness, their known association with outbreaks and severe disease, and the presence of mobile genetic elements such as LEE and the OI-122 PAI (Morabito et al., 2003). SPT A comprises STEC strains of serotypes O157:H7 and O157:NM, associated with both outbreaks and sporadic cases of HUS. SPT B includes STEC strains capable of causing HUS and outbreaks but less commonly than STEC O157, such as those belonging to serotypes O26:H11, O103:H2, O111:NM, O121:H19, and O145:NM, while SPT C contains strains belonging to serotypes O91:H21 and O113:H21, which lack LEE and have been sometimes associated with HUS. STEC serotypes associated with diarrhea but not with HUS or STEC strains circulating only in the animal reservoirs have been included in SPT D or E, respectively (Karmali et al., 2003).

Reservoirs

STEC strains have been isolated from a variety of domestic animals including pigs, poultry, cats, dogs, and horses and from wild animals including deer, raccoons, flies, and birds (Beutin et al., 1993; Wallace et al., 1997). However, the bulk of data suggest that prevalence of STEC is greater in ruminants than in other animals and that domestic ruminants, especially cattle, represent one of the largest reservoirs for STEC pathogenic to humans.

Transmission of STEC infections usually occurs via consumption of contaminated food of bovine origin, although several outbreaks have been associated with other food commodities including produce, apple cider, raw milk, or milk products (European Food Safety Authority, 2007). Other vehicles of infection such as contaminated drinking or recreational water and direct contact with infected humans or animals are also reported (European Food Safety Authority, 2007).

The detection of STEC in the vehicles of infection is hampered by the following factors: (i) due to the low infectious dose, methods with a high sensitivity are required; (ii) detection in food is complicated by the low proportion of STEC compared to the total *E. coli* flora, requiring protocols that allow discrimination of STEC from the ubiquitous nonpathogenic *E. coli*; (iii) biochemical differentiation of the STEC group from nonpathogenic *E. coli* is not reliable; (iv) STEC strains that expose only a partial virulence genes profile are found, especially in food and the environment.

Sometimes, these strains belong to the *E. coli* serogroups most involved in severe human disease. Therefore, a suitable method for

STEC detection in food should be able to detect the presence of a high number of targets: Shiga toxins or Shiga toxin-coding genes, serogroup or serogroup-associated genes, and accessory virulence genes. In addition, it should demonstrate that these genes are simultaneously present in the same viable bacterial cell.

STEC DETECTION IN FOOD

Since the minimal infectious dose of STEC is very low, usually qualitative procedures with enrichment steps are established. Due to significant differences in methods for detection of STEC O157 and non-O157, these two groups will be discussed separately.

Different food categories and types of samples are recommended to be tested for STEC: carcass samples, fresh meat, minced meat and meat preparations, ready-to-eat dried or fermented meat products (e.g., salami), fresh vegetables and salads, raw and low-heat-treated milk and dairy products.

Detection of O157 STEC in Food Samples

Cultural Methods for Detection of STEC O157:H7

Cultural methods for detection of STEC O157:H7 in food samples basically comprise a combination of an enrichment step and a serogroup-specific concentration with plating on selective agar and/or chromogenic solid media, followed by confirmation of presumptive positive colonies by biochemical and serological testing as well as by molecular methods.

Enrichment Step. Growth experiments by Doyle and Schoeni (1994) in tryptone soy broth (TSB) have shown that *E. coli* O157:H7 grows best between 30°C and 42°C. However, incubation at 42°C might hamper the recovery of injured cells. In addition to the incubation temperature, the composition of the enrichment broth has an important role on the growth capacity of STEC O157. In the literature, several different enrichment broths are described (Table 1).

In a recently published study (Jasson et al., 2009), lag phases and growth rates of healthy, stressed, and sublethally injured *E. coli* O157 cells in 10 enrichment broths were performed. TSB provided the fastest growth but failed to recover oxidation-stressed *E. coli* O157. TSB-yeast extract medium, TSB supplemented with 8 mg/liter novobiocin plus 16 mg/liter vancomycin (TSB+), buffered peptone water (BPW), and BPW supplemented with 8 mg/liter vancomycin (BPW+V) enabled resuscitation of *E. coli* O157 cells regardless of the preculture conditions.

Modified TSB plus 10 mg/liter novobiocin (mTSB+N), EC medium, EC reduced bile salts medium (ECred), and TSB supplemented with 8 mg/liter novobiocin plus 16 mg/liter vancomycin plus 2 mg/liter rifampin plus 1 mg/liter K-tellurite plus 1.5 g/liter bile salts

TABLE 1 Enrichment broths described for STEC O157 detection

Enrichment broth[a]	Supplement	Reference
mTSB	None	Doyle and Schoeni, 1987
mTSB	Novobiocin	Doyle and Schoeni, 1987
	Acriflavin	Kim and Doyle, 1992
EEB	Vancomycin, cefixime, cefsulodin	Weagant et al., 1995
LST	Novobiocin	Hitchins et al., 1998
mEC	Novobiocin	Okrend et al., 1990
BPW	Vancomycin, cefixime, cefsulodin	Chapman et al., 1994

[a] Abbreviations: EEB: *Enterobacteriaceae* enrichment broth; LST, lauryl tryptose broth; EC, *E. coli* broth; mEC, modified *E. coli* broth.

no. 3 (TSB++) all failed to recover *E. coli* O157 cells for at least one type of stress. The use of TSB, TSB+, BPW, and BPW+V was compared with that of mTSB+N (as described in the International Organization for Standardization [ISO] 16654:2001 standard method) for reliable detection of low numbers of healthy, stressed, and sublethally injured *E. coli* O157 (approximately 10 CFU/10 g) from foods (sprouted seeds, fermented sausage, raw milk, and raw ground beef).

When low numbers of healthy cells were inoculated, BPW, BPW+V, TSB, TSB+, and mTSB+N enabled growth to detectable numbers within 6 h of enrichment at 41.5°C. The results showed that mTSB+N failed to recover detectable numbers of *E. coli* O157 cells sublethally injured by freeze and food-processing stresses, in contrast to what was obtained with BPW and BPW+V. This study highlights that using mTSB+N for recovery of *E. coli* O157 from foods may yield false-negative results.

Separation and Concentration Techniques. Studies have shown that beef products may contain as few as 0.3 to 15 CFU of *E. coli* O157:H7 per gram (Bolton et al., 1996; Tuttle et al., 1999). To improve the sensitivity of the detection procedures, various separation and concentration techniques have been described (Bassel et al., 1983; Payen et al., 1992; Miles et al., 1995). The most widely used is the immunomagnetic separation technique. Basically, the system uses magnetic beads coated with anti-O157 antibodies, and the strains of *E. coli* O157 are bound to the beads, which are then separated by means of a magnetic field. Despite the possible cross-reaction with other bacterial antigens, the detection limit can be improved up to 100 times for *E. coli* O157 (Wright et al., 1994), and 1 CFU of *E. coli* O157:H7 per 25 g of food may be detected.

A different system intended for the selective concentration of *E. coli* O157 uses labeled recombinant phage tail fiber protein as a specific and sensitive tool for detection and capture of STEC O157.

Selective Media. STEC O157:H7 strains generally do not ferment sorbitol and are β-D-glucuronidase negative. Based on these biochemical characteristics, the enriched samples can be streaked onto sorbitol-containing agar plates or media supplemented with MUG (4-methylumbelliferyl-β-D-glucuronide), a fluorogenic substrate of β-D-glucuronidase.

False-positive and false-negative results may occur when the STEC O157 strains are streaked onto agar media that allow only the assessment of sorbitol fermentation. False-negative results may occur due to sorbitol-fermenting *E. coli* O157:H- strains. On the other hand, the colony morphology of many other organisms such as *Proteus* spp. or *Hafnia* spp. cannot be distinguished from that of *E. coli* O157 on MacConkey agar with sorbitol, cefixime, and tellurite (CT-SMAC). Therefore, all presumptive positive colonies have to be confirmed as *E. coli* by biochemical tests.

Other aspects may complicate the detection of STEC O157 by relying on the classical CT-SMAC medium. McCleery and Rowe (1995) showed that a 2-hour enrichment at 25°C in TSB is necessary for the detection of sublethally injured (52°C/60 min) *E. coli* O157 on sorbitol-MacConkey agar. Another study reported that the growth of nonmotile O157: H- strains was inhibited by the tellurite supplement in the CT-SMAC agar (Karch et al., 1996).

Different chromogenic media have been developed, although several problems may be encountered; The CHROMagar was found not to be sufficiently selective if foods with higher background flora were tested (Bettelheim, 1998; Restaino et al., 1999). In this case, typical *E. coli* O157 colonies were overgrown by the background flora. For the detection of heat-stressed O157 cells from minced meat, Taormina et al. (1998) recommend the use of another O157 agar.

Serological Identification of Isolated Strains. Isolated strains have to be confirmed as *E. coli* O157. However, when using polyclonal antibodies, cross-reactions with *Escherichia hermannii, Brucella abortus, Brucella melitensis, Yersinia enterocolitica* O:9, *Kebsiella* spp., *Enterobacter* spp., *Hafnia* spp., *Salmonella* spp., or *Pseudomonas* spp. may occur. All presumptive positive O157 colonies must therefore be confirmed biochemically as *Escherichia coli*. Different latex agglutination-based systems for O157 and O157:H7 are available.

Testing of Isolated Colonies for the Presence of the *stx* Genes

Once an *E. coli* strain is confirmed as O157, the presence of the *stx* genes or the capability to produce Stx must be assessed. Several methods based on conventional or real-time PCR that aim at testing isolated colonies for the presence of the *stx* genes are described in the literature. Examples of primers used for the detection of *stx* genes are given in Table 2. As an example, by using primer pairs stx1F/stx1R and stx2F/stx2R (Paton and Paton, 1998a), *stx1* and *stx2* genes can be detected. The latter primer pair can recognize all the variants of *stx2* except *stx2f*. The amplification of the *stx2f* variant can be obtained by an individual primer set, 128-1/128-2 (Schmidt et al., 2000).

As far as the assessment of the Stx production is concerned, this can be done either by in vitro identification of the cytopathic effect induced by the supernatant of STEC cultures onto Vero cell monolayers or by immunochemical detection methods. The latter include several commercially available immunoassays (see "Shiga Toxin Detection by ELISA" below.

Immunochemical Detection

All immunological detection procedures indicated include a preceding enrichment step. Doyle and Schoeni (1987) described a membrane technology, an immunoblot technique using nitrocellulose membranes, for the detection of STEC O157 in foods. Using polyclonal antibodies, these authors detected 1.5 CFU of *E. coli* O157:H7 per gram minced meat. The major drawback of this procedure is that it is complex and thus not suitable for the screening of food samples in routine laboratories.

Several immunochemical systems, mainly enzyme immunoassays, are commercially available. The detection limit for most of these kits is indicated by the suppliers to be as low as 1 CFU/g food. The advantages of using such approaches are that these systems can be easy to use and some of them are now validated. The drawbacks are the need, for some systems, to possess dedicated instruments and the restriction of their use to *E. coli* O157 detection only.

Biosensors

A further, rapid approach for STEC O157 detection is based on bio- or immunosensors (Mujika et al., 2008; Subramanian et al., 2006). Such biosensors are designed to be capable of detecting and quantifying small mag-

TABLE 2 Primer sequences for *stx1* and *stx2* detection

Target genes	Primers	Primer sequences	Amplicon size (bp)	Reference
stx1 group	stx1F	ATAAATCGCCATTCGTTGACTAC	180	Paton and Paton, 1998a
	stx1R	AGAACGCCCACTGAGATCATC		
stx2 group	stx2F	GGCACTGTCTGAAACTGCTCC	255	Paton and Paton, 1998a
	stx2R	TCGCCAGTTATCTGACATTCTG		
stx2f	128-1	AGA TTG GGC GTC ATT CAC TGG TTG	428	Schmidt et al., 2000
	128-2	TAC TTT AAT GGC CGC CCT GTC TCC		

netic field variations caused by the presence of superparamagnetic beads bound to the antigens previously immobilized on the sensor surface via an antibody-antigen reaction.

PCR

Feng et al. (1993) showed that *E. coli* O157:H7 can be specifically detected by a single modification in the 5′-end of the *vidA* structural gene. Since then, several methods have been described in the literature as conventional as well as real-time PCR systems targeting this point mutation. All these methods have in common the need to use an internal control in order to exclude the presence, in the sample, of possible substances inhibiting the amplification reaction.

Different real-time PCR kits are available for food testing, but all these systems include an enrichment step preceding the PCR screening.

Detection of Non-O157 STEC

Culture Methods for Detection of Non-O157 STEC

Culture methods for the detection of non-O157 STEC in food samples are based on a combination of an enrichment step followed by plating of the enrichment culture onto an agar-containing medium. Non-O157 STEC strains display a heterogeneous range of phenotypic properties, making it difficult to find selective or differential media to recover these pathogens. Nevertheless, a set of differential and confirmation plating media for STEC of serogroups O26, O103, O111, and O145, as well as sorbitol-positive and -negative O157 STEC, have been recently developed (Possé et al., 2008a, 2008b). Typical or suspected colonies have thereafter to be identified as STEC, e.g., by dot blot or PCR analysis, and further characterized by molecular, biochemical, and serological testing.

Shiga Toxin Detection by ELISA

A variety of enzyme-linked immunosorbent assay (ELISA) kits or latex agglutination kits are available, with which different toxin subtypes are detected simultaneously or separately. These kits use antibodies to the Shiga toxins or ganglioside Gb3 receptors to capture the toxins into an enrichment broth. The complexes are then revealed by secondary antibodies conjugated with the detection system.

Although these assays are generally less sensitive than traditional culture techniques, they offer advantages in terms of rapidity, reduced labor costs, and high-volume throughput. On the other hand, due to differences in the binding antibody affinities, some of these systems cannot detect all Stx variants (in particular, Stx2d and Stx2e). Recently, Beutin et al. (2010) published the results of two interlaboratory studies aiming at the detection of STEC in minced beef samples. A strikingly low sensitivity was observed when using an Stx-ELISA as detection method. This finding could be attributed to the use of the Stx-ELISA that showed deficiencies for detection of Stx1 and some variants of Stx2. It is therefore recommended that only detection systems that have been independently and thoroughly evaluated for their specificity and sensitivity be used.

PCR Detection of Non-O157 STEC

Several methods have been described in the literature as conventional as well as real-time PCR systems for detection of non-O157 STEC. All these procedures are based on an enrichment step preceding the PCR screening. One real-time PCR-based protocol for STEC detection is part of a method for STEC detection and isolation from food developed by the STEC ad hoc group in Working Group 6 of the Technical Committee 275 of the European Committee for Normalisation. The method has been discussed and approved by Working Group 6 as a "Technical Specification" and is now in publication in the form of an ISO standard.

In detail, the screening step (Table 3) of this method targets *stx1*, *stx2* (Perelle et al., 2004), and the intimin-coding *eae* genes (Nielsen and Andersen, 2003) as well as genes specific for serogroups O26, O103, O111, O145, and, in the complete form, O157 (Perelle et al., 2004,

TABLE 3 Degenerate primers and probes used in 5′ nuclease PCR assays for the detection of different genes

Target gene (serotype)	Forward primer, reverse primer, and probe sequences (5′–3′)	Reference
stx1	TTTGTYACTGTSACAGCWGAAGCYTTAC CCCCAGTTCARWGTRAGRTCMACRTC Probe: CTGGATGATCTCAGTGGGCGTTCTTATGTAA	Perelle et al., 2004
stx2	TTTGTYACTGTSACAGCWGAAGCYTTACG CCCCAGTTCARWGTRAGRTCMACRTC Probe: TCGTCAGGCACTGTCTGAAACTGCTCC	Perelle et al., 2004
eae	CAT TGA TCA GGA TTT TTC TGG TGA TA CTC ATG CGG AAA TAG CCG TTA Probe: ATAGTCTCGCCAGTATTCGCCACCAATACC	Nielsen and Andersen, 2003
rfbE (O157)	TTTCACACTTATTGGATGGTCTCAA CGATGAGTTTATCTGCAAGGTGAT Probe: AGGACCGCAGAGGAAAGAGAGGAATTAAGG	Perelle et al., 2004
wbdI (O111)	CGAGGCAACACATTATATAGTGCTTT TTTTTGAATAGTTATGAACATCTTGTTTAGC Probe: TTGAATCTCCCAGATGATCAACATCGTGAA	Perelle et al., 2004
wzx (O26)	CGCGACGGCAGAGAAAATT AGCAGGCTTTTATATTCTCCAACTTT Probe: CCCCGTTAAATCAATACTATTTCACGAGGTTGA	Perelle et al., 2004
ihp1 (O145)	CGATAATATTTACCCCACCAGTACAG GCCGCCGCAATGCTT Probe: CCGCCATTCAGAATGCACACAATATCG	Perelle et al., 2004
wzx (O103)	CAAGGTGATTACGAAAATGCATGT GAAAAAAGCACCCCCGTACTTAT Probe: CATAGCCTGTTGTTTTAT	Perelle et al., 2005

2005). This procedure aims at facilitating the screening for negative samples, but it requires a confirmation step with isolation of the STEC strains responsible for the positive PCRs.

STEC Isolation and Characterization

Besides immunomagnetic separation techniques for some non-O157 STEC serogroups, the isolation of non-O157 STEC strains can be achieved by colony dot blot hybridization or by Shiga toxin immunoblot techniques. Both these techniques are laborious and mainly used by reference laboratories.

Several methods (conventional as well as real-time PCR systems) are described in the literature for testing isolated colonies for the presence of the *stx* genes. Examples of primers used for the detection of *stx* genes are mentioned in Table 2 and have been discussed above. Some PCR protocols for STEC identification and typing can be found in the "laboratory methods" section of the EU Reference Laboratory for *E. coli* including the VTEC website (http://www.iss.it/vtec/work/cont.php?id=152&lang=2&tipo=3).

Classical antibody-based serotyping of *E. coli* is an important method in diagnostic microbiology for epidemiological purposes, as well as for a rough virulence assessment. However, serotyping is tedious, and its use is restricted to a few reference laboratories. To improve this aspect, Ballmer et al. (2007) developed and validated a molecular approach for serotyping based on microarray technology. The genes encoding the O-antigen flippase (*wzx*) and the O-antigen polymerase (*wzy*) were selected as target sequences identifying the different O antigens, whereas *fliC* and related genes, which code for the flagellar monomer, were chosen since differences in their DNA sequences represent the different H types. The microarray contains oligonucleotide DNA probes, each in duplicate, representing 24 of the epidemiologically most relevant STEC O antigens (O4, 6 to 9, 15, 26, 52, 53, 55, 79, 86, 91, 101, 103, 104, 111, 113, 114, 121, 128, 145, 157, and 172) as well as 47 of the 53 different H antigens (H antigens 1 to 12, 14 to 16, 18 to 21, 23 to 34, 37 to 43, 45, 46, 48, 49, 51 to 54, and 56). Evaluation of the microarray with a set of defined strains representing all O and H serotypes covered revealed that it has a high sensitivity and a high specificity (Ballmer et al., 2007). All of the conventionally typed 24 O groups and all of the 47 H serotypes were correctly identified. Moreover, strains that were nonmotile or nontypeable by immunological serotyping assays yielded unequivocal results, which may be a useful alternative to classical serotyping.

Bugarel et al. (2010) showed that the presence of *eae, ent/espL2, nleB, nleE*, and *nleH1-2* genes can be used as the signature for STEC strains with high virulence for humans, and they designed an assay for simultaneous detection of genes encoding Stx1 and Stx2 (stx_1 and stx_2), intimins (*eae*), enterohemolysin (*ehxA*), and six different *nle* genes derived from genomic islands OI-71 and OI-122, which permits discrimination between STEC causing serious illness in humans and other *E. coli* strains (Bugarel et al., 2010). This study provides important information on the virulence genes of STEC and could be exploited for the development of arrays aiming at the identification of the more virulent strains. Such a complex system may be used to examine the virulence gene set from isolated strains offering a genetic "barcoding" strategy useful to screen for different determinants.

CONCLUSIONS AND FUTURE TRENDS

Reliable detection and characterization of STEC is only possible through a combination of culture, molecular, and immunological methods. In order to assess the potential health hazard for humans, strain isolation and characterization of further virulence factors are required. Moreover, future efforts should include the establishment of standard enrichment protocols that can be used universally regardless of the serotypes tested or the matrix used.

State-of-the-art technologies such as high-throughput microarray or miniaturized PCR systems are increasingly being developed for the detection and characterization of bacterial pathogens. Although these approaches are still reserved for reference laboratories, they are becoming cheaper and easier to handle and may eventually be used in the future in routine diagnostic laboratories.

The establishment of the international standard method agreed upon at a worldwide level in the framework of ISO is the essential requirement for the technical implementation of monitoring programs in livestock animals and foodstuffs. At the European level, this approach has been laid down in a recently published technical specification by the European Food Safety Authority depicting the monitoring program for STEC in animals and food (European Food Safety Authority, 2009), bringing the STEC issue into the correct perspective for determining the microbiological criteria to be included in future regulations.

REFERENCES

Ballmer, K., B. M. Korczak, P. Kuhnert, P. Slickers, R. Ehricht, and H. Hächler. 2007. Fast DNA serotyping of *Escherichia coli* by use of an oligonucleotide microarray. *J. Clin. Microbiol.* **4:** 5370–5379.

Bassel, R. M., E. R. Ritcher, and G. J. Banwart. 1983. Monitoring microbial numbers in food by density centrifugation. *Appl. Environ. Microbiol.* **45:** 1156–1159.

Bettelheim, K. A. 1998. Reliability of CHROMagar O157 for the detection of enterohaemorrhagic *E. coli* O157 but not EHEC belonging to other serogroups. *J. Appl. Microbiol.* **85:**425–428.

Beutin, L., D. Geier, H. Steinrück, S. Zimmermann, and F. Scheutz. 1993. Prevalence and some properties of verotoxin (Shiga-like toxin) producing *Escherichia coli* in seven different species of healthy domestic animals. *J. Clin. Microbiol.* **31:** 2483–2488.

Beutin, L., G. Krause, S. Zimmermann, S. Kaulfuss, and K. Gleier. 2004. Characterization of Shiga toxin-producing *Escherichia coli* strains isolated from human patients in Germany over a 3-year period. *J. Clin. Microbiol.* **42:**1099–1108.

Beutin, L., A. Martin, G. Krause, K. Steege, S. Haby, K. Pries, N. Albrecht, A. Miko, and S. Jahn. 2010. Results, conclusions and recommendations of two ring trials for the detection and isolation of STEC from minced beef samples. *J. Verb. Lebensm.* **5:**21–34.

Bolton, F. J., L. Crozier, and J. K. Williamson. 1996. Isolation of *E. coli* O157 from raw meat products. *Lett. Appl. Microbiol.* **23:**317.

Bugarel, M., L. Beutin, and P. Fach. 2010. A low-density macroarray targeting non-locus of enterocyte effacement effectors (*nle* genes) and major virulence factors of Shiga toxin-producing *Escherichia coli* (STEC): a new approach for molecular risk assessment of STEC isolates. *Appl. Environ. Microbiol.* **76:**203–211.

Chapman, P. A., D. J. Wright, and C. A. Siddons. 1994. A comparison of immunomagnetic separation and direct culture for isolation of verocytotoxin-producing *Escherichia coli* O157 from bovine faeces. *J. Med. Microbiol.* **40:**424–427.

Coombes, B. K., M. E. Wickham, M. Mascarenhas, S. Gruenheid, B. B. Finlay, and M. A. Karmali. 2008. Molecular analysis as an aid to assess the public health risk of non-O157 Shiga toxin-producing *Escherichia coli* strains. *Appl. Environ. Microbiol.* **74:**2153–2160.

Doyle, M. P., and S. L. Schoeni. 1987. Isolation of *Escherichia coli* O157:H7 from retail fresh meats and poultry. *Appl. Environ. Microbiol.* **53:**2394–2396.

Doyle, M. P., and S. L. Schoeni. 1994. Survival and growth characteristics of *E. coli* associated with hemorrhagic colitis. *Appl. Environ. Microbiol.* **48:** 855–856.

European Food Safety Authority. 2007. Scientific opinion of the panel on biological hazards on a request from EFSA on monitoring of verotoxigenic *Escherichia coli* (VTEC) and identification of human pathogenic types. *EFSA J.* **579:**1–61.

European Food Safety Authority. 2009. Scientific report of EFSA: technical specifications for the monitoring and reporting of verotoxigenic *Escherichia coli* (VTEC) on animals and food (VTEC surveys on animals and food). *EFSA J.* **7:**1–43.

Feng, P. 1993. Identification of *Escherichia coli* serotype O157:H7 by DNA probe specific for an allele of *uid* A gene. *Mol. Cell. Probes* **7:**151–154.

Friedrich, A. W., M. Bielaszewska, W. Zhang, M. Pulz, T. Kuczius, A. Ammon, and H. Karch. 2002. *Escherichia coli* harboring Shiga toxin 2 gene variants: frequency and association with clinical symptoms. *J. Infect. Dis.* **185:**74–84.

Hitchins, A. D., P. Feng, W. D. Watkins, S. R. Rippey, and L. A. Chandler. 1998. *Escherichia coli* and the coliform bacteria, p. 4.01–4.29. *In Bacteriological Analytical Manual.* U.S. Food and Drug Administration, Washington, DC.

Jasson, V., A. Rajkovic, L. Baert, J. Debevere, and M. Uyttendaele. 2009. Comparison of enrichment conditions for rapid detection of low numbers of sublethally injured *Escherichia coli* O157 in food. *J. Food Prot.* **72:**1862–1868.

Johnson, K. E., C. M. Thorpe, and C. L. Sears. 2006. The emerging clinical importance of non-O157 Shiga toxin-producing *Escherichia coli*. *Clin. Infect. Dis.* **43:**1587–1595.

Karch, H., C. Janetzki-Mittmann, S. Aleksic, and M. Datz. 1996. Isolation of enterohemorrhagic *Escherichia coli* O157 strains from patients with hemolytic-uremic-syndrome by using immunomagnetic separation, DNA-based methods, and direct culture. *J. Clin. Microbiol.* **34:**516–519.

Karmali, M. A., M. Mascarenhas, S. Shen, K. Ziebell, S. Johnson, R. Reid-Smith, J. Isaac-Renton, C. Clark, K. Rahn, and J. B. Kaper. 2003. Association of genomic O island 122 of *Escherichia coli* EDL 933 with verocytotoxin-producing *Escherichia coli* seropathotypes that are linked to epidemic and/or serious disease. *J. Clin. Microbiol.* **41:**4930–4940.

Kim, J., J. Nietfeldt, and A. K. Benson. 1999. Octamer-based genome scanning distinguishes a unique subpopulation of *Escherichia coli* O157:H7 strains in cattle. *Proc. Natl. Acad. Sci. USA* **96:** 13288–13293.

Kim, M. S., and M. P. Doyle. 1992. Dipstick immunoassay to detect enterohemorrhagic *Escherichia coli* O157:H7 in retail ground beef. *Appl. Environ. Microbiol.* **58:**1764–1767.

Konczy, P., K. Ziebell, M. Mascarenhas, A. Choi, C. Michaud, A. M. Kropinski, T. S. Whittam, M. Wickham, B. Finlay, and M. A. Karmali. 2008. Genomic O island 122, locus for enterocyte effacement, and the evolution of virulent verocytotoxin-producing *Escherichia coli*. *J. Bacteriol.* **190:**5832–5840.

McCleery, D. R., and M. T. Rowe. 1995. Development of a selective plating technique for the recovery of *Escherichia coli* O157:H7 after heat stress. *Lett. Appl. Microbiol.* **21:**252–256.

Miles, C. A., M. J. Morley, W. R. Hudson, and B. M. Mackey. 1995. Principles of separating microorganisms from suspensions using ultrasound. *J. Appl. Bacteriol.* **78:**47–54.

Nielsen, E. M., and M. T. Andersen. 2003. Detection and characterization of verocytotoxin-producing *Escherichia coli* by automated 5′ nuclease PCR assay *J. Clin. Microbiol.* **41:**2884–2893.

Morabito, S., R. Tozzoli, E. Oswald, and A. Caprioli. 2003. A mosaic pathogenicity island made up of the locus of enterocyte effacement and a pathogenicity island of *Escherichia coli* O157:H7 is frequently present in attaching and effacing *E. coli*. *Infect. Immun.* **71:**3343–3348.

Mujika, M., S. Arana, E. Castano, M. Tijerno, R. Vilares, J. M. Ruano-Lopez, A. Cruz, L. Sainz, and J. Berganza. 2008. Magnetoresistive immunosensor for the detection of *Escherichia coli* O157:H7 including a microfluidic network. *Biosens. Bioelectron.* **24:**1253–1258.

Okrend, A. J. G., B. E. Rose, and B. Bennett. 1990. A screening method for the isolation of *Escherichia coli* O157:H7 from ground beef. *J. Food Prot.* **53:**249–252.

Paton, A. W., and J. C. Paton. 1998a. Detection and characterization of Shiga-toxigenic *Escherichia coli* by using multiplex PCR assays for *stx1, stx2, eaeA*, enterohemorrhagic *E. coli hlyA, rfbO111*, and *rfbO157*. *J. Clin. Microbiol.* **36:**598–602.

Paton, J. C., and A. W. Paton. 1998b. Pathogenesis and diagnosis of Shiga toxin-producing *Escherichia coli* infections. *Clin. Microbiol. Rev.* **11:**450–479.

Payen, M. J., S. Cambell, R. A. Patchett, and R. G. Kroll. 1992. The use of immobilised lectins in the separation of *Staphylococcus aureus, Escherichia coli, Listeria* and *Salmonella* spp. from pure cultures and foods. *J. Appl. Bacteriol.* **73:**41–52.

Perelle, S., F. Dilasser, J. Grout, and P. Fach. 2004. Detection by 5′-nuclease PCR of Shiga-toxin producing *Escherichia coli* O26, O55, O91, O103, O111, O113, O145 and O157:H7, associated with the world's most frequent clinical cases. *Mol. Cell. Probes* **18:**185–192.

Perelle, S., F. Dilasser, J. Grout, and P. Fach. 2005. Detection of *Escherichia coli* serogroup O103 by real-time polymerase chain reaction. *J. Appl. Microbiol.* **98:**1162–1168.

Possé, B., L. De Zutter, M. Heyndrickx, and L. Herman. 2008a. Novel differential and confirmation plating media for Shiga toxin-producing *Escherichia coli* serotypes O26, O103, O111, O145 and sorbitol-positive and -negative O157. *FEMS Microbiol. Lett.* **282:**124–131.

Possé, B., L. De Zutter, M. Heyndrickx, and L. Herman. 2008b. Quantitative isolation efficiency of O26, O103, O111, O145 and O157 STEC serotypes from artificially contaminated food and cattle faeces samples using a new isolation protocol. *J. Appl. Microbiol.* **105:**227–235.

Restaino, L., E. W. Frampton, K. W. Turner, and D. Allison. 1999. A chromogenic plating medium for isolating *E. coli* O157:H7 from beef. *Lett. Appl. Microbiol.* **29:**26–30.

Schmidt, H., J. Scheef, S. Morabito, A. Caprioli, L. Wieler, and H. Karch. 2000. A new Shiga toxin 2 variant (Stx2f) from *Escherichia coli* isolated from pigeons. *Appl. Environ. Microbiol.* **66:**1205–1208.

Subramanian, A., J. Irudayaraj, and T. Ryan. 2006. A mixed self-assembled monolayer-based surface plasmon immunosensor for detection of *E. coli* O157:H7. *Biosens. Bioelectron.* **15:**998–1006.

Taormina, P. J., M. Rocelle, S. Clavero, and L. R. Beuchat. 1998. Comparison of selective agar media and enrichment broths for recovering heat-stressed *E. coli* O157:H7 from ground beef. *Food Microbiol.* **15:**631–638.

Tarr, P. I., C. A. Gordon, and W. L. Chandler. 2005. Shiga-toxin-producing *Escherichia coli* and haemolytic uraemic syndrome. *Lancet* **365:**1073–1086.

Tuttle, J., T. Gomez, M. P. Doyle, J. G. Wells, T. Zhao, R. V. Tauxe, and P. M. Griffin. 1999. Lessons from a large outbreak of *E. coli* O157:H7 infections: insights into the infectious dose and method of widespread contamination of hamburger patties. *Epidemiol. Infect.* **122:**185–192.

Wallace, J. S., T. Cheasty, and K. Jones. 1997. Isolation of Vero cytotoxin-producing *Escherichia coli* O157 from wild birds. *J. Appl. Microbiol.* **82:** 399–404.

Weagant, S. D., J. L. Bryant, and K. G. Jinneman. 1995. An improved rapid technique for isolation of *Escherichia coli* O157:H7 from foods. *J. Food Prot.* **58:**7–12.

Wick, L. M., W. Qi, D. W. Lacher, and T. S. Whittam. 2005. Evolution of genomic content in the stepwise emergence of *Escherichia coli* O157:H7. *J. Bacteriol.* **187:**1783–1791.

Wickham, M. E., C. Lupp, M. Mascarenhas, A. Vazquez, B. K. Coombes, N. F. Brown, B. A.

Coburn, W. Deng, J. L. Puente, M. A. Karmali, and B. B. Finlay. 2006. Bacterial genetic determinants of non-O157 STEC outbreaks and hemolytic-uremic syndrome after infection. *J. Infect. Dis.* **194:**819–827.

Wright, D. J., P. A. Chapman, and C. A. Siddons. 1994. Immunomagnetic separation as a sensitive method for isolation *Escherichia coli* O157 from food samples. *Epidemiol. Infect.* **113:**31–39.

RAPID SCREENING OF ANIMAL FEEDS FOR MYCOTOXINS AND *SALMONELLA* CONTAMINATIONS

C. Wolf-Hall, H. Zhao, P. Häggblom, and J. Hoorfar

17

The global market for animal feed ingredients is ever expanding, and large volumes of materials are handled with an increasing complexity. Animal feed primarily originates from plants, but some ingredients are of animal, fungal (e.g., spent oyster mushroom substrate), or mineral origin. The variety of feed and feed ingredients is large, and hazards such as prions and industrial, environmental, bacterial, and fungal contaminants as well as insect and plant toxins may be transmitted to and through animals, ultimately entering the human food chain.

Mycotoxins and *Salmonella* are major and well-known food safety risks linked to animal feed. Mycotoxins can affect the health and susceptibility to infectious agents of animals, contributing to food safety issues in the human food supply, and mycotoxins can potentially be transmitted to the human food supply through meat, milk, or eggs (U.S. Department of Agriculture–Grain Inspection, Packers and Stockyards Administration, 2006). *Salmonella* is a contaminant of animal feed that can be transmitted through the animals and their byproducts to humans (Wales et al., 2010; Wierup and Haggblom, 2010). Feed is suspected to be a major source of *Salmonella* in poultry flocks (Jarquin et al., 2009).

The United Nations Food and Agriculture Organization (FAO) posts international information about feed safety issues (Food and Agriculture Organization, 2001a). An analysis of data reported to the European Community's Rapid Alert System for food and feed indicated that "various subspecies of *Salmonella* account for all but one of the 197 reports on microbiological hazards in animal feed" (9% of total reports in food and feed), while mycotoxins accounted for less than 1% of total reports for food and feed, with 14 reports for feed specifically (Kleter et al., 2009).

For a recent review of mycotoxins of agronomical importance, see Wolf-Hall (2010). The most significant mycotoxins of feed safety concern include aflatoxins, ergot alkaloids, fumonisins, ochratoxins, trichothecenes, and zearalenone. Of the trichothecenes, deoxynivalenol (DON), also known as vomitoxin, is

C. Wolf-Hall, Department of Veterinary and Microbiological Sciences, Great Plains Institute of Food Safety, 1523 Centennial Blvd., P.O. Box 6050-Dept. 7690, North Dakota State University, Fargo, ND 58108. *H. Zhao*, Great Plains Institute of Food Safety, 1523 Centennial Blvd., PO Box 6050-Dept. 7690, North Dakota State University, Fargo, ND 58108. *P. Häggblom*, Department of Chemistry, Environment and Animal Feed, National Veterinary Institute, SE-75189 Uppsala, Sweden. *J. Hoorfar*, National Food Institute, Technical University of Denmark, Mørkhøj Bygade 19, Blok H, DK-2860 Søborg, Denmark.

Rapid Detection, Identification, and Quantification of Foodborne Pathogens, ed. by J. Hoorfar

the most abundantly and commonly produced form and therefore the most likely to be tested for in feed. However, highly toxic trichothecenes such as T2 toxin are also screened for.

The Food and Drug Administration (FDA) has been developing a comprehensive animal feed safety program, which is to be risk based. The FDA has stated in draft documentation (Food and Drug Administration, 2010) the description for a risk-based approach as follows.

> A risk-based approach for the animal feed safety system identifies and assesses the risks to animal and human health posted by biological, chemical and physical hazards in animal feed. In this context, risk is a function of the likelihood of human or animal exposure to deleterious amounts of such hazards in feeds, and the significance of the health consequences in response to those exposures. Analysis of the risks posed by feed hazards will help the agency identify the appropriate regulatory approach for each hazard, and will permit the agency to make effective and efficient use of regulatory resources.

Therefore, accurate data for the occurrence, quantities, and validation of intervention strategies for high-risk chemical and biological hazards including mycotoxins and *Salmonella* in animal feed are much needed. Rapid methods for collection of these data are necessary.

The FDA has established limits on select contaminants in food and feed, including aflatoxin action levels (Food and Drug Administration, 1994), guidance levels for fumonisin (Food and Drug Administration, 2001), guidance levels for patulin (Food and Drug Administration, 2005), guidance for DON (Food and Drug Administration, 1993), and regulatory limits for *Salmonella* (Food and Drug Administration, 1989). The FDA has no action, guidance, or advisory levels for ochratoxin A or zearalenone and handles incidences of contamination by such toxins on a case-by-case basis. For these two mycotoxins, the FDA is collecting data to determine if limits should be established. The European Union, through the Codex Alimentarius Commission, has similar concerns (Codex Alimentarius Commission, 2009). Murphy et al. (2006) compiled a list of specific guidance and regulations for mycotoxins in food and feed for the United States and the European Union, including type of food/feed, action level, and source of information. "In order to ensure reliable determination of the toxin content at the tolerance levels, routine analytical methods must have detection limits of less than the tolerance level" (Schneider et al., 2004).

Hazards such as mycotoxins in feed affect both safety and international trade (Dohlman, 2003). The impact of testing information can affect global economies and food security and safety, with particularly devastating consequences for developing countries that may export their highest-quality commodities and be left with contaminated materials for domestic use. Therefore, accurate and rapid tests are vital to both feed safety and international trade.

FLOW OF FEED AND HACCP

Coffey and Cummins (2008) reviewed feed-to-food risk assessment focusing on mycotoxins in bovine feed. They described key parameters involved in the development of a feed chain risk assessment approach to identify risk management strategies to control transmission of mycotoxins through the feed and food chain. With the tremendous variety of feed stocks, the number of processing types and stages, and the multiple mycotoxins that can be present, control practice results can vary widely. "Mycotoxins may never be completely eradicated from the chain but with good agricultural practice, effective crop drying and storage and efficient processing procedures it is possible to keep their level, and potential human exposure to a minimum" (Coffey and Cummins, 2008).

The FAO has published a manual on the application of the hazard analysis and critical control points (HACCP) system in mycotoxin prevention and control (Food and Agriculture Organization, 2001b). The publication serves as an introduction to what mycotoxins are, gives an overview of HACCP, and includes examples of HACCP applied to specific commodities in different countries. Appendices include definitions of terms, descriptions of tasks involved in developing a HACCP system, a sample form for description and identified use of product, an example of a decision tree to identify critical control points, and an example of a HACCP worksheet.

For *Salmonella* testing, challenges include sensitivity to prevent false negatives and recovering representative samples from large consignments of feed ingredients. Jarquin et al. (2009) reviewed the processing of feeds, possible points where introduction of *Salmonella* into feed may occur, and current detection methods and discussed the needs for advancements in testing methods. Contamination points may include interactions of animal pests with raw ingredients and/or finished product, contaminated animal tissue such as fish meal, environmental contamination of the feed mill, contaminated feed transmission to a breeder flock, spread to chick progeny, spread to production houses, and environmental contamination of production houses. Heat processing of feed material such as extrusion pelleting can be an effective intervention dependent on time, temperature, and moisture content.

Because of the higher value of poultry breeding flocks, process flow is designed for one-way movement from clean to dirty. Due to the very low infectious dose for contamination of day-old juveniles, sensitive testing methods are required for detection of one cell/25 g of poultry feed.

Wierup and Haggblom (2010) studied the incidence of *Salmonella* in pig feed ingredients before introduction to feed mills and from HACCP-based surveillance of feed mills. They found *Salmonella* to be frequently isolated from imported consignments of soybean meal and rapeseed meal. They concluded that effective HACCP-based controls are needed in feed mills and that hygiene in crushing plants is critically important to prevent downstream contamination. They highlighted the challenge to the European Union for screening feed ingredients, some of which (up to 98%) are imported from nonmember states, with unknown *Salmonella* status.

Wales et al. (2010) reviewed interventions of chemical treatment and water in feed for control of *Salmonella*. Topics covered included organic acids and derivatives, acid treatment of drinking water, formaldehyde, essential oils, and chlorate. Challenges to intervention studies included masking, strain variation, natural versus artificial contamination, effects of background flora, and unknown. The effectiveness of intervention treatments needs to be checked before treated batches of feed material are used. Reliance on single interventions was discouraged. It was recommended that chemical treatments be part of a package including careful testing, selection of ingredients, mill hygiene, and control of recontamination.

SAMPLING AND TESTING

Sampling for Mycotoxins

"A sampling procedure is a multistage process and consists of three distinct phases: sampling, sample preparation and analysis" (Cheli et al., 2009). The European Commission designated sample methods and performance criteria for the analytical methods to be used for managing mycotoxins in feed (European Communities Commission, 2006). The *Codex Alimentarius* describes sample selection requirements for aflatoxin analysis (Codex Alimentarius Commission, 2009). Sampling information includes materials to be sampled, incremental sample selection, number of incremental samples for lots of varying weight, weight of the incremental sample, static lots, dynamic lots, packaging and transportation of samples, and sealing and labeling of samples. Sample preparation includes precautions and homogenization (grinding) information. Shephard et al. (2010) recently reviewed developments in mycotoxin sampling and analysis as published in 2008 and 2009, emphasizing new methodology. In their recent review of the evolution of legislation dealing with sampling plans, Cheli et al. (2009) described the critical importance of sampling protocols as fungal development and mycotoxin production are "spot processes," which are influenced by crop variety, agronomic practices, weather conditions, storage, and processing, and concluded that "The combination of representative non-destructive sampling procedures with validated rapid methods of analysis represents the future challenge in mycotoxin control measures."

While standards are gradually becoming harmonized globally, there are impacts due to differences in sampling plans. For example, Whitaker et al. (2010) describe disparities in sampling plans between the Codex Alimentarius Commission and the U.S. almond industry. The dual-sampling plan designed by the Codex Alimentarius Commission for almonds resulted in significant economic impact to the industry. A comparison of performances of the sampling plans indicated that the sampling plan with dual limits resulted in more rejections of almond lots than a similar sampling plan with a single limit. The information from this Monte Carlo method-based .study was intended to help the U.S. almond industry adapt their sampling plans to ensure that the testing results of the exporters would be more consistent with the results of importers.

Monitoring for *Salmonella*

Monitoring for *Salmonella* in animal feed is very challenging, as indicated in the review by Jarquin et al. (2009). When sampling for salmonellae in animal feed or in feed raw materials, the analyses usually give qualitative results, in contrast to mycotoxin analyses, while the acceptance limit is zero (no positive samples detected). A sampling plan often includes many incremental samples, and a single positive sample will sometimes cause rejections, a situation that puts high demands on aseptic work in order to prevent false-positive consignments. Thus, the detection probability and reported prevalence of *Salmonella* contamination is highly dependent on the sampling plan used (Anonymous, 2008). The aim of the monitoring may also be to check on the environmental hygiene of the premises, equipment, and trucks. Depending on the purpose of the sampling, different sample types are used such as dust samples, swabs, or scrapings, especially when following up on a suspicion of contamination in a plant. Because of the generated qualitative data and also because the majority of investigated samples are negative, sampling plans for microorganisms are typically based on theoretical assumptions (International Commission on Microbial Specifications in Food, 2002). Typically, specification limits for *Salmonella* are "absence in 25g," but the implementation could vary between countries. According to the International Commission on Microbial Specifications in Food (2002), five negative 25-g samples would be sufficient to conclude with 95% certainty that less than 50% of all possible 25-g samples are positive, provided a uniform distribution of bacteria in the lot. With nonrandom or spotwise distribution, the number of samples required to achieve a similar level of confidence may be considerably higher. Operation characteristics curves may be used to reflect the uncertainty of a microbiological sampling plan (Fig. 1) (International Commission on Microbial Specifications in Food, 2002). Detection limits as low as one cell in 10 tons of feed may be needed to control for flock infection due to nonuniform distribution. Detection can be hampered by stress damage contributing to the viable but not culturable state of the target organism. The sampling uncertainty of salmonellae and mycotoxins in feed was reviewed in a recent article (Andersson and Haggblom, 2009).

Analytical Methods

Goryacheva et al. (2007) describe two main groups of methods for mycotoxin analysis: "laborious methods for determination of mycotoxins with high sensitivity and precision, and screening methods for rapid detection in a non-laboratory environment." The laborious methods include liquid and gas chromatography methods, often with mass spectrometry of fluorescence detection. The challenges for detecting and quantifying mycotoxins include the need for low detection limits and common needs of sample cleanup

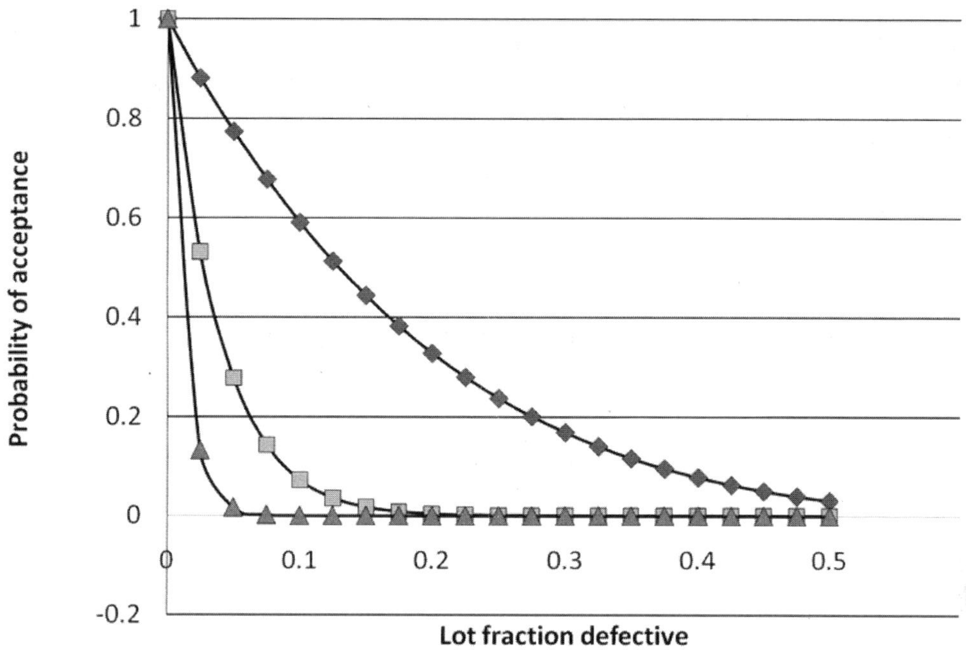

FIGURE 1 Operations characteristics curve showing the effect of sample size (i.e., number of increments) on the probability of detecting salmonellae based on calculations using the International Commission on Microbial Specifications in Food model (two-class sampling plan). *x* axis, true fraction contaminated with *Salmonella*; *y* axis, probability that the batch is classified as salmonella free (%).

to prevent background interference. The laborious methods have advantages when it comes to the ability to detect and quantify multiple mycotoxins simultaneously.

Regarding ochratoxin A testing, Murphy et al. (2006) stated the following:

> Because of the large variety of food matrices in which ochratoxin has been found, there is no universally suitable method of analysis. Differences in extraction conditions and clean-up are as varied as the foods. Analysis is typically accomplished by high performance liquid chromatography (HPLC), although liquid chromatography—mass spectrometry—is an important secondary confirmation technique.

The concern over feed matrix differences is significant for all analyses and requires testing methods to be individually validated for the specific matrix to be tested.

Reiter et al. (2009) reviewed the topic of sample preparation strategies for aflatoxin analysis of food and feed. Adequate removal of interfering substances is an important step for many analytical methods used for aflatoxin testing. The choice of solvents has impact on disposal and environmental concerns, which has led to the ban on chlorinated solvents for routine testing. Common solvents used include acetonitrile, methanol, and acetone. Liquid-liquid extraction protocols have been replaced with solid-phase extraction or immunoaffinity extraction methods to further reduce the use of organic solvents and speed up the cleanup process. Supercritical fluid extraction using carbon dioxide has also been used for extraction purposes. The development of more efficient solid-phase extraction columns such as aluminum oxide columns helps speed sample cleanup and utilize less solvent. Immunoaffinity columns provide high selectivity during purification and work by use of antibodies that bind the target analyte and then release the analyte when the antibodies are destroyed with an organic solvent. Some immunoaffinity columns can be reused by regeneration with phosphate-buffered saline, and this process helps lower costs of analysis. Solid-phase extraction and immunoaffinity strategies can be automated and set up within HPLC systems but can lead to challenges with pressure instability.

There are several sources of information about mycotoxin and *Salmonella* methods for feed. The European Mycotoxin Awareness Network (2010) has a comprehensive website with fact sheets for various mycotoxin analytical methods, including information for rapid test kits. Methods manuals make available established and validated methods information. Because analytical methods results can vary between labs and analysts, it is critical for methods to be verified through interlaboratory collaborations. Table 1 lists examples of sources for validated mycotoxin and *Salmonella* analytical methods. The *Codex Alimentarius* allows for modifications and development of methods for analysis of aflatoxins, providing specific requirements with which methods of analysis should comply, which include recovery, precision for reproducibility, and precision for repeatability (Codex Alimentarius Commission, 2009).

Most mycotoxin methods detect single toxins, resulting in the need for multiple tests to thoroughly screen feed for safety. Simultaneous multiple mycotoxin testing involves more complicated sample extraction, cleanup, and processing (Goryacheva et al., 2007). This can result in lower sensitivities. Chromatography methods capable of simultaneous multiple my-

TABLE 1 Examples of validated methods for mycotoxins and *Salmonella*

Analyte	Method	Reference(s)[a]
Aflatoxin	Thin-layer chromatography	AOAC and AACC
	Mass spectrometry	AOAC
	Immunoaffinity column	AOAC-IUPAC
	Qualitative minicolumn	AOAC-AACC
	Rapid test kits	AOAC Rapid
Ochratoxin	Thin-layer chromatography	AOAC-IUPAC
	Liquid chromatography	AOAC
	Immunoaffinity by column HPLC	AOAC
Ergotamine	Chromatography	AOAC
Patulin	Liquid chromatography	AOAC
Fumonisin	Liquid chromatography	AOAC-IUPAC and AACC
	Competitive direct ELISA	AOAC
DON	Thin-layer chromatography	AOAC-AACC
	Gas chromatography	AOAC
T2 toxin	Rapid test kits	AOAC Rapid
Zearalenone	Thin-layer chromatography	AOAC-AACC
	Liquid chromatography	AOAC and AACC
	ELISA kit	AOAC
Salmonella	Qualitative microbiological/immunoassay	AOAC, AACC, FDA-BAM, FSIS, and ISO
	Rapid test kits	AOAC Rapid and FDA-BAM

[a] Hyphens indicate cross-referenced methods. Abbreviations: AOAC, Association of Analytical Communities, AOAC International (2005); AACC, American Association of Cereal Chemists, AACC International (2010); IUPAC, International Union of Pure and Applied Chemistry (2010); AOAC Rapid, AOAC *Performance Tested Methods Validated Methods* (AOAC International, 2010); FDA-BAM, Food and Drug Administration *Bacteriological Analytical Manual* (Food and Drug Administration, 2009); FSIS, Food Safety and Inspection Service *Microbiology Laboratory Guidebook* (U.S. Department of Agriculture, Food Safety and Inspection Service, 2008); ISO, *Microbiology of Food and Animal Feeding Stuffs—Horizontal Method for the Detection of* Salmonella *spp*. (International Organization for Standardization, 2002).

cotoxin determination have been reported. Sulyok et al. (2006) reported a method for detection of 39 mycotoxins in wheat and maize, using a single extraction step followed by liquid chromatography with electrospray ionization–triple-quadrupole mass spectrometry without cleanup. To date, these types of tests have not gained official method status, but they are likely to do so in the future.

Rapid, nondestructive methods such as near-infrared reflectance have advantages over chromatography or antibody-based methods. The lack of need for solvents or reagents is the greatest advantage. Shephard et al. (2010) reviewed several recent developments in mycotoxin analysis, including near-infrared reflectance methods, and indicated that viable methods had been reported for aflatoxin and ochratoxin in various grains and spices.

RAPID METHODS FOR DETECTION OF MYCOTOXINS IN ANIMAL FEED

Chromatography methods remain the most common types of standard methods for quantification of mycotoxins. However, rapid methods based on antibody technology are becoming more widely adopted, especially for screening of samples to save on labor and equipment costs and to test in the field. Table 2 summarizes types of rapid testing methods verified by the U.S. Department of Agriculture—Grain Inspection, Packers and Stockyards Administration (2010) for specific commodities.

Goryacheva et al. (2007) reviewed immunochemical methods for detection of mycotoxins, with emphasis on simultaneous multiple-analyte determination. The main trends in research and development include sensitivity improvement, matrix effect reduction, simplification, shorter time of analysis, and multimycotoxin screening. Formats used for multimycotoxin screening include microtiter plate immunoassays and sensors. The two main ways to detect multimycotoxins were described as "using antibodies with suitable cross-reactivities for the determination of groups of related compounds and also by application of antibodies to different compounds in separate rows in one ELISA plate" (Goryacheva et al., 2007). Table 3 summarizes the types of immunoassay methods reviewed by Goryacheva et al. (2007); see the reference for details about limits of detection. The development of array immunoassay systems was the most common approach to detecting multiple mycotoxins. Challenges for future development include increased sensitivity and an adequate supply of high-quality antibodies or synthetic alternatives.

Another emerging issue with mycotoxin detection is the need to account for masked mycotoxins (Shephard et al., 2010; Zachariasova et al., 2010). Masked mycotoxins are conjugated forms of the mycotoxins bound to matrix materials. Special steps are needed to ensure that bound mycotoxins are accounted for, as they may be released upon ingestion and hydrolysis due to digestive enzymes.

Maragos (2009) reviewed the topic of fluorescence polarization immunoassays (FPIA) for mycotoxins. The primary advantage of this homogenous methodology is speed, where there is not a need for separation of free and bound tracer. "Development of competitive FPIA for mycotoxins requires a fluorescent tracer (usually a fluorescein conjugate of the toxin), a high molecular weight material to bind the toxin (generally an antibody), and an instrument to measure the polarization" (Maragos, 2009). A summary of FPIA methods described in the literature for most of the major mycotoxins was provided, including specific mycotoxin form, tracers used, matrices, sample cleanup methods, and limits of detection. Most of the methods described took from 2 to 15 minutes for completion, making them the leading format for rapidity of detection. However, the sequential-solution-based methodology limits adaptability to field applications.

RAPID METHODS FOR DETECTION OF *SALMONELLA* IN ANIMAL FEED

Many of the same challenges that exist for mycotoxin monitoring are also relevant to *Salmonella.* These include the need for sensitive detection limits (1 CFU/25-g sample) and ad-

TABLE 2 U.S. Department of Agriculture–Grain Inspection, Packers and Stockyards Administration-verified rapid methods for mycotoxins in various commodities[a]

Mycotoxins	Type(s) of method	Commodity[b]
Aflatoxins	Microtiter well plate assay	Corn, corn germ meal, corn gluten meal, cornmeal, corn bran, corn flour, corn screenings, cracked corn, corn/soy blend, flaking corn grits, cottonseed, ddg, ddgs, oats, milled rice, rough rice, brown rice, popcorn, rice bran, rye, sorghum, soybeans, soybean meal, wheat
	ELISA-based antibody-coated tubes	Wheat, barley, malted barley, corn, rice, wheat flour, wheat middlings
	Lateral flow strip	Corn, corn flour, corn germ meal, corn gluten meal, cornmeal, corn screenings, corn soy blend, cracked corn, flaking corn grits, ddg, ddgs, milled rice, oats, popcorn, rough rice, rye, sorghum, soybeans, wheat
	Fluorescence polarization	Corn, wheat
	Flowthrough rapid test	Corn
	Affinity column immunoassay	Corn, condensed distiller's solubles, corn bran, corn flour, corn germ meal, corn gluten feed, corn gluten meal, cornmeal, corn soy blend, ddg, ddgs, flaking corn grits, milled rice, brown rice, rice bran, rough rice, popcorn, sorghum, soybeans, wheat
	Rapid, fluorometric, solid-phase cleanup	Corn, cornmeal, corn/soy blend, milled rice, popcorn, sorghum, soybeans, wheat, corn screenings, ddgs
Fumonisins	Lateral flow strip	Corn
	Fluorescence polarization	Corn
	Microtiter well plate assay	Corn, cornmeal, popcorn, rough rice, wheat, corn soy blend, corn germ meal, sorghum, flaking corn grits
	Direct competitive ELISA	Corn, cornmeal, ddgs, popcorn, rice, sorghum, wheat
	Rapid, fluorometric, solid phase cleanup	Corn, cornmeal, corn soy blend, corn germ meal, sorghum, popcorn, flaking corn grits
Ochratoxins	Lateral flow strip	Wheat, barley, malted barley
	Rapid enzyme immunoassay	Wheat
	Rapid, fluorometric, solid-phase cleanup	Wheat

(*continued*)

TABLE 2 (*Continued*)

Mycotoxins	Type(s) of method	Commodity[b]
DON	Lateral flow strip	Barley, corn, malted barley, milled rice, oats, rough rice sorghum, wheat, wheat flour, wheat middlings
	Fluorescence polarization	Wheat
	Microtiter well plate assay	Wheat, barley, malted barley, oats, corn, rice, wheat germ, wheat flour, wheat middlings, sorghum
	Rapid enzyme immunoassay	Wheat, barley, malted barley, corn, rice, wheat flour
	ELISA-based antibody-coated tubes	Wheat, barley, malted barley, corn, oats, rice, wheat flour, wheat middlings
Zearalenone	Lateral flow strip	Barley, corn, ddgs, flaking corn grits, milled rice, oats, rough rice, sorghum, wheat, wheat flour
	Microtiter well plate assay	Corn, wheat

[a] Data from U.S. Department of Agriculture–Grain Inspection, Packers and Stockyards Administration, 2010).
[b] Abbreviations: ddg, distiller's dried grains; ddgs, distiller's dried grains with solubles.

dressing background interferences that may cause false-positive or false-negative results. Major differences between mycotoxin testing and *Salmonella* testing are (i) the zero tolerances set for *Salmonella* for many matrices, whereas mycotoxins have allowable limits; (ii) the need to distinguish living from dead cells for *Salmonella*; and (iii) the need to isolate and characterize *Salmonella* from positive samples for epidemiological purposes. Although a tremendous amount of research and development has been done and is still in progress, the new technologies still require further study in order to become validated methods (Koyuncu et al., 2010).

For *Salmonella* testing, the culture-based methods used to detect salmonellae in animal feed are laborious, time-consuming, and often not specific enough, leading to the need for development of rapid methods such as PCR methods (Löfström et al., 2004). The authors described the need for simplified sample preparation. They also indicated the advantages of culture enrichment prior to analysis.

Alakomi and Saarela (2009) in a review described traditional detection and isolation as "a multi-step protocol with nonselective preenrichment, followed by a selective enrichment step, isolation on selective agar media, and a preliminary biochemical and serological confirmation," with a requirement of 3 working days for negative results and 5 working days for confirmed positive results. They continue describing several chromogenic and fluorogenic growth media developed for faster detection based on specific enzyme activities and then lead into the description of more rapid methods including those based on metabolic and enzymatic properties, antibodies, nucleic acids, or filtration.

Fierens and Huyghebaert (1996) evaluated five screening methods for *Salmonella* in naturally contaminated feed samples. The methods included selective microbiological media, enzyme-linked immunosorbent assay (ELISA), and immunomagnetic separation. They found variable results, which were thought to be due to differences in sensitivity and specificity. En-

TABLE 3 Summary of types of immunoassays for mycotoxin analysis[a]

Type of method	Method	Description	Mycotoxins
Instrumental immunosensors		Detection of analyte-antibody interactions causing changes in physicochemical parameters detected with electrical sensors	
	Colorimetric and luminescent sensors	Based on light (visible or UV) transformation into signal	Aflatoxin, DON, citrinin, fumonisin, ochratoxin, zearalenone
	Surface plasmon resonance sensors	Based on changes on the surface of thin metal films	Aflatoxin, DON, fumonisin, zearalenone
	Eletrochemical sensors	Usually based on a competitive ELISA format	Aflatoxin, DON, fumonisin, ochratoxin, zearalenone
	Fluorescence polarization immunoassay	A homogenous method, with the analyte labeled with a fluorophore. Labeled and nonlabeled analytes compete for antibody-binding sites. The signal is the value of fluorescence polarization of the label, which is inversely proportional to the amount of free analyte in the sample.	Aflatoxin, DON, zearalenone
	Capillary electrophoretic immunoassay	Combines separation of analytes with high specificity of antibodies and sensitive detection	Fumonisin
Noninstrumental	Lateral flow tests	Immunochromatographic strip tests with conjugate pad, porous membrane, and absorbent pad. The solubilized analytes move through the pad by capillary flow.	Aflatoxin, DON, T2 toxin, fumonisin, ochratoxin, zearalenone
	Dipstick tests	Usually based on membrane spotted with specific antibody and requires several steps including sample binding, analyte-enzyme conjugation, and chromogenic substrate reaction	
	Flowthrough tests	Also known as immunofiltration assay or enzyme-linked immunofiltration assay. Based on a direct competitive ELISA format with antibodies attached to a membrane. Utilizes chromogenic detection.	Aflatoxin, DON, T2 toxin, roridin A, diacetoxyscirpenol, fumonisin, ochratoxin
	Tandem column tests	Integration of solid-phase cleanup and immunoassay	Aflatoxin, ochratoxin

[a] Data from Goryacheva et al., 2007.

richment stages seemed to influence the number of positive results, emphasizing the importance of enrichment to ensure detectable levels. Although accuracy, sensitivity, and specificity of methods may be consistent, detection limits can be affected by different matrices independent of the detection method used (Koyuncu and Haggblom, 2009; Koyuncu et al., 2010). Significant differences in detection limits for *Salmonella* were seen between palm kernel meal and rapeseed, soybean meal, and pelleted feed, which is a problem when prevalence data are compared as well as when feed ingredients are tested for conformance with microbiological criteria.

Critical steps for rapid methods include capturing the target microbe, removal of PCR inhibitors, the use of robust polymerases, and rapid identification. Also important is the isolation of the target microbe for validation and epidemiological comparisons. Rapid methods were described as methods that can provide positive results in less than 24 hours. The newer and future methods discussed included real-time PCR, RNA-based methods, quantitative multiplex PCR assays, capillary electrophoresis, microarrays, and biosensors.

Malorny et al. (2009) reviewed the topic of PCR methodologies for detection and serovar identification of *Salmonella* in food and feed. They included information on recommended enrichment procedures for food and feed matrices and sample preparation to minimize interfering substances that affect the reaction and discussed the organizations for validation and standardization of PCR for the detection of *Salmonella*. Real-time PCR was described as an effective screening tool. Limitations to PCR technology include the remaining need for culture-based isolation in the case of positive samples and needs for validation across a wide variety of sample matrices.

RAPID ENUMERATION OF *SALMONELLA*

In a recent study (C. Löfström, P. Häggblom, and J. Hoorfar, unpublished data), a PCR-based method for detection of low numbers of *Salmonella* in animal feed samples was developed, validated, and combined with an optimized most-probable-number (MPN) protocol to allow enumeration. The PCR method included an enrichment step in buffered peptone water at 37°C for 18 ± 2 h, followed by centrifugation of a withdrawn 1-ml buffered peptone water aliquot. DNA was extracted by an automated procedure from the pellet and subjected to real-time PCR. The qualitative PCR was compared to a reference culture method using modified semisolid Rappaport-Vassiliadis agar plates, according to ISO 6579 (International Organization for Standardization, 2002). Of 81 naturally or artificially contaminated samples tested (soya meal, rapeseed meal, rapeseed cake, and pellets), only 3 samples gave results that differed between the PCR and modified semisolid Rappaport-Vassiliadis methods. Cycle threshold values for naturally contaminated samples were higher than those of samples artificially contaminated with low numbers (approximately 2 CFU/25 g feed) of stressed *Salmonella* cells. To allow quantification of low numbers of *Salmonella* in feed, the developed PCR method was combined with an MPN approach. The novel MPN scheme was modified in order to make the procedure less laborious, time-consuming, and costly, as well as better adjusted to enumerate *Salmonella* cells in feed samples. This was achieved by two different approaches: (i) the dilution scheme was adjusted to better enumerate the small quantity of organisms presumably found in feed, and (ii) the selective enrichment steps were replaced by the qualitative PCR method. In conclusion, the developed PCR method can be used as an alternative method for detecting low numbers of *Salmonella* in feed samples. In combination with the novel MPN scheme, it can also be employed to generate quantitative data of feed contamination and in addition support the development of sampling plans for feed. Studies are in progress to further validate the performance with more naturally contaminated samples.

CONCLUSION AND FUTURE TRENDS

Food safety and a rapidly changing global market for feed and feed ingredients increase the need and urgency for development of sensitive, accurate, precise, fast, and economically feasible testing methods for multiple mycotoxins and *Salmonella*. Rapid methods are entering the market and being developed at an astounding rate, as evidenced by the amount of recent review information covered in this chapter. Traditional methods are still widely used, but kits for rapid detection are gaining validation status by standardizing organizations. There are still many issues to resolve including very significant challenges related to representative sample collection to ensure no false negatives occur. Sampling plans will need to be globally harmonized for consistency but must be balanced with equity concerns for economic impact and food security in the sense of an adequate and safe food supply for developing countries. Sample processing and analysis still present limitations for detection of multiple mycotoxins, masked mycotoxins, and metabolically injured (viable but not culturable) *Salmonella*.

REFERENCES

Alakomi, H. L., and M. Saarela. 2009. *Salmonella* importance and current status of detection and surveillance methods. *Qual. Assur. Saf. Crops Foods* **3:** 142–152.

American Association of Cereal Chemists. 2010. *Approved Methods of Analysis*, 11th ed. American Association of Cereal Chemists, AACC International, St. Paul, MN. http://www.aaccnet.org/ApprovedMethods/default.aspx

Andersson, M. G., and P. Haggblom. 2009. Sampling for contaminants in feed. *Feed Int.* **March:** 16–19.

Anonymous. 2008. Scientific opinion of the Panel on Biological Hazards on a request from the Health and Consumer Protection, Directorate General, European Commission on Microbiological Risk Assessment in feedingstuffs for food producing animals. *EFSA J.* **720:**1–84.

AOAC International. 2005. *Official Methods of Analysis of AOAC International (OMA)*, 18th ed. Association of Analytical Communities, AOAC International, Gaithersburg, MD. http://www.eoma.aoac.org/.

AOAC International. 2010. *Performance Tested Methods^{sm} Validated Methods*. Association of Analytical Communities, AOAC International, Gaithersburg, MD. http://www.aoac.org/testkits/testedmethods.html.

Cheli, F., A. Campagnoli, L. Pinottie, E. Fusi, and V. Dell'Orto. 2009. Sampling feed for mycotoxins: acquiring knowledge from food. *Ital. J. Anim. Sci.* **8:**5–22.

Codex Alimentarius Commission 2009. *Codex Alimentarius Current Official Standards*. Food and Agriculture Organization/World Health Organization, Rome, Italy. http://www.codexalimentarius.net/web/standard_list.do?lang=en.

Coffey, R., and E. Cummins. 2008. Feed to food risk assessment, with particular reference to mycotoxins in bovine feed. *Int. J. Risk Assess. Manag.* **8:** 266–286.

Dohlman, E. 2003. Mycotoxin hazards and regulations: impacts on food and animal feed crop trade. *In* J. C. Buzby (ed.), *International Trade and Food Safety: Economic Theory and Case Studies. U.S. Department of Agriculture Economic Research Service, Agricultural Economic Report No. AER828.* U.S. Department of Agriculture, Washington, DC. http://www.ers.usda.gov/publications/aer828/.

European Communities Commission. 2006. Commission recommendation of 17 August 2006 on the presence of deoxynivalenol, zearalenone, ochratoxin a, T-2 and HT-2 and fumonisins in products intended for animal feeding, 2006/576/EC. *Off. J. Eur. Union* L **229:**7–9.

European Mycotoxin Awareness Network. 2010. *European Mycotoxin Awareness Network*. http://www.mycotoxins.org/.

Fierens, H., and A. Huyghebaert. 1996. Screening of *Salmonella* in naturally contaminated feeds with rapid methods. *Int. J. Food Microbiol.* **31:**301–309.

Food and Agriculture Organization. 2001a. *Feed and Food Safety Gateway*. United Nations Food and Agriculture Organization, Rome, Italy. http://www.fao.org/ag/AGa/AGAP/FRG/Feedsafety/feedsafety.htm.

Food and Agriculture Organization. 2001b. *Manual on the Application of the HACCP System in Mycotoxin Prevention and Control*. United Nations Food and Agriculture Organization, Rome, Italy. http://www.fao.org/docrep/y1390e/y1390e00.HTM.

Food and Drug Administration. 2010. *Animal Feed Safety System*. Food and Drug Administration, Rockville, MD. http://www.fda.gov/AnimalVeterinary/SafetyHealth/AnimalFeedSafetySystemAFSS/default.htm.

Food and Drug Administration. 2009. *Bacteriological Analytical Manual*. Food and Drug Administration, Rockville, MD. http://www.fda.gov/Food/ScienceResearch/LaboratoryMethods/BacteriologicalAnalyticalManualBAM/default.htm.

Food and Drug Administration. 1989. Animal feeds contaminated with *Salmonella* microorganisms. *CFR 500.35. Title 21—Food and Drugs,* chapt. I. U.S. Food and Drug Administration, Department of Health and Human Services, Subchapter E – Animal drugs, feeds, and related products. Food and Drug Administration, Rockville, MD. http://www.accessdata.fda.gov/scripts/cdrh/cfdocs/cfcfr/CFRSearch.cfm.

Food and Drug Administration. 2005. *Apple Juice, Apple Juice Concentrates, and Apple Juice Products—Adulteration with Patulin. CPG 510.150. Food and Drug Administration Compliance Policy Guide.* U.S. Food and Drug Administration, Rockville, MD. http://www.fda.gov/ICECI/ComplianceManuals/CompliancePolicyGuidanceManual/ucm074427.htm.

Food and Drug Administration. 1994. *Action Levels for Aflatoxins in Animal Feeds. U.S. Food and Drug Administration Compliance Policy Guide. CPG 683.100.* Food and Drug Administration, Rockville, MD. http://www.fda.gov/ICECI/ComplianceManuals/CompliancePolicyGuidanceManual/ucm074703.htm.

Food and Drug Administration. 1993. *Guidance for Industry and FDA Letter to State Agricultural Directors, State Feed Control Officials, and Food, Feed, and Grain Trade Organizations Final Guidance. U.S. Food and Drug Administration Compliance Policy Guide.* Food and Drug Administration, Rockville, MD. http://www.fda.gov/Food/GuidanceCompliance RegulatoryInformation/GuidanceDocuments/NaturalToxins/ucm120184.htm.

Food and Drug Administration. 2001. *Guidance for Industry: Fumonisin Levels in Human Foods and Animal Feeds; Final Guidance. Guidance for Industry no. 112.* U.S. Food and Drug Administration, Rockville, MD. http://www.fda.gov/Food/GuidanceComplianceRegulatoryInformation/GuidanceDocuments/ChemicalContaminantsandPesticides/ucm109231.htm.

Goryacheva, I. Y., S. De Saeger, S. A. Eremin, and C. Van Peteghem. 2007. Immunochemical methods for rapid mycotoxin detection: evolution from single to multiple analyte screening: a review. *Food Addit. Contam.* **24:**1169–1183.

International Commission on Microbial Specifications in Food. 2002. *Microbiological testing of Food Safety Management.* Kluwer Academic Publishing/Plenum Press, New York, NY.

International Organization for Standardization. 2002. *Microbiology of Food and Animal Feeding Stuffs—Horizontal Method for the Detection of* Salmonella *spp. ISO 6579:2002.* International Organization for Standardization, Geneva, Switzerland. http://www.iso.org/iso/iso_catalogue/catalogue_tc/catalogue_detail.htm?csnumber=29315&commid=47858.

International Union of Pure and Applied Chemistry. 2010. International Union of Pure and Applied Chemistry. http://www.iupac.org/.

Jarquin, R., I. Hanning, S. Ahn, and S. C. Ricke. 2009. Development of rapid detection and genetic characterization of *Salmonella* in poultry breeder feeds. *Sensors* **9:**5308–5323.

Kleter, G. A., A. Prandini, L. Filippi, and H. J. P. Marvin. 2009. Identification of potentially emerging food safety issues by analysis of reports published by the European Community's Rapid Alert System for Food and Feed (RASFF) during a four-year period. *Food Chem. Toxicol.* **47:** 932–950.

Koyuncu, S., and P. Haggblom. 2009. A comparative study of cultural methods for the detection of *Salmonella* in feed and feed ingredients. *BMC Vet. Res.* **5:**6–15.

Koyuncu, S., M. G. Andersson, and P. Haggblom. 2010. Accuracy and sensitivity of commercial PCR-based methods for detection of *Salmonella* enteric in feed. *Appl. Environ. Microbiol.* **76:**2815–2822.

Löfström, C., R. Knutsson, C. E. Axelsson, and P. Radstrom. 2004. Rapid and specific detection of *Salmonella* spp. in animal feed samples by PCR after culture enrichment. *Appl. Environ. Microbiol.* **70:**69–75.

Malorny, B., S. Huchn, R. Dieckmann, N. Kramer, and R. Helmuth. 2009. Polymerase chain reaction for the rapid detection and serovar identification of *Salmonella* in food and feeding stuff. *Food Anal. Methods* **2:**81–95.

Maragos, C. 2009. Fluorescence polarization immunoassay of mycotoxins: a review. *Toxins* **1:**196–207.

Murphy, P. A., S. H. Hendrich, C. Landgren, and C. M. Bryant. 2006. Food mycotoxins: an update. *J. Food Sci.* **71:**R51–R65. http://members.ift.org/NR/rdonlyres/62024337-FB0E-4952-AD9F-55E860551809/0/mycotoxins.pdf.

Reiter, E., J. Zentek, and E. Razzazl. 2009. Review on sample preparation strategies and methods used for the analysis of aflatoxins in food and feed. *Mol. Nutr. Food Res.* **53:**508–524.

Schneider, E., V. Curtui, C. Seidler, R. Dietrich, E. Usleber, and E. Martlbauer. 2004. Rapid methods for deoxynivalenol and other trichothecenes. *Toxicol. Lett.* **153:**113–121.

Shephard, G. S., F. Berthiller, J. Dorner, R. Krska, G. A. Lombaert, B. Malone, C. Maragos, M. Sabina, M. Solfrizzo, M. W. Trucksess, H. P. van Egmond, and T. B. Whitaker. 2010. Developments in mycotoxins analysis: an update for 2008–2009. *World Mycotoxin J.* **3:**3–23.

Sulyok, M., F. Berthiller, R. Krska, and R. Schuhmacher. 2006. Development and validation of a liquid chromatography/tandem mass spectrometric method for the determination of 39 mycotoxins in wheat and maize. *Rapid Commun. Mass Spectrom.* **20:**2649–2659.

U.S. Department of Agriculture–Food Safety Inspection Service. 2008. *Microbiology Laboratory Guidebook.* U.S. Department of Agriculture, Washington, DC. http://www.fsis.usda.gov/Science/Microbiological_Lab_Guidebook/index.asp.

U.S. Department of Agriculture–Grain Inspection, Packers and Stockyards Administration. 2006. Mycotoxins in grain and feed, chapt. 3. *In Grain Fungal Diseases & Mycotoxin Reference.* U.S. Department of Agriculture, Washington, DC. http://archive.gipsa.usda.gov/pubs/mycobook.pdf.

U.S. Department of Agriculture–Grain Inspection, Packers and Stockyards Administration. 2010. *Mycotoxins.* U.S. Department of Agriculture, Washington, DC. http://www.gipsa.usda.gov/GIPSA/webapp?area=home&subject=grpi&topic=rd-my.

Wales, A. D., V. M. Aallen, and R. H. Davies. 2010. Chemical treatments of animal feed and water for the control of *Salmonella* (report). *Foodborne Pathog. Dis.* **1:**3–15.

Whitaker, T. B., A. B. Slate, J. G. Adams, T. Birmingham, and F. G. Giesbrecht. 2010. Comparing the performance of sampling plans that use a single regulatory limit based upon total aflatoxins to sampling plans that use dual limits based upon B_1 and total aflatoxins. *World Mycotoxin J.* **3:** 35–44.

Wierup, M., and P. Haggblom. 2010. An assessment of soybeans and other vegetable proteins as source of *Salmonella* contamination in pig production. *Acta Vet. Scand.* **52:**15–23.

Wolf-Hall, C. 2010. Mold and mushroom toxins, p. 275–285. *In* V. Juneja and J. Sofos (ed.), *Pathogens and Toxins in Foods: Challenges and Interventions.* ASM Press, Washington, DC.

Zachariasova, M., O. Lacinaa, A. Malachovaa, M. Kostelanskaa, J. Poustkaa, M. Godulab, and J. Hajslova. 2010. Novel approaches in analysis of *Fusarium* mycotoxins in cereals employing ultra performance liquid chromatography coupled with high resolution mass spectrometry. *Anal. Chim. Acta* **662:**51–61.

DAIRY PRODUCTION CHAIN

IV

LISTERIA MONOCYTOGENES IN MILK, CHEESE, AND THE DAIRY ENVIRONMENT

A. D. Hitchins, K. N. Jordan, M. Sanaa, and M. Wagner

18

The realization that listeriosis was a foodborne disease was largely due to the involvement of *Listeria monocytogenes*-contaminated dairy products in the earliest recognized foodborne outbreaks (Norton and Braden, 2007). Thus, dairy sample matrices are routinely included in developing and validating rapid methods for *L. monocytogenes* in all food categories.

It has to be acknowledged that these rapid methods are generally not specific to dairy foods. The methods, both commercial and noncommercial, have generally been developed for use with samples from several categories of foods. As with any food, dairy products' specific characteristics are particularly important at the initial steps of testing processes, i.e., sampling and analytical portion preparation. These may be crucial in recovering *Listeria* cells from the matrices. Dairy samples represent a full range of physical characteristics from fluid (milk) through semisolid (yogurt) to soft solid (butter, cream, some cheeses) and hard solid (other cheeses).

In the solid phase, the cheese may present physical heterogeneity in the form of rind and core fractions. Fluid milk products will vary compositionally with the fat content according to the various types of raw milk processing. Environmental sampling can be relatively simple physically but can possibly be complicated by the degree of attachment to surfaces and sequestration in biofilm matrices.

SAMPLING PLANS, SAMPLE SIZE, SAMPLE TREATMENT, AND SAMPLE STORAGE

The control of *L. monocytogenes* in dairy food products is mainly ensured by a preventive approach, such as implementation of good hygiene practice and application of procedures based on hazard analysis and critical control point (HACCP) principles. Sampling of the production and processing environment can be a useful tool to identify and prevent the presence of *L. monocytogenes* in dairy products and also for validating and verifying the correct functioning of their procedures based on HACCP principles and good hygiene practice. As an example, Table 1 presents sampling

A. D. Hitchins, U.S. Food and Drug Administration, Center for Food Safety and Applied Nutrition, 5100 Paint Branch Parkway, College Park, MD 20740-3835. *K. N. Jordan*, Teagasc, Moorepark Food Research Centre, Moorepark, Fermoy, Co. Cork, Ireland. *M. Sanaa*, Agence Française de Sécurité Sanitaire des Aliments, Directorate of Risk Assessment, 27-31, avenue du Général Leclerc, BP 19-94701 Maisons-Alfort Cédex, France. *M. Wagner*, Institute of Milk Hygiene, Milk Technology and Food Science, University of Veterinary Medicine, Veterinärplatz 1, 1210 Vienna, Austria.

Rapid Detection, Identification, and Quantification of Foodborne Pathogens, ed. by J. Hoorfar

TABLE 1 Sampling plans for *Listeria monocytogenes* applied in Europe (EC 2073/2005)[a]

Products[b]	*n*	*c*	*m* (CFU/g)
Ready-to-eat foods intended for infants and ready-to-eat foods for special medical purposes	10	0	<0.04
Ready-to-eat foods able to support the growth *of L. monocytogenes*, other than those intended for infants and for special medical purposes	5	0	100[c]
Ready-to-eat foods unable[d] to support the growth of *L. monocytogenes*, other than those intended for infants and for special medical purposes	5	0	100

[a] *n*, number of units comprising the sample; *c*, number of sample units giving values over *m*; *m*, acceptable bacterial cell concentration in a sample unit. Note: *m* is used in 2 class plans. In 3 class plans, *M*, the marginally acceptable bacterial concentration in a sample unit, is also used.

[b] For all products, criteria apply to the duration of the shelf life of the product placed on the market.

[c] This criterion applies if the manufacturer is able to demonstrate, to the satisfaction of the competent authority, that the product will not exceed the limit of 100 CFU/g throughout the shelf life. The operator may fix intermediate limits during the process that should be low enough to guarantee that the 100-CFU/g limit is not exceeded at the end of the shelf life.

[d] Products with pH of ≤4.4 or water activity of ≤0.92, products with pH of ≤5.0 and water activity of ≤0.94, and products with a shelf life of less than 5 days are automatically considered to belong to this category. Other categories of products can also belong to this category, subject to scientific justification.

plans applied in Europe (European Union Commission, 2005).

The quality of the sample received for analysis is of primary importance. Samples should be correctly collected and handled. If samples are not representative of the sampled lot and are not stored in adequate conditions, the laboratory results will be useless.

Sampling

In general, there are two levels of sampling; the first concerns the selection of a certain number of unit products (*n*) from the whole lot, which are then submitted to a laboratory for testing, and the second concerns the laboratory's sampling from the set of subsamples, such as sets of sticks of butter or ice cream bars. Commonly representative parts of the subsamples are homogenized, and then a portion of the homogenate is taken for analysis. Less commonly, the analytical portion may be just from the homogenate of a single subsample unit.

At the first level, the objective is to select certain production units that are most representative of the lot and to assess the lot status with acceptable accuracy. A representative sample is one that reflects, as far as possible, the composition of the lot from which it is drawn. This is best achieved by random sampling. Sample accuracy is associated with the sample size and the characteristics of the lot contamination. An accurate sample is one that has a high probability of detecting a contaminated lot with a stated proportion of contaminated units.

To ensure a representative subsample (analytical unit), one should agitate liquids or free-flowing materials until the contents are homogeneous. If the sample is taken from a solid material, the analytical unit could be obtained by taking a portion from several locations within the sample unit. To reduce the workload, the analytical units may be combined for analysis (International Commission on Microbiological Specifications for Foods, 1986): conventionally, it is suggested that a composite contain not more than five analytical units, but under certain circumstances, the lot inspection is based on a sample of 25 g from five different units (e.g., cheeses).

The sampling procedure is essential when pathogens such as *L. monocytogenes* are not homogenously distributed within the food matrix and between the food units.

If the bacteria are present in a liquid food, such as milk, and the cells are randomly distributed without clustering (no mutual attraction or repulsion), then their count conforms to a Poisson distribution. So, the sample could be taken from the containers without any specific procedure. The sample volume will depend mainly on the target concentration and the accepted risk of finding a sterile sample when in fact the true bacterial concentration (or density) is greater than zero. In the case of a nonhomogeneous distribution of bacterial cells, particularly in a solid matrix, their count could be described by the negative binomial distribution (http://en.wikipedia.org/wiki/Negative_binomial_distribution) with two parameters, p and k (the scale and shape parameters, respectively).

The parameter k could be interpreted as a clustering index. To compare the two situations (homogeneous versus heterogeneous distribution), one could consider a certain amount (λ) of bacterial cells present in the same unit of product (e.g., cheese) with different gradients of clustering. The negative binomial distribution could easily be parameterized by using λ and k. When k tends to infinity, the negative binomial tends to the Poisson distribution. Indeed, k is a parameter measuring the heterogeneity of bacterial cell distribution.

For a fixed value of the mean concentration, the probability of a sample being uncontaminated is higher if the negative binomial distribution is assumed than if the Poisson distribution is assumed. Thus, the sample size and sample mass would need to be larger for contamination following the negative binomial distribution than for contamination described by the Poisson distribution if one wants to reduce the risk of finding all samples uncontaminated when in fact the true bacterial concentration is greater than zero or a certain limit (m or M) (Table 1, footnote *a*; Tripathi, 1985). For example, assuming that the true concentration is 10 CFU/g, the probability of a sample being uncontaminated is close to zero for the Poisson distribution and is 0.54, 0.34, and 0.10 for negative binomial distributions with k equal to 0.1, 0.2, and 0.5, respectively, based on one sample of 5 g.

The composition and nature of each lot affect the homogeneity and uniformity of the total sample mass. The proper statistical sampling procedure must consider the following: (i) whether the dairy food is solid, semisolid, viscous, or liquid; and (ii) whether the pathway of contamination is assumed to support homogeneous or heterogeneous distribution of *L. monocytogenes* cells within and between food units.

When composite samples are used for the laboratory analysis, it is recommended to conduct the pooling at the laboratory after the homogenization step whenever possible.

Sample Handling and Preparation

The isolation and identification of *L. monocytogenes* from dairy products or environmental samples necessitate the use of selective and enrichment procedures that maintain the level of other kinds of microorganisms present at reasonable numbers and allow multiplication of *L. monocytogenes* to levels that are enough for its detection (Donnelly and Nyachuba, 2007). The level of the other kinds of microorganisms depends on the nature of the sampled material (food at a particular stage in its processing or environmental samples such as swabs of dairy floors or walls, rinsing water, etc.) and the conditions of sample storage and transport to the laboratory.

Whenever possible, samples should be submitted to the laboratory in their original unopened packaging. If products are in bulk or in containers too large for submission to the laboratory, representative portions should be transferred to sterile containers under aseptic conditions. Sterile sampling equipment and aseptic techniques must be used.

Sample refrigeration at a temperature between 0°C and 4°C is recommended for handling, storing, and shipping materials to be analyzed for *L. monocytogenes*. Depending on the food matrix, in certain circumstances *L. monocytogenes* can grow at temperatures close to 4°C (Junttila et al., 1988); however, the growth rate is considered low with no significant influence on the population count. Low temperature has been used as a selective procedure (Gray et al., 1948), because the majority of strains of dairy food microflora do not grow at low temperature; that is, they are not psychrotrophic (Jay, 1986). Unless the sample is received frozen, it is strongly recommended that freezing be avoided. If the sample is already frozen, it should not be thawed until analysis (Food and Drug Administration, 2010).

In the case of composite samples analysis, they should be prepared under laboratory conditions observing aseptic techniques. One should take care to make subsamples representative of a food's outer surface as well as its interior. To increase the probability of detection, it is useful to combine the subsamples after they are blended or stomached. If composite samples are not required, 25-g analytical portions of food are simply blended or stomached in 225 ml of basal enrichment medium and preenriched/enriched as described later. Usually, a 25-g sample is retained for possible enumeration. It must be stored at 4°C if it is not frozen or, if frozen, in a nondefrosting freezer (International Organization for Standardization, 2004a).

Enumeration of *L. monocytogenes* in contaminated food is now required. As in general few numbers of samples are expected to be contaminated and then most often at low concentration (1 to 10 CFU/g), it is preferable to conduct detection first and then, when positive results are obtained, to proceed to enumeration. Two samples from the same unit of product may not have the same amount of *L. monocytogenes* because of possible clustering and heterogeneous distribution of cells. Various degrees of nonuniform distribution of microorganisms are to be expected in any food sample. To ensure more even distribution, it is important that mixing procedures be conducted thoroughly.

RAPID METHODS FOR DETECTION, ENUMERATION, CHARACTERIZATION, AND TYPING

Analytical Portion Preparation

As noted above, proper sample preparation is critical for rapid methods of listeria detection and enumeration. Only guidelines are presented here for the various general types of dairy samples.

Dairy Samples for Detection Methods

Typically, in detection methods 25 ml/g is diluted into 225 ml of selective enrichment broth and homogenized. For fermented dairy products, neutralization with sodium hydroxide may be advisable if the enrichment is only weakly buffered.

Dairy Samples for Enumeration Methods

For kit methods, use the diluent specified by the manufacturer. A good general-purpose diluent is Butterfield's phosphate solution (Butterfield, 1932; Food and Drug Administration, 2010).

Fluid Dairy Samples

Fluid dairy samples include light, whipping, and heavy creams; half and half; ice cream; whole, chocolate, skim, and other forms of milk, etc.; and milk-based infant formulas. For enumeration, 2 ml of sample is used directly for capture enumeration. For more viscous fluids like ice cream, chocolate milk, and heavy cream, 5 ml is diluted with 5 ml sterile water and 2 ml is used for immunomagnetic capture. For greater sensitivity, more sample and diluent may be concentrated by centrifugation (10,000 × *g* for 15 min). The fat layer is removed, the remaining supernatant is decanted, and the pellet is recovered for analysis.

Solid Dairy Samples

For solid dairy samples, routine sample preparation is described; but for greater sensitivity, more sample and diluent may be used for concentration by centrifugation as with fluid dairy samples.

For hard cheeses like mozzarella, 10 g is added to 90 ml buffer in a stomacher bag with a mesh insert and homogenized. The homogenate is filtered, and 2 ml is used for capture enumeration. Soft cheeses like ricotta are treated similarly but with only a 1 in 4 dilution. Hard grated cheeses like Parmesan are diluted more (1 in 20). Milk chocolate is shaved with a grater and stomached at a 1 in 3 dilution. Use of trisodium citrate (2%, wt/vol) in the first dilution of a cheese sample is recommended to help solubilize the cheese and release listeriae from the matrix.

Butter is melted in a container placed in a 30°C water bath. A volume of 5 ml is mixed with 5 ml of buffer and vortexed for 1 min. The mixture is then cooled to solidify the fat and the fat is removed with a sterile implement. Yogurt and milk puddings are diluted up to 1 in 6 in buffer and stomached in a bag with a mesh insert. The homogenate is filtered for the capture process.

Environmental Samples for Enumeration

For enumeration in environmental samples, a defined area (typically 10 by 10 cm) of product contact surfaces, machinery, walls, and floors, etc., is wiped with premoistened sterile swabs, sponges, or gauze pads. Dey/Engley (D/E) neutralizing broth is typically used for moistening. Samples are swabbed in a standard pattern of movements. Used collection implements are placed in an appropriate additional volume (5 to 10 ml) of D/E broth at 4 to 5°C. Swabs are vortexed, and pads/sponges are squeezed to obtain the supernatant. Supernatant (3 ml) is added to Petrifilm plates, or 0.1 to 0.2 ml is spread onto chromogenic agar plates. For cell capture, 2 ml is mixed with excess immunomagnetic beads (10^8) to ensure the presence of less than 1 CFU per bead on average. Loaded beads are recovered and plated.

Detection

Traditional methods for detection of *L. monocytogenes* include enrichment of the sample, plating on a semiselective agar, e.g., PALCAM (polymyxin-acriflavine-lithium chloride-ceftazidime-esculin-mannitol agar), isolation of suspect colonies, and finally confirmation of *L. monocytogenes* by biochemical tests or by an API test strip, taking a total time of several days (International Organization for Standardization, 2004a; Food and Drug Administration, 2010; U.S. Department of Agriculture—Food Safety and Inspection Service, 2010). Alternatively, direct plating without enrichment can be used, again followed by confirmation of isolates (International Organization for Standardization, 2004b). These agar-based methods are the "gold standard" for isolation of *L. monocytogenes* and are important when an isolate is required for governmental regulatory actions. These are detailed in another chapter. On the other hand, rapid methods for detection of *L. monocytogenes* offer the possibility of faster screening for negative results and the possibility of automation. As a result, a considerable amount of research has been directed towards rapid methods for detection of *L. monocytogenes*.

Different Rapid Methods for Detection of *L. monocytogenes*

The ability to detect low levels of *L. monocytogenes* in dairy foods is essential, as the contaminated foods are generally ready-to-eat foods that are not further processed before consumption. The detection of such sporadic low-level contamination is hampered by a high level of background competing microflora and by interference from the food matrix. These difficulties must be balanced with the need for a timely result so that product can be released as *Listeria*-free.

Rapid methods for detection of *L. monocytogenes* can be interpreted as a reduction in the number of days to obtain a result, but the process may still take a relatively long time, i.e., days rather than hours. There are many instruments and test kits on the market (Jantzen et al., 2006). However, it is not the purpose of this chapter to review the different instruments and kits, just the basis on which they work. These rapid methods include (i) agar-based methods including chromogenic agars, (ii) immunoassay-based methods, (iii) concentration methods, and (iv) molecular methods.

Agar-Based Methods

The addition of chromogenic substrates to a traditional agar base is a useful method for the differentiation of specific, closely related species in mixed populations. Several different media based on this principle of chromogens are available for detection of *L. monocytogenes*. Incorporation of the chromogenic substrate 5-bromo-4-chloro-3-indolyl-β-D-glucopyranoside, which can be cleaved by β-D-glucosidase, present in all *Listeria* species, demonstrates the presence of *Listeria* spp. Pathogenic *Listeria* spp. including *L. monocytogenes* and *L. ivanovii* can be detected by the presence of a phospholipase C enzyme that cleaves the substrate phosphatidylinositol phospholipase C. Typical *Listeria* colonies are blue-green, while colonies of pathogenic *Listeria* have an additional white precipitate around the colony. However, a few other non-*Listeria* strains can result in false positives, and suspect colonies must be confirmed as *L. monocytogenes* (to distinguish them from *L. ivanovii* and other false positives) by either biochemical tests or PCR.

A comprehensive study of the performance of different chromogenic agars was performed by Stessl et al. (2009). They showed that for food samples, a 48-h incubation on chromogenic agar is necessary for identification of *L. monocytogenes* and confirmation of suspect colonies is necessary. A more complete validation of chromogenic agars, including naturally contaminated samples and challenge tests with stressed cells and low inocula, is required.

Immunoassay-Based Methods

The antibody-antigen reaction can be used as a means of detecting the presence of *Listeria*. In detection of *L. monocytogenes*, advantage can be taken of this specific reaction if *L. monocytogenes* is considered an antigen. Antibodies reacting specifically with *Listeria* spp. or with *L. monocytogenes* can be produced. These can be used in different ways to detect the presence of *Listeria* spp. or *L. monocytogenes*. For example, immunomagnetic separation, latex agglutination, enzyme-linked immunosorbent assays, and enzyme-linked fluorescent assay techniques are among those that have been developed for detection of *Listeria* after enrichment. Various instruments and test kits are available, and many have been tested for use with dairy products (Gangar et al., 2000; Food and Drug Administration, 2010; http://www.rapidmicrobiology.com/test-methods/Listeria.php).

After enrichment, there can be a heat treatment to inactivate cells and promote the immune reaction, leading to increased safety and greater sensitivity.

Concentration Methods

In order to improve the limit of detection, the cells present in a sample can be concentrated into a smaller volume. Such systems use the antibody-antigen reaction whereby paramagnetic beads are coated with an antibody that specifically attracts *L. monocytogenes* cells (antigen), which are then concentrated by attracting the beads to a magnet. A concentration of cells of up to 1,000-fold can be achieved, and the concentrated beads can be assayed (by plating, PCR, etc.) for detection (Avoyne et al., 1997).

Molecular Methods

There are many different PCR assays that can be used to detect specific genes in *L. monocytogenes* (Levin, 2003). However, the need for complex sample preparation, the inability to clearly distinguish live and dead cells, and the use of gel electrophoresis (or alternative expensive instrumentation) have limited the transfer of this methodology from research to widespread applied use. Despite these limitations, a number of commercial kits using PCR have been developed. The various reagents, primers, and instructions are supplied with the kit in order to simplify the procedure as much as possible. Currently, however, all the available commercial kits are qualitative and require an enrichment step prior to use. Quantification and direct detection are confined to research methods. In this regard, the application of real-time PCR (Rodríguez-Lázaro et al., 2004) may have potential future use. At present, screening samples for negative results is probably the widest use of PCR-based analyses. This will speed up the process of product release because a negative result can be confirmed rapidly.

Validation and Standardization

One of the major drawbacks to the widespread use of rapid methods for detection of *L. monocytogenes* in the dairy industry is lack of validation and standardization by international bodies. Although various rapid methods have been validated (by such organizations as AOAC and AFNOR) and tested in collaborative studies (D'Agostino et al., 2004; Silbernagel et al., 2004), to date none have been accepted by the International Organization for Standardization as international standards. Half a dozen *Listeria* genus kits have been validated by the AOAC International. For details, consult the *Bacteriological Analytical Manual* (Food and Drug Administration, 2010). All of the methods have been validated collaboratively or precollaboratively for various dairy products from among the following: 2% fat milk, chocolate milk, nonfat dry milk, Brie cheese, soft cheese, cheddar cheese, and ice cream.

Combinations of Rapid and Conventional Methods

In practice, many of the rapid methods are used in combination with conventional methodologies. In order to detect low numbers, enrichment is a necessary first step in analysis of *L. monocytogenes* in dairy products. This is often followed by PCR or immunoassay methods as a confirmatory test in place of biochemical tests, resulting in a saving of 2 days for a confirmed result (Fig. 1).

Enumeration

Qualitative versus Quantitative Data

For the most part, qualitative data, i.e., the presence or absence of *L. monocytogenes*, are required for dairy products. In some cases, European Union regulations allow for <100 CFU/g (European Union Commission, 2005). However, the detection limit of a di-

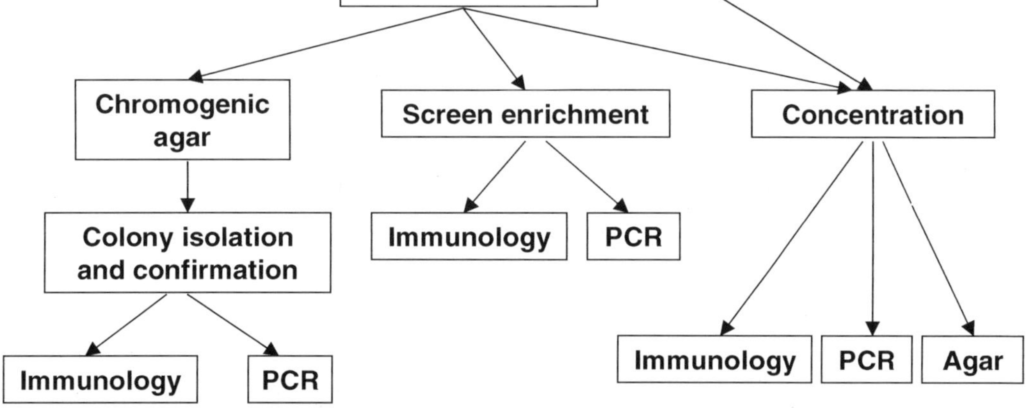

FIGURE 1 Alternative modes of complementation of standard cultural detection methods with rapid methods.

rect count is 100 CFU/g, and in the absence of reliable quantitative rapid methods below this value, a positive/negative result applies. All of the above rapid methods will provide qualitative data with good sensitivity (<1 CFU/g detected) if preceded by an enrichment step whereby low numbers of *Listeria* are increased. However, many modern methodologies, such as risk assessment and biotracing (Barker et al., 2009), require quantitative data, i.e., numbers of *L. monocytogenes* cells present. At about 10^2 to 10^3 cells, the limit of detection of rapid methods is not sensitive enough to provide such quantitative data by direct detection.

Therefore, enumeration of *Listeria* and specifically *L. monocytogenes* in dairy products is useful for several reasons: (i) assessment of contamination in the dairy food production environment; (ii) routine surveillance of the levels of listeriae in products for risk assessment purposes; (iii) estimation of foodborne doses consumed in epidemiological outbreaks; and (iv) determining compliance with an official tolerance level. It should be pointed out that enumeration per se is not mandatory for determining tolerance level compliances. That can be achieved with the usual qualitative presence and absence methods by simply choosing an appropriate analytical portion size. The portion(s) should come from a properly homogenized sample of sufficient amount to be representative of the product as is done in current practice.

The classical enumeration methods for *Listeria* enumeration in dairy products, i.e., the *Listeria* plate count and the *Listeria* most-probable-number (MPN) procedure, are more or less slow because they are limited by the inherent growth rate of listeriae, the usual need for identification of listeria isolates to the species level, and interference by food components such as particulates and microflora. Improvements in enumeration are mainly due to molecular techniques that address, more or less, at least some of these limitations. Currently at least, the important advantage of the classical listeria enumeration methods over the molecular methods remains that their dependence on growth ensures that only viable cells are being counted.

Direct Cell Counts. The fastest way to enumerate bacteria like *L. monocytogenes* in dairy foods would be to do a direct microscopic count of a quantitative dilution. In practice, this has not proved feasible. It is difficult to distinguish viable from dead cells and to distinguish *L. monocytogenes* from other *Listeria* species, other rod-shaped members of the food microflora, and food matrix particulates in homogenates. A 6-hour direct count of a culture was achieved by using aseptate filamentous growth induced by an antibiotic to distinguish the viable cells, which were stained with a fluorescent dye (Frank et al., 1992). The method could detect more surviving viable cells after starvation than could a conventional nonselective plate count.

In the absence of a differential reaction, the method was only applicable to pure cultures. Fluorescein-labeled polyclonal anti-*Listeria* antibody was used to count listeriae in fresh vegetables (Tortorello et al., 1997) after suitable filtration. However, antibody cross-reaction was a potential problem. The authors surmised that the method would be applicable to milk after suitable enzymatic digestion to ensure filterability.

Colony Counts. Total microbial or aerobic plate counts have been traditionally useful to determine the microflora concentrations in milk and other dairy products. The spiral plate method has been an important innovation for colony enumeration in dairy foods (Donnelly et al., 1976). A circulating plate of agar culture medium is inoculated with a test fluid (milk or diluted food) in a controlled quantitative way from the center to the edge of the rotating plate. The plate is incubated in the usual fashion. A density range of 1,000-fold above a 500-CFU/ml minimum is achievable with a consequent saving of time, dilutions, and plates. It can be used to enumerate listeriae on selective agars (Loessner et al., 1988) and has been applied to enumeration of listeriae in nondairy foods. Dairy food samples do not appear to have been tested.

Listeriae in dairy foods can be enumerated by CFU counts on esculin containing selective agars. A proportion of the colonies have to be confirmed as *L. monocytogenes* or not, in order to estimate the number of *L. monocytogenes* present. This is time-consuming when done by the classical identification tests. Initial ways to reduce this time have included use of a nonradioactive DNA probe with spiked Brie cheese (Datta et al., 1993) and use of enhanced hemolysis agar to enumerate spiked Brie, provolone, Muenster, and Camembert cheeses (Heisick et al., 1995). The latter method still required an additional test of xylose catabolism to distinguish *L. monocytogenes* from *L. seeligeri*. The introduction of differential selective agars, the so-called chromogenic agars, represents the state of the art in colony counting. These agars (see "Detection" above) allow counting of presumptive *L. monocytogenes* colonies. The only other *Listeria* species likely to be confused with *L. monocytogenes* is *L. ivanovii*, which is fortunately not common in foods (Huang et al., 2007). The International Organization for Standardization uses the ALOA medium in its *L. monocytogenes* colony counting method, but it was chosen over other chromogenic agars only because its formulation is nonproprietary. Hitchins found (A. D. Hitchins, unpublished data) that recovery of quantitative *L. monocytogenes* spikes was good, from a variety of hard cheeses (Cheddar, Monterey Jack, provolone, Gruyere, and Cheshire) and soft cheeses (Brie, Camembert, Chimay, Reblochon, Gorgonzola, Pont L'Eveque, Raclette, Roquefort, and Taleggio) including mold-ripened and some smoked varieties, when using CHROMagar and Rapid L'mono. These two agars generally had good microflora suppression with only slight reduction of plating efficiency relative to nonselective agar. Although the ALOA medium performed well in enumerating *L. monocytogenes* in Mexican-style cheese samples connected to a North Carolina outbreak (Hitchins, unpublished), it was not tested with other cheeses because it became commercially unavailable in the United States. The imported component bovine peptides were considered to be potentially contaminated with prions.

Real-time PCR can be used to specifically enumerate CFU of *L monocytogenes* in milk products. Nogva et al. (2000) used *hlyA* as a target and tested skim and raw whole milks, using nonspecific binding to paramagnetic beads in sample preparation. The detection limit was 6 to 60 CFU per PCR. Hein et al. (2001) used the specific allele of *iap* and tested with ultrahigh-temperature-treated supermarket milk (3.8% fat). In both methods, DNA was enzymatically extracted from concentrated cells. For cell concentration, Nogva et al. (2000) used nonspecific binding magnetic beads, while Hein et al. used centrifugation. When Hein et al. (2001) quantitatively spiked ultrahigh-temperature-treated milk with stationary-phase cells, they found that nonviable and lysed cell DNA can cause significant discrepancies between viable and real-time PCR counts (cycle threshold values), which can only be regarded as potentially presumptive pending later confirmation by CFU or MPN counts.

MPN Enumeration

As the required detection limit of *L. monocytogenes* is <1 cell/25 g, enrichment is needed to boost the detection limit. Therefore, rapid methods are best applied after enrichment (see "Detection," above). They do not have the required detection limit for a direct count. When enrichment is applied, only qualitative data are possible, as enumeration after stationary-phase enrichment is meaningless. Levels of competing bacteria, different growth rates, etc., will influence the numbers, so that one enriched culture cannot be directly compared to another to imply initial numbers. Direct plating on chromogenic agars, direct PCR, or immunomagnetic separation is not a viable option, as none of these methods are sensitive enough to detect the numbers of *L. monocytogenes* generally found in food.

MPN combined with rapid methods is currently the best approach to the problem of enumerating the low levels of *L. monocytogenes* that are typically found in dairy foods when contamination does occur.

MPN enumeration estimates are slow and inconvenient and have wide 95% confidence intervals. The interval widths can be reduced by using many more than the usual three replicates (tubes) per food dilution level, but this is more inconvenient.

Spread and pour plate colony counts of listeriae in foods are hampered by limitation of the spreadable volume or by food particulates in 10^{-1} dilutions, respectively. These limitations can be reduced by suitable enzymatic digestion of particulates coupled with filtration onto hydrophobic grid membrane filters (HGMF), which are incubated on selective and differential culture media. Entis and Lerner (2000) have developed such an MPN method, in which grid squares replace traditional MPN tubes, which conveniently reduces the experimental equipment needed. They found that MPN values of spiked foods (pasteurized milk, raw milk, and Mexican-style cheese) by the HGMF method and traditional three-replicate MPN were statistically comparable in the range of 5 to 500 MPN/g or ml.

Traditional listeria MPN methodology requires that the growth in positive-culture tubes be identified to species level if the goal is to enumerate *L. monocytogenes* specifically. This is time-consuming. Use of PCR with *L. monocytogenes*-specific primers reduces species identification confirmation time. This is even more dramatic with real-time PCR, as demonstrated with Brie cheese, fresh cheese, ice cream, pasteurized milk (De Martinis et al., 2007) (Table 2), and Cheddar cheese (De Jesus and Hitchins, 2008). The total time for enumerating *L. monocytogenes* was about 24 h. Interference by target DNA from dead or autolyzed target cells is less of a problem in PCR-MPN enumeration than in PCR direct enumeration. Enumeration of viable *L. monocytogenes* cells by PCR with propidium monoazide in the presence of dead cells is possible within certain boundary concentrations but has apparently not been applied to dairy products (Pan and Breidt, 2007).

TABLE 2 Real-time PCR MPN enumeration of viable *L. monocytogenes* scheme[a]

Sample: 1 to 10 subsamples from a lot of dairy food
Composite sample (optional): 5 subsamples per composite
Homogenize: single subsample or composite by blending or pummeling
Selectively enrich MPN tubes: 1.0, 0.1, 0.01 ml or g per 10 ml BLEB[b] in triplicate
Incubate MPN culture: 35°C for 24 ± 2 h
Concentrate culture sediment: 10,000 × *g* for 10 min at 5°C.
Wash culture sediment: Resuspend pellet in 5 ml PBS[b]
Reconcentrate washed sediment: 10,000 × *g* for 10 min at 5°C.
Extract DNA: Heat pellet 10 min in 2 ml molecular biological grade water at 95 to 100°C
Separate and decant supernatant: 10,000 × *g* for 10 min at 5°C.
Dilute supernatant: 1:20 in molecular biological grade water
Real-time PCR assay: Diluted extract (4 μl); specific primers for 16S-rRNA gene; PCR mixture; SYBR Green; total volume, 25 μl.

[a] Based on De Martinis et al., 2007.
[b] Abbreviations: BLEB, buffered Listeria enrichment broth; PBS, phosphate-buffered saline.

Environmental Sample Enumeration

L. monocytogenes can persist in the dairy processing environment and from there contaminate products. Environmental swabs can be taken to assess the presence of *L. monocytogenes*. In these cases, a qualitative result for the presence/absence of *L. monocytogenes* is required; therefore, enrichment of the sample is necessary. This can be followed by any of the above-mentioned rapid methods for detection. While qualitative rapid methods would be satisfactory for most environmental surveillance, a more quantitative approach appears to be developing.

Some of the methods already mentioned for enumeration of *L. monocytogenes* in dairy foods have also been applied to its enumeration in samples from dairy plant-related environmental surfaces such as stainless steel. The HGMF-MPN enumeration method (Entis and Lerner, 2000) mentioned above can also be applied to environmental swab samples.

Enumeration of live *L. monocytogenes* in a mixture of live and dead cells obtained from sanitized biofilm on stainless steel coupons was achieved by the propidium iodide PCR method (Pan and Breidt, 2007). *L. monocytogenes* cells dried on stainless steel and recovered by swabbing into D/E neutralizing broth were enumerable by immunomagnetic capture on beads (Mitchell et al., 1994). This method is generic for *Listeria* species. The value of quantitative listeria testing by this method was reported for a food processing plant (Kohn et al., 1997). In regard to the enumeration of *L. monocytogenes* in environmental samples, media for storage and transport of samples at different temperatures (−4 to 15°C) were evaluated (Bazaco et al., 2007). Neutralizing buffer and Copan SRK maintained bacterial concentrations at all the tested temperatures. D/E neutralizing broth permitted significant growth at 10°C and above.

A potentially convenient method for environmental listeria detection (Nyachuba and Donnelly, 2007) or enumeration is Listeria (EL) Plate. This is a sample-ready culture medium containing selective agents, nutrients, a cold-water-soluble gelling agent, and a chromogenic dye to indicate which of the colonies are *Listeria* (red-violet color). It can detect all the species of *Listeria* except *L. seeligeri* and *L. ivanovii*, which, however, are not environmentally very significant. The method has been validated by the AOAC Research Institute for stainless steel, plastic surfaces, ceramic tile, and sealed concrete surfaces (Performance Tested Method no. 030601). The method is summarized in Table 3.

TABLE 3 Enumeration scheme for environmental *Listeria*[a]

Sampling step: Sponge/swab defined areas of dairy food contact surfaces, walls, floors, equipment, etc., with up to 10 ml moisturizing agent[b].
Repair step: Add prewarmed sterile buffered peptone water (5 ml) and manually/mechanically mix (about 1 min); incubate for 1 h at 20 to 30°C.
Plating step: Peel open proprietary plate and pipette 3 ml of repair-step fluid onto the lower film; close top film by rolling action to remove possible air bubbles; gently spread contents per manufacturer's instructions; medium gels in ≥10 min.
Incubation step: 28 ± 2 h at 36 ± 2°C in a humid atmosphere.
Interpretative step: Count red-violet colonies, which are presumptive *Listeria*.
Confirmation step: Colonies are accessible for optional genus and species determination.

[a] The Petrifilm Environmental Listeria Plate is basically equivalent to a pour plate that accommodates a larger-than-usual sample volume.

[b] For example, sterile water, buffered peptone water, or D/E neutralizing broth.

Current Status of Enumeration Methods

Plate count and MPN estimate methods for *L. monocytogenes* in dairy foods still involve a substantial growth period and are therefore relatively slow. However, as mentioned above, they now are faster than the classical culture methods due to faster identification techniques that are also less presumptive than previous ones. A commercial environmental sample enumeration scheme has already been summarized. While it is not strictly rapid, the saving of medium preparation time is certainly convenient.

Identification and Typing

Typing *Listeria* is an inevitable task in all laboratories dealing with the protection of the food chain on an advanced level. Typing includes not only the discrimination of apathogenic from nonpathogenic species but also advanced discrimination on serotype or subserotype level. Molecular typing has revolutionized the potential of typing information and has allowed substantial progress in the elucidation of contamination chains and outbreak investigations. Furthermore, typing is used to unravel the phylogeny of listeria species in basic science.

Dairy food chain investigations especially require the support of *Listeria* typing, since some products such as smear-ripened soft cheeses are known to support the growth of *Listeria* to high numbers, which may lead to outbreaks of listeriosis. Dairy farm environments have been found to be heavily contaminated in many studies and pose a risk if raw milk cheese is made, if underpasteurization occurs, or if postpasteurization cross-contamination has not been prevented (Oliver et al., 2005).

A phenomenon of particular concern is the repetitive recovery of genetically indistinguishable *Listeria* isolates from dairy process environments, which was initially described in 1996 (Wagner et al., 1996; Unnerstad et al., 1996). Long-term follow-ups of occurrence of such *Listeria* persisters show the tremendous capability of such clones to survive in-house, obviously in biofilm formation in niches where the hygiene management is insufficient to eradicate the pathogen. Since biofilm formation is stimulated not only by polysaccharide production but also by DNA abundance (Harmsen et al., 2010), research should be directed toward showing the impact of pathogen killing through disinfection on the persistence phenomenon. Whether persisters share genetic traits making them highly adapted to food processing environments is not elucidated yet.

A recently published European Centre for Disease Prevention and Control/European Food Safety Authority paper on the disturbing increase of cases of listeriosis in many countries worldwide (Anonymous, 2009; Allerberger and Wagner, 2010) suggests that strengthening of research on *Listeria* transmission is needed. Studying *Listeria* transmission routes, however, requires reproducible and rapid subtyping techniques. To detect outbreak episodes, subtyping of *L. monocytogenes* is mandatory in human disease surveillance. Since the incubation period in listeriosis can be more than 70 days, clusters of associated cases are usually spread out in countries, provinces, or federal states. This is illustrated by the recent Austrian outbreak in which cases occurred in five European Union member states and fatalities have been recorded from at least two states. Typing provided the final proof that an acid-curd cheese was implicated, thus facilitating targeted recall actions (Fretz et al., 2010). Several outbreaks in the United States have been confirmed only on the basis of molecular typing applied to centrally stored and processed isolate collections. These studies have revealed that neither the clinical manifestation nor the outcome of infection can be predicted on the basis of the lineage assignments of the isolates involved (Sauders et al., 2006). Others have even suggested that outbreak clones, such as the one having caused the California Jalisco cheese outbreak in 1985, are still prevailing in farm environments and causing sporadic cases of human listeriosis in the United States (Borucki et al., 2004).

Species Discrimination

The genus *Listeria* is currently comprised of the six species *L. monocytogenes, L. ivanovii, L. innocua, L. seeligeri, L. welshimeri*, and *L. grayi*, which were recognized decades ago. Two novel species, *L. rocourti* and *L. marthii*, have been recently described, which suggests that more species can be expected when environmental isolates are more systematically characterized in the future (Graves et al., 2009; Leclercq et al., 2009). Only two of them, *L. monocytogenes* and *L. ivanovii*, which are distinguishable by hemolysis and phospholipase activity on blood and chromogenic agars, respectively (Stessl et al., 2009), are pathogenic to humans and mammals. *L. monocytogenes* must be mainly discriminated from the closely related, frequently appearing apathogenic species *L. innocua*.

PCR assays have certainly improved the field of species discrimination, and some of them are also used to further discriminate the nonpathogenic species (Bubert et al., 1999). Molecular assays are helpful in those cases where pathogenic and nonpathogenic species occur together. *L. innocua* occurred for years in the plant involved in the recent Austrian outbreak of listeriosis. Latterly, *L. monocytogenes* has become the predominant contaminant for unknown reasons (I. Rückerl, personal communication). In cases where a pathogenic and a nonpathogenic species occur in the same sample, research has shown that the apathogenic species often grows more rapidly than the pathogenic species does (Oravcová et al., 2008). This results in the predominance of the nonpathogenic species on culture plates used for isolation. In such cases, only the use of chromogenic agars or PCR combined with plate swabbing unravels the real dimension of the contamination event, and only typing of many clones from the same plate can demonstrate that the outbreak clone hides somewhere among the competing flora.

Serotyping

Traditional serotyping of *Listeria monocytogenes* will remain important as long as the phenomenon that three serovars (1/2a, 1/2b, and 4b) cause the majority of human cases remains unexplained. Neither the proliferation of sequence information nor outcomes of virulence and in vivo studies have provided a satisfying explanation for this observation, nor have they resulted in identification of a biomarker(s) explaining the serovar compliance with pathogenicity in *L. monocytogenes*. Serotyping also retains its importance in that there is a match between the serovar distribution and the evolvement of three evolutionary lineages in *Listeria* (Rasmussen et al., 1995). In evolutionary terms, isolates of serovars 1/2b and 4b have been evolved separately from isolates of serovars 1/2a and 4a. Serotyping involves the detection of O and H antigens, and 13 serovars are described in the species *L. monocytogenes*. From dairy environments, listeriae are usually of serovar 1/2a or, more rarely, of serovar 1/2b (Waak et al., 2002). Serovar 4b isolates are rarely found. A recent study has reported serovar 4b clones isolated from the processing environments of small cheese plants to be highly related (Wagner et al., 2006). Serotyping schemes for the nonpathogenic species are incomplete. Such typing information would be helpful in understanding the transmission of these bacteria into food chains as models for the transmission routes of the pathogenic species.

Rapid methods aside, serology is quite rapid itself and is still useful not the less because sets of the serogroups closely mimic *L. monocytogenes* genetic lineages. The disadvantages of serotyping are the high cost of antisera and the not-infrequent existence of untypeable strains. Serotyping by PCR has been applied with only partial success. The O antigens are mainly if not totally cell wall teichoic acids. They have different molecular weights according to serotype. This is due to different chain lengths and different side chain substituents. It may be possible to exploit this in a direct physicochemical assay.

The antibody-based typing was recently complemented by a PCR-based serotyping scheme (Doumith et al., 2004). However, the PCR scheme only discerns the most prevalent serovars and is not capable of detecting all serovars described so far.

Molecular Typing

Molecular typing has become the dominant typing method in recent years and has fueled research on *Listeria* transmission routes. In most cases, nucleic acids are the target analytes. Nevertheless, any technology that provides a fingerprint of isolate-specific traits is potentially useful in subtyping, whether it is genomic, proteomic, or metabolomic.

Nucleic acid-based molecular subtyping has been developed and standardized based on random amplification of polymorphic DNA (Wagner et al., 1999), pulsed-field gel electrophoresis (PFGE) (Graves and Swaminathan, 2001), variable numbers of tandem repeats (Wagner et al., 2000), multiple-locus variable numbers of tandem repeats (Murphy et al., 2007; Sperry et al., 2008), and multilocus sequence typing (Cai et al., 2002). Very few studies have used amplified fragment length polymorphism (Aarts et al., 1999) due to its complexity. It should be emphasized that the various methods provide different types of information. Due to insufficient interlaboratory reproducibility, random amplification of polymorphic DNA is mainly useful for quick screens on outbreak investigations that are run under standardized conditions in the same laboratory. With PFGE, a database (PulseNet Europe/USA; www.pasteur.fr) has been made accessible to wider research communities for comparing the profiles of their isolates with those centrally stored. Universal databases are essential in *Listeria* epidemiology, since the majority of cases are sporadic. While whole-genome-based subtyping technologies such as PFGE provide epidemiological information, sequence-based subtyping technologies are mainly used for phylogenic studies. Current advances in typing technologies include whole-genome microarrays that reveal the presence/absence of the total gene content from an isolate. This technology has the advantage that it combines screening for biomarkers with phylogenetic analysis. Phenotype-based typing technologies such as Fourier transform infrared spectroscopy can also be used for early-response investigations, since huge numbers of isolates can be processed cheaply. Results of such phenotypic techniques still need follow-up fine-tuning with genetic subtyping.

Trends

A public health focus on listeriosis will still be necessary in the future due to an increasing incidence reported from many countries worldwide (Allerberger and Wagner, 2010). More emphasis should be given to elucidating the complexities of transmission routes such as environmental niches in food processing, food animal carriage, and consumer behavior in regard to handling and preparing food domestically. The ecophysiology of saprophytic organisms such as *Listeria* is largely unknown and should be a matter of intensified research. The availability of efficient typing protocols will play a pivotal role in current and future epidemiological studies. Currently, PFGE is the gold standard in studies on *Listeria* epidemiology worldwide. PFGE is, however, time-consuming and requires trained personnel to run the analysis on a reproducible level. The capabilities of spectroscopic techniques are under investigation, and the discriminatory power of these technologies seems to be dependent on the microbe analyzed. Advances by spectroscopic techniques such as Fourier transform infrared spectroscopy (Rebuffo-Scheer et al., 2007) or mass spectroscopy (Barbuddhe et al., 2008) have shown that they are useful in species identification. However, they usually require pure cultures and their discrimination at the subtype level has been difficult to assess so far.

Novel subtyping approaches in *Listeria* research can develop based on the advances in parallel sequencing. The merits of sequencing are that the readouts are very easy to analyze and to compare and that the information can be easily stored in databases and shipped around for supranational alignments and comparisons.

CONCLUSION AND FUTURE DEVELOPMENTS

In summary, rapid methods offer the advantages of high specificity, the possibility of automation, reduced labor input, and the possibility of rapidly screening for negative results, enabling earlier release of product. However, they do suffer from the disadvantage of lack of international validation/standardization, issues of sensitivity that need to be overcome to enable better direct detection, the requirement for expertise, and the expense of capital cost for equipment. If some of these disadvantages can be overcome, use of instrument/kit-based rapid methods will become widespread.

The rapidity of dairy detection methods of a given sensitivity is largely limited by the growth rate of *L. monocytogenes* under selective enrichment conditions. However, additional contributing negative factors can be the dairy food microflora competitiveness (in the case of nonpasteurized milk products) and any inhibitory properties of the particular dairy food matrix (e.g., the acidity of fermented products). One advantage of the preliminary enrichment culture is that it helps to ensure that the detection target is from a viable cell. This is critical, as many rapid detection methods cannot distinguish dead from live cells. As a result, this has tended to limit these methods to a screening function with the consequent regulatory requirement of confirming that viable cells containing the target are present. Capitalization on the short half-life of mRNA molecules in living cells would be helpful in alleviating the problem of viability.

A major advantage of the rapid methods is that they can be specific for *L. monocytogenes*. In practice, at least in the United States, *Listeria* genus methods have been mostly used as screening methods for dairy foods. The rationale is that species like *L. innocua* are harbingers of potential *L. monocytogenes* contamination in dairy plants. Thus, although *L. monocytogenes* may be absent from samples, preemptive sanitization is still deemed necessary in a dairy plant if *L. innocua* has been found in product or environmental samples.

The greatest advances in dairy listeriology have been in the area of *L. monocytogenes* isolate typing. Perhaps there are too many methods, but a trend toward using DNA sequencing methods is developing fast.

Detection of *L. monocytogenes* in the dairy processing plant environment is a very important measure in controlling contamination. Food contact surfaces, equipment, and other parts of processing plants that are producing listeria-containing products are often environmentally sampled by swabbing/sponging. In the United states, about 30 to 100 environmental samples may be collected in a single plant inspection. Enumeration of *L. monocytogenes* on food contact surfaces might be more informative than mere recording of presence or absence. Recoverability of *L. monocytogenes* from biofilm matrix is an area that needs investigation in view of the known persistence of particular strains in food processing plants for years.

REFERENCES

Aarts, H. J. M., L. E. Hakemulder, and A. M. A. Van Hoef. 1999. Genomic typing of *Listeria monocytogenes* strains by automated laser fluorescence analysis of amplified fragment length polymorphism fingerprint patterns. *Int. J. Food Microbiol.* **49:**95–102.

Allerberger, F., and M. Wagner. 2010. Listeriosis: a resurgent foodborne infection. *Clin. Microbiol. Infect.* **16:**16–23.

Anonymous. 2009. *The Community Summary Report on Foodborne Outbreaks in the European Union in 2007.* European Food Safety Authority, Parma, Italy.

Avoyne, C., M. Butin, J. DeLaval, and J.-L. Bind. 1997. Detection of *Listeria* spp. in food samples by immunomagnetic capture: ListerScreen method. *J. Food Prot.* **60:**377–384.

Barbuddhe, S. B., T. Maier, G. Schwarz, M. Kostrzewa, H. Hof, E. Domann, T. Chakraborty, and T. Hain. 2008. Rapid identification and typing of listeria species by matrix-assisted laser desorption ionization–time of flight mass spectrometry. *Appl. Environ. Microbiol.* **74:**5402–5407.

Barker, G. C., N. Gomez, and J. Smid. 2009. An introduction to biotracing in food chain systems. *Trends Food Sci. Technol.* **20:**220–226.

Bazaco, M. C., J. D. Eifert, R. C. Williams, and S. Kathariou. 2007. Quantitative recovery of *Listeria monocytogenes* and select *Salmonella* serotypes from environmental sample media. *J. AOAC Int.* **90:**250–257.

Borucki, M. K., J. Reynolds, C. C. Gay, K. L. McElwain, S. H. Kim, D. P. Knowles, and J. Hu. 2004. Dairy farm reservoir of *Listeria monocytogenes* sporadic and epidemic strains. *J. Food Prot.* **67:**2496–2499.

Bubert, A., I. Hein, M. Rauch, A. Lehner, B. Yoon, W. Goebel, and M. Wagner. 1999. Detection and differentiation of *Listeria* spp. by a single reaction based on multiplex PCR. *Appl. Environ. Microbiol.* **65:**4688–4692.

Butterfield, C. T. 1932. The selection of a dilution water for bacterial examinations. *J. Bacteriol.* **23:** 355–368.

Cai, S., D. Y. Kabuki, A. Y. Kuaye, T. G. Cargioli, M. S. Chung, R. Nielsen, and M. Wiedmann. 2002. Rational design of DNA sequence-based strategies for subtyping *Listeria monocytogenes*. *J. Clin. Microbiol.* **40:**3319–3325.

D'Agostino, M. Wagner, J. A. Vazquez-Boland, T. Kuchta, R. Kapriskova, J. Hoorfar, S. No-

vella, M. Scortti, J. Ellison, A. Murray, I. Fernandez, M. Kuhn, J. M. Pazlarova, A. Heuvelink, and N. Cook. 2004. A validated PCR method to detect *Listeria monocytogenes* using raw milk as a food model—towards an international standard. *J. Food Prot.* **67:**1646–1655.

Datta, A. R., M. A. Moore, B. A. Wentz, and J. Lane. 1993. Identification and enumeration of *Listeria monocytogenes* by nonradioactive DNA probe colony hybridization. *Appl. Environ. Microbiol.* **59:** 144–149.

De Jesus, A., and A. D. Hitchins. 2008. *Abstr. 122nd AOAC Annu. Meet. Expo.*, abstr. P-W-402. AOAC International, Gaithersburg, MD.

De Martinis, E. C. P., R. E. Duvall, and A. D. Hitchins. 2007. Real-time PCR detection of 16S rRNA genes speeds most-probably-number enumeration of foodborne *Listeria monocytogenes*. *J. Food Prot.* **70:**1650–1655.

Donnelly, C. B., J. E. Gilchrist, J. T. Peeler, and J. E. Campbell. 1976. Spiral plate count method for the examination of raw and pasteurized milk. *Appl. Environ. Microbiol.* **32:**21–27.

Donnelly, C. W., and D. G. Nyachuba. 2007. *Conventional methods to detect and isolate* Listeria monocytogenes, p. 216–256. *In* E. T. Ryser, and E. M. Marth (ed.), *Listeria, Listeriosis and Food Safety*, 3rd ed. CRC Press, Boca Raton, FL.

Doumith, M., C. Buchrieser, P. Glaser, C. Jacquet, and P. Martin. 2004. Differentiation of the major *Listeria monocytogenes* serovars by multiplex PCR. *J. Clin. Microbiol.* **42:**3819–3822.

Entis, P., and I. Lerner. 2000. Twenty-four hour direct presumptive enumeration of *Listeria monocytogenes* in food and environmental samples using the ISO-GRID method with LM-137 agar. *J. Food Prot.* **63:**354–363.

European Union Commission. 2005. Regulation (EC) No 2073 on microbiological criteria for foodstuffs. *Off. J. Eur. Union* L **338:**1–26.

Food and Drug Administration. 2010. *Bacteriological Analytical Manual.* http://www.fda.gov/Food/ScienceResearch/LaboratoryMethods/BacteriologicalAnalyticalManualBAM/UCM061208. Accessed 24 May 2010.

Frank, J. F., M. A. Gassem, and R. A. N. Gillett. 1992. A direct viable count method suitable for use with *Listeria monocytogenes*. *J. Food Prot.* **55:**697–700.

Fretz, R., U. Sagel, W. Ruppitsch, A. Pietzka, A. Stoger, S. Huhulescu, S. Heuberger, J. Pichler, P. Much, G. Pfaff, K. Stark, R. Prager, A. Flieger, O. Feenstra, and F. Allerberger. 2010. Listeriosis outbreak caused by acid curd cheese 'Quargel', Austria and Germany 2009. *Euro Surveill.* **15**(5):pii:19477.

Gangar, V., M. S. Curiale, A. D'Onorio, A. Schultz, R. L. Johnson, and V. Atrache. 2000. VIDAS enzyme-linked immunofluorescent assay for detection of *Listeria* in foods: collaborative study. *J. AOAC Int.* **83:**903–918.

Graves, L. M., and B. Swaminathan. 2001. PulseNet standardized protocol for subtyping *Listeria monocytogenes* by macrorestriction and pulse-field electrophoresis. *Int. Food Microbiol.* **65:**55–62.

Graves, L. M., L. O. Helsel, A. G. Steigerwalt, R. E. Morey, M. I. Daneshvar, S. E. Roof, R. H. Orsi, E. D. Fortes, S. R. Milillo, H. C. den Bakker, M. Wiedmann, B. Swaminathan, and B. D. Sauders. 2009. *Listeria marthii* sp. nov., isolated from the natural environment, Finger Lakes National Forest. *Int. J. Syst. Evol. Microbiol.* **60:** 1280–1288.

Gray, M. L., H. J. Stafseth, F. Thorp, Jr., L. B. Sholl, and W. F. Riley, Jr. 1948. A new technique for isolating listerellae from the bovine brain. *J. Bacteriol.* **55:**471–476.

Harmsen, M., M. Lappann, S. Knøchel, and S. Molin. 2010. Role of extracellular DNA during biofilm formation by *Listeria monocytogenes*. *Appl. Environ. Microbiol.* **76:**2271–2279.

Hein, I., D. Klein, A. Lehner, A. Bubert, E. Brandl, and M. Wagner. 2001. Detection and quantification of the *iap* gene of *Listeria monocytogenes* and *Listeria innocua* by a new real-time quantitative PCR assay. *Res. Microbiol.* **152:**37–46.

Heisick, J. E., L. I. Rosas-Marty, and S. R. Tatini. 1995. Enumeration of viable *Listeria* species and *Listeria monocytogenes* in foods. *J. Food Prot.* **58:** 733–736.

Huang, B., S. Eglezos, B. A. Heron, H. Smith, T. Graham, J. Bates, and J. Savil. 2007. Comparison of multiple PCR with conventional biochemical methods for the identification of *Listeria* spp. isolates from food and clinical samples in Queensland, Australia. *J. Food Prot.* **70:**1874–1880.

International Commission on Microbiological Specifications for Foods. 1986. *Microorganisms in Foods. 2. Sampling for Microbiological Analysis: Principles and Specific Applications*, 2nd ed. University of Toronto Press, Toronto, Ontario, Canada.

International Organization for Standardization. 2004a. *ISO Normative 11290-1:2004 Horizontal Method for the Detection and Enumeration of* Listeria monocytogenes. *Part 1. Detection Method.* International Organization for Standardization, Geneva, Switzerland.

International Organization for Standardization. 2004b. *ISO Normative 11290-2:2004 Horizontal Method for the Detection and Enumeration of* Listeria monocytogenes. *Part 2. Enumeration Method.* International Organization for Standardization, Geneva, Switzerland.

Jantzen, M. M., J. Navas, A. Corujo, R. Moreno, V. Lopez, and J. V. Martinez-Suarez. 2006. Specific detection of *Listeria monocytogenes* in foods using commercial methods: from chromogenic media to real-time PCR. *Span. J. Agric. Res.* **4:**235–247.

Jay, J. J. 1986. *Modern Food Microbiology*, 3rd ed. Van Nostrand Reinhold Co., New York, NY.

Junttila, J. R., S. I. Niemela, and J. Hirn. 1988. Minimum growth temperatures of *Listeria monocytogenes* and non-hemolytic *Listeria*. *J. Appl. Bacteriol.* **65:**321–327.

Kohn, B. A., K. Costello, and A. B. Phillips. 1997. HACCP verification procedures made easier by quantitative *Listeria* testing. *Dairy Food Environ. Sanit.* **17:**76–80.

Leclercq, A., D. Clermont, C. Bizet, P. A. Grimont, A. Le Flèche-Matéos, S. M. Roche, C. Buchrieser, V. Cadet-Daniel, A. Le Monnier, M. Lecuit, and F. Allerberger. 2010. *Listeria rocourtiae* sp. nov. *Int. J. Syst. Evol. Microbiol.* **60**(Pt. 9)**:**2210–2214.

Levin, R. E. 2003. Application of the polymerase chain reaction for detection of *Listeria monocytogenes* in foods: a review of methodology. *Food Biotechnol.* **17:**99–116.

Loessner, M. J., R. H. Bell, J. M. Jay, and L. A. Shelef. 1988. Comparison of seven plating media for enumeration of *Listeria* spp. *Appl. Environ. Microbiol.* **54:**3003–3007.

Mitchell, B. A., J. A. Milbury, A. M. Brookins, and B. J. Jackson. 1994. Use of immunomagnetic capture on beads to recover *Listeria* from environmental samples. *J. Food Prot.* **57:**743–745.

Murphy, M., D. Corcoran, J. F. Buckley, M. O'Mahony, P. Whyte, and S. Fanning. 2007. Development and application of a multilocus-locus variable-number of tandem-repeat analysis (MLVA) to subtype a collection of *Listeria monocytogenes*. *Int. J. Food Microbiol.* **115:**187–194.

Nogva, H. K., K. Rudi, K. Naterstad, A. Holck, and D. Lillehaug. 2000. Application of 5′-nuclease PCR for quantitative detection of *Listeria monocytogenes* in pure cultures, water, skim milk, and unpasteurized whole milk. *Appl. Environ. Microbiol.* **66:**4266–4271.

Norton, D. M., and C. R. Braden. 2007. Foodborne listeriosis, p. 305–356. *In* E. T. Ryser and E. M. Marth (ed.), *Listeria, Listeriosis and Food Safety*, 3rd ed. CRC Press, Boca Raton, FL.

Nyachuba, D. G., and C. W. Donnelly. 2007. Comparison of 3M™ Petrifilm™ environmental *Listeria* plates against standard enrichment methods for the detection of *Listeria monocytogenes* of epidemiological significance from environmental surfaces. *J. Food Sci.* **72:**M346–M354.

Oliver, S. P., B. M. Jayarao, and R. A. Almeida. 2005. Foodborne pathogens in milk and the dairy farm environment: food safety and public health implications. *Foodborne Pathog. Dis.* **2:**15–129.

Oravcová, K., T. Trncíková, T. Kuchta, and E. Kaclíková. 2008. Limitation in the detection of *Listeria monocytogenes* in food in the presence of competing *Listeria innocua*. *J. Appl. Microbiol.* **104:**429–437.

Pan, Y., and F. Breidt, Jr. 2007. Enumeration of viable *Listeria monocytogenes* cells by real-time PCR with propidium monoazide and ethidium monoazide in the presence of dead cells. *Appl. Environ. Microbiol.* **73:**8028–8031.

Rasmussen, O. F., P. Skouboe, L. Dons, L. Rossen, and J. E. Olsen. 1995. *Listeria monocytogenes* exists in at least three evolutionary lines: evidence from flagellin, invasive associated protein and listeriolysin O genes. *Microbiology* **141:**2053–2061.

Rebuffo-Scheer, C. A., J. Schmitt, and S. Scherer. 2007. Differentiation of *Listeria monocytogenes* serovars by using artificially neural network analysis of Fourier-transformed infrared spectra. *Appl. Environ. Microbiol.* **73:**1036–1040.

Rodríguez-Lázaro, D., M. Hernández, M. Scortti, T. Esteve, J. A. Vázquez-Boland, and M. Pla. 2004. Quantitative detection of *Listeria monocytogenes* and *Listeria innocua* by real-time PCR: assessment of *hly*, *iap*, and *lin02483* targets and AmpliFluor technology. *Appl. Environ. Microbiol.* **70:**1366–1377.

Sauders, B. D., Y. Schukken, L. Kornstein, V. Reddy, T. Bannerman, E. Salehi, N. Dumas, B. J. Anderson, J. P. Massey, and M. Wiedmann. 2006. Molecular epidemiology and cluster analysis of human listeriosis cases in three U.S. states. *J. Food Prot.* **69:**1680–1689.

Silbernagel, K. M., C. N. Carver, R. P. Jechorek, and R. L. Johnson. 2004. Evaluation of VIDAS *Listeria monocytogenes* II (LMO2) immunoassay method for the detection of *Listeria monocytogenes* in foods: collaborative study. *J. AOAC Int.* **87:**1123–1132.

Sperry, K. E., S. Kathariou, J. S. Edwards, and L. G. Wolf. 2008. Multiple-locus variable-number tandem-repeat analysis as a tool for subtyping *Listeria monocytogenes* strains. *J. Clin. Microbiol.* **46:**1435–1450. (Erratum, **47:**1984, 2009.)

Stessl, B., W. Luf, M. Wagner, and D Schoder. 2009. Performance testing of six chromogenic ALOA-type media for the detection of *Listeria monocytogenes*. *J. Appl. Microbiol.* **106:**651–659.

Tortorello, M. L., K. F. Reineke, and D. Stewart. 1997. Comparison of antibody-direct epifluorescent filter technique with the most probable number procedure for rapid enumeration of *Listeria* in vegetables. *J. AOAC Int.* **80:**1208–1214.

Tripathi, R. C. 1985. Negative binomial distribution, p. 169–177. *In* S. Kotz and N. C. Johnson (ed.), *The Encyclopedia of Statistical Science*, vol. 6. John Wiley & Sons, New York, NY.

Unnerstad, H., E. Bannerman, J. Bille, M. L. DanielssonTham, E. Waak, and W. Tham. 1996. Prolonged contamination of a dairy with *Listeria monocytogenes. Neth. Milk Dairy J.* **50:**493–499.

U.S. Department of Agriculture—Food Safety and Inspection Service. 2010. *Microbiological Laboratory Guidebook.* http://www.fsis.usda.gov/Science/Microbiological_Lab_Guidebook/index.asp. Accessed 4 June 2010.

Waak, E., W. Tham, and M. L. Danielsson-Tham. 2002. Prevalence and fingerprinting of *Listeria monocytogenes* strains isolated from raw whole milk in farm bulk tanks and in dairy plant receiving tanks. *Appl. Environ. Microbiol.* **68:**3366–3370.

Wagner, M., A. Maderner, and E. Brandl. 1996. Random amplification of polymorphic DNA for tracing and molecular epidemiology of *Listeria* contamination in a cheese plant. *J. Food Prot.* **59:**384–389.

Wagner, M., A. Maderner, and E. Brandl. 1999. Development of a multiple primer RAPD assay as a tool for phylogenetic analysis in *Listeria* spp. strains isolated from milk product associated epidemics, sporadic cases of listeriosis and dairy environments. *Int. J. Food Microbiol.* **52:**29–37.

Wagner, M., F. Eliskases-Lechner, P. Rieck, I. Hein, and F. Allerberger. 2006. Characterization of *Listeria monocytogenes* isolated from fifty Austrian small cheese making factories. *J. Food Prot.* **96:** 1297–1303.

Wagner, M., A. Lehner, D. Klein, and A. Bubert. 2000. Single-stranded conformation polymorphisms (SSCP) in the *hly* gene and PCR analysis of a repeat region in the *iap* gene to identify and type *Listeria monocytogenes. J. Food Prot.* **63:**332–336.

BACILLUS CEREUS IN MILK AND DAIRY PRODUCTION

M. Ehling-Schulz, U. Messelhäusser, and P. E. Granum

19

INTRODUCTION TO THE PATHOGEN AND COMMODITY

Bacillus cereus is a gram-positive, spore-forming, motile, aerobic rod that also grows well anaerobically. It is a common soil saprophyte and is easily spread to many types of foods, especially of plant origin, but is frequently isolated from dairy products, meat, and eggs. There are several members in the *B. cereus* group: *B. cereus, B. thuringiensis, B. mycoides*, and *B. weihenstephanensis*.

Members of the *B. cereus* group are a special problem for the dairy industry and are frequently found in pasteurized milk and milk-derived products, such as milk powder and infant formulas (Becker et al., 1994; Shaheen et al., 2006). Due to the heat and acid resistance of its spores, *B. cereus* is not eliminated by pasteurization or normal sanitation procedures, and without ultrahigh temperature (UHT) treatment, it is not possible to completely avoid *B. cereus* in some dairy products.

M. Ehling-Schulz, Food Microbiology Unit, Clinic of Ruminants, Department for Farm Animals and Veterinary, Public Health, Veterinaerplatz 1, A-1210 Vienna, Austria. *U. Messelhäusser*, Bavarian Health and Food Safety Authority, Veterinärstr. 2, D-85764 Oberschleißheim, Germany. *P. E. Granum*, Department of Food Safety and Infection Biology, Norwegian School of Veterinary Science, N-0033 Oslo, Norway.

The psychrotolerant members of the *B. cereus* group, *B. mycoides* and *B. weihenstephanensis*, are characterized by their ability to grow at low temperatures and have frequently been reported to cause spoilage of refrigerated foods (Mayr et al., 1999). Milk, the food matrix from which *B. weihenstephanensis* was originally isolated (Lechner et al., 1998), is a common source of contamination with psychrotolerant *B. cereus* group members (Te Giffel et al., 1996). These bacteria, growing at low temperatures, can in addition cause milk spoilage due to the activity of proteases and phospholipases.

Besides its spoilage potential, *B. cereus* is of special consumer concern because of its toxigenic potential. Very different types of toxins are associated with two types of gastrointestinal disease caused by *B. cereus*, namely, diarrhea and emesis (Ehling-Schulz et al., 2004a; Stenfors Arnesen et al., 2008). The emetic syndrome, characterized by intensive vomiting, is caused by a single heat-stable peptide toxin, called cereulide, which is preformed in food.

On the other hand, different enterotoxins contributing to diarrhea have been described. The emetic syndrome is mainly characterized by vomiting a few hours after ingestion of the contaminated food. In diarrhea, symptoms ap-

Rapid Detection, Identification, and Quantification of Foodborne Pathogens, ed. by J. Hoorfar

pear usually 8 to 16 hours after ingestion and include abdominal pain and diarrhea. The emetic type of disease is the result of intoxication by cereulide-contaminated food, whereas the diarrheal type is thought to be the consequence of a foodborne infection with enterotoxic *B. cereus* (Ehling-Schulz et al., 2011; Schoeni and Wong, 2005; Stenfors Arnesen et al., 2008). Emetic *B. cereus* food poisoning is frequently associated with rice and pasta dishes; however, emetic strains have also been isolated from milk, pudding, dairy-based baby food, and ice cream (Messelhäusser et al., 2010). In a widespread incident of emetic poisoning in Japan, milk was reported to have been the source; and in a recent outbreak in Germany, ricotta cheese has been identified as the source of emetic *B. cereus*. The diarrheal type of foodborne disease was first described in connection with the consumption of contaminated vanilla sauce in 1948. Now, at least three types of potential enterotoxins resulting in symptoms ranging from mild to occasionally bloody diarrhea have been described (Stenfors Arnesen et al., 2008).

The true incidence of *B. cereus* food poisoning is unknown for a number of reasons, including misdiagnosis of the illness, which may be symptomatically similar to other types of food poisoning. While the diarrheal syndrome caused by *B. cereus* resembles the symptoms of a *Clostridium perfringens* infection, the emetic type of food poisoning caused by *B. cereus* may be mixed with the symptoms caused by mild types of *Staphylococcus aureus* enterotoxins (see also chapter 20).

The incidence of foodborne intoxications caused by bacterial toxins has been steadily increasing in Europe throughout the last years. For instance, in April 2009 the European Food Safety Authority reported an increase of 42% in food-related intoxications by *Bacillus*, mainly *B. cereus*, toxins (Anonymous, 2009). Beside a general overview on novel methods for detection and enumeration of *B. cereus*, this chapter will focus on recently developed methods for the detection and quantification of the *B. cereus* toxins. It will briefly discuss novel methods to determine the toxigenic potential of a strain and present the current methods for subtyping of *B. cereus* group organisms. However, detection and quantification of dormant spores are dealt with in another chapter.

CHAIN FLOW, SAMPLE TYPES, SAMPLE SIZE, SAMPLE TREATMENT, AND SAMPLE STORAGE

As *B. cereus* is found in quite different niches on farms, contamination of raw milk often occurs during milking via soil-contaminated udders and teats and the milking equipment (Christiansson et al., 1999). Although farms are important sources of *B. cereus* spores, additional contamination might occur at dairy plants. For instance, silo tanks are often contaminated with *B. cereus* spores, but also pasteurizers as well as filling machines have been reported to be involved in contamination of milk and processed milk with *B. cereus*. Since *B. cereus* spores might occur in low numbers at certain steps of the milk production, a simple membrane filtration step could facilitate enumeration and isolation of *B. cereus* spores from milk. In brief, milk is heat treated at 75°C for 5 minutes and filtrated through a membrane filter with 0.8-μm pore size. This filter is incubated on a blood agar plate containing 10 ppm polymyxin at 20°C for 48 hours (for details, see Christiansson et al., 1997).

Infant formulas are of special concern, because even low-dose *B. cereus* contamination in these products can cause intoxications. Contamination can occur at different steps of the production chain. A typical production chain of infant formula is depicted in Fig. 1, indicating the major contamination entry points as well as possible sampling points. In the European Union, a mandatory sampling plan exists for the detection of presumptive *B. cereus* in infant formula at the end of the production process. Because of the nonhomogeneous distribution of *B. cereus* in milk powder and dried products, like infant formulas, Regulation (EC) No. 2073/2005 (Anonymous,

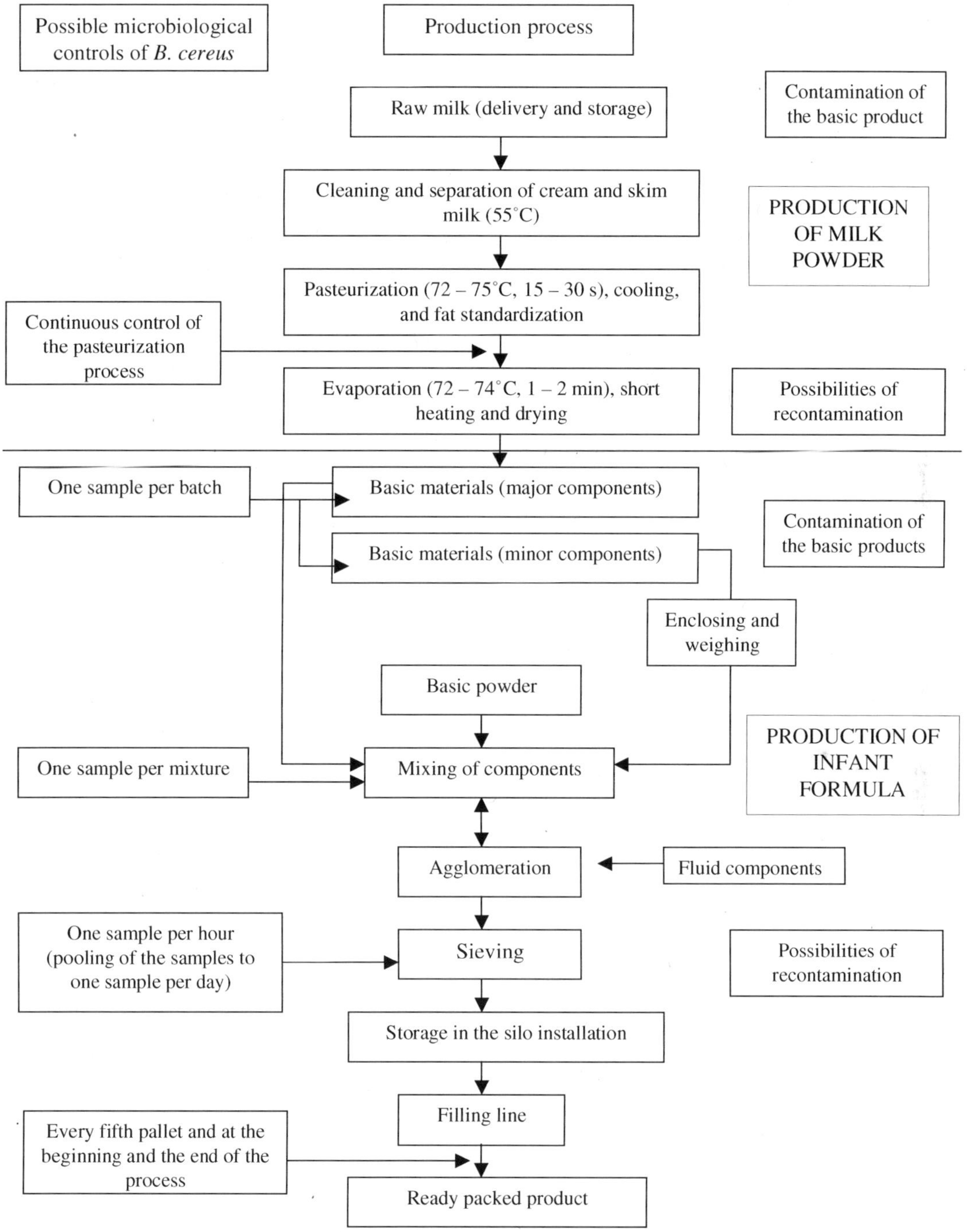

FIGURE 1 Production chain of infant formula (indicating possible sampling points).

2005) requires five samples of each production batch and determination of the number of presumptive *B. cereus* organisms in every sample according to EN/ISO 7932. The production batches are then classified into three categories: satisfactory, acceptable, and unsatisfactory. This three-class-sampling plan is part of the hazard analysis and critical control points concept of the infant formula production to avoid food poisonings (Fig. 1).

Sample collection must be carried out aseptically, and samples must be transported directly to the lab, where analyses are started immediately. Samples must be kept at temperatures below 6°C. When stored at refrigerated temperatures, it must be kept in mind that some *B. cereus* group strains are able to grow at low temperature while others are not. Thus, strains with a potentially higher food poisoning risk might be outcompeted (Guinebretiere et al., 2008).

If analysis cannot begin directly, samples shall be frozen at −20°C to prevent a shift in the microbial community towards psychrotolerant bacteria (Fricker et al., 2010). Before examination, samples shall be thawed at room temperature and processed immediately.

For the detection of *B. cereus* in food, usually two different sample amounts are used for qualitative or quantitative detection. All methods for quantification of *B. cereus* use 10 g/10 ml as the sample size. For quantification of heat-resistant spores, heat-shock treatment of the food samples is performed at 75°C for 20 minutes before plating on agar plates. By comparing results of direct plating and plating after heat shock, the ratio between vegetative cells and spores can be determined. Qualitative detection methods to test for the presence or absence of toxigenic strains of *B. cereus* might be especially useful in outbreak situations in order to exclude "negative" samples from further laborious toxin testing. In addition, a qualitative detection provides a higher diagnostic reliability, which is necessary in outbreak investigations. The usual sample size for qualitative detection is 25 g or 10 g, both diluted 1:10 in enrichment medium.

RAPID DETECTION

Isolation and enumeration of *B. cereus* from foods are routinely performed using conventional selective plating media. For instance, the polymyxin-egg yolk-mannitol-bromothymol blue agar and mannitol-egg yolk-polymyxin agar are endorsed by the International Organization for Standardization and the Food and Drug Administration (FDA) for the identification of presumptive *B. cereus* colonies. However, some *B. cereus* strains do not show the typical reactions and might therefore be misidentified or missed on these types of media, especially if complex matrices like food are analyzed (Fricker et al., 2008). Molecular methods provide more accurate as well as more rapid detection and enumeration of *B. cereus* than the currently used agar plate methods.

Preanalytical Sample Preparation

Preanalytical sample preparation is a critical step in the detection and enumeration of *B. cereus* from milk and milk products for two reasons. (i) *B. cereus* frequently occurs in milk and milk products as spores or a mixture of spores and vegetative cells. Due to their stability and stress resistance, *Bacillus* spores could be difficult to lyse. (ii) Milk and milk components like chocolate could interfere with the PCR. A preenrichment step is therefore mostly used for the qualitative detection of especially pathogenic *B. cereus* in different food matrices before PCR for the detection of the different toxin genes. The advantages of a preenrichment step are that (i) the probability of detecting low numbers of toxigenic strains in the sample increases with a preenrichment step; and (ii) possible PCR inhibitors in the sample are diluted by preenrichment, thereby minimizing the risk of false-negative PCR results. An overview of the different possibilities for detection and enumeration of *B. cereus* in food samples is provided in Fig. 2.

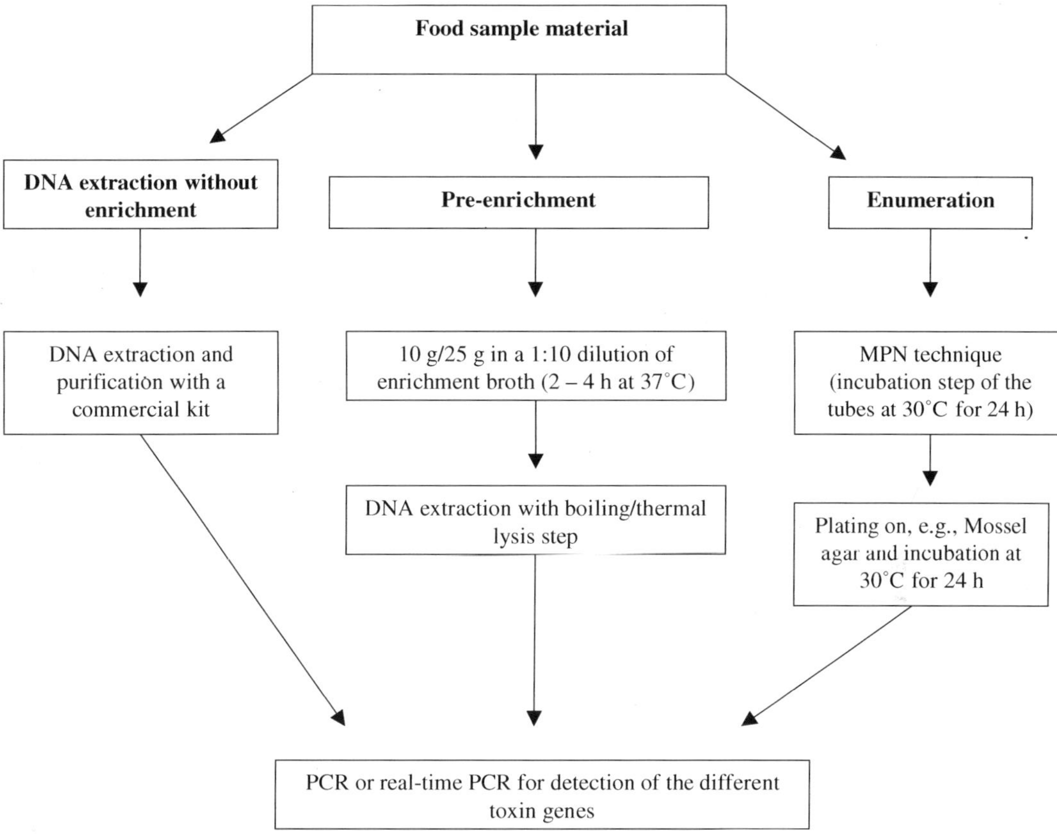

FIGURE 2 Work flow for detection and enumeration of pathogenic *B. cereus* in food samples.

Various methods have been described in the literature for DNA extraction from foods for real-time PCR. Several DNA extraction methods, including commercially available DNA isolation kits, have been tested for their suitability for *B. cereus* DNA extraction from milk and milk products as well as from other foodstuffs during the development of molecular detection systems for emetic *B. cereus*. An overview of the detection limits for DNA prepared by different extraction methods and subjected to standard PCR and real-time PCR is given in Table 1. Phenol-chloroform extraction yields the greatest amount of pure and high-molecular-weight genomic DNA, but it is a quite laborious technique not suitable for routine analysis. Based on our survey, we recommend using either (i) one of the commmercially available DNA extraction kits according to the manufacturer's instructions or (ii) the simple boiling methods described below in combination with a short nonselective enrichment step in brain heart infusion glucose broth at 37°C.

Protocol: Simple Boiling Method for DNA Extraction from *B. cereus* after Preenrichment

Required equipment includes incubators, water bath, a centrifuge, and pipettes. In brief, 25-g samples are mixed with 225 ml of brain heart infusion glucose broth (Merck, Ger-

TABLE 1 Comparison of DNA extraction methods for *B. cereus* from milk products[a]

Method	Duration (h)[b]	Standard PCR (EM1-EM2)	Real-time PCR TaqMan *ces*[c]	Real-time PCR SYBR Green *ces*[c]	Reference or source
Phenol-chloroform extraction	6	10^0	ND	ND	Ehling-Schulz et al., 2004b
Puregene DNA isolation kit	3	10^6	ND	ND	Gentra, United States
First magnetic milk kit	1.5	10^4	ND	ND	Gen-ial GmbH, Germany
AquaPure genomic DNA isolation kit	1.5	ND	10^2 (10^0 after 4-h enrichment)	10^1 (10^0 after 2-h enrichment)	Biorad, Germany
NucleoSpin food kit	1.3	10^6	10^2 (10^0 after 4-h enrichment)	10^1 (10^0 after 2-h enrichment)	Macherey-Nagel, Germany
Automatic sample preparation work station (QIAcube)	~1, depending on the kit	ND	10^2 (10^0 after 4-h enrichment)	ND	Qiagen GmbH, Germany
Automatic sample preparation work station (Maxwell)	0.75	ND	10^2 (10^0 after 4-h enrichment)	ND	Promega, Germany
Boiling method	0.3	$>10^6$	10^5 (10^0 after 6-h enrichment)	10^3 (10^0 after 4-h enrichment)	Fricker et al., 2007

[a] Similar detection limits were found for rice and milk rice pudding. ND, not determined.
[b] Time needed for five samples to be extracted in parallel.
[c] Detection limits after 2 to 6 h of unspecific enrichment are given in parentheses.

many) supplemented with 0.1% glucose) and incubated statically at 37°C for 2 to 6 hours. Subsequently, cells from 1-ml enriched samples are harvested by centrifugation, and cell pellets are resuspended in 300 μl sterile MilliQ, boiled for 10 min, and then cooled on ice. After pelleting cell and food residues, a 5-μl aliquot of the supernatant can be used as the PCR template (for further details, see Fricker et al., 2007).

Detection and Enumeration

In recent years, a wide range of molecular methods for the detection of *B. cereus* have been developed and published. PCR and real-time PCR methods are mainly used for the detection of the different toxin genes of emetic and diarrheal *B. cereus* strains. For the enumeration of these pathogens in food, classical culture methods, especially the most-probable-number (MPN) or colony count techniques on various solid agars, are still in use. The culture methods can be combined with a molecular differentiation of the isolates. This procedure allows an analytical estimate of the number of pathogenic *B. cereus* cells in the food sample investigated.

Enumeration of *B. cereus* Cells

There are two main enumeration methods for *B. cereus* that are applied in routine diagnostics. The MPN technique is mainly used for the enumeration of low concentrations of organisms (<100/g). For example, 10 g of sample material is added into 90 ml of sterile peptone diluent (1:10 dilution), and decimal dilutions are prepared. One-milliliter portions are transferred to 3, 5, or 10 tubes containing enrichment broth (e.g., Trypticase-peptone-glucose-yeast extract broth) with at least three consecutive dilutions for each dilution step (Blodgett, 2003). Depending on the number of tubes used for one dilution step, the calculated detection limit of the method is <3.0 MPN/g (three tubes), <1.8 MPN/g (five tubes), or <0.9 MPN/g. The tubes are incubated at 30°C ± 2°C under aerobic conditions for 24 h ± 2 h. After the incubation period, one loopful of every tube is plated onto, e.g., Mossel agar or blood agar, which is incubated as described above. When using MPN for enumeration of *B. cereus* strains with pathogenic potential, *B. cereus* colonies on the agar can be examined for the presence of the different toxin genes by PCR. Then, the MPN of emetic or diarrheal *B. cereus* in the sample is calculated based on the PCR results (Messelhäusser et al., 2010). In the same way, colony count techniques in combination with molecular methods can be used to calculate the concentration of pathogenic *B. cereus* in food samples. For quantification of *B. cereus*, PCR methods have been developed (for examples, see Reekmans et al., 2009, and Yang et al., 2007), and it is expected that in the coming years quantitative PCR and real-time PCR will be more commonly used as an alternative to the classical culture enumeration methods. However, there are still some challenges to overcome, concerning the comparability of culture and molecular results, especially with regard to the risk of detecting dead cells with PCR but also regarding appropriate sample preparation.

PCR

For the detection of the different toxin genes of emetic and diarrheal *B. cereus* strains, PCR and real-time PCR methods are widely used in food diagnostic laboratories. Used mainly in outbreak situations, molecular methods are a rapid tool to screen a higher number of samples (before or after enrichment) for the presence of enterotoxigenic or emetic *B. cereus*. But PCR users should always keep in mind that molecular methods detect only the toxin-encoding genes and not the toxins. Therefore, molecular detection of pathogenic *Bacillus cereus* should always be combined with the detection of the toxin in the sample material or with epidemiological data from outbreak investigations. A concise overview of different PCR and real-time-PCR methods detecting the diarrheal or emetic toxin genes published so far is given in Table 2. For the detection of diarrheal *B. cereus*, conventional PCR methods are more common, because of the multiplex capacity of this technology. A multiplex PCR system allows, e.g., the detection of genes encoding the major toxins Nhe (nonhemolytic enterotoxin), Hbl (hemolysin BL), and CytK (cytotoxin K) in a single PCR run. Examples for PCR systems detecting the different toxin genes of enterotoxigenic *B. cereus* are described in Table 2. Depending on the thermal cycler, the multiplex capacity of real-time PCR allows only the simultaneous detection of four or five toxin genes in one single-tube reaction in comparison to conventional PCR. However, due to the advantages of real-time PCR, especially the absence of any post-PCR analysis, there has been an increasing number of reports on real-time PCR systems for the detection of emetic as well as diarrheal *B. cereus*. For example, a real-time

TABLE 2 Methods for detection of *B. cereus* toxins

Method	Toxin(s) (genes) detected									Reference(s)
	NHE			*HBL*				CytK	Ces	
	nheA	*nheB*	*nheC*	*hblA*	*hblB*	*hblC*	*hblD*			
PCR	×	×	×	×	×	×	×	×		Guinebretiere et al., 2002
									×	Ehling-Schulz et al., 2004b
	×	×		×			×	×	×	Ehling-Schulz et al., 2006
Real-time PCR									×	Fricker et al., 2007
	×						×	×	×	Wehrle et al., 2010
Quantitative real-time PCR	Detection and quantification of enterotoxigenic *B. cereus*									Yang et al., 2007
Enzyme-linked immunosorbent assay		×				×				Moravek et al., 2006
HPLC-MS									×	Bauer et al., 2010; Haggblom et al., 2002

PCR assay for the detection of the cereulide synthetase gene (*ces*) in combination with an internal amplification control (IAC) is described by Fricker et al. (2007).

Protocol: Detection of the *ces* Gene of *B. cereus* by Real-Time PCR

Depending on the sample material, total bacterial DNA is extracted from food as described in "Preanalytical Sample Preparation" above, or DNA is extracted directly from a *B. cereus* colony by thermal lysis (briefly, one *B. cereus* colony from an agar plate is suspended in 1.5 ml of 0.9% NaCl and heated for 10 minutes at 95°C). The detection system described by Fricker et al. (2007) is a probe-based duplex real-time PCR using TaqMan technology. The heterologous IAC system based on the pUC 19 plasmid is used with a separate primer and probe system targeting the pMB1 replicon of the plasmid (Fricker et al., 2007; Messelhäusser et al., 2010). The assay is run with a 25-μl PCR mixture, containing 12.5 μl of a commercial multiplex real-time PCR master mixture consisting of 0.5 μM of each primer (ces_for 5′-CGCCGAAAGTGATTATACCAA-3′ and ces_rev 5′-TATGCCCCGTTCTCAAACTG-3′ for emetic *B. cereus* and IAC_2_fw 5′-TGT-GAA-ATA-CCG-CAC-AGA-TG-3′ and IAC_2_re 5′-AGC-TGG-CGT-AAT-AGC-GAA-G-3′ for the IAC), 0.2 μM for each probe (ces_probe FAM-GGGAAAATAACGAGAAATGCA-TAMRA for emetic *B. cereus* and IAC_2_probe HEX-GAG-AAA-ATA-CCG-CAT-CAG-GC-TAMRA for the IAC), 1 fg of plasmid DNA pUC19, and 5 μl DNA template. The temperature-time program is a typical two-step TaqMan profile, starting with an initial denaturation step of 95°C for 10 minutes followed by 45 cycles containing a denaturation step of 95°C for 15 seconds and an amplification step of 55°C for 60 seconds.

Besides the probe-based real-time PCR assays, SYBR Green assays using an intercalating dye for detection of the PCR product are also described in the literature (Fricker et al., 2007;

Wehrle et al., 2010). The advantages of using SYBR Green assays are the lower costs of the assays and mostly a higher sensitivity for the detection of the target genes. In contrast, probe-based systems give a higher security concerning the specificity of the reaction using not only target-specific primers but also a target-specific probe.

MALDI-TOF (MS) Technology

Matrix-assisted laser desorption ionization–time of flight (mass spectrometry) (MALDI-TOF [MS]) is a relatively new technique used in microbiology to identify bacteria mainly to the species level. Isolates are cultured under standardized conditions, e.g., on blood agar for 24 h at 30°C. Then, a small amount of culture material is spread on a carrier matrix, overlaid with an organic acid, and passed into the MALDI-TOF (MS). The MALDI-TOF creates a characteristic mass spectrum fingerprint and compares the result with the reference database. The whole procedure including sample preparation takes less than 15 min per sample. In comparison to other identification methods, MALDI-TOF (MS) is a fast and reliable technique for identification of pure bacterial colonies. One of the disadvantages of the technique is that pure cultures are needed for the identification. So far, detection of bacterial cells directly from sample material has not been possible. Another drawback is that the differentiation of emetic, enterotoxigenic, and apathogenic *B. cereus* and *B. thuringiensis* isolates is not possible. Currently, MALDI allows only the identification of *B. cereus* sensu lato. However, MALDI-TOF (MS) combined with chemometric analysis may be one of the possibilities not only for a rapid identification but also for the differentiation of *B. cereus* isolates from food in the future (Lasch et al., 2009).

Detection and Quantification of Toxins

The spectrum of potential *B. cereus* toxicity ranges from strains used as probiotics in animal feed to highly toxic strains already reported as being responsible for fatalities (for reviews, see Ehling-Schulz et al., 2004a, and Stenfors Arnesen et al., 2008). It is therefore of utmost importance to detect and quantify not only the pathogens but also their toxins.

Detection and Quantification of Cereulide

The toxin responsible for the emetic type of poisoning is a small heat- and acid-resistant peptide called cereulide, which is preformed in food. Due to its extreme heat and pH stability, it might be possible that preformed cereulide would still be present in a final food product, whereas the bacterium itself would be inactivated during processing. Therefore, molecular analysis by real-time PCR (see "Enumeration of *B. cereus* Cells" above) should be complemented by toxin analysis methods. The actual minimal dose of cereulide causing illness is unknown. However, through feeding experiments of primates and cell culture-based toxicity tests, it is estimated to be about 8 μg toxin/kg of body weight. We have analyzed samples from food poisoning cases in Germany and found the minimum dose to range between 3 and 20 μg cereulide/g food (M. Ehling-Schulz, M. Fricker, and U. Messelhäusser, unpublished data). Unfortunately, a rapid and economic method for quantification of cereulide in complex matrices like food is not yet available. In addition, laborious and time-consuming analyses are currently required to determine the toxigenic potential of a certain *B. cereus* contaminant in foods. However, knowledge of the toxicity of contaminants would be of special importance for the food industry to implement the appropriate actions in cases of *B. cereus* contaminations, which cannot be completely avoided in non-UHT-treated dairy products. Routine methods for the matrix-independent quantification of cereulide and for the rapid detection of highly toxigenic strains are urgently needed.

Currently, there are three possibilities for detecting the emetic toxin: via cell culture-based assays using HEp-2 cells, via a sperm-based bioassay, and by high-performance liquid chromatography connected to ion trap mass spectrometry (HPLC-MS) (for a review, see Ehling-Schulz et al., 2004a). With the exception of HPLC-MS, these assays do not specifically detect cereulide; e.g., the bioassay is also sensitive to other mitochondrial toxins like valinomycin. All the assays mentioned above were hampered by the lack of appropriate reference material. Since cereulide is not commercially available, valinomycin is commonly used as a surrogate standard. Therefore, reference material of cereulide as well as its $^{13}C_6$-isotopologue has been generated by in vivo production in *B. cereus*. The ^{13}C-labeled cereulide has been used to establish a highly sensitive, accurate, and robust stable isotope dilution analysis (SIDA) for the quantitative analysis of cereulide in foods by LC-MS/MS detection (Bauer et al., 2010). Cross-validation of the SIDA has shown that the MS response of valinomycin is significantly influenced by the food matrix, questioning its appropriateness as a surrogate for cereulide (Bauer et al., 2010). Current research focuses on the adaptation of the aforementioned SIDA to be implemented in routine diagnostic labs carrying out MS-based analysis. Meanwhile, the most practical assay might still be the HEp-2-based cytotoxicity assay, keeping in mind its drawbacks, e.g., not fully quantitative results. Generally, HEp-2 cells are easy to cultivate, and toxin extraction can be carried out by the procedure described below.

Protocol: Cytotoxicity Assay in Hep-2 Cells

Required equipment includes incubators (including CO_2 atmosphere), an autoclave, access to cell culture lab and HEp-2 cell lines, and a microtiter plate reader. In brief, 25-g samples are mixed with 225 ml of Luria-Bertani broth and incubated for 24 hours at 30°C, 150 rpm. After enrichment, food samples are autoclaved (15 min at 120°C) to denature heat-labile toxins, and samples are serially diluted (twofold) in 96-well microtiter plates with Eagle minimum essential medium (MEM) with Earle salts and supplements and 2% ethanol as a solvent for cereulide. The HEp-2 assay is carried out as described previously (Dietrich et al., 1999; Lücking et al., 2009).

Detection and Quantification of Enterotoxins

There are three different potential enterotoxins produced by *B. cereus*: Nhe, Hbl, and CytK. All *B. cereus* strains tested so far harbor the genes for Nhe, about 50 to 60% harbor the genes for Hbl, and 40% carry the genes for CytK (Stenfors Arnesen et al., 2008). However, a few strains of *B. cereus* do not transcribe fully active toxins and can therefore be nontoxigenic. Since both Nhe and Hbl are three component toxins, all three components must be transcribed and folded correctly. For both these toxins, several variants have been described, but we know little about the variation in toxicity.

Since these toxins all must be produced in the small intestine of the host in order to cause food infections, it is of little relevance to detect these toxins in foods. Neither of the toxins is stable enough to pass through the stomach acid barrier followed by proteolysis; hence, the toxins must be produced in the intestinal tract. Not only do the different strains of *B. cereus* vary in the amount of toxin produced, but also there are huge differences in optimal production temperatures. As expected, the psychrotolerant *B. weihenstephanensis* usually has optimal toxin production at a much lower temperature (20 to 30°C) than the mesophilic *B. cereus* strains (Stenfors Arnesen et al., 2007) and will probably not be able to cause food poisoning. This means that testing of toxin potential for each different strain should be conducted at 37°C, although toxin production is regulated by several environmental factors, not temperature alone (Duport et al., 2006; Messaoudi et al., 2010).

In light of what has been outlined above, we suggest that the toxin production be tested after growth of the bacterial strain in brain heart infusion supplemented with 1% glucose at 37°C for 6 h and harvested by centrifugation. The supernatants can then be tested for toxin production. It would be preferable to know the toxin genes harbored by the tested strain prior to testing. This can be done by multiplex PCR (see Ehling-Schulz et al.,

2006, and Wehrle et al., 2010). Toxicity can be tested by using tissue cultures. Vero cells are frequently used since they are easily grown and maintained.

Protocol: Cytotoxicity Assay in Vero Cells

Vero cells are grown in MEM supplemented with 5% fetal calf serum. Cells are seeded into 24-well plates 2 to 3 days before testing. Before use, check that the growth of the Vero cells is confluent. If so, remove the medium and wash the cells once with 1 ml preheated (37°C) MEM.

1. Add 1 ml preheated (37°C) low-leucine medium to each well and then add the toxin to be tested (1 to 100 μl of *B. cereus* supernatant); incubate the cells for 2 hours at 37°C.
2. Remove the low-leucine medium with the toxin, and wash each well once with 1 ml preheated (37°C) low-leucine medium. Mix 8 ml preheated low-leucine medium with 16 μl [^{14}C]leucine, and add 300 μl of this mixture to each well; incubate the cells for 1 hour at 37°C.
3. Remove the radioactive medium, add 1 ml 5% trichloroacetic acid (TCA) to each well, and incubate at room temperature for 10 minutes. Remove the TCA, and wash the wells twice with 1 ml 5% TCA.
4. After removing the TCA, add 300 μl 0.1 M KOH, and incubate at room temperature for 10 minutes. Transfer the content of each well to liquid scintillation tubes with 2 ml liquid scintillation cocktail. Vortex the tubes, and count the radioactivity in a scintillation counter for 1 minute.
5. The percentage of inhibition of protein synthesis is calculated using the following formula: [(negative control − sample)/negative control] × 100; the negative control is Vero cells from wells without addition of sample.

An alternative method is the propidium iodide uptake in Vero cells. This is performed with a spectrofluorimeter. Two-day-old confluent monolayers of Vero cells are used as described above. Cell suspensions with a final concentration of about 106 cells in 2 ml EC buffer containing propidium iodide (5 μg ml^{-1}) are held in a thermostatically controlled (37°C) 1-cm-path length quartz cuvette, to which the toxin is then added. Cells are continuously mixed by use of a magnetic stirrer and "flea." Fluorescence should be monitored every 30 seconds by using excitation/emission wavelengths of 575/615 nm and 5-nm slits for both. Results are used without subtraction of background fluorescence.

Neither of these tests will tell the relative influence of the different toxins causing the cytotoxicity. An alternative, apart from PCR to give the genetic toxin background, is to have access to neutralizing antibodies directed against each of the toxins. The relative amounts of the different toxins produced by the *B. cereus* strains may also be tested (semi-quantification) by using Western blots. However, for two of these toxins (Nhe and Hbl) three different antibodies are needed for each toxin. The last alternative is to use a commercial rapid immunochemical test.

However, since the kits only test for the first of the toxin components in each of the two operons, it is difficult to estimate the activity of each toxin, although for many strains the relative amount of either L2 or NheA will give an indication of toxin amounts produced.

At present, the best approach to detect all the components of the three toxins is to develop an enzyme-linked immunosorbent assay, which makes quantification possible. However, seven (preferably) monoclonal antibodies are then needed. Finally, another approach might be to use an LC-MS/MS method to quantify the toxins, but the methods are not fully developed and the equipment is quite expensive.

RAPID CHARACTERIZATION AND SUBTYPING

Genotyping Methods

In order to understand epidemiological processes and to trace and track *B. cereus* contaminants, suitable typing methods are required. Several independent genotypic studies, using multilocus enzyme electrophoresis, multilocus sequence typing (MLST), and amplified fragment length polymorphism markers, showed that the *B. cereus* group of organisms can be grouped in three major clusters (for a review, see Ehling-Schulz et al., 2010). Clinical *B. cereus* isolates tended to cluster in clonal groups within clade I, which also contains *B. anthracis*, while dairy and environmental isolates were mainly scattered throughout the other two clades. The psychrotolerant *B. mycoides* and *B. weihenstephanensis* form a group (clade III) clearly separated from the other two clades. Psychrotolerant strains belonging to the latter clade have rarely been connected to food poisoning, although these strains frequently carry enterotoxin genes.

During recent years, MLST has become the "gold standard" for typing of *B. cereus* group strains, and several MLST schemes have been developed that are now combined to form a supertree in the SuperCAT database (http://mlstoslo.uio.no/). However, since MLST requires intensive sequencing efforts, its application as a rapid tool for large population analysis is limited. The sporulation stage III AB gene (*spoIIIAB*) has been shown to be a suitable chromosomal marker resembling the structure of MLST-based trees (Ehling-Schulz et al., 2005) while being less expensive, as only a single gene has to be sequenced compared to the seven housekeeping genes used for MLST. The suitability of *spoIIIAB* as a single genetic marker was also verified with the sliding-window-based approach to compare whole genomes with the same isolates clustering close to *B. anthracis* as described recently (Segerman et al., 2010). It might therefore be a useful tool for a rapid and high-throughput analysis of *B. cereus* populations.

Protocol: Sequence Typing Using *spoIIIAB* as Single Chromosomal Marker

Required equipment includes a thermocycler, pipette, and reaction tubes. In brief, the primers Spo2F (5′-CGACGAGGATAACCCAATTTGC-3′) and Spo2R (5′-CAGTGAGAGACCGAGGCAAC-3′) are used to amplify a part of the *spoIIIAB* gene locus from *B. cereus* group strains. The PCR protocol starts with a denaturation step for 5 min at 95°C, followed by 30 cycles of 20 s at 95°C, 1 min at 59°C, and 2 min at 72°C each, and ends with a final elongation step at 72°C for 5 min. PCR amplicons are purified, sequenced, and analyzed as described previously (Ehling-Schulz et al., 2005).

Metabolic Profiling by FTIR

An interesting alternative to the genetic method for population and epidemiological studies requiring high-throughput capacities might be Fourier transform infrared (FTIR) spectroscopy. This technique has been found to successfully differentiate microorganisms on different genetic levels (Wenning et al., 2008) and has been applied to study the population structure of foodborne and clinical *B. cereus* (Ehling-Schulz et al., 2005). Figure 3 shows that FTIR is also a suitable tool in contamination route analysis, as it does not require substantial sample preparation like genotypic methods. The assay can normally be carried out within 24 hours. In one occasion, milk powder from a European country contaminated with low levels of *B. cereus* was exported and used to produce a milk concentrate outside Europe. The final product contained high levels of *B. cereus* organisms. However, FTIR clearly showed that the *B. cereus* that was found at a low level in the exported milk powder was not the source for the high *B. cereus* contamination observed in the final product (Fig. 3).

CONCLUSION AND FUTURE TRENDS

B. cereus is a ubiquitous sporeformer that cannot be totally avoided in most food products, especially in non-UHT-treated dairy products. Since the pathogenic potential of members of this species can be quite variable, ranging from strains used as probiotics to strains causing deadly diseases, methods to discriminate hazardous strains from harmless isolates may be the challenge in *B. cereus*

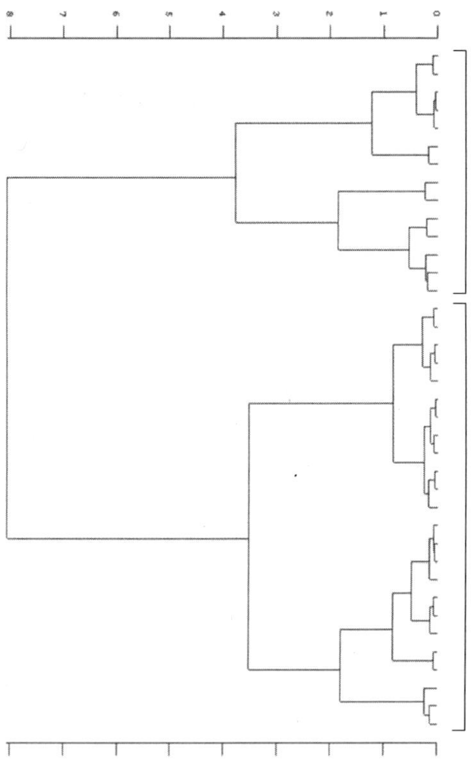

FIGURE 3 Contamination route analysis of *B. cereus*-contaminated milk concentrate, using FTIR spectroscopy. Exported milk powder from Europe containing low levels of *B. cereus* was used to produce a milk concentrate outside Europe. The final product contained high levels of *B. cereus* organisms. Hierarchical cluster analysis of FTIR spectra recorded from isolates obtained from both raw material and end product clearly showed that the *B. cereus* organisms found in the raw material were not the source for the high *B. cereus* contamination observed in the final product.

diagnostics for the next years. The knowledge of *B. cereus* pathogenicity gained during the last decade may render it possible to develop detection and typing systems to differentiate between hazardous and harmless strains. The different multiplex PCR systems for toxin gene profiling developed recently (see, for example, Ehling-Schulz et al., 2006, and Wehrle et al., 2010) might be a first step towards a new diagnostic strategy, taking into consideration the pathogenic potential rather than taxonomy. Another approach in this direction is genome typing techniques. A recent report (Zwick et al., 2008) presented a mixed whole-genome microarray for discrimination of closely related *B. anthracis* strains and typing of *B. cereus*. Such genotyping systems might also be suitable for the classification of foodborne *B. cereus*. However, the sole presence or absence of individual virulence factors does not fully explain the pathogenic potential of a certain strain. In addition, specific and sensitive systems for rapid and accurate quantification of toxins must supplement molecular detection and typing methods.

ACKNOWLEDGMENT

We thank Martina Fricker for providing us with unpublished data from evaluation studies of DNA extraction methods.

REFERENCES

Anonymous. 2005. Commission regulation (EC) No 2073/2005 of 15 November 2005 on microbiological criteria for foodstuffs. *Off. J. Eur. Union* L **338:**1–26.

Anonymous. 2009. *The Community Summary Report on Food-Borne Outbreaks in the European Union in 2007.* European Food Safety Authority, Parma, Italy.

Bauer, T., T. Stark, T. Hofmann, and M. Ehling-Schulz. 2010. Development of a stable isotope dilution analysis for the quantification of the *Bacillus cereus* toxin cereulide in foods. *J. Agric. Food Chem.* **58:**1420–1428.

Becker, H., G. Schaller, W. von Wiese, and G. Terplan. 1994. *Bacillus cereus* in infant foods and dried milk products. *Int. J. Food Microbiol.* **23:**1–15.

Blodgett, R. 2003. Most probable number from serial dilutions. *In Bacteriological Analytical Manual*, Appendix 2. Food and Drug Administration, Rockville, MD.

Christiansson, A., J. Bertilsson, and B. Svensson. 1999. *Bacillus cereus* spores in raw milk: factors affecting the contamination of milk during the grazing period. *J. Dairy Sci.* **82:**305–314.

Christiansson, A., K. Ekelund, and H. Ogura. 1997. Membrane filtration method for enumeration and isolation of spores of *Bacillus cereus* from milk. *Int. Dairy J.* **7:**743–748.

Dietrich, R., C. Fella, S. Strich, and E. Märtlbauer. 1999. Production and characterization of monoclonal antibodies against the hemolysin BL enterotoxin complex produced by *Bacillus cereus*. *Appl. Environ. Microbiol.* **65:**4470–4474.

Duport, C., A. Zigha, E. Rosenfeld, and P. Schmitt. 2006. Control of enterotoxin gene expression in *Bacillus cereus* F4430/73 involves the redox-sensitive ResDE signal transduction system. *J. Bacteriol.* **188:**6640–6651.

Ehling-Schulz, M., M. Fricker, and S. Scherer. 2004a. *Bacillus cereus*, the causative agent of an emetic type of food-borne illness. *Mol. Nutr. Food Res.* **48:**479–487.

Ehling-Schulz, M., M. Fricker, and S. Scherer. 2004b. Identification of emetic toxin producing *Bacillus cereus* strains by a novel molecular assay. *FEMS Microbiol. Lett.* **232:**189–195.

Ehling-Schulz, M., M. H. Guinebretiere, A. Monthan, O. Berge, M. Fricker, and B. Svensson. 2006. Toxin gene profiling of enterotoxic and emetic *Bacillus cereus*. *FEMS Microbiol. Lett.* **260:**232–240.

Ehling-Schulz, M., R. Knutsson, and S. Scherer. 2011. *Bacillus cereus*, p. 147–164. *In* P. Fratamico, S. Kathariou, and Y. Liu (ed.), *Genomes of Food- and Water-Borne Pathogens*. ASM Press, Washington, DC.

Ehling-Schulz, M., B. Svensson, M.-H. Guinebretiere, T. Lindback, M. Andersson, A. Schulz, M. Fricker, A. Christiansson, P. E. Granum, E. Märtlbauer, C. Nguyen-The, M. Salkinoja-Salonen, and S. Scherer. 2005. Emetic toxin formation of *Bacillus cereus* is restricted to a single evolutionary lineage of closely related strains. *Microbiology* **151:**183–197.

Fricker, M., U. Messelhausser, U. Busch, S. Scherer, and M. Ehling-Schulz. 2007. Diagnostic real-time PCR assays for the detection of emetic *Bacillus cereus* strains in foods and recent food-borne outbreaks. *Appl. Environ. Microbiol.* **73:** 1892–1898.

Fricker, M., R. Reissbrodt, and M. Ehling-Schulz. 2008. Evaluation of standard and new chromogenic selective plating media for isolation and identification of *Bacillus cereus*. *Int. J. Food Microbiol.* **121:**27–34.

Fricker, M., B. Skånseng, K. Rudi, B. Stessl, and M. Ehling-Schulz. 20 September 2010. Shift from farm to dairy tank microbiota revealed by a polyphasic approach is independent from geographical origin. *Int. J. Food Microbiol.* (Epub ahead of print.)

Guinebretiere, M. H., V. Broussolle, and C. Nguyen-The. 2002. Enterotoxigenic profiles of food-poisoning and food-borne *Bacillus cereus* strains. *J. Clin. Microbiol.* **40:**3053–3056.

Guinebretiere, M. H., F. L. Thompson, A. Sorokin, P. Normand, P. Dawyndt, M. Ehling-Schulz, B. Svensson, V. Sanchis, C. Nguyen-The, M. Heyndrickx, and P. De Vos. 2008. Ecological diversification in the *Bacillus cereus* group. *Environ. Microbiol.* **10:**851–865.

Haggblom, M. M., C. Apetroaie, M. A. Andersson, and M. S. Salkinoja-Salonen. 2002. Quantitative analysis of cereulide, the emetic toxin of *Bacillus cereus*, produced under various conditions. *Appl. Environ. Microbiol.* **68:**2479–2483.

Lasch, P., W. Beyer, H. Nattermann, M. Stämmler, E. Siegbrecht, R. Grunold, and D. Naumann. 2009. Identification of *Bacillus anthracis* by using matrix-assisted laser desorption ionization–time of flight mass spectrometry and artificial neural networks. *Appl. Environ. Microbiol.* **75:**7229–7242.

Lechner, S., R. Mayr, K. P. Francis, B. M. Pruss, T. Kaplan, E. Wiessner-Gunkel, G. S. Stewart, and S. Scherer. 1998. *Bacillus weihenstephanensis* sp. nov. is a new psychrotolerant species of the *Bacillus cereus* group. *Int. J. Syst. Bacteriol.* **48**(Pt. 4)**:**1373–1382.

Lücking, G., M. K. Dommel, S. Scherer, A. Fouet, and M. Ehling-Schulz. 2009. Cereulide synthesis in emetic *Bacillus cereus* is controlled by the transition state regulator AbrB, but not by the virulence regulator PlcR. *Microbiology* **155:**922–931.

Mayr, R., F. von Stetten, K. P. Francis, and S. Scherer. 1999. Significance of psychrotolerant aerobic sporeformers in food spoilage and methodologies for their detection and identification. *Mitt. Lebensm. Hyg.* **90:**41–61.

Messaoudi, K., T. Clavel, P. Schmitt, and C. Duport. 2010. Fnr mediates carbohydrate-dependent regulation of catabolic and enterotoxin genes in *Bacillus cereus* F4430/73. *Res. Microbiol.* **161:**30–39.

Messelhäusser, U., P. Kampf, M. Fricker, M. Ehling-Schulz, R. Zucker, B. Wagner, U. Busch, and C. Holler. 2010. Prevalence of emetic *Bacillus cereus* in different ice creams in Bavaria. *J. Food Prot.* **73:**395–399.

Moravek, M., R. Dietrich, C. Buerk, V. Broussolle, M. H. Guinebretiere, P. E. Granum, C.

Nguyen-The, and E. Märtlbauer. 2006. Determination of the toxic potential of *Bacillus cereus* isolates by quantitative enterotoxin analyses. *FEMS Microbiol. Lett.* **257:**293–298.

Reekmans, R., P. Stevens, T. Vervust, and P. De Vos. 2009. An alternative real-time PCR method to detect the *Bacillus cereus* group in naturally contaminated food gelatine: a comparison study. *Lett. Appl. Microbiol.* **48:**97–104.

Schoeni, J. L., and A. C. L. Wong. 2005. *Bacillus cereus* food poisoning and its toxins. *J. Food Prot.* **68:**636–648.

Segerman, B., D. De Medici, M. Ehling Schulz, P. Fach, L. Fenicia, M. Fricker, P. Wielenga, B. Van Rotterdam, and R. Knutsson. 2010. Use of whole genome sequencing as a high resolution diagnostic typing tool when tracing bioterror organisms in the food and feed chain. *Int. J. Food Microbiol.* doi:10.1016/j.ijfoodmicro.2010.06.027. Epub ahead of print.

Shaheen, R., M. A. Andersson, C. Apetroaie, A. Schulz, M. Ehling-Schulz, V. M. Ollilainen, and M. S. Salkinoja-Salonen. 2006. Potential of selected infant food formulas for production of *Bacillus cereus* emetic toxin, cereulide. *Int. J. Food Microbiol.* **107:**287–294.

Stenfors Arnesen, L. P., A. Fagerlund, and P. E. Granum. 2008. From soil to gut: *Bacillus cereus* and its food poisoning toxins. *FEMS Microbiol. Rev.* **32:**579–606.

Stenfors Arnesen, L. P., K. O'Sullivan, and P. E. Granum. 2007. Food poisoning potential of *Bacillus cereus* strains from Norwegian dairies. *Int. J. Food Microbiol.* **116:**292–296.

Te Giffel, M. C., R. R. Beumer, S. Leijendekkers, and F. M. Rombouts. 1996. Incidence of *Bacillus cereus* and *Bacillus subtilis* in foods in the Netherlands. *Food Microbiol.* **13:**53–58.

Wehrle, E., A. Didier, M. Moravek, R. Dietrich, and E. Märtlbauer. 2010. Detection of *Bacillus cereus* with enteropathogenic potential by multiplex real-time PCR based on SYBR Green I. *Mol. Cell. Probes* **24:**124–130.

Wenning, M., S. Scherer, and D. Naumann. 2008. Infrared spectroscopy in the identification of microorganisms, p. 71–96. *In* M. Diem, P. R. Griffiths, and J. M. Chalmers (ed.), *Vibrational Spectroscopy for Medical Diagnosis.* John Wiley & Son Ltd., Chichester, United Kingdom.

Yang, I. C., D. Y. Shih, J. Y. Wang, and T. M. Pani. 2007. Development of rapid real-time PCR and most-probable-number real-time PCR assays to quantify enterotoxigenic strains of the species in the *Bacillus cereus* group. *J. Food Prot.* **70:**2774–2781.

Zwick, M. E., M. P. Kiley, A. C. Stewart, A. Mateczun, and T. D. Read. 2008. Genotyping of *Bacillus cereus* strains by microarray-based resequencing. *PLoS One* **3:**e2513.

STAPHYLOCOCCUS AUREUS IN THE DAIRY CHAIN

B. Stessl, I. Hein, M. Wagner, and M. Ehling-Schulz

20

GENERAL ASPECTS

Staphylococcus aureus is an important pathogen causing a variety of diseases. In human and veterinary medicine, this pathogen is one of the major causes of infections of various regions of the body including the mammary glands (Amir, 2002; Adam and Southwood, 2006; Moret-Stalder et al., 2008). Bovine mastitis accounts for substantial economical loss in the dairy industry and is often caused by infection with *S. aureus*. Biofilm formation of *S. aureus* in the udder has been proposed to be an important factor for long-term infection of animals and recurrent cases of mastitis (Melchior et al., 2006). Early detection of affected animals is necessary for an effective herd management. In addition, some strains are able to produce staphylococcal enterotoxins (SEs) causing food intoxications (Lee et al., 2007).

Symptoms develop rapidly (1 to 6 h after eating) and include mainly emesis but also stomach cramps. Since the emetic syndrome caused by *S. aureus* cannot be differentiated symptomatically from intoxications with *Bacillus cereus* toxins (see chapter 19), methods for rapid differentiation between these two pathogens and their heat-resistant toxins are important for accurate and reliable diagnostics (Granum and Lund, 1997).

Thus, two different aspects are of importance concerning the presence of *S. aureus* in the dairy chain: (i) clinical and subclinical infections of cows, not only posing a health risk via shedding of *S. aureus* in their milk but also being an economic factor in dairy farming; and (ii) transmission into the dairy chain and production of enterotoxins leading to human food poisoning.

There are five major classical SE types causing food poisoning, i.e., SEA, SEB, SEC, SED, and SEE, and several newly identified SEs. Currently, 18 different SEs are recognized (Chiang et al., 2008). The involvement of the newly identified SEs in food poisoning is not clear, and it has been proposed that only those that induce emesis after oral administration in a monkey model should be designated SEs, whereas the others should be designated staphylococcal enterotoxin-like superantigens (Lina et al., 2004). Concerning the classical toxins, for monkeys the 50% emetic doses vary between 5 and 20 μg per animal, but outbreak

B. Stessl, I. Hein, and M. Wagner, Institute of Milk Hygiene, Milk Technology and Food Safety, Department for Farm Animals and Veterinary Public Health, Veterinaerplatz 1, A-1210 Vienna, Austria. *M. Ehling-Schulz,* Food Microbiology Unit, Clinic of Ruminants, Department for Farm Animals and Veterinary Public Health, Veterinaerplatz 1, A-1210 Vienna, Austria.

Rapid Detection, Identification, and Quantification of Foodborne Pathogens, ed. by J. Hoorfar

data suggest that humans may be more sensitive (Bergdoll, 1979; Jørgensen et al., 2005a).

The upper temperature limit for production of SEs varies between 35°C and 44°C (Schmitt et al., 1990). Because SEs are stable with respect to heat and storage, they may be present in milk and milk products even if viable *S. aureus* cells are absent (Jablonski and Bohach, 1997). Foods with levels of SE-producing *S. aureus* greater than 10^5 CFU per g are considered a risk (Bergdoll, 1979). European legislation requires testing of cheese, milk powder, and whey powder for the presence of SEs if levels above 10^5 CFU per g are detected. In addition, three levels of classification apply to the presence of *S. aureus* in milk products. The *m* value separates acceptable from marginally acceptable samples, and the *M* value separates marginally acceptable from defective-quality samples (Table 1). The limits for the last two categories of cheese apply to that stage during processing in which the highest numbers are to be expected (Anonymous, 2007a).

Contamination of milk and milk products with *S. aureus* may occur either at the farm level (due to a priori contaminated raw milk, cross-contamination during milking, or contaminated milking equipment) or during food processing and handling (due to contaminated environment or human carriers). The nasopharynx is considered to be an important ecological niche for *S. aureus*. Approximately 20% of the healthy human population is persistently and 80% is intermittently colonized with *S. aureus* in the nose (van Belkum et al., 2009; Burian et al., 2010).

In addition to hygiene problems leading to contamination at some point during the chain, a process failure is often associated with staphylococcal food poisoning. In 2000, a massive outbreak with more than 10,000 cases involved occurred in Japan and was caused by the consumption of reconstituted milk (see below) that contained SEA and SEH (Ikeda et al., 2005). Due to a power failure, the recovery milk warmed to 40°C was stored for more than 9 h without refrigeration, which presumably resulted in extensive production of SEs (Soejima et al., 2007). Another outbreak involving SEH in mashed potato made with raw milk was reported in Norway in 2003. The mashed potato was a leftover from a Christmas party held the night before and had been reheated (Jørgensen et al., 2005a). In June 2007, an outbreak involving SEA and SED occurred in Austria after consumption of pasteurized milk products. Risk analysis suggested that toxin production might have occurred during storage of surplus pasteurized milk for 3 days, followed by repasteurization and further processing (Schmid et al., 2009). Another outbreak, in Switzerland in July 2008, involved consumption of SED in raw goat milk originating from goats suffering from mastitis (Giezendanner et al., 2009). SEE is rarely involved in food poisoning. In France, an outbreak occurred at the end of 2009 involving SEE toxin after consumption of soft cheese made from unpasteurized milk (Ostyn et al., 2010).

Traditional microbiological methods used for the detection and quantification of *S. aureus* are time-consuming, and therefore, faster high-throughput methods have been evaluated and used as alternative tools. The main field of application of these methods is in-house process control, since for many countries (including those in the European Union) the official guidelines still refer to the traditional microbiological methods (Anonymous, 2007b). A vast array of rapid methods is also available for the detection of SEs. The following sections include discussions of these methods.

SAMPLING IN THE PRODUCTION CHAIN

The entrance of *S. aureus* into the dairy chain is multifactorial (Fig. 1). *S. aureus* is frequently associated with dairy cows and is a leading cause of intramammary infection, which is often subclinical (Akineden et al., 2001). *S. aureus* can easily be shed into the milk during collection and enter the milk processing chain (Heidinger et al., 2009; Schmid et al., 2009). Particularly, contamination of raw milk is possible during milking (via milking equipment

TABLE 1 Food safety and process hygiene criteria of the European Regulation EC 2005/2073 for cheese and milk products[a]

Food category	Microorganisms/toxins	Sampling plan		Limits (CFU/g)		Point of application
		n	*c*	*m*	*M*	
Food safety criteria						
Cheese, milk powder, and whey powder	SEs	5	0	Not detected in 25 g		End product
Food hygiene criteria						
Cheeses made from raw milk	Coagulase-positive staphylococci	5	2	10^4	10^5	Time during the manufacturing process when the number of staphylococci is expected to be highest[b]
Cheeses made from milk that has undergone a lower heat treatment than pasteurization and ripened cheeses made from milk or whey that has undergone pasteurization or a stronger heat treatment	Coagulase-positive staphylococci	5	2	100	1,000	
Unripened soft cheeses (fresh cheeses) made from milk or whey that has undergone pasteurization or a stronger heat treatment	Coagulase-positive staphylococci	5	2	10	100	End of the manufacturing process[b]

[a] Abbreviations: *n*, number of units comprising the sample; *c*, number of sample units giving values between *m* and *M*; *m*, the acceptable microbiological level in a sample unit; *M*, the level which, when exceeded in one or more samples, would cause the lot to be rejected.

[b] If values of >10^5 CFU/g are detected, the cheese batch has to be tested for SEs.

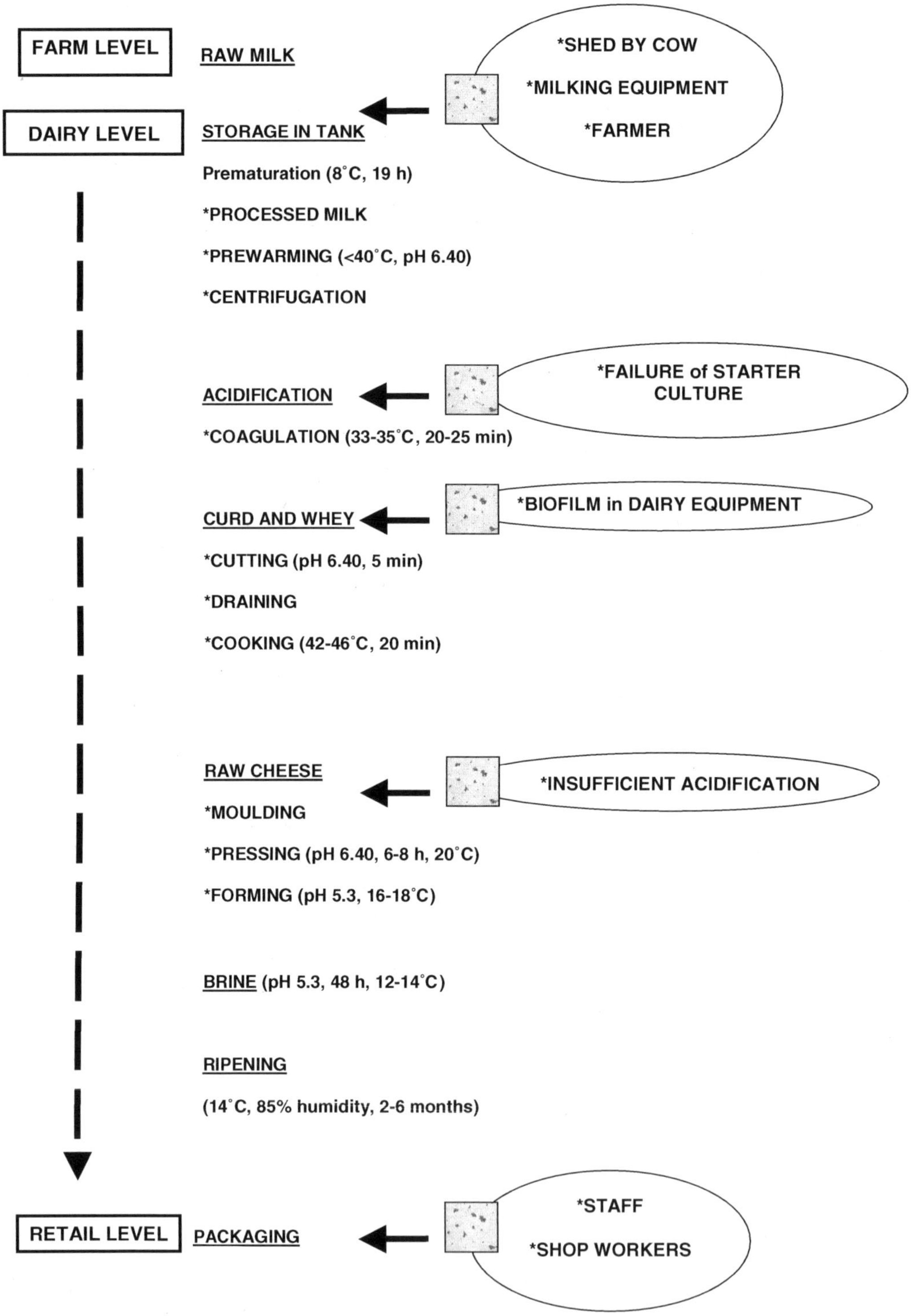

FIGURE 1 Possible entrance of *S. aureus* in the dairy chain.

or farm workers) and storage of milk in the farm tank. Each further processing step facilitates the entrance/growth of *S. aureus* associated with bovine, human, or environmental origin. The use of bulk tanks and other methods of rapid cooling of raw milk on the farm limits the growth of *S. aureus* strains (Henning et al., 2004).

Subsequent factors may have a major influence on the production of enterotoxins by *S. aureus* in the cheese production. Generally, a growth of enterotoxigenic staphylococci to 1 to 5 million per gram of cheese during the cheese-making process is necessary to produce amounts of enterotoxin detectable by enzyme-linked immunosorbent assay (ELISA) methods. Mild heat treatment (<40°C), which is usual during the processing of raw milk cheeses, may leave a residual population of coagulase-positive staphylococci in the curd. In addition, recontamination by an *S. aureus*-produced biofilm in difficult-to-access niches of the dairy may affect the milk after heat treatment. The most important factor permitting the growth of staphylococci and other potentially pathogenic bacteria, e.g., *Listeria*, during cheese making is a weak/insufficient acidification of the lactic starter culture. *S. aureus* tends to increase 1.5 to 3 log units in semihard or soft cheeses made from raw milk under normal conditions during cheese making. In the case of a starter failure, it can increase up to even 5 log cycles (Zangerl and Ginzinger, 2006). The most frequent causes for deficient pH decrease usually are an infection of the starter culture by bacteriophage or the presence of inhibitory substances in the milk. Defined mesophilic cultures (such as *Lactococcus lactis* subsp. *lactis*, *Lactococcus lactis* subsp. *cremoris*, and *Leuconostoc lactis*) with a temperature optimum of 30°C and defined thermophilic starter cultures (such as *Streptococcus thermophilus*, *Lactobacillus delbrueckii*, and *Lactobacillus helveticus*) with an optimum growth temperature of 42°C are normally used for the acidification process. Additionally, undefined "natural starters" are used in the production of traditional cheeses such as Emmentaler and Bergkäse. Fluctuations in culture composition can result in variable cheese processing. Natural starters are considered to be highly tolerant to phage infection because they are reproduced in the presence of phages, which leads to the dominance of resistant or tolerant strains (Carminati et al., 2010). The epidemiology of cheese-related outbreaks demonstrated that surface-ripened soft cheeses generally represented a greater risk for the transmission of pathogens than other cheeses (Zottola and Smith, 1991).

In Brazil, a food poisoning outbreak involving 50 individuals was related to the consumption of traditional handmade Minas cheese, made from raw milk, which was contaminated by enterotoxigenic strains of *S. aureus*. In this study, the density of the microorganism was quite high, up to 2.0×10^8 CFU/g cheese (Carmo et al, 2002). Additionally, Poli et al. (2007) investigated Monte Veronese, a semihard cheese from raw cow milk, for the detection of enterotoxin-producing *S. aureus*. *S. aureus* numbers in cheese were higher than the limit tolerated by the Italian food legislation in 78% of instances. Longer ripening did not always result in a numerical decline of this pathogen. Thus, the most efficient remedy is reducing the initial contamination by improving the hygiene level of milking and manufacturing. Delbes et al. published in 2006 an investigation of *S. aureus* in Saint-Nectaire, an uncooked, semihard cheese from raw cow milk. Coagulase-positive staphylococci grew rapidly during the first 6 h. Thus, the initial level of coagulase-positive staphylococci in raw milk should be maintained below 100 CFU/ml and optimally below 40 CFU/ml. To limit growth, acidification should be managed to obtain pH values around or below 5.8 at 6 h in Saint-Nectaire cheeses. Lindqvist et al. (2002) identified several data gaps in their model on quantitative microbial risk assessment exemplified by *S. aureus* in unripened cheese made from raw milk. Due to limited data and the absence of dose-response models, a complete risk assessment was not possible. The initial level of *S. aureus*, followed by storage temperature, had the largest influence on the probability of an unsatisfactory cheese at the two pH values (5.2 and 6.5) investigated.

The investigation of dairy samples for the detection and quantification of coagulase-positive staphylococci should be performed according to the standard guidelines delineated in ISO 6888 (Anonymous, 1999). Coagulase-positive staphylococci are enumerated after inoculation of a specific quantity of the test sample onto a solid selective medium (Baird-Parker agar) in duplicate. A selection of typical and atypical colonies suspicious for *S. aureus* are tested in the coagulase test with rabbit plasma fibrinogen. According to the European Commission (EC) Regulation 2073, five samples of 25 g, representing different batches of cheese, should be investigated for the detection and enumeration of coagulase-positive staphylococci (Anonymous, 2005). Sampling from semihard cheese productions should be taken, particularly before ripening, where the total bacterial content is assumed to be at the maximum. Therefore, samples should be taken before and after brining from the edge of the cheese and from unripened cheese. This should be initiated by the producer and is considered monitoring of food hygiene. Hence, not more than two of five samples should exceed limits between the m value and the M value. In the case of a semihard cheese made from raw milk, these values should be $m = 10,000$ and $M = 100,000$. In addition, in the case of a food safety breach, five samples must be taken by a two-class plan at the end of ripening. Therefore, the number of units comprising the sample should be 5, and the number of sample units giving values between m and M must be zero. Additionally, for microbiological investigation of dairy samples, the thermonuclease test is recommended to assess staphylococcal growth and potential enterotoxin production (Lachica, 1980). This test is, however, a screening method with a risk of false-negative or false-positive results. Therefore, thermonuclease test-positive cheeses must always be subjected to enterotoxin assays before they are declared unsafe for consumption. Milk and cheese samples should be transported immediately under cooled conditions (5 to 8°C) to the laboratory and should be investigated on the same day.

RAPID DETECTION METHODS

Methods To Detect and Quantify the Pathogen *S. aureus*

Rapid detection methods are applied at different stages of the microbiological analysis: species identification of colonies on agar plates and detection of the pathogen in the enrichment or directly in the sample, which enables quantification also. Since analysis time is a major issue in food processing and the toxin-producing ability of some strains requires a risk assessment approach based on the numbers of pathogen present in the sample, direct toxin quantification in the sample is preferred for later stages of the chain, whereas classical microbiological investigation is mainly used for samples at the farm level, e.g., for mastitis diagnostics.

Generally, either compounds of the surface, metabolites produced by *S. aureus*, RNA, or—most commonly—DNA is the target for detection. Sometimes, a combination of different targets is applied, e.g., separation of the bacteria with paramagnetic beads targeted at surface compounds followed by DNA detection (Aprodu et al., 2010).

Most methods start with a pretreatment of the sample, which is aimed at concentrating *S. aureus* cells or DNA and separating these from the sample for further analysis (Stevens and Jaykus, 2004; Aprodu et al., 2010). When techniques were compared for separation efficiency, i.e., paramagnetic beads coated with recombinant *S. aureus* phage endolysin-derived, cell wall-binding domain proteins, cell density-based flotation, and a combination of solvents, detergents, and centrifugation for separation of *S. aureus* from milk, the method using sample preparation based on a solvents-detergents combination was found to be superior to the beads method (Aprodu et al., unpublished). Graber et al. (2007) added *Lactobacillus casei* to their protocol to prevent attachment of *S. aureus* to milk fat globules and to displace staphylococci already stuck to the globule surface. Another interesting approach, which might be helpful when investigating milk from herds with mastitis, involves pretreatment of milk with amino-silica to remove somatic cells (Kubota et al., 2007). Fewer methods are available for separation of whole cells of *S. aureus* or DNA from cheese. A DNA extraction method based on the lysing and nuclease-inactivating properties of the chaotropic agent guanidinium thiocyanate, together with the nuclease-binding properties of silica particles, achieved a detection limit of 100 CFU/g raw milk cheese when combined with PCR detection (Cremonesi et al., 2007). The application of a method incorporating pronase for digestion of milk proteins, lysostaphin, and proteinase K yielded similar detection limits for semihard cheese, Camembert-style soft cheese, and fresh goat's cheese when combined with real-time PCR analysis (Hein et al., 2001). Most methods applied in the dairy chain are targeted toward the DNA of *S. aureus*. For PCR and real-time PCR, commonly used targets are the genes coding for thermonuclease (*nuc*), coagulase (*coa*), 23S rRNA, and a heat shock protein (*htrA*) (Hein et al., 2005; Chiang et al., 2007; Cremonesi et al., 2007; Graber et al., 2007; Poli et al, 2007). A duplex real-time PCR system for a rapid one-step differentiation between *S. aureus* and emetic *B. cereus* has been described (Fricker et al., 2007). Enterotoxin-producing *S. aureus* and emetic *B. cereus* cause a similar type of emetic disease; their differential diagnostic is therefore of special importance for food authorities and epidemiological surveillance (see chapter 19). Multiplex PCR approaches are mainly used at the farm level to detect different pathogens in mastitis milk (Phuketes et al., 2003; Gillespie and Oliver, 2005). Combinations of multiplex PCR with biochip-based detection of the amplicons or universal PCR targeting the 16S rRNA gene with DNA ligation reaction detection and a universal tag array, incorporating two specific probes for each target sequence, have also been used for that purpose (Lee et al., 2008; Cremonesi et al., 2009).

A major issue is to remove PCR inhibitors present in milk and cheese from the isolated DNA. Thus, it is mandatory to include internal amplification controls to identify false-negative results (Hoorfar et al., 2004). Well-performing methods achieve a detection limit of about 10 to 100 CFU/ml milk (Hein et al., 2005; Goto et al., 2007). Concerning the detection limit, the bottleneck for many methods is the limited amount of isolated DNA (2 to 10 μl) that is eventually analyzed in the PCR. In addition, DNA-based methods suffer from the inability to discriminate DNA of viable *S. aureus* from that of dead *S. aureus*. DNA binding agents ethidium monoazide (EMA) and propidium monoazide (PMA), which can be covalently linked to the DNA by exposure to visible light, have been tested for their ability to block PCR amplification of DNA from dead *S. aureus*. Here, PMA, but not EMA, was able to selectively enter (mainly) dead *S. aureus* cells, thus being useful for viable-versus-dead discrimination in PCR or real-time PCR, whereas EMA blocked PCR amplification of DNA from viable *S. aureus* also (Kobayashi et al., 2009a, 2009b). The high-copy-number RNA species tmRNA provides a new target, which might also be suitable for detection of viable *S. aureus*. A real-time nucleic acid sequence-based amplification assay targeting *S. aureus*-specific tmRNA was able to detect 1 to 10 CFU of *S. aureus*/ml milk in 3 to 4 h, without culture enrichment (O'Grady et al., 2009).

An immunosensor targeted at the cell-bound protein A of *S. aureus* was able to detect 4.8×10^3 *S. aureus* cells/ml milk (Escamilla-Gómez et al., 2008). *S. aureus* cells and anti-rabbit immunoglobulin G labeled with alkaline phosphatase compete for the binding sites of rabbit immunoglobulin G, which is coimmobilized with tyrosinase on a mercaptopropionic acid, self-assembled, monolayer modified, gold electrode. Phenol generated in the enzyme reaction with alkaline phosphatase is detected amperometrically at the tyrosinase-modified electrode through the electrochemical reduction of the *o*-quinone formed. Recently, matrix-assisted laser desorption ionization–time of flight (mass spectroscopy) (MALDI-TOF [MS]) has been explored for the identification of microorganisms (Seng et al., 2009). This technique uses protein fingerprints combined with chemometric analysis for identification of bacteria. The major advantages of this technique are the simplicity of sample preparation and its open platform characteristics, whereas the major drawback is the cost of the MS instrument and software required for analysis. Several groups have tested the applicability of MALDI-TOF (MS) as well as the variant technique, surface-enhanced laser desorption ionization–TOF (MS) for identification and characterization of staphylococci (Dubois et al., 2010; Rajakaruna et al., 2009; Spanu et al., 2010). Sensitivity and specificity were reported to be around 99%. In a recent study, Dubois et al. (2010) typed 152 staphylococcal strains representing 22 staphylococcal species by MALDI-TOF (MS) and found a specificity of 99.3%. Of 152 strains, 151 were correctly identified. Due to the simplicity of sample preparation (see work flow in Fig. 2) and the speed of MALDI-TOF (MS) analysis, numerous samples can be processed per day. Compared to the costs for an API 20Staph or real-time PCR, the costs for a MALDI-TOF (MS) analysis are quite low. In addition, for standard plating media like Columbia blood agar for cultivation of the bacteria, only a few inexpensive chemicals are required (Liu et al., 2007). Although still an upfront technique, MALDI-TOF (MS) technology is expected to be increasingly implemented not only in clinical diagnostic labs but also in food microbiology labs along with other metabolic fingerprinting methods. Other methods still in the developmental stage, with no field data available for application in milk and cheese, include detection of mastitis pathogens by analysis of volatile bacterial metabolites and multicolored fluorescent resonance energy transfer silica nanoparticles, labeled with specific antibodies for direct detection of bacteria including *S. aureus* (Wang et al., 2007; Hettinga et al., 2008).

Detection and Quantification of Toxins

The presence of higher numbers of *S. aureus* (>10^5 staphylococci per g) in the dairy chain has to be confirmed according to EC regulation 2073/2005 by SE detection. According to the FDA *Bacteriological Analytical Manual*, eight enterotoxins (types A, B, C1, C2, C3, D, E, and H) have been identified as distinct serological entities (Bennett, 2001). Tracer-labeled immunoassays such as ELISA and enzyme-linked fluorescent immunoassays (ELFI) are recommended for routine laboratory confirmation of enterotoxins. Most of the ELISA/ELFI kits use specific polyclonal or monoclonal antibodies that are absorbed, e.g., on microtiter wells or polystyrene balls (Park et al., 1994; Bennett and McClure, 1994; Hennekinne et al., 2007).

The sensitivity for different ELISA or ELFI assays varies from 0.1 ng SE per g to >0.5 SE ng per g reference material (Hennekinne et al., 2007). A recent study published by Boyle et. al. (2010) reports the detection of enterotoxin B with lateral flow immunochromatographic devices in milk and milk products. The described advantage of this method is the omission of the extraction step that is needed for SEB concentrations below 5 and 0.25 ng/ml detection levels. An electrical protein array chip on the basis of enzyme-linked sandwich immunoassay was lately developed by Quiel et al. (2010). The limit of detection of SEB in different liquid matrices was 1 ng/ml, which was detectable in less than 23 min. Sasaki et al. (2005) developed an avidin-biotin sandwich ELISA for the quantification of SEA, SEB, and SEC1 using monoclonal antibodies. The linearity of this system was 0.78 to 25 ng/ml for each SE. A fluorescent latex microparticle immunoassay for the detection of 0.5 ng/ml whole milk was introduced (Medina,

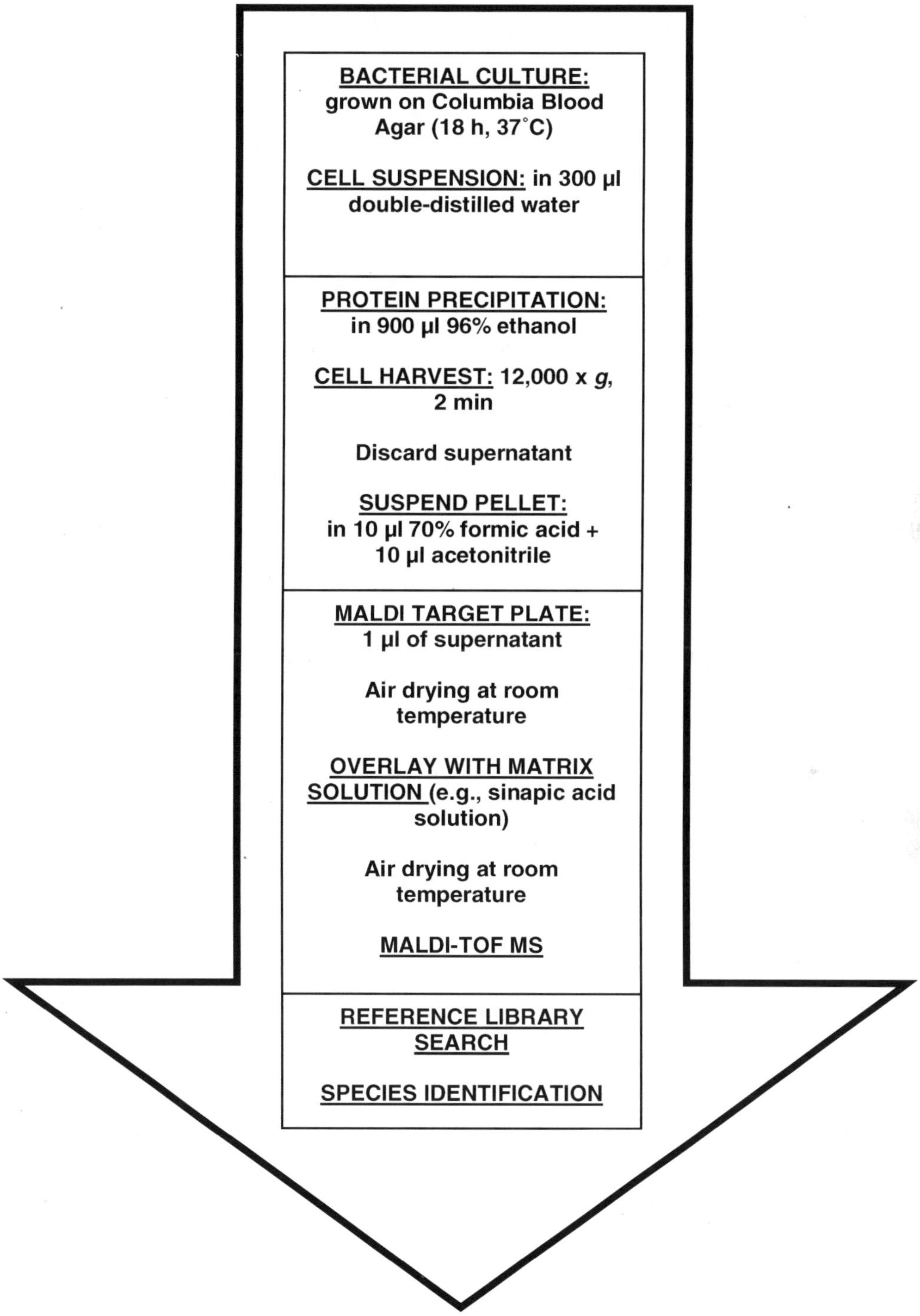

FIGURE 2 Work flow for identification of staphylococci by MALDI-TOF (MS).

2006), whereby anti-SEB was immobilized onto carboxylated polystyrene microparticles. Mishra et. al. (2008) reported that with a silicon nanowire transistor as a basis for the electronic detection of bacterial toxins, the detection limit of SEB was in the range of 10 to 35 fM. The ultrasensitive detection was demonstrated by electrochemical impedance spectroscopy. A further method for the detection of enterotoxin B was described by Khreich et al. (2008). This immunochromatographic test used immunoliposomes as labels for the detection of SEB with a detection limit of 0.5 ng/ml whole milk. A spectral surface plasmon resonance biosensor was developed by Homola et al. (2002) for the detection of enterotoxin B in milk. The sandwich detection mode, in which secondary antibodies are bound to the already captured toxin molecules, showed a detection limit of 0.5 ng/ml milk. Another approach by Chen (2008) introduced conductometric immunosensors for the detection of enterotoxin B based on a bioelectrocatalytic reaction on microcomb electrodes. The recovery for SEB spiked in milk was on average 116%. A novel tool for quantification of staphylococcal enterotoxin gene expression during cheese manufacturing by reverse transcription-quantitative PCR was published by Duquenne et al. (2010). Expression of these SE genes was monitored during the first 72 h of the cheese-making process, and mRNA data were correlated with enterotoxin production.

Rapid Characterization and Subtyping of *S. aureus*

Characterization and subtyping of *S. aureus* in the dairy chain highlight a possible contamination flow within the chain. It may also reveal the source association of the strains found in the dairy chain with strains present in the environment or in humans and other animals. Usually, pure cultures derived from single colonies are required for this purpose. If the flow of distinct genes such as SE genes or the *mecA* gene of methicillin-resistant *S. aureus* (MRSA) coding for a specific penicillin-binding protein (PBP2) and resulting in beta-lactam antibiotic resistance is the topic of interest, direct detection of these genes at different stages of the chain can be performed. However, in most cases specific traits of distinct *S. aureus* strains isolated from the chain are of interest. Phenotypic characterization includes metabolic fingerprinting and resistance profiles to different antibiotics (Jørgensen et al., 2005b; Gonano et al., 2009; Morandi et al., 2010). Most methods are based on genotypic traits. The gold standard for typing of *S. aureus* is pulsed-field gel electrophoresis with the rare cutting restriction enzyme SmaI, as was used, for example, in the study by Zadoks et al. (2002). Most MRSA isolated from livestock belong to sequence type 398, which is nontypeable with SmaI (Bosch et al., 2010). In a recent study, nearly 10% of Belgian farms reporting *S. aureus* mastitis had an MRSA problem, with sequence type 398 confirmed in all cases (Vanderhaeghen et al., 2010). Recently, a neoschizomer of SmaI, designated Crf9I, was shown to be insensitive for the DNA methylation, leading to typing problems with SmaI (Bosch et al., 2010). PCR-based methods followed by gel electrophoresis or sequencing require less hands-on time than pulsed-field gel electrophoresis. Polymorphisms in the coagulase (*coa*) gene yield distinct fingerprints when PCR amplicons of that gene are digested with frequent-cutting restriction enzymes such as AluI or HaeIII and separated on agarose gels (Saei et al., 2009; Morandi et al., 2010). Multiple-locus variable-number tandem repeat analysis involves PCR amplification of genes with variable-number tandem repeats followed by separation of the amplicons. Genes commonly used are *spa, clfA, clfB, sdr, sspA, coa, fnb, sdrC,* and *sdrD* (Gilbert et al., 2006; Morandi et al., 2010). The presence of SE genes is often used in a multiplex PCR format as a subtyping method (Jørgensen et al., 2005b; Gonano et al., 2009). However, correlation with other typing methods is moderate (Morandi et al., 2010). Multilocus sequence typing (MLST), *spa*-typing, and staphylococcal cassette chromosome *mec*

(SCC*mec*) typing involve PCR amplification and sequence analysis of one (*spa*-typing) or multiple genes. These methods yield objective sequence data that can be easily compared to international databases. MLST includes sequence analysis of seven housekeeping genes: *arcC, aroE, glpF, gmk, pta, tpi,* and *yqiL* (Jørgensen et al., 2005b; Smith et al., 2005). A comparison of MLST types of 258 bovine isolates from the United States, Chile, and the United Kingdom showed that 87.4% belonged to one clonal complex, supporting the theory of the host specificity of *S. aureus* (Smith et al., 2005). *spa*-typing involves sequence analysis of the protein A gene variable-repeat region and is usually used in a clinical context (Koreen et al., 2004). Thus, rapid genotyping in the dairy chain is applied at the farm level to type *S. aureus* isolated from cases of clinical or subclinical mastitis (Aires-de-Sousa et al., 2007; Vanderhaeghen et al., 2010). Variations in the structure and sequence of the large mobile genetic element SCC*mec*, carrying the *mecA* gene, are explored for typing of MRSA. This method has been used for typing of MRSA isolated from bovine milk in Korea. The isolates harbored a unique SCC*mec* subtype, which did not correlate with the emergence of community-acquired MRSA in human infection (Kwon et al., 2005).

CONCLUSION AND FUTURE TRENDS

Rapid detection and quantification methods could serve as valuable tools for early indication of process failures, and as part of hazard analysis and critical control points concepts. The trend in *S. aureus* enterotoxin detection and quantification is enhanced by the desire to reduce the time from sampling to identification and confirmation in the laboratory. Results should be obtained within hours instead of days. These commercially available rapid methods (e.g., classical ELISA) can detect approximately 1.0 nanogram of toxin/g of food. The bottleneck of the rapid methods is the high detection level (>10^5 *S. aureus* cells/g are needed) and the elaborative sample preparation in the case of semihard and hard cheeses. Most noncommercialized rapid methods have lower limits of detection (0.1 ng SE per g to >0.5 SE ng per g food) but are only established for the detection of SEB in liquid samples. A further recent development is a real-time immunoquantitative PCR method for the detection of SE. Many immunological and biochemical assays often work well in buffer systems but are hindered when tested in the complex chemical environment of foods. To overcome these biases and improve the limits of detection, we implemented an immunomagnetic PCR signal amplification assay for recovery and detection of SEA and SEB in foods (Panneerseelan and Muriana, 2009). For further characterization of enterotoxigenic *S. aureus*, a number of techniques have been developed lately, including, for example, DNA microarrays and subtyping methods (e.g., MLST, *spa*-typing, multiple-locus variable-number tandem-repeat analysis). Characterization and subtyping methods facilitate tracking and tracing the contamination and transmission routes of *S. aureus* in the dairy chain.

REFERENCES

Adam, E. N., and L. L. Southwood. 2006. Surgical and traumatic wound infections, cellulitis, and myositis in horses. *Vet. Clin. N. Am. Equine Pract.* **22:**335–361.

Aires-de-Sousa, M., C. E. S. R. Parente, O. Vieira-da-Motta, I. C. F. Bonna, D. A. Silva, and H. de Lencastre. 2007. Characterization of *Staphylococcus aureus* isolates from buffalo, bovine, ovine, and caprine milk samples collected in Rio de Janeiro State, Brazil. *Appl. Environ. Microbiol.* **73:** 3845–3849.

Akineden, Ö., C. Annemüller, A. A. Hassan, C. Lämmler, W. Wolter, and M. Zschöck. 2001. Toxin genes and other characteristics of *Staphylococcus aureus* isolates from milk of cows with mastitis. *Clin. Diagn. Lab. Immunol.* **8:**959–964.

Amir, L. 2002. Breastfeeding and *Staphylococcus aureus:* three case reports. *Breastfeed. Rev.* **10:**15–18.

Anonymous. 1999. *EN ISO 6888-1. Microbiology of Food and Animal Feeding Stuffs—Horizontal Method for the Enumeration of Coagulase-Positive Staphylococci (Staphylococcus aureus and Other Species). Part 1. Technique using Baird-Parker Agar Medium.* International

Organization for Standardization, Geneva, Switzerland.

Anonymous. 2005. Commission Regulation 2073/2005/EC of 15 November 2005 on microbiological criteria for foodstuffs. *Off. J. Eur. Commun.* L **338:**1–26.

Anonymous. 2007a. Commission Regulation (EC) No 1441/2007 of 5th December 2007 on microbiological criteria for foodstuffs. European Commission, Brussels, Belgium.

Anonymous. 2007b. *EN ISO 7218:2007. Microbiology of Food and Animal Feeding Stuff—General Requirements of Guidance for Microbiological Examination.* International Organization for Standardization, Geneva, Switzerland.

Aprodu, I., G. Walcher, J. Schelin, I. Hein, B. Norling, P. Rådström, A. Nicolau, and M. Wagner. 2010. Advanced sample preparation for the molecular quantification of *Staphylococcus aureus* in artifically and naturally contaminated milk. *Int. J. Food Microbiol.* (Epub ahead of print.)

Bennett, R. W., and F. McClure. 1994. Visual screening with enzyme immunoassay for staphylococcal enterotoxins in foods: collaborative study. *J. AOAC Int.* **77:**357–364.

Bennett, R. W. 2001. Staphylococcal enterotoxins, chapt. 13 A. *Bacteriological Analytical Manual.* Food and Drug Administration, Rockville, MD. http://www.fda.gov/Food/ScienceResearch/LaboratoryMethods/BacteriologicalAnalyticalManualBAM/default.htm.

Bergdoll, M. S. 1979. Staphylococcal intoxications, p. 443–494. *In* H. Riemann and F. L. Bryan (ed.), *Food-Borne Infections and Intoxications.* Academic Press, Inc., New York, NY.

Bosch, T., A. J. de Neeling, L. M. Schouls, K. W. van der Zwaluw, J. A. J. W. Kluytmans, H. Grundmann, and X. W. Huijsdens. 2010. PFGE diversity within the methicillin-resistant *Staphylococcus aureus* clonal lineage ST398. *BMC Microbiol.* **10:**40.

Boyle, T., J. M. Njoroge, R. L. Jones, Jr., and M. Principato. 2010. Detection of staphylococcal enterotoxin B in milk and milk products using immunodiagnostic lateral flow devices. *J. AOAC Int.* **93:**569–575.

Burian, M., C. Wolz, and C. Goerke. 2010. Regulatory adaptation of *Staphylococcus aureus* during nasal colonization of humans. *PLoS ONE* **5**(4): e10040. doi:10.1371/journal.pone. 0010040.

Carminati, D., G. Giraffa, A. Quiberoni, A. Binetti, V. Suarez, and J. Reinheimer. 2010. Advances and trends in starter cultures for dairy fermentations, p. 177–192. *In* F. Mozzi, R. R. Raya, and G. M. Vignolo (ed.), *Biotechnology of Lactic Acid Bacteria: Novel Applications.* Blackwell Publishing, New York, NY.

Carmo, L. S., R. Souza Dias, V. Roberto Linardi, M. J. de Sena, D. Aparecidados Santos, M. E. de Faria, E. Castro Pena, M. Jett, and L. Guilherme Heneine. 2002. Food poisoning due to enterotoxigenic strains of *Staphylococcus* present in Minas cheese and raw milk in Brazil. *Food Microbiol.* **19:**9–14.

Chen, Z. G. 2008. Conductometric immunosensors for the detection of staphylococcal enterotoxin B based bio-electrocatalytic reaction on micro-comb electrodes. *Bioprocess. Biosyst. Eng.* **31:**345–350.

Chiang, Y. C., C. M. Fan, W. W. Liao, C. K. Lin, and H. Y. Tsen. 2007. Real-time PCR detection of *Staphylococcus aureus* in milk and meat using new primers designed from the heat shock protein gene *htrA* sequence. *J. Food Prot.* **70:**2855–2859.

Chiang, Y. C., W. W. Liao, C. M. Fan, W. Y. Pai, C. S. Chiou, and H. Y. Tsen. 2008. PCR detection of staphylococcal enterotoxins (SEs) N, O, P, Q, R, U, and survey of SE types in *Staphylococcus aureus* isolates from food-poisoning cases in Taiwan. *Int. J. Food Microbiol.* **121:**66–73.

Cremonesi, P., G. Perez, G. Pison, P. Moroni, S. Morandi, M. Luzzana, M. Brasca, and B. Castiglioni. 2007. Detection of enterotoxigenic *Staphylococcus aureus* isolates in raw milk cheese. *Lett. Appl. Microbiol.* **45:**586–591.

Cremonesi, P., G. Pisoni, M. Severgnini, C. Consolandi, P. Moroni, M. Raschetti, and B. Castiglioni. 2009. Pathogen detection in milk samples by ligation detection reaction-mediated universal array method. *J. Dairy Sci.* **92:**3027–3039.

Delbes, C., J. Alomar, N. Chougui, J. F. Martin, and M. C. Montel. 2006. *Staphylococcus aureus* growth and enterotoxin production during the manufacture of uncooked, semihard cheese from cows' raw milk. *J. Food Prot.* **69:**2161–2167.

Dubois, D., D. Leyssene, J. P. Chacornac, M. Kostrzewa, P. O. Schmit, R. Talon, R. Bonnet, and J. Delmas. 2010. Identification of a variety of *Staphylococcus* species by matrix-assisted laser desorption ionization–time of flight mass spectrometry. *J. Clin. Microbiol.* **48:**941–945.

Duquenne, M., I. Fleurot, M. Aigle, C. Darrigo, E. Borezée-Durant, S. Derzelle, M. Bouix, V. Deperrois-Lafarge, and A. Delacroix-Buchet. 2010. Tool for quantification of staphylococcal enterotoxin gene expression in cheese. *Appl. Environ. Microbiol.* **76:**1367–1374.

Escamilla-Gómez, V., S. Campuzano, M. Pedrero, and J. M. Pingarrón. 2008. Immunosensors for the determination of *Staphylococcus aureus* using a tyrosinase-mercaptopropionic acid modified electrode as an amperometric transducer. *Anal. Bioanal. Chem.* **391:**837–845.

Fricker, M., U. Messelhäusser, U. Busch, S. Scherer, and M. Ehling-Schulz. 2007. Diagnostic real-time PCR assays for the detection of emetic *Bacillus cereus* strains in foods and recent food-borne outbreaks. *Appl. Environ. Microbiol.* **73:** 1892–1898.

Giezendanner, N., B. Meyer, M. Gort, P. Müller, and C. Zweifel. 2009. Raw milk-associated *Staphylococcus aureus* intoxication in children. *Schweiz. Arch. Tierheilkd.* **151:**329–331.

Gilbert, F. B., A. Fromageau, L. Gélineau, and B. Poutrel. 2006. Differentiation of bovine *Staphylococcus aureus* isolates by use of polymorphic tandem repeat typing. *Vet. Microbiol.* **117:**297–303.

Gillespie, B. E., and S. P. Oliver. 2005. Simultaneous detection of mastitis pathogens, *Staphylococcus aureus, Streptococcus uberis,* and *Streptococcus agalactiae* by multiplex real-time polymerase chain reaction. *J. Dairy Sci.* **88:**3510–3518.

Gonano, M., I. Hein, P. Zangerl, A. Rammelmayr, and M. Wagner. 2009. Phenotypic and molecular characterization of *Staphylococcus aureus* strains of veterinary, dairy and human origin. *Epidemiol. Infect.* **137:**688–699.

Goto, M., H. Takahashi, Y. Segawa, H. Hayashidani, K. Takatori, and Y. Hara-Kudo. 2007. Real-time PCR method for quantification of *Staphylococcus aureus* in milk. *J. Food Prot.* **70:**90–96.

Graber, H. U., M. G. Casey, J. Naskova, A. Steiner, and W. Schaeren. 2007. Development of a highly sensitive and specific assay to detect *Staphylococcus aureus* in bovine mastitic milk. *J. Dairy Sci.* **90:**4661–4669.

Granum, P. E., and T. Lund. 1997. *Bacillus cereus* and its food poisoning toxins. *FEMS Microbiol. Lett.* **157:**223–228.

Heidinger, J. C., C. K. Winter, and J. S. Cullor. 2009. Quantitative microbial risk assessment for *Staphylococcus aureus* and *Staphylococcus* enterotoxin A in raw milk. *J. Food Prot.* **72:**1641–1653.

Hein, I., A. Lehner, P. Rieck, K. Klein, E. Brandl, and M. Wagner. 2001. Comparison of different approaches to quantify *Staphylococcus aureus* cells by real-time quantitative PCR and application of this technique for examination of cheese. *Appl. Environ. Microbiol.* **67:**3122–3126.

Hein, I., H. J. Jorgensen, S. Loncarevic, and M. Wagner. 2005. Quantification of *Staphylococcus aureus* in unpasteurised bovine and caprine milk by real-time PCR. *Res. Microbiol.* **156:**554–563.

Hennekinne, J. A., F. Guillier, S. Perelle, M. L. De Buyser, S. Dragacci, S. Krys, and B. Lombard. 2007. Intralaboratory validation according to the EN ISO 16 140 standard of the Vidas SET2 detection kit for use in official controls of staphylococcal enterotoxins in milk products. *J. Appl. Microbiol.* **102:**1261–1272.

Henning, D. R., R. Flowers, R. Reiser, and E. T. Ryser. 2004. Pathogens in milk and milk products, p. 103–152. *In* M. Wehr and J. F. Frank (ed.), *Standard Methods for the Examination of Dairy Products.* American Public Health Association, Washington, DC.

Hettinga, K. A., H. J. F. van Valdenberg, T. J. G. M. Lam, and A. C. M. van Hooijdonk. 2008. Detection of mastitis pathogens by analysis of volatile bacterial metabolites. *J. Dairy Sci.* **91:**3834–3839.

Homola, J., J. Dostálek, S. Chen, A. Rasooly, S. Jiang, and S. S. Yee. 2002. Spectral surface plasmon resonance biosensor for detection of staphylococcal enterotoxin B in milk. *Int. J. Food Microbiol.* **75:**61–69.

Hoorfar, J., N. Cook, B. Malorny, M. Wagner, D. De Medici, A. Abdulmawjood, and P. Fach. 2004. Diagnostic PCR: making internal amplification control mandatory. *Lett. Appl. Microbiol.* **38:**79–80.

Ikeda, T., N. Tamate, K. Yamaguchi, and S. Makino. 2005. Mass outbreak of food poisoning disease caused by small amounts of staphylococcal enterotoxins A and H. *Appl. Environ. Microbiol.* **71:** 2793–2795.

Jablonski, L. M., and G. A. Bohach. 1997. *Staphylococcus aureus,* p. 353–375. *In* L. R. Doyle, L. R. Beuchat, and T. J. Montville (ed.), *Food Microbiology: Fundamentals and Frontiers.* ASM Press, Washington, DC.

Jørgensen, H. J., T. Mathisen, A. Løvseth, K. Omoe, K. S. Qvale, and S. Loncarevic. 2005a. An outbreak of staphylococcal food poisoning caused by enterotoxin H in mashed potato made with raw milk. *FEMS Microbiol. Lett.* **252:**267–272.

Jørgensen, H. J., T. Mørk, D. A. Caugant, A. Kearns, and L. M. Rørvik. 2005b. Genetic variation among *Staphylococcus aureus* strains from Norwegian bulk milk. *Appl. Environ. Microbiol.* **71:** 8352–8361.

Khreich, N., P. Lamourette, H. Boutal, K. Devilliers, C. Créminon, and H. Volland. 2008. Detection of *Staphylococcus* enterotoxin B using fluorescent immunoliposomes as label for immunochromatographic testing. *Anal. Biochem.* **377:**182–188.

Kobayashi, H., M. Oethinger, M. J. Tuohy, G. S. Hall, and T. W. Bauer. 2009a. Unsuitable distinction between viable and dead *Staphylococcus aureus* and *Staphylococcus epidermidis* by ethidium bromide monoazide. *Lett. Appl. Microbiol.* **48:**633–638.

Kobayashi, H., M. Oethinger, M. J. Tuohy, G. S. Hall, and T. W. Bauer. 2009b. Improving clinical significance of PCR: use of propidium monoazide to distinguish viable from dead *Staph-*

ylococcus aureus and *Staphylococcus epidermidis*. *J. Orthop. Res.* **27:**1243–1247.

Koreen, L., S. V. Ramaswamy, E. A. Graviss, S. Naidich, J. M. Musser, and B. N. Kreiswirth. 2004. *spa* typing method for discriminating among *Staphylococcus aureus* isolates: implications for use of a single marker to detect genetic micro- and macrovariation. *J. Clin. Microbiol.* **42:**792–799.

Kubota, M., T. Hayashi, K. Iwasaki, H. Ohtsuka, M. Kohiruimaki, S. Kawamura, K. Sakaguchi, and R. Abe. 2007. Rapid and effective method for separation of *Staphylococcus aureus* from somatic cells in mastitis milk. *J. Dairy Sci.* **90:**4100–4107.

Kwon, N. H., K. T. Park, J. S. Moon, W. K. Jung, S. H. Kim, J. M. Kim, S. K. Hong, H. C. Koo, Y. S. Joo, and Y. H. Park. 2005. Staphylococcal cassette chromosome *mec* (SCC*mec*) characterization and molecular analysis for methicillin-resistant *Staphylococcus aureus* and novel SCC*mec* subtype IVg isolated from bovine milk in Korea. *J. Antimicrob. Chemother.* **56:**624–632.

Lachica, R. V. 1980. Accelerated procedure for the enumeration and identification of food-borne *Staphylococcus aureus*. *Appl. Environ. Microbiol.* **39:** 17–19.

Lee, K. H., J. W. Lee, S. W. Wang, L. Y. Liu, M. F. Lee, S. T. Chuang, Y. M. Shy, C. L. Chang, M. C. Wu, and C. H. Chi. 2008. Development of a novel biochip for rapid multiplex detection of seven mastitis-causing pathogens in bovine milk samples. *J. Vet. Diagn. Investig.* **20:** 463–471.

Lee, Y. D., B. Y. Moon, J. H. Park, H. I. Chang, and W. J. Kim. 2007. Expression of enterotoxin genes in *Staphylococcus aureus* isolates based on mRNA analysis. *J. Microbiol. Biotechnol.* **17:**461–467.

Lina, G., G. A. Bohach, S. P. Nair, K. Hiramatsu, E. Jouvin-Marche, and R. Mariuzza. 2004. Standard nomenclature for the superantigens expressed by *Staphylococcus*. *J. Infect. Dis.* **189:**2334–2336.

Lindqvist, R., S. Sylvén, and I. Vågsholm. 2002. Quantitative microbial risk assessment exemplified by *Staphylococcus aureus* in unripened cheese made from raw milk. *Int. J. Food Microbiol.* **78:**155–170.

Liu, H., Z. Du, J. Wang, and R. Yang. 2007. Universal sample preparation method for characterization of bacteria by matrix-assisted laser desorption ionization–time of flight mass spectrometry. *Appl. Environ. Microbiol.* **73:**1899–1907.

Medina, M. B. 2006. Development of a fluorescent latex microparticle immunoassay for the detection of staphylococcal enterotoxin B (SEB). *J. Agric. Food Chem.* **54:**4937–4942.

Melchior, M. B., H. Vaarkamp, and J. Fink-Gremmels. 2006. Biofilms: a role in recurrent mastitis infections? *Vet. J.* **171:**398–407.

Mishra, N. N., W. C. Maki, E. Cameron, R. Nelson, P. Winterrowd, S. K. Rastogi, B. Filanoski, and G. K. Maki. 2008. Ultra-sensitive detection of bacterial toxin with silicon nanowire transistor. *Lab. Chip* **8:**868–871.

Morandi, S., M. Brasca, R. Lodi, L. Brusetti, C. Andrighetto, and A. Lombardi. 2010. Biochemical profiles, restriction fragment length polymorphism (RFLP), random amplified polymorphic DNA (RAPD) and multilocus variable number tandem repeat analysis (MLVA) for typing *Staphylococcus aureus* isolated from dairy products. *Res. Vet. Sci.* **88:**427–435.

Moret-Stalder, S., C. Fournier, R. Miserez, S. Albini, M. G. Doherr, M. Reist, W. Schaeren, M. Kirchhofer, H. U. Graber, A. Steiner, and T. Kaufmann. 2008. Prevalence study of *Staphylococcus aureus* in quarter milk samples of dairy cows in the Canton of Bern, Switzerland. *Prev. Vet. Med.* **88:**72–76.

O'Grady, J., L. Katrina, B. Glynn, T. J. Smith, T. Barry, and M. Maher. 2009. tmRNA—a novel high-copy-number RNA diagnostic target—its application for *Staphylococcus aureus* detection using real-time NASBA. *FEMS Microbiol. Lett.* **301:** 218–223.

Ostyn, A., M. L. De Buyser, F. Gullier, J. Groult, B. Félix, S. Salah, G. Delmas, and J. A. Hennekinne. 2010. First evidence of a food poisoning outbreak due to staphylococcal enterotoxin type E, France, 2009. *Euro Surveill.* **15**(13): pii=19528.

Panneerseelan, L., and P. M. Muriana. 2009. An immunomagnetic PCR signal amplification assay for sensitive detection of *Staphylococcus aureus* enterotoxins in foods. *J. Food Prot.* **72:**2538–2546.

Park, C. E., M. Akhtar, and M. K. Rayman. 1994. Evaluation of a commercial enzyme immunoassay kit (RIDASCREEN) for detection of staphylococcal enterotoxins A, B, C, D, and E in foods. *Appl. Environ. Microbiol.* **60:**677–681.

Phuketes, P., G. F. Browning, G. Anderson, and P. D. Mansell. 2003. Multiplex polymerase chain reaction as a mastitis screening test for *Staphylococcus aureus, Streptococcus agalactiae, Streptococcus dysgalactiae* and *Streptococcus uberis* in bulk milk samples. *J. Dairy Res.* **70:**149–155.

Poli, A. E. Guglielmini, S. Sembeni, M. Spiazzi, F. Dellaglio, F. Rossi, and S. Torriani. 2007. Detection of *Staphylococcus aureus* and enterotoxin genotype diversity in Monte Veronese, a protected designation of origin Italian cheese. *Lett. Appl. Microbiol.* **45:**529–534.

Quiel, A., B. Jürgen, G. Piechotta, A. P. Le Foll, A. K. Ziebandt, C. Kohler, D. Köster, S. Engelmann, C. Erck, R. Hintsche, J. Wehland, M. Hecker, and T. Schweder. 2010. Electrical protein array chips for the detection of staphylococcal virulence factors. *Appl. Microbiol. Biotechnol.* **85:**1619–1627.

Rajakaruna, L., G. Hallas, L. Molenaar, D. Dare, H. Sutton, V. Encheva, R. Culak, I. Innes, G. Ball, A. M. Sefton, M. Eydman, A. M. Kearns, and H. N. Shah. 2009. High throughput identification of clinical isolates of *Staphylococcus aureus* using MALDI-TOF-MS of intact cells. *Infect. Genet. Evol.* **9:**507–513.

Saei, H. D., M. Ahmadi, K. Mardani, and R. A. Batavani. 2009. Molecular typing of *Staphylococcus aureus* isolated from bovine mastitis based on polymorphism of the coagulase gene in the north west of Iran. *Vet. Microbiol.* **137:**202–206.

Sasaki, T., Y. Terano, T. Shibata, H. Kawamoto, T. Kuzuguchi, E. Kohyama, T. Watanabe, T. Ohyama, and M. Gemba. 2005. Establishment of highly specific and quantitative immunoassay systems for staphylococcal enterotoxin A, B, and C using newly-developed monoclonal antibodies. *Microbiol. Immunol.* **49:**589–597.

Schmid, D., R. Fretz, P. Winter, M. Mann, G. Höger, A. Stöger, W. Ruppitsch, J. Ladstätter, N. Mayer, A. de Martin, and F. Allerberger. 2009. Outbreak of staphylococcal food intoxication after consumption of pasteurized milk products, June 2007, Austria. *Wien. Klin. Wochenschr.* **121:**125–131.

Schmitt, M., U. Schuler-Schmid, and W. Schmidt-Lorenz. 1990. Temperature limits of growth, TNase and enterotoxin production of *Staphylococcus aureus* strains isolated from foods. *Int. J. Food Microbiol.* **11:**1–19.

Seng, P., M. Drancourt, F. Gouriet, B. La Scola, P. E. Fournier, J. M. Rolain, and D. Raoult. 2009. Ongoing revolution in bacteriology: routine identification of bacteria by matrix-assisted laser desorption ionization time-of-flight mass spectrometry. *Clin. Infect. Dis.* **149:**543–551.

Smith, E. M., L. E. Green, G. F. Medley, H. E. Bird, L. K. Fox, Y. H. Schukken, J. V. Kruze, A. J. Bradley, R. N. Zadoks, and C. G. Dowson. 2005. Multilocus sequence typing of intercontinental bovine *Staphylococcus aureus* isolates. *J. Clin. Microbiol.* **43:**4737–4743.

Soejima, T., E. Nagao, Y. Yano, H. Yamagata, H. Kagi, and K. Shinagawa. 2007. Risk evaluation for staphylococcal food poisoning in processed milk produced with skim milk powder. *Int. J. Food Microbiol.* **115:**29–34.

Spanu, T., E. De Carolis, B. Fiori, M. Sanguinetti, T. D'Inzeo, G. Fadda, and B. Posteraro. 2011. Evaluation of matrix-assisted laser desorption ionization-time-of-flight mass spectrometry in comparison to *rpoB* gene sequencing for species identification of bloodstream infection staphylococcal isolates. *Clin. Microbiol. Infect.* **17:**44–49.

Stevens, K. A., and L. A. Jaykus. 2004. Bacterial separation and concentration from complex samples matrices: a review. *Crit. Rev. Microbiol.* **30:**7–24.

van Belkum, A., N. J. Verkaik, C. P. de Vogel, H. A. Boelens, J. Verveer, J. L. Nouwen, H. A. Verbrugh, and H. F. Wertheim. 2009. Reclassification of *Staphylococcus aureus* in nasal carriage types. *J. Infect. Dis.* **199:**1820–1826.

Vanderhaeghen, W., T. Cerpentier, C. Adriaensen, J. Vicca, K. Hermans, and P. Butaye. 2010. Methicillin-resistant *Staphylococcus aureus* (MRSA) ST398 associated with clinical and subclinical mastitis in Belgian cows. *Vet. Microbiol.* doi: 10.1016/j.vet-mic.2009.12.044.

Wang, L., W. Zhao, M. B. O'Donoghue, and W. Tan. 2007. Fluorescent nanoparticles for multiplexed bacteria monitoring. *Bioconjug. Chem.* **8:** 297–301.

Zadoks, R. N., W. B. van Leeuwen, D. Kreft, L. K. Fox, H. W. Barkema, Y. H. Schukken, and A. Van Belkum. 2002. Comparison of *Staphylococcus aureus* isolates from bovine and human skin, milking equipment, and bovine milk by phage typing, pulsed-field gel electrophoresis, and binary typing. *J. Clin. Microbiol.* **40:**3894–3902.

Zangerl, P., and W. Ginzinger. 2006. Staphylococcus aureus *in Cheese—a Review.* Trade Association of Food Industry, Bundesanstalt für Alpenländische Milchwirtschaft, Rotholz, Austria.

Zottola, E. A., and L. B. Smith. 1991. Pathogens in cheese. *Food Microbiol.* **8:**171.

CRONOBACTER SPECIES IN POWDERED INFANT FORMULA

K. N. Jordan and S. Fanning

21

Cronobacter species, formerly known as *Enterobacter sakazakii*, was classified into a new genus comprising six species, one of which has three subspecies (Iversen et al., 2007, 2008). These six species include *Cronobacter sakazakii, C. dublinensis, C. malonaticus, C. muytjensii, C. turiensis*, and genomospecies I. They are gram-negative bacteria, members of the family *Enterobacteriaceae*. Although these bacteria are widespread in the environment, a specific reservoir remains to be identified. Nonetheless, *Cronobacter* species have been detected from a wide variety of sources, including different food types, animal feed, and the environment (Kandhai et al., 2004). The frequency at which *Cronobacter* species can be isolated varies with the corresponding source but in general is low. A notable exception, however, is in the case of dried-food matrices including powdered infant formula (PIF) and other similar medically formulated foods. The bacteria appear to be adapted to survive in low water activity matrices/environments, sometimes extending over considerable time periods (Edelson-Mammel et al., 2005).

Infection with *Cronobacter* species is linked with cases of necrotizing enterocolitis, bacteremia, and meningitis in vulnerable neonates. Based on the available data, mortality rates have been estimated to be up to 80% (Lai, 2001). PIF and the environment in which this food is produced have been epidemiologically identified as an important vehicle or source for transmission to neonates. The clinical nature of the infection in this vulnerable group of infants, allied with the negative commercial impacts linked to a particular PIF brand and its manufacturer, represents unique challenges to the infant formula industry and to public health professionals alike.

MICROBIOLOGICAL CRITERIA

Cronobacter species are occasional contaminants of PIF, and consequently, this particular food type presents a significant health risk to vulnerable neonates (Food and Agriculture Organization/World Health Organization [FAO/WHO], 2004, 2006; Chap et al., 2009). Microbiological criteria are designed to control the concentration and prevalence of bacteria in foods. These criteria specify the microorganism of concern, the analytical

K. N. Jordan, Teagasc, Moorepark Food Research Centre, Moorepark, Fermoy, Co. Cork, Ireland. *S. Fanning*, WHO-Collaborating Centre for Research, Reference & Training on *Cronobacter*, Centre for Food Safety, UCD Veterinary Sciences Centre, University College Dublin, Belfield, Dublin 4, Ireland.

Rapid Detection, Identification, and Quantification of Foodborne Pathogens, ed. by J. Hoorfar

method for their detection/quantification, a sampling plan defining the number of samples to be taken and the size of the analytical unit, the microbiological limits deemed appropriate at a specified point in the food chain, and the number of analytical units that should conform to those limits (Codex Alimentarius Commission, 2003).

The sampling plans specified in microbiological criteria for *Cronobacter* species can be adjusted in a number of ways to achieve an acceptable level of control of the pathogen in PIF. For example, the number of samples and/or the analytical unit size and/or the number of analytical units that must conform to a microbiological limit can be increased. These changes in turn increase the stringency of the microbiological criteria and thereby reduce the risk of *Cronobacter* species contaminating the PIF available on the market.

In the past few years, FAO/WHO via their Joint Meetings on Microbiological Risk Assessment developed a risk assessment model for *Cronobacter* species in PIF (Food and Agriculture Organization/World Health Organization, 2008). The model is freely available, and the PIF industry and other interested food producers are encouraged to make use of this analytical tool.

The model allows risk managers to explore the impact on the relative risk of different microbiological criteria in the form of two-class sampling plans, either in isolation or in combination with different PIF handling and preparation strategies.

The user is required to specify the within- and between-batch prevalence of *Cronobacter* species in PIF. If sampling plans are specified, their impact on the prevalence of positive batches is calculated assuming that the tested batches that are found positive are removed from the market. Finally, the user can also specify different handling and preparation practices for the PIF remaining on the market after testing and evaluate the impact different practices may have on the reduction of risk. The model calculates an overall relative risk reduction compared to a baseline scenario that is predefined but can be modified. In this way, the impact of implementing a PIF sampling plan and/or guidelines on safe PIF preparation practices can be evaluated.

CONVENTIONAL MICROBIOLOGICAL METHODS TO DETECT *CRONOBACTER* SPECIES

An outbreak of *Cronobacter* spp. in April 2001 was the catalyst for the early development of culture methods to detect this pathogen (Himelright et al., 2002). This prompted the U.S. Food and Drug Administration (FDA) to devise a detection method for *Cronobacter* (Food and Drug Administration, 2002, 2009). The method gained widespread acceptance, despite the fact that it was never validated or included in the *Bacteriological Analytical Manual*. This protocol was based on a method developed earlier to enumerate *Enterobacteriaceae* in dried foods (Mossel, 1970) and PIF (Muytjens et al., 1988). A schematic representation of the steps involved is presented in Fig. 1a. More recently, the International Organization for Standardization (ISO) and the International Dairy Federation devised a protocol, outlined in Fig. 2a, specifically designed to improve the detection of *Cronobacter* species in milk-based powdered formula (ISO, 2006). In this protocol, buffered peptone water was used to preenrich samples, followed by incubation in a modified lauryl sulfate broth containing vancomycin. Chromogenic agars developed during this time were then used to aid in the identification, followed by confirmation using biochemical tests. The method takes about 5 days for a positive confirmation.

FIGURE 1 Flow diagram for detection of *Cronobacter* species by the original FDA method (a) and the revised FDA method (b). EE broth, *Enterobacteriaceae* enrichment broth; PBS, phosphate-buffered saline; TSA, tryptic soy agar; VRBG, violet red bile agar with glucose.

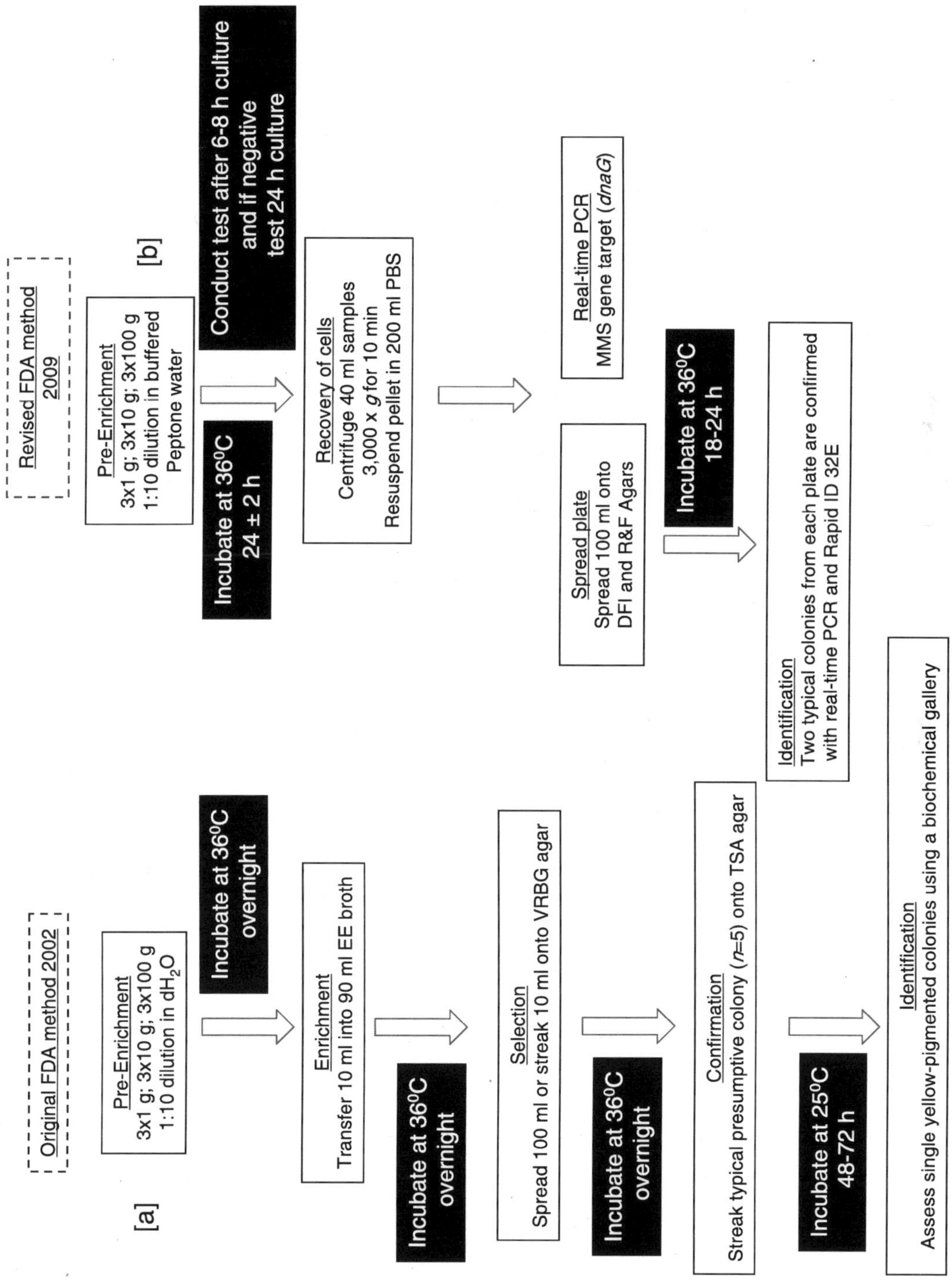

[a]
Original FDA method 2002
Pre-Enrichment
3x1 g; 3x10 g; 3x100 g
1:10 dilution in dH_2O
Incubate at 36^{0}C overnight
Enrichment
Transfer 10 ml into 90 ml EE broth
Incubate at 36^{0}C overnight
Selection
Spread 100 ml or streak 10 ml onto VRBG agar
Incubate at 36^{0}C overnight
Confirmation
Streak typical presumptive colony (n=5) onto TSA agar
Incubate at 25^{0}C 48-72 h
Identification
Assess single yellow-pigmented colonies using a biochemical gallery
[b]
Revised FDA method 2009
Pre-Enrichment
3x1 g; 3x10 g; 3x100 g
1:10 dilution in buffered Peptone water
Incubate at 36^{0}C 24 ± 2 h
Conduct test after 6-8 h culture and if negative test 24 h culture
Recovery of cells
Centrifuge 40 ml samples
3,000 x *g* for 10 min
Resuspend pellet in 200 ml PBS
Real-time PCR
MMS gene target (*dnaG*)
Spread plate
Spread 100 ml onto DFI and R&F Agars
Incubate at 36^{0}C 18-24 h
Identification
Two typical colonies from each plate are confirmed with real-time PCR and Rapid ID 32E

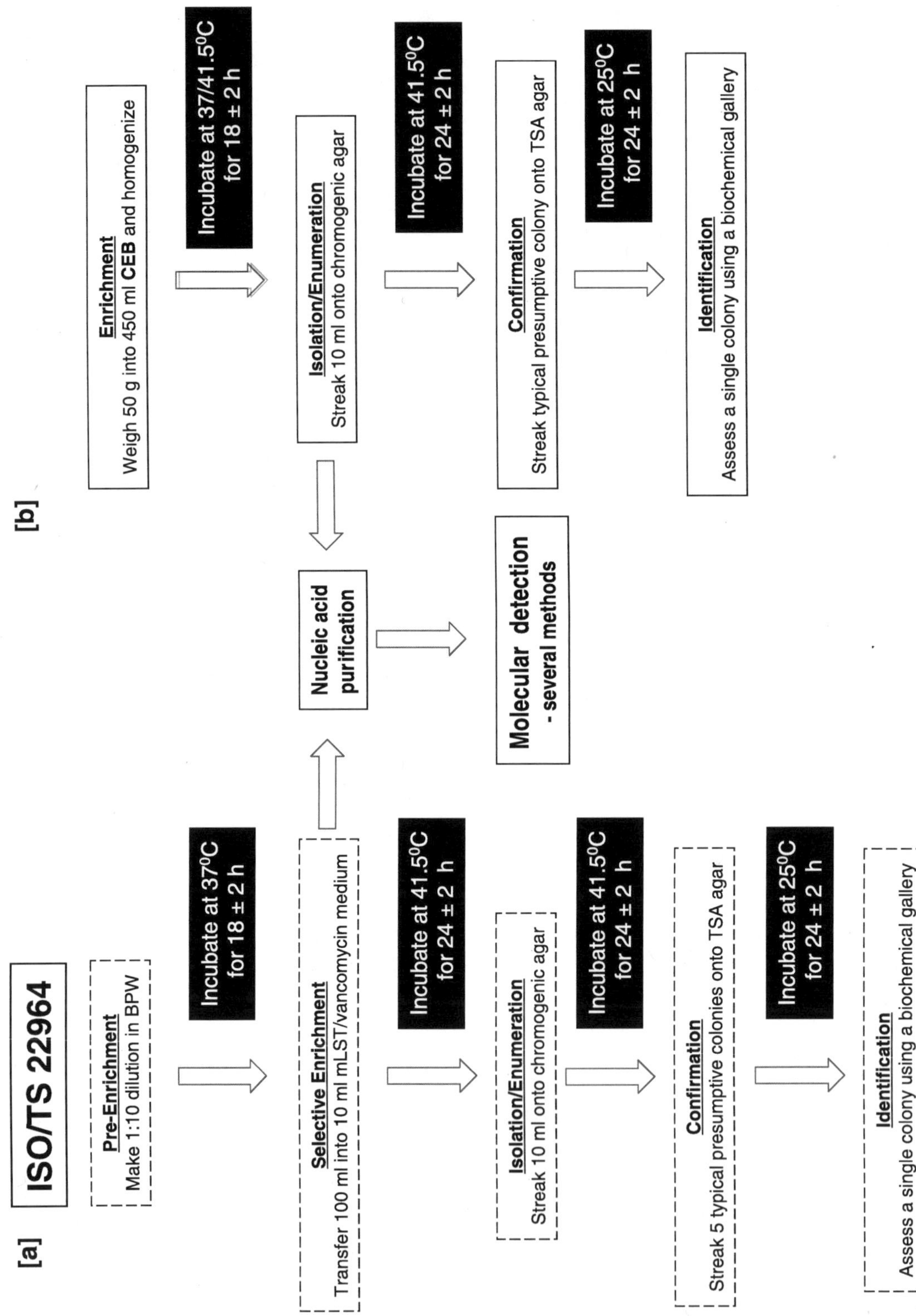
[a]
ISO/TS 22964
Pre-Enrichment
Make 1:10 dilution in BPW
Incubate at 37°C for 18 ± 2 h
Selective Enrichment
Transfer 100 ml into 10 ml mLST/vancomycin medium
Incubate at 41.5°C for 24 ± 2 h
Isolation/Enumeration
Streak 10 ml onto chromogenic agar
Incubate at 41.5°C for 24 ± 2 h
Confirmation
Streak 5 typical presumptive colonies onto TSA agar
Incubate at 25°C for 24 ± 2 h
Identification
Assess a single colony using a biochemical gallery
Nucleic acid purification
Molecular detection - several methods
[b]
Enrichment
Weigh 50 g into 450 ml CEB and homogenize
Incubate at 37/41.5°C for 18 ± 2 h
Isolation/Enumeration
Streak 10 ml onto chromogenic agar
Incubate at 41.5°C for 24 ± 2 h
Confirmation
Streak typical presumptive colony onto TSA agar
Incubate at 25°C for 24 ± 2 h
Identification
Assess a single colony using a biochemical gallery

RAPID DETECTION PROTOCOLS FOR *CRONOBACTER* SPECIES

Generally, rapid methods are understood to mean the application of a molecular or instrument-based technique to enable pathogen detection in hours rather than days. In common with *Listeria monocytogenes* (see chapter 18), rapid detection of *Cronobacter* species, in this context, can be interpreted to mean improving conventional plating methods, with or without molecular methods for confirmation.

Improved Conventional Methods

Muytjens et al. (1984) reported that *Cronobacter* species could metabolize the substrate *o*-nitrophenyl-β-D-glucopyranoside (ONPG), thereby producing a yellow color following hydrolysis. This relatively unique biochemical characteristic, based on the enzymatic activity of β-glucosidase, was used in the formulation of many of the currently available commercial agars designed for *Cronobacter* species. Inclusion of the chromogenic substrate 5-bromo-4-chloro-3-indolyl-D-β-glucopyranoside (β-XGlu) can be used to determine whether an organism possesses the corresponding β-glucosidase activity. Once this substrate is cleaved, a blue-green colony is produced that is clearly distinguished on the agar medium. This approach forms the basis for commercially available agars such as *Enterobacter sakazakii* isolation agar, used as part of the ISO/TS 22964 protocol (Fig. 2a) and Druggan-Forsythe-Iversen agar (Iversen et al., 2004). The latter was modified subsequently as described by Druggan and Iversen (Druggan and Iversen, 2009). The development of chromogenic agar was a considerable advance towards the provision of quick analytical protocols to aid in the reliable detection of *Cronobacter* species. Recently, Lampel and Chen (2009) modified the original FDA method (Fig. 1a) by making use of developments in selective agars and molecular detection strategies for more reliable identification (as outlined in Fig. 1b).

A number of *Cronobacter* isolates that could not be recovered by the ISO/International Dairy Federation protocol were identified, and this prompted a reevaluation of the early steps designed to recover and provide enrichment of the bacterium. In a novel approach, a one-step protocol for the recovery of *Cronobacter* species was reported by O'Brien et al. (2009). In this protocol (Fig. 2b), a combined preenrichment/enrichment broth, known as *Cronobacter* enrichment broth was developed and applied in conjunction with several selective agars. Compared to the other methods outlined, this protocol facilitates a faster detection of *Cronobacter* species. A comprehensive review of these developments in selective enrichment broths to support the reliable recovery of *Cronobacter*, together with a detailed description of selective/differential agars that are now commercially available, is presented elsewhere (Druggan and Iversen, 2009; Healy et al., 2010).

Once obtained, an isolate can be subjected to further characterization by phenotypic and genotypic methods, up to and including genome sequencing (Fig. 3).

Concentration of Cells

Mullane et al. (2006) described the development and application of a simple protocol based on the use of cationic-magnetic beads, designed to capture and concentrate *Cronobacter* organisms from large volumes of PIF suspended in buffered peptone water. This protocol could reliably detect 1 to 5 CFU of *Cronobacter* species in 500-g samples of PIF in less than 24 h. Concentration was followed by direct plating onto commercially available

FIGURE 2 The ISO method for detection and enumeration of *Cronobacter* species. Abbreviations: BPW, buffered peptone water; CEB, *Cronobacter* enrichment broth; mLST, modified lauryl sulfate-tryptose; TSA, tryptic soy agar.

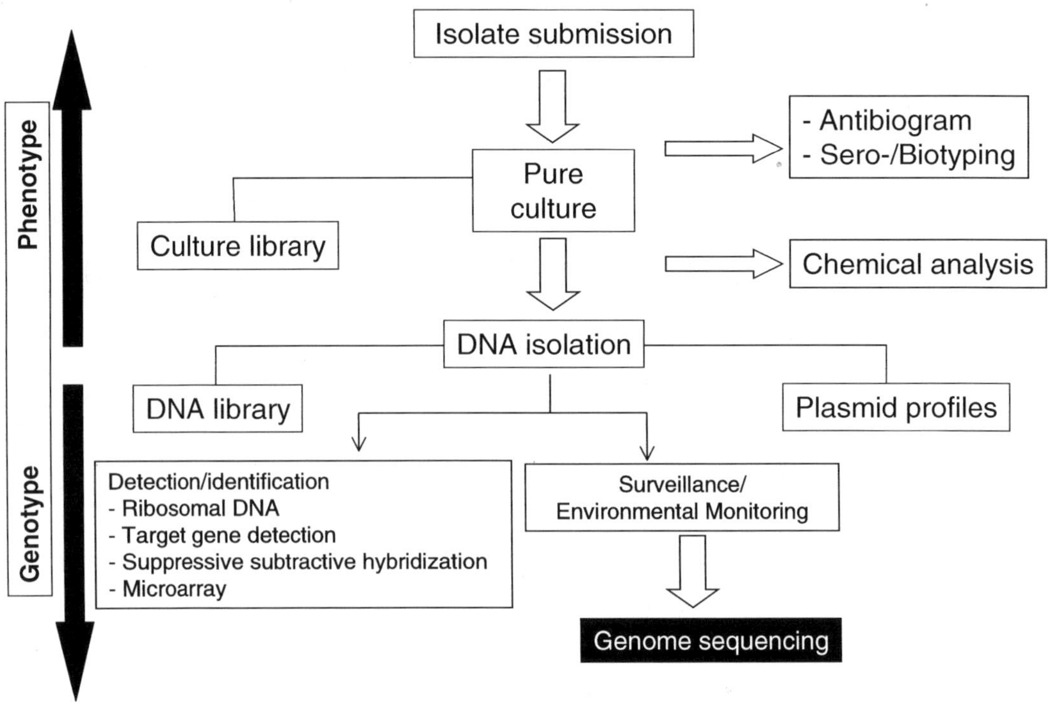

FIGURE 3 Genotypic and phenotypic methods for characterization of *Cronobacter* species.

chromogenic agar together with a molecular confirmatory real-time PCR modified from the original method described by Seo and Brackett (2005).

Molecular Methods

The development of accurate, sensitive, and specific molecular methods for the detection of *Cronobacter* species is important in ensuring that product can be released in a timely manner, without compromising on safety. Compared to traditional methods, molecular approaches, including those making use of PCR technology, have the capacity to speed up the detection process. However, as the detection limit of PCR assays remains high (approximately 10^2 to 10^3 CFU/g), an enrichment step continues to be essential. Similar to detection strategies used for *L. monocytogenes*, a combination of traditional culture enrichment and PCR and subsequent amplicon detection may yield the best results.

In general, the first step in any PCR assay is the extraction of the target DNA from the food/environmental matrix (see http://www.protocol-online.org/prot/Microbiology/Bacteria/Nucleic_Acid_Extraction/index.html). As milk powder and other food matrices may contain inhibitors that can directly interfere with the activity of *Taq* DNA polymerase used to amplify the target gene, great care is needed in the application of these protocols. In order to eliminate any possibility of misdiagnosis, the inclusion of an internal control is strongly advised.

Several PCR detection strategies (using both conventional and real-time PCR approaches) targeting different genes have been developed (Table 1). These are based on the use of single targets such as the 16S-23S rRNA gene intergenic spacer region (ITS) (Liu et al., 2006), the *ompA* gene (Mohan Nair and Venkitanarayanan, 2006), or more than one target gene/region such as the 16S rRNA

TABLE 1 PCR primers and target genes used for detection of *Cronobacter* species by conventional, duplex, and real-time PCR

Primers	Sequence	Target gene(s)	Amplicon size (bp)	Reference
SG-F	GGGTTGTCTGCGAAGCGAA	ITS	282	Liu et al., 2006
SG-R	GTCTTCGTGCTGCGAGTTG			
ESS-F	GGATTTAACCGTGAACTTTTTCC	*ompA*	469	Mohan Nair and Venkitanarayanan, 2006
ESS-R	CGCCAGCGATGTTAGAAGA			
E-F	GGCGGAGCCGAATAACTG	*gluA*	673	Ye et al., 2007
E-R	CGTGCCTGCATGAGAAAA			
MMS-F	GGGATATTGTCCCCTGAAACAG	MMS operon	78	Seo and Brackett, 2005
MMS-R	CGAGAATAAGCCGCGCATT			
Probe	6FAM-AGAGTAGTAGTTGTAGAGGCCGTGCTTCCGAAAG-TAMRA			
ES-16s-F	CAAGTCGAACGGTAACAGGG	16S rRNA	148	Malorny and Wagner, 2005
ES-16s-R	GTCCCCCACTTTGGTCCG			
Probe	FAM-TGCTGCTCTGCTGACGAGTGGC-DarkQuencher			
F	GGGTTGTCTGCGAAAGCGAA	ITS	282	Zhou et al., 2008
R	GTCTTCGTGCTGCGAGTTTG			
F	GGATTTAACGTGAACTTTTCC		469	
R	CGCCAGCGATGTTAGAAGA	*ompA*		
616-V	AGAGTTTGATYMTGGCTC	16s rRNA	1,500	Lehner et al., 2004
630-R	CAKAAAGGAGGTGATCC			
ESAG-F	TGAAAGCAATCGACAAGAAG	α-Glucosidase	1,680	Lehner et al., 2006
ESAG-R	ACTCATTACCCCTCCTGATG			

and α-1,4-glucosidase (*gluA*) genes (Malorny and Wagner, 2005; Lehner et al., 2004, 2006), the *ompA* gene and ITS region (Zhou et al., 2008), the *gluA* and *ompA* genes (Ye et al., 2007), and the *dnaG* gene within the macromolecular synthesis (MMS) operon (Seo and Brackett, 2005). A comparison of many of these methods (Ye et al., 2009) reported 10 of 243 samples positive for *Cronobacter* species by the ISO method and by all the PCR methods tested. Four false positives were detected by PCR targeting the ITS region, one false positive was obtained by PCR targeting the *ompA* gene, and two false-negative results were obtained with the ISO method. Based on these and other data, PCR methods are now considered as alternatives to traditional ISO methods, although standardization and acceptance by standardization bodies remain to be completed.

Several commercial kits are now available and include PCR assays.

CHARACTERIZATION OF *CRONOBACTER* SPECIES BY MOLECULAR SUBTYPING

Molecular subtyping methods have been used to fingerprint clinical *Enterobacteriaceae* isolates associated with nosocomial infections and foodborne outbreaks (Swaminathan et al., 2001; Block et al., 2002). These protocols contribute towards extending our understanding of microbial ecology and epidemiology, and they are regarded as useful tools to monitor foodborne disease. There are three main approaches to the use of molecular subtyping, as follows.

1. Quality control in food production wherein bacterial strains are used, such as the subtyping of probiotic bacteria added to foods

2. Monitoring the production environment to identify areas of contamination and identification of prevalent bacterial strain types (that may be adapted to the production environment)

3. Investigation of foodborne outbreaks

The principle of these subtyping methods relies on the fact that identical isolates share highly similar or indistinguishable DNA profile patterns or fingerprints, allowing individual strains to be linked in outbreak investigations while excluding other non-case-related isolates.

Pulsed-field gel electrophoresis (PFGE) has been used to investigate the relationship between *Cronobacter* strains isolated from PIF and those cultured from infected infants (Smeets et al., 1998; Block et al., 2002). Other molecular subtyping protocols, including ribotyping and random amplification of polymorphic DNA have also been applied (Nazarowec-White and Farber, 1999). A summary of the currently available molecular subtyping methods is shown in Table 2; many of these have been applied to *Cronobacter* species. Despite the frequent use of PFGE in clinical settings, the food industry has been slower to embrace this approach due to the need for high-throughput, cost-effective test protocols. Nonetheless, PFGE is beginning to form an important part of the food industry's surveillance plan, whereby it can provide a useful

TABLE 2 Summary of molecular subtyping techniques applicable to *Cronobacter* species[a]

Methods	Techniques
First-generation methods	Plasmid DNA profiling
	Restriction digests of purified plasmids
Second-generation methods	Ribotyping
Third-generation methods	PFGE
	PCR-based amplification:
	Rapid amplification of polymorphic DNA
	PCR ribotyping
	Repetitive PCR
	Enterobacterial repetitive intergenic consensus PCR
	Amplified fragment length polymorphism
Fourth-generation methods	Multilocus variable-number tandem-repeat analysis
	Multilocus sequence typing
	DNA sequencing

[a] Modified from Dijkshoorn and Towner, 2003.

addition to hazard analysis of critical control point plans aimed at limiting transmission of spoilage and pathogenic foodborne organisms. Mullane et al. (2007) conducted a surveillance program to identify points of contamination with *E. sakazakii* (*Cronobacter* species) within a PIF manufacturing facility, investigate clonal persistence among strains, and identify possible dissemination routes of *Cronobacter* species. Indistinguishable PFGE profiles were identified among *Cronobacter* isolates cultured from the manufacturing environment, from sampling points, along the processing chain, and from intermediate and final products. These findings demonstrate that the environment of the PIF manufacturing facility represents an important contamination route of the final product. In other related studies, similar approaches were adopted, wherein molecular subtyping was used as an additional aid to facilitate the control of the microbial ecology of low-moisture manufacturing environments (Mullane et al., 2008a; Baumgartner et al., 2009).

Using a PCR-based approach, a semiautomated subtyping method making use of repetitive sequences found dispersed throughout bacterial chromosomes was described (Healy et al., 2008). Variable tandem repeat sequences (Mullane et al., 2008b) and a multilocus sequence typing-based method (Kuhnert et al., 2009) have also been developed as alternative approaches to PFGE.

OTHER PATHOGENS

While *Cronobacter* species remain one of the key organisms of most concern to the PIF and other low-moisture food manufacturers, other pathogens may also be present. These include typical *Enterobacteriaceae* such as *Pantoea* species, *Citrobacter* species, *Klebsiella* species, and *Enterobacter* species, in addition to *Salmonella* species, some strains of *Bacillus* species, and *Staphylococcus aureus*, all of which are monitored closely by traditional microbiological culture-based methods. The approaches outlined above for detecting *Cronobacter* species could also be applied in a similar manner to these pathogens. In fact, it would be possible to combine the detection of all pathogens by using a carefully designed multiplex-based PCR approach.

FUTURE PERSPECTIVES

Recently, the genome sequence of *Cronobacter sakazakii* BAA-894 was published (Kucerova et al., 2010), and subsequent mining of these data may facilitate the development of novel PCR and subtyping approaches along with the refinement of existing methods. In the future, it is to be expected that these developments will lead to an improvement in molecular analytical methods and a more widespread acceptance and use of these protocols for detection and subtyping of *Cronobacter* species and other food safety bacterial pathogens of concern. Developments in the provision of user-friendly microarray methodology for detection of multiple pathogens on a single platform (Wang et al., 2009) will result in more cost-effective pathogen detection protocols. Allied with these developments, validation and standardization of these newer methodologies will increase their acceptance and use among those in the broader food industry and the regulatory bodies.

ACKNOWLEDGMENTS

We acknowledge the assistance of Niall Mullane (Danone, Wexford, Ireland) and Wayne Anderson (Food Safety Authority of Ireland) in the preparation of the manuscript.

REFERENCES

Baumgartner, A., M. Grand, M. Liniger, and C. Iversen. 2009. Detection and frequency of *Cronobacter* spp. (*Enterobacter sakazakii*) in different categories of ready-to-eat foods other than infant formula. *Int. J. Food Microbiol.* **136:**189–192.

Block, C., O. Peleg, N. Minster, B. Bar-Oz, A. Simhon, I. Arad, and M. Shapiro. 2002. Cluster of neonatal infections in Jerusalem due to unusual biochemical variant of *Enterobacter sakazakii*. *Eur. J. Clin. Microbiol. Infect. Dis.* **21:**613–616.

Chap, J., P. Jackson, R. Siqueira, N. Gaspar, C. Quintas, J. Park, T. Osaili, R. Shaker, Z. Jaradat, S. H. P. Hartantyo, N. Abdullah Sani, S. Estuningsih, and S. J. Forsythe. 2009. International survey of *Cronobacter sakazakii* and other *Cronobacter* spp. in follow up formulas and infant foods. *Int. J. Food Microbiol.* **136:**185–188.

Codex Alimentarius Commission. 2003. Principles for the establishment and application of microbiological criteria for foods (CAC/GL 21-1997), p. 45–52. *In Food Hygiene Basic Texts*, 3rd ed. Food and Agriculture Organization/World Health Organization, Rome, Italy.

Dijkshoorn, L., and K. Towner. 2003. An introduction to the generation and analysis of microbial typing data, p. 1–30. *In* L. Dijkshoorn, K. Towner, and M. Strulens (ed.), *New Approaches for the Generation and Analysis of Microbial Typing Data.* Elsevier BV, Amsterdam, The Netherlands.

Druggan, P., and C. Iversen. 2009. Culture media for the isolation of *Cronobacter* spp. *Int. J. Food Microbiol.* **136:**169–178.

Edelson-Mammel, S. G., M. K. Porteous, and R. L. Buchanan. 2005. Survival of *Enterobacter sakazakii* in a dehydrated powdered infant formula. *J. Food Prot.* **68:**1900–1902.

Food and Agriculture Organization/World Health Organization. 2004. Enterobacter sakazakii *and Other Micro-Organisms in Powdered Infant Formula: Meeting Report. Microbiological Risk Assessment Series, No. 6.* Food and Agriculture Organization/World Health Organization, Rome, Italy. ftp://ftp.fao.org/docrep/fao/007/y5502e/y5502e00.pdf. Accessed July 2010.

Food and Agriculture Organization/World Health Organization. 2006. Enterobacter sakazakii *and* Salmonella *in Powdered Infant Formula: Meeting Report. Microbiological Risk Assessment Series, No. 10.* Food and Agriculture Organization/World Health Organization, Rome, Italy. ftp://ftp.fao.org/docrep/fao/007/y5502e/y5502e00.pdf. Accessed July 2010.

Food and Agriculture Organization/World Health Organization. 2008. *Risk Assessment for* Cronobacter sakazakii *in Powdered Infant Formula.* Food and Agriculture Organization/World Health Organization, Rome, Italy. http://www.mramodels.org/ESAK/default.aspx. Accessed July 2010.

Food and Drug Administration. 2002. Isolation and enumeration of *Enterobacter sakazakii* from dehydrated powdered infant formula. *In Bacteriological Analytical Manual*, 8th ed. Food and Drug Administration, Gaithersburg, MD.

Food and Drug Administration. 2009. *Questions and Answers on Method for* E. sakazakii *in Powdered Infant Formula.* Food and Drug Administration, Gaithersburg, MD. http://www.fda.gov/Food/ScienceResearch/LaboratoryMethods/ucm114668.htm. Accessed July 2010.

Healy, B., S. Cooney, S. O'Brien, C. Iversen, P. Whyte, J. Nally, J. J. Callanan, and S. Fanning. 2010. *Cronobacter* (*Enterobacter sakazakii*): an opportunistic foodborne pathogen. *Foodborne Pathog. Dis.* **7:**339–350.

Healy, B., N. Mullane, V. Collin, S. Mailler, C. Iversen, S. Chatellier, M. Storrs, and S. Fanning. 2008. Evaluation of an automated repetitive sequence-based PCR system for subtyping *Enterobacter sakazakii. J. Food Prot.* **71:**1273–1278.

Himelright, I., E. Harris, V. Lorch, M. Anderson, T. Jones, A. Craig, M. Kuehnert, T. Forster, M. Arduino, B. Jensen, and D. Jernigan. 2002. *Enterobacter sakazakii* infections associated with the use of powdered infant formula—Tennessee 2001. *JAMA* **17:**2204–2205.

International Organization for Standardization. 2006. *Milk and Milk Products—Detection of* Enterobacter sakazakii. *Technical Specification ISO/TS 22964. ISO/TS22964:2006(E) and IDF/RM 210: 2006(E)*, 1st ed. International Organization for Standardization, Geneva, Switzerland.

Iversen, C., M. Lane, and S. J. Forsythe. 2004. The growth profile, thermotolerance and biofilm formation of *Enterobacter sakazakii* grown in infant milk formula. *Lett. Appl. Microbiol.* **38:**378–382.

Iversen, C., A. Lehner, N. Mullane, J. Marugg, S. Fanning, R. Stephan, and H. Joosten. 2007. Identification of *Cronobacter* spp. (*Enterobacter sakazakii*). *J. Clin. Microbiol.* **45:**3814–3816.

Iversen, C., N. Mullane, B. McCardell, B. D. Tall, A. Lehner, S. Fanning, R. Stephan, and H. Joosten. 2008. *Cronobacter* gen. nov., a new genus to accommodate the biogroups of *Enterobacter sakazakii*, and proposal of *Cronobacter sakazakii* gen. nov., comb. nov., *Cronobacter malonaticus* sp. nov., *Cronobacter turicensis* sp. nov., *Cronobacter muytjensii* sp. nov., *Cronobacter dublinensis* sp. nov., *Cronobacter* genomospecies 1, and of three subspecies, *Cronobacter dublinensis* subsp. *dublinensis* subsp. nov., *Cronobacter dublinensis* subsp. *lausannensis* subsp. nov. and *Cronobacter dublinensis* subsp. *lactaridi* subsp. nov. *Int. J. Syst. Evol. Microbiol.* **58:**1442–1447.

Kandhai, M. C., M. Reij, L. G. M. Gorris, O. Guillaume-Gentil, and M. van Schothorst. 2004. Occurrence of *Enterobacter sakazakii* in food production environments and households. *Lancet* **363:**39–40.

Kucerova, E., S. W. Clifton, X. Q. Xia, F. Long, S. Porwollik, L. Fulton, C. Fronick, P. Minx, K. Kyung, W. Warren, R. Fulton, D. Feng, A. Wollam, N. Shah, V. Bhonagiri, W. E. Nash, K. Hallsworth-Pepin, R. K. Wilson, M. McClelland, and S. J. Forsythe. 2010. Genome sequence of *Cronobacter sakazakii* BAA-894 and comparative genomic hybridization analysis with other *Cronobacter* species. *PLoS ONE* **5:**e9556.

Kuhnert, P., B. M. Korczak, R. Stephan, H. Joosten, and C. Iversen. 2009. Phylogeny and prediction of genetic similarity of *Cronobacter* and related taxa by multilocus sequence analysis (MLSA). *Int. J. Food Microbiol.* **136:**152–158.

Lai, K. K. 2001. *Enterobacter sakazakii* infections among neonates, infants, children, and adults. Case reports and a review of the literature. *Medicine* (Baltimore) **80:**113–122.

Lampel, K. A., and Y. Chen. 2009. Method for the isolation and detection of *Enterobacter sakazakii* (Cronobacter) from powdered infant formula. *Int. J. Food Microbiol.* **136:**179–184.

Lehner, A., T. Tasara, and R. Stephan. 2004. 16S rRNA gene based analysis of *Enterobacter sakazakii* strains from different sources and development of a PCR assay for identification. *BMC Microbiol.* **4:**43.

Lehner, A., S. Nitzsche, P. Breeuwer, B. Diep, K. Thelen, and R. Stephan. 2006. Comparison of two chromogenic media and evaluation of two molecular based identification systems for *Enterobacter sakazakii* detection. *BMC Microbiol.* **6:**15.

Liu, Y., X. Cai, X. Zhang, Q. Gao, X. Yang, Z. Zheng, M. Luo, and X. Huang. 2006. Real time PCR using TaqMan and SYBR Green for detection of *Enterobacter sakazakii* in infant formula. *J. Microbiol. Methods* **65:**21–31.

Malorny, B., and M. Wagner. 2005. Detection of *Enterobacter sakazakii* strains by real-time PCR. *J. Food Prot.* **68:**1623–1627.

Mohan Nair, M. K., and K. S. Venkitanarayanan. 2006. Cloning and sequencing of the *ompA* gene of *Enterobacter sakazakii* and development of an *ompA*-targeted PCR for rapid detection of *Enterobacter sakazakii* in infant formula. *Appl. Environ. Microbiol.* **72:**2539–2546.

Mossel, D. A. A., and M. Alina Ratto. 1970. Rapid detection of sublethally impaired cells of *Enterobacteriaceae* in dried foods. *Appl. Environ. Microbiol.* **20:**273–275.

Mullane, N., B. Healy, J. Meade, P. Whyte, P. G. Wall, and S. Fanning. 2008a. Dissemination of *Cronobacter* spp. (*Enterobacter sakazakii*) in a powdered milk protein manufacturing facility. *Appl. Environ. Microbiol.* **74:**5913–5917.

Mullane, N., J. Murray, D. Drudy, N. Prentice, P. Whyte, P. G. Wall, A. Parton, and S. Fanning. 2006. Detection of *Enterobacter sakazakii* in dried infant milk formula by cationic-magnetic-bead capture. *Appl. Environ. Microbiol.* **72:**6325–6330.

Mullane, N. R., M. Ryan, C. Iversen, M. Murphy, P. O'Gaora, T. Quinn, P. Whyte, P. G. Wall, and S. Fanning. 2008b. Development of multiple-locus variable-number tandem-repeat analysis for the molecular subtyping of *Enterobacter sakazakii*. *Appl. Environ. Microbiol.* **74:**1223–1231.

Mullane, N. R., P. Whyte, P. G. Wall, T. Quinn, and S. Fanning. 2007. Application of pulsed-field gel electrophoresis to characterise and trace the prevalence of *Enterobacter sakazakii* in an infant formula processing facility. *Int. J. Food Microbiol.* **116:**73–81.

Muytjens, H. L., H. Roelofs-Willemse, and G. H. Jaspar. 1988. Quality of powdered substitutes for breast milk with regard to members of the family *Enterobacteriaceae*. *J. Clin. Microbiol.* **26:**743–746.

Muytjens, H. L., J. van der Ros-van de Repe, and H. A. van Druten. 1984. Enzymatic profiles of *Enterobacter sakazakii* and related species with special reference to the alpha-glucosidase reaction and reproducibility of the test system. *J. Clin. Microbiol.* **20:**684–686.

Nazarowec-White, M., and J. M. Farber. 1999. Phenotypic and genotypic typing of food and clinical isolates of *Enterobacter sakazakii*. *J. Med. Microbiol.* **48:**559–567.

O'Brien, S., B. Healy, C. Negredo, S. Fanning, and C. Iversen. 2009. Evaluation of enrichment methods for detection of *Cronobacter* spp. (*Enterobacter sakazakii*) from powdered infant formula. *J. Food Prot.* **72:**1472–1475.

Seo, K. H., and R. E. Brackett. 2005. Rapid, specific detection of *Enterobacter sakazakii* in infant formula using a real-time PCR. *J. Food Prot.* **68:**59–63.

Smeets, L. C., A. Voss, H. L. Muytjens, J. F. G. M. Meis, and W. J. G. Melchers. 1998. Genetische karakterisatie van *Enterobacter sakazakii*-isolaten van Nederlandse patiënten met neonatale meningitis. *Nederlands Tijdschrift voor Medische Microbiologie* **6:**113–115.

Swaminathan, B., T. J. Barrett, S. B. Hunter, R. V. Tauxe, and the CDC PulseNet Task Force. 2001. PulseNet: the molecular subtyping network for foodborne bacterial disease surveillance, United States. *Emerg. Infect. Dis.* **7:**382–389.

Wang, M., B. Cao, Q. Gao, Y. Sun, P. Liu, L. Feng, and L. Wang. 2009. Detection of *Enterobacter sakazakii* and other pathogens associated with infant formula powder by use of a DNA microarray. *J. Clin. Microbiol.* **47:**3178–3184.

Ye, Y., Q. Wu, L. Yao, X. Dong, K. Wu, and J. Zhang. 2009. A comparison of polymerase chain reaction and international organization for standardization methods for determination of *Enterobacter sakazakii* contamination of infant formulas from Chinese mainland markets. *Foodborne Pathog. Dis.* **6:**1229–1234.

Ye, Y.-W., Q.-P. Wu, W.-P. Guo, J.-M. Zhang, and X.-H. Dong. 2007. Rapid detection for *Enterobacter sakazakii* based on species-specific PCR in powdered milks. *Microbiology* **34:**1192–1197.

Zhou, Y., Q. Wu, X. Xu, X. Yang, Y. Ye, and J. Zhang. 2008. Development of an immobilization and detection method of *Enterobacter sakazakii* from powdered infant formula. *Food Microbiol.* **25:** 648–652.

FRESH PRODUCE, SEAFOOD, AND WATER

V

PATHOGEN TESTING IN FRESH PRODUCE AND IRRIGATION WATER

C. H. Rambo and S. D. Pillai

22

The reality today is that fresh produce is a food item that is viewed as highly vulnerable to pathogen contamination. There is a high risk of contamination because the produce is grown under natural conditions, packed, and consumed without extensive postharvest treatments or cooking. Fresh produce, by virtue of its cultivation, handling, and consumption practices, is prone to pathogen contamination and therefore becomes a vehicle for widespread foodborne outbreaks. The focus of this chapter is to highlight the challenges associated with monitoring for pathogens in fresh produce. We discuss some of the contemporary approaches of how certain regions in the United States are attempting to reduce the potential for outbreaks from fresh produce. The chapter includes a description of the key pathogens that have been associated with produce-related outbreaks and provides an overview of some of the indicator organisms that can be used to screen for the presence of fecal contamination.

C. H. Rambo and S. D. Pillai, Food Safety and Environmental Microbiology Program, Texas A&M University, Room 418B Kleberg Center, MS 2472, College Station, TX 77843-2472.

GLOBALIZATION OF FRESH FRUIT AND VEGETABLE SUPPLIES AND FOODBORNE ILLNESSES

The globalization of foods has resulted in fresh produce grown in one country being consumed in another country. This change can be attributed to rising consumer incomes, international trade agreements, and improved technologies. U.S. fresh fruit and vegetable imports between 1990–1992 and 2004–2006 increased from $2.7 billion to $7.9 billion (Huang and Huang, 2007). Tomatoes, peppers, and cucumbers account for 60% of the total volume of fresh vegetables that are imported into the United States from developed countries; Canada and the European Union were among the top sources of U.S. food imports; the greatest growth between 1998 and 2007 was among imports from the developing countries (Brooks et al., 2009). In the last decade, U.S. imports came primarily from Canada, Mexico, the European Union, Australia, Brazil, and China. Imports from China are steadily increasing. Today, 13% of vegetables consumed in the United States are grown outside the United States.

This globalization of food supplies brings along significant challenges, especially as it pertains to maintaining the microbiological safety of fruits and vegetables. This is especially

Rapid Detection, Identification, and Quantification of Foodborne Pathogens, ed. by J. Hoorfar

significant given that there are no harmonized international regulations ensuring the microbiological safety or quality of preharvest farming or postharvest processing practices. There is a significant lack of information globally on the number of outbreaks associated with imported fruits and vegetables. Other than the United States and the European Union, a majority of the countries have no established epidemiological surveillance or outbreak investigations focused on foodborne infections. There are, however, reports about how alfalfa sprout seeds that were produced in Denmark caused illnesses in the United States and Finland (Mahon et al., 1997). An outbreak of *Salmonella enterica* serotype Stanley infections occurred in the United States and Finland in 1995. Alfalfa sprout consumption was the only exposure associated with these *S.* Stanley infections. Alfalfa sprouts eaten by patients in six U.S. states and Finland were traced to seed shipped by a Dutch shipper. This is a classic example of how alfalfa sprouts grown from contaminated seed (produced in one country) caused an international outbreak of more than 200 *S.* Stanley infections in two different countries.

There are numerous examples of how other fresh produce commodities grown in Mexico and South American countries have caused outbreaks in the United States. The highly publicized outbreaks associated with *Cyclospora* and more recent hepatitis A virus (HAV) and *Salmonella* infections associated with green onions and jalapeño peppers are some of these examples.

PRODUCE-RELATED OUTBREAKS

The United States has been in the grip of a number of outbreaks linked to pathogens on produce as diverse as tomatoes, cantaloupe, spinach, and jalapeño peppers. These disease outbreaks continue to take a devastating toll on human health care costs. According to estimates, as many as 199 people were infected by the 2006 *Escherichia coli* O157:H7 contamination of spinach. In 2008, over 1,500 people were affected by *Salmonella* contamination of jalapeño peppers. *Salmonella* serovar Saintpaul contamination of alfalfa sprouts starting in February 2009 sickened over 235 people in over 14 states. These contamination events, outbreaks, recalls, and media exposures continue to economically devastate the fresh produce industry and retail food franchises. In England and Wales, between 1992 and 2006, of 2,274 food-related disease outbreaks, approximately 90 outbreaks were associated with the consumption of prepared salads (Little and Gillespie, 2008).

Federal agencies, state governments, and academic institutions have spent considerable amounts of human and monetary resources in trying to identify the sources of contamination, evaluating different pathogen intervention strategies, and developing extensive industry and consumer outreach activities. Unfortunately, the return on investment that a grower obtains in terms of his/her investment into a particular intervention strategy or program is quite small at the farm level compared to the return on investment at the retail stage. The value of a commodity increases significantly as it moves through the supply chain from the farm to the retail level. Thus, the increased value of the commodity at the retail level provides the necessary impetus for ensuring that the commodity meets the food safety expectations of the customer. The fact that there are published and validated pathogen detection technologies does not automatically imply that the fresh produce industry will adopt these technologies.

It is not uncommon to hear academic scientists lament that the industry is unwilling to adopt rapid molecular pathogen detection technologies. In order to understand who in the fresh produce industry would require rigorous microbial pathogen monitoring and who would be forced to institute these monitoring programs, it is necessary to understand the inherent market forces at play, since they determine who would be willing to invest in education and technology adoption. For example, the distributors and the retail industry have the necessary incentive to invest in education and adoption of new technologies to protect their brand name and customer base and avoid recalls and associated economic ramifications. The United Fresh Produce Association estimated that losses to processors alone reached $50 million to $100 million during the 2006 spinach outbreak.

PRODUCTION CHAINS

The food supply in the United States is one of the safest in the world. Yet, outbreaks related to food products are still occurring. Foodborne diseases cause approximately 76 million illnesses, 325,000 hospitalizations, and 5,000 deaths in the United States per year (Mead et al., 1999). Fresh produce is at risk of getting contaminated with pathogens not only during the preharvest stage but also during the postharvest, packing/further processing, and transportation steps. While the commodity is still in the field, it can get contaminated with microbial pathogens through feces from wildlife, pathogen-contaminated irrigation water, improper manure additions, aerosol contamination, rainfall runoff, and improper human activities. Produce is especially prone to contamination from workers during harvest. Contamination can occur through dirty harvest equipment, dirty rinse water, unclean transport vehicles, infected personnel, lack of personal hygiene, and improper storage. During processing, fresh produce can become contaminated through tainted processing equipment, improper recycling of wash water, and cross-contamination.

In recent years, fresh produce has attracted widespread attention because of the numbers and type of outbreaks that have been attributed to contaminated produce. *Salmonella* has been found on tomatoes, lettuce, sprouts, and cantaloupe. *Shigella* has been found on several vegetables, including lettuce. *E. coli* was also found on lettuce, and enteric viruses have been linked to different kinds of raw produce. *Campylobacter, Listeria*, and a variety of protozoans have also caused produce-associated foodborne outbreaks (Beuchat, 1998).

Even though disease outbreaks associated with fresh produce are becoming commonplace, the United States does not have a mandatory sampling and monitoring system. The only ongoing surveillance program is the U.S. Department of Agriculture's Agricultural Marketing Service Microbiological Data Program (MDP). The MDP was implemented in 2001 and is a national foodborne pathogen database program. It is a collaboration between the federal agencies and state agricultural departments. The MDP "manages the collection, analysis, data entry, and reporting of foodborne pathogens on selected agricultural commodities" (http://www.ams.usda.gov/mdp). Data collection is carried out in 11 states representing about 50% of the nation's population and all regions of the country. Samples are obtained by federal or state employees close to the point of consumption (terminal markets, large chain store, and wholesale distribution centers) where lot numbers and grower/packer/distributor information are still available. The samples are, however, not collected at retail outlets. The data collected through the MDP are available to federal and state public health agencies and industry for food safety decision-making purposes.

In 2009, the MDP was sampling and testing alfalfa sprouts, peanut butter, lettuce (loose and bagged), spinach, hot peppers, green onions, and tomatoes. The types of produce sampled and the pathogens that are screened for vary slightly from year to year. In 2009, the MDP tested for pathogenic *E. coli* (Shiga-toxigenic *E. coli* and enterotoxigenic *E. coli*), *E. coli* O157:H7, and *Salmonella* spp. In 2008, for example, it also tested for *Shigella*. The normal monthly sampling rate is 64 samples per commodity. Samples are tested either at state laboratories or by the Agricultural Marketing Service National Science Laboratory. All isolated pathogens are then sent for pulsed-field gel electrophoresis (PFGE) analysis and the resulting patterns are uploaded into the Centers for Disease Control and Prevention's (CDC's) PulseNet database and matched against human isolates or outbreak patterns. Even though the MDP is considered to be a voluntary data collection program, it is the most extensive monitoring system for fresh produce in the United States.

The U.S. Food and Drug Administration (FDA) in 1998 developed the *Guide To Minimize Microbial Food Safety Hazards for Fresh Fruits and Vegetables* (Food and Drug Administration, 1998). This document outlined a list of Good Agricultural Practices (GAPs) for growing, harvesting, transporting, cooling, packing, and storage of fresh produce. Many retailers started employing third-party verification for suppliers adhering to the GAPs. The GAPs for the most part were common-sense recommendations to prevent the commodities from exposure to microbial pathogens. In 2004, the FDA sent a letter to the tomato and lettuce industries to make them aware of the FDA's concerns regarding outbreaks and to strongly encourage the adherence to the FDA's GAP document. The outbreaks still continued.

In spring 2007, due to the growing number of outbreaks associated with fresh produce grown in California, a group of California handlers of leafy greens established the Leafy Greens Products Handler Marketing Agreement (LGMA) in response to the September 2006 *E. coli* outbreak that was attributed to spinach grown in the Salinas Valley. Since that time, approximately 99% of the fresh produce industry in California has joined the LGMA. The LGMA is a voluntary program and has already been widely accepted by the produce industry, grocers, and food service firms. However, since this was a voluntary program that was developed in California, there are no regulatory requirements for other states in the United States to join the LGMA. However, from a marketing perspective, other states and growers that do not join the LGMA may be at a competitive disadvantage. Presently, only California and Arizona have adopted the LGMA. Florida will probably be the next state adopting the LGMA. Many of the LGMA compliance requirements fall upon growers to implement, as it requires signatory members to source their leafy greens solely from growers found to be in compliance with a set of food safety provisions called "best practices."

Currently, the LGMA has the most specific microbial testing requirements for irrigation water, soil amendments, nonsynthetic crop treatments (e.g., manure, compost teas, and biofertilizers) (Table 1). However, there are no specific microbial standards for the commodities per se.

The CDC has established the Foodborne Diseases Active Surveillance Network, FoodNet, which monitors all foodborne diseases. It is a collaboration between the CDC, the U.S. Department of Agriculture, and the FDA. The network monitors foodborne diseases and conducts related epidemiologic studies designed to help public health officials better understand the epidemiology of foodborne diseases in the United States. FoodNet conducts population-based surveillance in 10 states for laboratory-confirmed cases of infection caused by *Campylobacter, Cryptosporidium, Cyclospora, Listeria, Salmonella, E. coli, Shigella, Vibrio*, and *Yersinia*. Even though this system monitors human infections rather than microbiological loads on foods, it is crucial for linking outbreaks to food sources, such as fresh produce.

GLOBAL REGULATIONS

For the global market, the most important guidelines stem from the Codex Alimentarius and the World Health Organization (WHO). The Codex Alimentarius has two codes that pertain to fresh produce, the *Code of Hygienic Practice for Fresh Fruits and Vegetables* (*CAC/RCP 53-2003*) and the *Recommended International Code of Practice—General Principles of Food Hygiene* (*CAC/RCP 1-1969, Rev. 4-2003*) (Codex Alimentarius Commission, 2003a, 2003b). Neither of these codes specifies any kind of sampling plan. However, the Code of Hygienic Practice recommends that "potential sources of contamination from the environment should be identified" (Codex Alimentarius Commission, 2003a), and the Recommended International Code of Practice states that adequate monitoring should be performed, of both fresh produce and inputs, especially water sources (Codex Alimentarius Commission, 2003b). Yet it does not give

TABLE 1 Microbial testing requirements of the LGMA

Sample	Sampling frequency	Target organism	Recommended test method	Acceptance criteria
Water (preharvest foliar application)	One sample per water source and tested prior to use within 60 days of use	*E. coli*	15-Tube MPN[a] or other U.S. accredited method	≤126 MPN/100 ml (geometric mean) or ≤235 MPN/100 ml (single sample)
Water (preharvest nonfoliar application) (e.g., drip irrigation)	One sample per water source and tested prior to use within 60 days of use	*E. coli*	15-Tube MPN or other U.S. accredited method	≤126 MPN/100 ml (geometric mean) or ≤576 MPN/100 ml (single sample)
Postharvest (direct contact surface)	One sample tested prior to use within 60 days of use	*E. coli*	15-Tube MPN or other U.S. accredited method	≤126 MPN/100 ml (geometric mean) or ≤235 MPN/100 ml (single sample)
Soil amendment (e.g., composted soil containing animal manure)	Each lot (≤5,000 cubic yards) before field application	Fecal coliforms; *Salmonella* spp.; *E. coli* O157:H7	Fecal coliforms: 9-tube MPN; *Salmonella*: U.S. EPA Method 1682; *E. coli* O157: H7: any lab-validated method	Fecal coliforms: negative or below detection limit/gram; *Salmonella*: <1/30 g; *E. coli* O157:H7 <1/30 g
Soil amendment (nonsynthetic crop treatment)	Each lot (≤5,000 cubic yards) before field application	Fecal coliforms; *Salmonella* spp.; *E. coli* O157:H7	Fecal coliforms: 9-tube MPN; *Salmonella*: EPA Method 1682; *E. coli* O157:H7: any lab-validated method	Fecal coliforms: negative or below detection limit/gram; *Salmonella*: <1/30 gram; *E. coli* O157:H7 <1/30 g

[a] MPN, most probable number.

specific monitoring/sampling plans but rather leaves it up to individual operations to determine what is appropriate. WHO also does not specify monitoring regimes, but rather refers to the Codex Alimentarius for food standards. It does have publications of the risk of certain food- and waterborne pathogens, but again no sampling plans are specified.

Globally, there are only guidelines for sampling and testing fresh produce. However, given the global nature of produce supply and marketing, major retailers are requiring growers to undergo third-party audits or certifications. Many of these third-party certifications do use the LGMA guidelines as an unofficial checklist. There are no Codex Alimentarius-based mandatory sampling plans. Everything is on a voluntary basis, and most often the respective governments implement some sort of monitoring program rather than the industry itself.

WATER-RELATED ISSUES

For both food- and waterborne pathogens, food hygiene guidelines go hand in hand with agricultural water guidelines. Probably, the single most important source of contamination for fresh produce is water, either irrigation water or water used during postharvest processing. In the United States, agriculture depends heavily on surface waters for irrigation purposes. In many parts of the country, however, municipal wastewater treatment plants release secondary effluent into these surface waters. In addition, wildlife, cattle-feeding operations, and septic tanks also contribute to the pathogen load in these waters. Unfortunately, there are no federally mandated microbial acceptance levels for irrigation water that growers can follow. As mentioned earlier, the FDA provides only guidance for the industry, the *Guide To Minimize Microbial Food Safety Hazards for Fresh Foods and Vegetables* (Food and Drug Administration, 1998)

In terms of microbial testing of agricultural water, it states that there are a number of gaps in the science and that microbial testing of agricultural water may be of limited usefulness. It recommends that "growers focus on good agricultural practices, such as manure management and runoff controls, to maintain and protect the quality of their water sources" (Food and Drug Administration, 1998).

In the Codex Alimentarius guidelines, monitoring input sources is mentioned, but no sampling plan is proposed. The WHO also has guidelines on the safe use of wastewater in agriculture, which are very similar to the U.S. Environmental Protection Agency (EPA) standards for wastewater. It is always difficult for international bodies to be specific, because it is every country's prerogative to set its own standards. Yet, the guidelines put forth by these organizations are taken as the basis for most countries' standards. Since we do live in a global society now, uniformity with regards to such guidelines and standards is becoming the norm. Developed nations will typically have stricter standards and guidelines than developing ones, yet no country is immune to the risk of foodborne pathogens, which is why it is important to develop rapid detection methods for these pathogens.

BACTERIAL PATHOGENS

Salmonella

The genus *Salmonella* belongs to the family *Enterobacteriaceae*, which belongs to the order *Enterobacteriales* and the order *Gammaproteobacteria*. *Salmonella* species are gram-negative, non-spore-forming, motile rods. They are found in warm- and cold-blooded animals, humans, and nonliving habitats, such as water. They cause zoonotic infections (transmitted among species). In humans, the infection is called salmonellosis, which is characterized by diarrhea, fever, and abdominal cramps. *Salmonella* species have to be ingested to cause infection. Common sources of *Salmonella* include poultry and poultry products, unclean food, polluted surface waters, and reptiles. The species *Salmonella enterica* and its many serovars are frequently implicated in foodborne outbreaks. The most recent *Salmonella* outbreak in the United States occurred in 2009. Peanut butter and peanut butter-containing products were linked to this outbreak (Centers for Disease Control and Prevention, 2009a). An outbreak of *Salmonella* Saintpaul also occurred in the U.S. in 2009. Alfalfa sprouts were identified as the source in this outbreak (Centers for Disease Control and Prevention, 2009b). In 2008, a *Salmonella* Saintpaul outbreak in the U.S. was linked to jalapeño and serrano peppers (Centers for Disease Control and Prevention, 2008). Samples that are tested for *Salmonella* are typically enriched first. Enrichment broths vary depending on the sample. Some of the common enrichment broths used include tryptic soy broth and universal preenrichment broth. The most commonly used isolation media are Rappaport-Vassiliadis medium, xylose lysine desoxycholate agar, bismuth sulfite agar, and Hektoen enteric agar. Confirmation of *Salmonella* is accomplished through biochemical and/or serological testing. Common biochemical tests include triple sugar iron agar slants, lysine iron agar, and urea broth (U.S. Environmental Protection Agency, 2006). The most commonly used serological test is the polyvalent flagellar test. Subtyping is accomplished via PFGE following the *Bacteriological Analytical Manual* method (Food and Drug Administration, 2008).

E. coli O157:H7

The genus *Escherichia* also belongs to the family *Enterobacteriaceae*, like *Salmonella* and *Shigella*. Members of the genus are gram-negative, rod-shaped facultative anaerobes. *E. coli*, the type species, is found in the intestine of warm-blooded organisms, where it is an essential component of the microflora. Most strains are harmless, but some (e.g., *E. coli* O157:H7) can cause serious gastrointestinal illness. *E. coli* is typically transmitted via the fecal-oral route. Sources of contamination include raw or undercooked meat, unhygienic food preparation, manure, and irrigation water. Dairy and beef cattle tend to be primary reservoirs for *E. coli*; hence, outbreaks are very common in raw or undercooked ground beef. Other outbreak sources include vegetables (spinach and lettuce), raw milk, unpasteurized juice, and cheese. In 2006, an *E. coli* O157:H7 outbreak affecting multiple states in the United States was linked to fresh spinach (Centers for Disease Control and Prevention, 2006). In 2007, an *E. coli* O157:H7 outbreak in The Netherlands and Iceland was linked to shredded lettuce (Friesema et al., 2008). *E. coli* is commonly enriched from environmental and food samples in lauryl sulfate tryptose broth or *E. coli* broth. Selection media include eosin-methylene blue (EMB) agar, *E. coli*-MUG (4-methylumbelliferyl-β-D-glucuronide) agar, and tellurite cefixime sorbitol MacConkey (TSMAC) agar (for *E. coli* O157:H7). Common biochemical tests include indole production, gas from lactose, and citrate. The methyl red Voges-Proskauer test is also used to confirm the presence of *E. coli*. The main biochemical feature of *E. coli* is the formation of gas from lactose. The conventional method for subtyping *E. coli* species is PFGE (Food and Drug Administration, 2008).

Shigella

The genus *Shigella* belongs to the family *Enterobacteriaceae*, which belongs to the order *Enterobacteriales* and the class *Gammaproteobacteria*. It comprises four species: *S. dysenteriae* (subgroup A), *S. flexneri* (subgroup B), *S. boydii* (subgroup C), and *S. sonnei* (subgroup D). Members of the genus are gram-negative, non-spore-forming, nonmotile, facultative anaerobes. They cause shigellosis primarily in primates (humans and apes) but not in other mammals. Even though the disease is typically considered to be waterborne, it has been transmitted via foods as well. Foods that have been found to cause shigellosis include potato salad, shellfish, and raw vegetables. Infection normally occurs via ingestion through fecal-oral contamination. The most recent *Shigella* outbreaks occurred in Europe. In 2007, an outbreak of *S. sonnei* was reported in Denmark. Epidemiological studies revealed that the source was baby corn that had been imported from Thailand (Lewis et al., 2007). In 2009, an outbreak of *S. dysenteriae* was reported in Sweden. The suspected source was sugar peas (sugar snaps) (Lofdahl et al., 2009). Sugar peas were also the suspected source of an *S. sonnei* outbreak in Denmark in 2009 (Muller et al., 2009). The conventional approach used to isolate *Shigella* from environmental and food samples involves culture-based methods. It uses *Shigella* broth containing novobiocin for enrichment. After enrichment, samples are streaked onto MacConkey agar. Typical colonies, slightly pink and translucent, are then biochemically and/or serologically confirmed to be *Shigella*. Common biochemical tests include glucose broth, triple sugar iron agar slant, lysine decarboxylase broth, motility agar, tryptone, and urea broth. *Shigella* species typically do not produce gas from carbohydrates. They are negative for H_2S, glucose (gas), lysine decarboxylase, motility, sucrose, lactose, and urease. Overall, *Shigella* species are very difficult to characterize by conventional culture-based methods because they tend to be biochemically inert and very similar to *Escherichia coli*. Non-culture-based methods such as PCR or DNA hybridization are used more frequently now (Food and Drug Administration, 2008).

CHALLENGES ASSOCIATED WITH THE DETECTION OF FOODBORNE VIRUSES AND PROTOZOA

Detection methods involving food matrices begin with time-consuming and laborious methods for separation and concentration of viruses from food matrices, whereby multiple samples and large volumes are tested for adequate sample representation. Detection of enteric viruses in foods faces many challenges, including detecting low viral numbers usually found in foods, noncultivability of the human noroviruses and wild-type HAV, sampling of large volumes, and removal of inhibitory substances from the food that may coincidentally affect downstream detection. Although food and water matrices differ significantly in their chemical and physical characteristics, there are common issues and challenges when it comes to sample processing and detection of pathogenic protozoa. *Cryptosporidium* and *Giardia* are two protozoan pathogens that are of considerable importance to food and water supplies. These parasites are prevalent in both human and animal populations, are transmitted by the fecal-oral route, and can contaminate water, food, and surfaces. Foodborne viruses of epidemiological significance include HAV and the noroviruses.

Norovirus

Norovirus is a nonenveloped, positive-sense single-stranded RNA virus belonging to the genus *Norovirus* in the *Caliciviridae* family. The noroviruses currently comprise five genogroups: GI (prototype, Norwalk virus); GII (prototype, Snow Mountain Agent); GIII (prototype, bovine enteric calicivirus); GIV (prototype, Alphatron and Fort Lauderdale virus); and GV (prototype, murine norovirus) (Zheng et al., 2006). Three of these genogroups infect humans. Human noroviruses (in genogroups I, II, and IV) are the leading cause of nonbacterial gastroenteritis worldwide, occurring mainly in shellfish. Noroviruses have also been detected on farms (Cheong et al., 2009) and in market-ready produce (Allwood et al., 2004) and have been implicated in outbreaks (Bowen et al., 2006). Despite this significant disease burden, an efficient tissue culture system for noroviruses has not yet been developed. This is in large part because members of the *Caliciviridae* family do not replicate efficiently in tissue culture. So far, researchers have not been able to fully understand how noroviruses replicate, in large part because an efficient tissue culture system is not available. However, surrogate viruses, such as the murine norovirus, have been used to study and elucidate the biology of noroviruses. Since noroviruses are so difficult to culture, there is no conventional culture method yet. Researchers are still trying to develop an efficient one. In lieu of this, it is even more difficult to detect noroviruses in environmental samples by using culture methods, which is why researchers have moved to molecular methods.

Hepatitis A Virus

HAV is a positive-strand RNA virus belonging to the genus *Hepatovirus* in the family *Picornaviridae*. HAV causes infectious hepatitis in humans. Transmission of the virus occurs via the fecal-oral route through contaminated water and food, especially shellfish. In 2009, there was an HAV outbreak in Australia that was linked to semidried tomatoes. In 2010, HAV isolates from an outbreak in The Netherlands were found to be identical to the ones from the Australia outbreak in 2009. The source was suspected to be semidried tomatoes (Petrignani et al., 2010). Cell culture is the conventional culture method for HAV. Detection of the virus is rather difficult, however, even in clinical and biological samples, because it grows very slowly and poorly in cell culture. HAV replicates within days rather than hours like the other viruses in the *Picornaviridae* family. In addition, it does not interfere noticeably with the host cell macromolecular synthesis and hence does not produce a cytopathic effect. HAV is typically cultured in the fetal Rhesus monkey kidney cell line (Frhk). Since HAV is difficult to culture, reverse transcription PCR is most commonly used to detect HAV in environmental samples (Food and Drug Administration, 2008).

Cyclospora and *Cryptosporidium*

Cyclospora and *Cryptosporidium* are obligate, intracellular, coccidian protozoan parasites. Both cause severe diarrhea illness through infestation of the gastrointestinal tract of animals and humans. Both waterborne protozoa belong to the same order, namely, *Eucoccidiorida*. *Cyclospora* belongs to the *Eimeriidae* family, and *Cryptosporidium* belongs to the *Cryptosporidiidae* family. Both protozoans produce environmentally resistant oocysts, which are ingested by animals and humans either directly from contaminated water or from foods that have been treated with contaminated water. *Cryptosporidium* has been recognized as a waterborne pathogen, and *Cryptosporidium parvum* is the major cause of cryptosporidiosis in livestock and humans. However, in recent years *Cryptosporidium* has been found on fresh produce as well. The most recent *C. parvum* outbreak occurred in Finland in 2008. A salad mixture was suspected to be the source (Ponka et al., 2009). Another outbreak of cryptosporidiosis occurred in Sweden in 2008 at a hotel restaurant. The most probable source was chopped fresh parsley (Insulander et al., 2008). On the other hand, *Cyclospora*, especially *Cyclospora cayetanensis*, is most often associated with contaminated produce and fruits. From 1995 to 2000, yearly outbreaks of cyclosporiasis occurred in the United States. The source of these outbreaks was raspberries from Guatemala (Ho et al., 2002). One of the most recent *Cyclospora* outbreaks occurred in 2007 in British Columbia, Canada. The suspected source was imported basil from Mexico (Shah et al., 2009). In these instances, the major source of contamination is thought to be irrigation water. The conventional method for detection of *Cyclospora* and *Cryptosporidium* oocysts in water samples consists of three steps. The first one is concentration. This step is especially important for water samples, as it will have a large impact on subsequent detection. This step is typically accomplished through filtration. The second step is purification of the sample. The third step, the actual detection step, uses an immunofluorescent assay to visualize the oocysts microscopically (Quintero-Betancourt et al., 2002). The U.S. Environmental Protection Agency approved a fluorescent antibody procedure to detect *Cryptosporidium* oocysts in water samples (U.S. Environmental Protection Agency, 1996). More research has focused on the detection of *Cryptosporidium* oocysts, but all the techniques are also applicable to *Cyclospora*.

Toxoplasma

The genus *Toxoplasma* belongs to the *Sarcocystidae* family in the order *Eucoccidiorida*. *T. gondii*, the type species, causes toxoplasmosis in humans and other warm-blooded animals. Toxoplasmosis is typically asymptomatic in healthy adults but can be fatal in immunocompromised individuals. Transmission of the disease occurs via consumption of raw or undercooked meat (especially pork) containing tissue cysts or via ingestion of oocysts from environmental matrices (water or meat) (Dumetre and Darde, 2003). In 2001, an outbreak of toxoplasmosis in Brazil was linked to contaminated drinking water (de Moura et al., 2006). Even though *Toxoplasma* oocysts can persist for long periods of time in the environment, they have some characteristics that may hamper their detection. First, they are exclusively shed by felids, unlike other waterborne protozoa such as *Cryptosporidium*. They tend to be excreted in large amounts for only a short time. Hence, the probability of detecting *Toxoplasma* oocysts in a typical environmental sample is low. Second, they may be confused with other coccidian oocysts, such as *Hammondia* sp. and *Neospora* sp. Debris in environmental samples may also pose a problem. The conventional detection method of *Toxoplasma* in environmental samples is the mouse bioassay (Dumetre and Darde, 2003). Mice are treated with the oocysts recovered from the environmental samples. Prior to ingestion, the oocysts are sporulated in the laboratory through exposure to aeration and proper temperature to render them infectious. Mouse bioassays typically take several weeks and may be unsuited for sensitive and low detection. Microscopy is also often used to detect oocysts, especially in water samples. The procedure is very similar to the method used by the EPA to detect *Cryptosporidium* oocysts in water (Villena et al., 2004). Researchers are currently evaluating many different methods to detect *Toxoplasma* in environmental samples because the conventional method is expensive and time-consuming.

FECAL CONTAMINATION INDICATORS

Different organisms have been proposed as indicators of fecal contamination such as fecal coliform bacteria, *E. coli*, enterococci, bacteriophages, etc., over the past many years. However, studies have repeatedly shown that bacterial indicators are not truly representative of all possible fecal contaminants, especially the enteric viruses. There have been reports spanning over 2 decades suggesting that bacteriophages are useful as indicators of fecal pollution (Endley et al., 2003b; Gerba, 1987; Havelaar et al., 1993). *Bacteroides fragilis* is an obligate anaerobe found in high concentrations in human feces; thus, the presence or absence of phages that infect these bacteria is also considered to be indicative of fecal pollution (Duran et al., 2003; Gantzer et al., 2002; International Organization for Standardization, 1999; Puig et al., 1999). The use of coliphages as indicators of fecal pollution is based on the assumption that their presence in water samples denotes the presence of bacteria capable of supporting their replication. Coliphages have been proposed as indicators of enteric viral contamination because they are found in relatively high numbers in the environment, they do not readily multiply in the environment, and the coliphage assay cost is only a fraction of that of a typical pathogenic enteric virus assay. Coliphages have also been suggested as a reliable indicator of enteric viruses because of their relative similarities in size, transport, survival, or persistence patterns and their densities found in sewage or septic samples. Coliphages are valuable as contamination indicators because they are found in relatively high numbers in sewage and they do not, for the most part, readily multiply in the environment. They are also relatively persistent in the environment and are relatively resistant to disinfection treatments, and the methods to detect them are relatively straightforward. Studies in our laboratory have shown that coliphages can be readily detected in fresh produce such as carrots, cilantro, and parsley (Endley et al., 2003a, 2003b).

Coliphages can be broadly categorized into somatic coliphages and male-specific (or F^+) coliphages. Somatic coliphages are phages that infect *E. coli* by attaching directly to the cell wall. Male-specific coliphages, on the other hand, infect the *E. coli* host bacteria by initially attaching to specific bacterial appendages termed sex-pilus or F-pilus. The use of male-specific coliphages as fecal indicators circumvents the technical complexities and costs involved in screening for enteric viral pathogens and yet provides some assurance about the presence of viral fecal contaminants (Allwood et al., 2003; Havelaar et al., 1993). *Bacteroides fragilis* is an obligate anaerobic bacterium found in high concentrations in human feces, and thus, the presence or absence of phages that infect these bacteria is considered to be indicative of human fecal contamination.

Due to the increased interest in using coliphages and *Bacteroides* phages as contamination indicators, methods to detect them have been constantly improving in the last couple of years. One of the key issues that confront bacteriophage detection from water and foods is deciding on the appropriate sample volume for analysis. Since enteric viruses are normally recovered by using very large sample volumes, the current trend in bacteriophage analysis (at least in the research laboratories) is to employ large sample volumes as well. A variety of different methods have been reported for the recovery, detection, and enumeration of phages (Pillai and Nwachuku, 2002). The differences in the methods reflect the differences in the sample volumes (used for analysis), the sample processing methods, and the host bacteria that are employed in detecting the presence of the bacteriophages. The EPA has published validated protocols for the detection of coliphages in drinking water (Method 1601 and Method 1602). The International Organization for

Standardization has also published a standardized protocol for the detection of male-specific coliphages and phages that infect *Bacteroides fragilis*. In our laboratory, we have successfully modified EPA Method 1601 and EPA Method 1602 to screen for male-specific and somatic coliphages in fresh produce (Endley et al., 2003a, 2003b) and in produce packing sheds (Endley et al., 2003b).

REFERENCES

Allwood, P. B., Y. S. Malik, C. W. Hedberg, and S. M. Goyal. 2003. Survival of F-specific RNA coliphage, feline calicivirus, and *Escherichia coli* in water: a comparative study. *Appl. Environ. Microbiol.* **69:**5707–5710.

Allwood, P. B., Y. S. Malik, S. Maherchandani, K. Vought, L. A. Johnson, C. Braymen, C. W. Hedberg, and S. M. Goyal. 2004. Occurrence of *Escherichia coli*, noroviruses, and F-specific coliphages in fresh market-ready produce. *J. Food Prot.* **67:**2387–2390.

Beuchat, L. R. 1998. *Surface Decontamination of Fruits and Vegetables Eaten Raw: a Review. Food Safety Issues. WHO/FSF/FOS 98.2.* Food Safety Unit, World Health Organization, Geneva, Switzerland.

Bowen, A., A. Fry, G. Richards, and L. Beuchat. 2006. Infections associated with cantaloupe consumption: a public health concern. *Epidemiol. Infect.* **134:**675–685.

Brooks, N., A. Regmi, and A. Jerardo. 2009. *U.S. Food Import Patterns, 1998–2007. Report FAU-125.* USDA Economic Research Service, U.S. Department of Agriculture, Washington, DC.

Centers for Disease Control and Prevention. 2009a. Multistate outbreak of *Salmonella* infections associated with peanut butter and peanut butter-containing products—United States, 2008–2009. *MMWR Morb. Mortal. Wkly. Rep.* **58:**35–90.

Centers for Disease Control and Prevention. 2006. Ongoing multistate outbreak of *Escherichia coli* serotype O157:H7 infections associated with consumption of fresh spinach—United States, September 2006. *MMWR Morb. Mortal. Wkly. Rep.* **55:** 1045–1046.

Centers for Disease Control and Prevention. 2009b. Outbreak of *Salmonella* serotype Saintpaul infections associated with eating alfalfa sprouts—United States, 2009. *MMWR Morb. Mortal. Wkly. Rep.* **58:**500–503.

Centers for Disease Control and Prevention. 2008. Outbreak of *Salmonella* serotype Saintpaul infections associated with multiple raw produce items—United States, 2008. *MMWR Morb. Mortal. Wkly. Rep.* **57:**929–934.

Cheong, S., C. Lee, S. W. Song, W. C. Choi, C. H. Lee, and S. J. Kim. 2009. Enteric viruses in raw vegetables and groundwater used for irrigation in South Korea. *Appl. Environ. Microbiol.* **75:** 7745–7751.

Codex Alimentarius Commission. 2003a. *Code of Hygienic Practice for Fresh Fruits and Vegetables (CAC/RCP 53-2003).* Food and Agriculture Organization/World Health Organization, Rome, Italy.

Codex Alimentarius Commission. 2003b. *Recommended International Code of Practice—General Principles of Food Hygiene (CAC/RCP 1-1969, Rev. 4-2003).* Food and Agriculture Organization/World Health Organization, Rome, Italy.

de Moura, L., L. M. Bahia-Oliveira, M. Y. Wada, J. L. Jones, S. H. Tuboi, E. H. Carmo, W. M. Ramalho, N. J. Camargo, R. Trevisan, R. M. Graca, A. J. da Silva, I. Moura, J. P. Dubey, and D. O. Garrett. 2006. Waterborne toxoplasmosis, Brazil, from field to gene. *Emerg. Infect. Dis.* **12:**326–329.

Dumetre, A., and M. L. Darde. 2003. How to detect *Toxoplasma gondii* oocysts in environmental samples? *FEMS Microbiol. Rev.* **27:**651–661.

Duran, A. E., M. Muniesa, L. Moce-Llivina, C. Campos, J. Jofre, and F. Lucena. 2003. Usefulness of different groups of bacteriophages as model micro-organisms for evaluating chlorination. *J. Appl. Microbiol.* **95:**29–37.

Endley, S., E. Johnson, and S. D. Pillai. 2003a. A simple method to screen cilantro and parsley for fecal indicator viruses. *J. Food Prot.* **66:**1506–1509.

Endley, S., L. Lu, E. Vega, M. E. Hume, and S. D. Pillai. 2003b. Male-specific coliphages as an additional fecal contamination indicator for screening fresh carrots. *J. Food Prot.* **66:**88–93.

Food and Drug Administration. 1998. *Guide To Minimize Microbial Food Safety Hazards for Fresh Fruits and Vegetables.* Center for Food Safety and Applied Nutrition, Food and Drug Administration, Washington, DC. www.fda.gov/food/guidancecomplianceregulatoryinformation/guidancedocuments/produceandplanproducts/ucm064458.htm.

Food and Drug Administration. 2008. *Bacteriological Analytical Manual.* Food and Drug Administration, Gaithersburg, MD. www.fda.gov/Food/ScienceResearch/LaboratoryMethods/BacteriologicalAnalyticalManualBAM/default.htm.

Friesema, I., G. Sigmundsdottir, K. van der Zwaluw, A. Heuvelink, B. Schimmer, C. de Jager, B. Rump, H. Briem, H. Hardardottir, A. Atladottir, E. Gudmundsdottir, and W. van Pelt. 2008. An international outbreak of Shiga toxin-producing *Escherichia coli* O157 infection due to lettuce, September–October 2007. *Euro Surveill.* **13**(50)**:**pii:19065.

Gantzer, C., J. Henny, and L. Schwartzbrod. 2002. *Bacteroides fragilis* and *Escherichia coli* bacteriophages in human faeces. *Int. J. Hyg. Environ. Health* **205:**325–328.

Gerba, C. P. 1987. Phages as indicators of fecal pollution. *In* S. M. Goyal, C. P. Gerba, and G. Bitton (ed.), *Phage Ecology*. John Wiley & Sons, New York, NY.

Havelaar, A. H., M. van Olphen, and Y. C. Drost. 1993. F-specific RNA bacteriophages are adequate model organisms for enteric viruses in fresh water. *Appl. Environ. Microbiol.* **59:**2956–2962.

Ho, A. Y., A. S. Lopez, M. G. Eberhart, R. Levenson, B. S. Finkel, A. J. da Silva, J. M. Roberts, P. A. Orlandi, C. C. Johnson, and B. L. Herwaldt. 2002. Outbreak of cyclosporiasis associated with imported raspberries, Philadelphia, Pennsylvania, 2000. *Emerg. Infect. Dis.* **8:**783–788.

Huang, S., and K. Huang. 2007. *Increased U.S. Imports of Fresh Fruit and Vegetables. Report FTS-328-01.* USDA Economic Research Service, U.S. Department of Agriculture, Washington, DC.

Insulander, M., B. de Jong, and B. Svenungsson. 2008. A food-borne outbreak of cryptosporidiosis among guests and staff at a hotel restaurant in Stockholm county, Sweden, September 2008. *Euro Surveill.* **13**(51)**:**pii:19071.

International Organization for Standardization. 1999. *Water Quality—Detection and Enumeration of Bacteriophages. Part 4. Enumeration of Bacteriophages Infecting* Bacteroides fragilis. International Organization for Standardization, Geneva, Switzerland.

Lewis, H. C., S. Ethelberg, M. Lisby, S. B. Madsen, K. E. Olsen, P. Rasmussen, C. Kjelso, L. S. Vestergaard, K. Qureshi, M. Howitz, and K. Molbak. 2007. Outbreak of shigellosis in Denmark associated with imported baby corn, August 2007. *Euro Surveill.* **12:**E070830 1.

Little, C. L., and I. A. Gillespie. 2008. Prepared salads and public health. *J. Appl. Microbiol.* **105:** 1729–1743.

Lofdahl, M., S. Ivarsson, S. Andersson, J. Langmark, and L. Plym-Forshell. 2009. An outbreak of *Shigella dysenteriae* in Sweden, May–June 2009, with sugar snaps as the suspected source. *Euro Surveill.* **14**(28)**:**pii:19268.

Mahon, B. E., A. Ponka, W. N. Hall, K. Komatsu, S. E. Dietrich, A. Siitonen, G. Cage, P. S. Hayes, M. A. Lambert-Fair, N. H. Bean, P. M. Griffin, and L. Slutsker. 1997. An international outbreak of *Salmonella* infections caused by alfalfa sprouts grown from contaminated seeds. *J. Infect. Dis.* **175:**876–882.

Mead, P. S., L. Slutsker, V. Dietz, L. F. McCaig, J. S. Bresee, C. Shapiro, P. M. Griffin, and R. V. Tauxe. 1999. Food-related illness and death in the United States. *Emerg. Infect. Dis.* **5:**607–625.

Muller, L., T. Jensen, R. F. Petersen, K. Molbak, and S. Ethelberg. 2009. Imported fresh sugar peas as suspected source of an outbreak of *Shigella sonnei* in Denmark, April–May 2009. *Euro Surveill.* **14**(24)**:**pii:19241.

Petrignani, M., M. Harms, L. Verhoef, R. van Hunen, C. Swaan, J. van Steenbergen, I. Boxman, I. S. R. Peran, H. Ober, H. Vennema, M. Koopmans, and W. van Pelt. 2010. Update: a food-borne outbreak of hepatitis A in the Netherlands related to semi-dried tomatoes in oil, January–February 2010. *Euro Surveill.* **20**(15)**:**pii: 19572.

Pillai, S. D., and N. Nwachuku. 2002. Bacteriophage methodologies, p. 374–384. *In* G. Bitton (ed.), *Encyclopedia of Microbiology*. Wiley & Sons, New York, NY.

Ponka, A., H. Kotilainen, R. Rimhanen-Finne, P. Hokkanen, M. L. Hanninen, A. Kaarna, T. Meri, and M. Kuusi. 2009. A foodborne outbreak due to *Cryptosporidium parvum* in Helsinki, November 2008. *Euro Surveill.* **14**(28)**:**pii:19269.

Puig, A., N. Queralt, J. Jofre, and R. Araujo. 1999. Diversity of *Bacteroides fragilis* strains in their capacity to recover phages from human and animal wastes and from fecally polluted wastewater. *Appl. Environ. Microbiol.* **65:**1772–1776.

Quintero-Betancourt, W., E. R. Peele, and J. B. Rose. 2002. *Cryptosporidium parvum* and *Cyclospora cayetanensis*: a review of laboratory methods for detection of these waterborne parasites. *J. Microbiol. Methods* **49:**209–224.

Shah, L., L. MacDougall, A. Ellis, C. Ong, S. Shyng, and L. LeBlanc. 2009. Challenges of investigating community outbreaks of cyclosporiasis, British Columbia, Canada. *Emerg. Infect. Dis.* **15:** 1286–1288.

U.S. Environmental Protection Agency. 1996. National primary drinking regulation: monitoring requirements for public drinking water suppliers: *Cryptosporidium, Giardia*, viruses, disinfection byproducts, water treatment plant data and other information requirements. *Fed. Regist.* **61:**24354–24388.

U.S. Environmental Protection Agency. 2006. *Method 1682: Salmonella in Sewage Sludge (Biosolids) by Modified Semisolid Rappaport-Vassiliadis (MSRV) Medium. EPA-821-R-06-14.* U.S. Environmental Protection Agency, Washington, DC.

Villena, I., D. Aubert, P. Gomis, H. Ferte, J. C. Inglard, H. Denis-Bisiaux, J. M. Dondon, E. Pisano, N. Ortis, and J. M. Pinon. 2004. Evaluation of a strategy for *Toxoplasma gondii* oocyst detection in water. *Appl. Environ. Microbiol.* **70:**4035–4039.

Zheng, D. P., T. Ando, R. L. Fankhauser, R. S. Beard, R. I. Glass, and S. S. Monroe. 2006. Norovirus classification and proposed strain nomenclature. *Virology* **346:**312–323.

NOROVIRUS, HEPATITIS A VIRUS, AND INDICATOR MICROORGANISMS IN SHELLFISH, SOFT FRUITS, AND WATER

A. Bosch, S. Bidawid, F. S. Le Guyader, D. Lees, and L. Jaykus

23

Viruses such as hepatitis A virus (HAV), noroviruses (NoV), sapoviruses, enteroviruses, astroviruses, adenoviruses, rotaviruses, and hepatitis E virus have all been implicated in food- and/or waterborne outbreaks of illness. As a matter of fact, there is a potential for any virus that causes illness after ingestion to be transmitted by food. However, in practice most reported incidents of viral foodborne illness are due to human NoV and HAV.

In the European Union, viral agents were reported to be responsible for approximately 12% of the foodborne outbreaks reported to the European Food Safety Authority (http://www.efsa.europa.eu/EFSA/efsa_locale-1178620753812_1211902515341.htm) during 2007 and were identified as the second most common causative agent group, after *Salmonella*. In the United States, the importance of viral agents (i.e., NoV) to the overall burden of foodborne disease is increasingly being recognized based on results from outbreak investigations. In a report on outbreak-associated illnesses compiled by the Centers for Disease Control and Prevention for the period from 1998 to 2002 (Lynch et al., 2006), viral pathogens, predominantly NoV, caused 33% of outbreaks and 41% of cases; the proportion of outbreaks attributed to viral agents increased from 16% in 1998 to 42% in 2002. Viral pathogens accounted for an increased proportion of outbreaks each year during this reporting period and a higher proportion of outbreaks of known etiology during this reporting period than preceding reporting periods. This is probably reflecting the increased availability of improved viral diagnostic tests. In most countries, however, the extent of the problem is not known.

Human viruses can contaminate food either through contamination at source, principally through sewage pollution of the environment, or in association with food processing through inadequate hygiene practices of operatives or systems. Consequently, many different food items such as vegetables, shellfish, and a great

A. Bosch, Virus Entèrics, Department of Microbiology, University of Barcelona, Avda. Diagonal 645, 08028 Barcelona, Spain. *S. Bidawid*, Microbiology Research Division, Health Canada, Food Directorate, Bureau of Microbial Hazards, Sir F.G. Banting Research Centre, Tunney's Pasture, Ottawa, Ontario, Canada K1A 0K9. *F. S. Le Guyader*, Laboratoire de Microbiologie-LNR, Département EMP, IFREMER, 1 rue de l'Ile d'Yeu, BP 21105, 44311 Nantes cedex 03, France. *D. Lees*, European Community Reference Laboratory, CEFAS Laboratory, Barrack Road, The Nothe, Weymouth, Dorset, DT4 8UB, United Kingdom. *L. Jaykus*, Department of Food, Bioprocessing and Nutrition Sciences, North Carolina State University, Raleigh, NC 27695-7624.

Rapid Detection, Identification, and Quantification of Foodborne Pathogens, ed. by J. Hoorfar

variety of ready-to-eat foods like sandwiches, cold meat, pastries, etc., have been implicated in foodborne viral infections. Bivalve shellfish are commonly involved in outbreaks of foodborne viral diseases. Shellfish are filter feeders, and if shellfish-growing waters are polluted with human sewage, the shellfish accumulate viruses infectious for humans. The difficulties in detecting virus in shellfish pose further problems, as does the fact that the correlation between levels of bacterial indicator organisms and the extent of viral contamination is poor.

Contamination of food with fecal material can occur at any step during its production. For example, contamination of shellfish usually occurs prior to harvesting. Contamination of berries (e.g., raspberries) may occur prior to harvesting (e.g., due to irrigation with fecally contaminated water), during harvesting (contamination by infected field workers), or during processing prior to distribution (contamination in the factory by infected food handlers or by spraying with contaminated water). Other foods implicated in outbreaks, including salads, sandwiches, and deli meats, have been contaminated at the site of preparation by infected food handlers.

Secondary transmission of NoV infection is common (often >30%), allowing amplification of an outbreak, particularly in closed settings. Such outbreaks are commonly recognized in health care institutions (e.g., hospitals or nursing homes) and on cruise ships (Lopman et al., 2004; Verhoef et al., 2010). Because NoV are relatively resistant to inactivation by many common disinfectants, outbreaks in these settings may require closure of the unit or ship for more extensive cleaning and disinfection.

Water is another frequent vehicle for enteric virus infections. Since current water treatments do not ensure the complete removal of these pathogens, viruses become contaminants of the water environment in numbers high enough to represent a public health threat, although low enough to pose serious difficulties for their detection. Poor water quality continues to pose a major threat to human health. Billions of cases of gastrointestinal illness occur annually worldwide. The World Health Organization (WHO) declared that diarrheal disease alone contributes to an estimated 4.1% of the total DALY (disability adjusted life years) of the global burden of disease and is responsible for the deaths of 1.8 million people every year (World Health Organization, 2004). It was figured that 88% of that burden is attributable to unsafe water supply, sanitation, and hygiene, and it is mostly in children from developing countries. A significant amount of disease could be prevented, especially in developing countries, through better access to safe water supply, adequate sanitation facilities, and better hygiene practices.

VIRUSES IN FOOD AND WATER

Viruses are present in food and water samples at low to very low levels compared with human clinical samples. However, unlike many enteric bacteria causing foodborne disease, enteric viruses can initiate an infection in very low numbers; the NoV and HAV infectious doses are generally accepted to be around 10 to 100 virus particles (Teunis et al., 2008; http://www.fda.gov/food/foodsafety/foodborneillness/foodborneillnessfoodbornepathogensnaturaltoxins/badbugbook/ucm071294.htm). Thus, sufficient sensitivity is a significant issue in the development of virus detection methods for food and water samples.

An expert meeting convened under the auspices of WHO/FAO and the World Organization for Animal Health reviewed available evidence and grouped food and waterborne viruses according to their ability to cause high morbidity, severe disease, or a significant threat thereof.

Subsequently, available evidence for a specific hazard-food commodity combination was reviewed by considering available information on estimates of the incidence of foodborne disease linked to a specific commodity and the level of evidence for the importance of that commodity in causing viral foodborne illness. This resulted in several virus-commodity combinations for which prevention and control measures should be considered:

- NoV and HAV in bivalve molluscan shellfish
- NoV and HAV in fresh produce
- NoV and HAV in prepared foods
- Rotaviruses in water for food preparation
- Emerging viruses in selected commodities

The present chapter will only deal with NoV and HAV detection in bivalve mollusks, soft fruits, and water.

FECAL POLLUTION INDICATORS

Fecal pollution indicators have a long history, dating back to the early 20th century, of use for monitoring and managing the pathogen risk associated with fecal pollution. Worldwide, legislative standards controlling the sanitary production of bivalve mollusks and drinking water utilize fecal indicators. For bivalve mollusks, fecal indicators are measured either in the shellfish themselves (European Union system) or in their growing waters (United States system).

Both systems require the grading of areas according to the degree of fecal pollution which determines their sanitary quality and hence determines the approval to harvest, the marketability, and the extent of food processing required for the product (Lees, 2000). The acceptability of bivalve mollusks in international trade, as set out by Codex Alimentarius, is also judged by their compliance with fecal indicator end product standards. *Escherichia coli* is now generally targeted for shellfish flesh monitoring (e.g., European Union system), whereas coliforms or fecal coliforms are analyzed for water monitoring (e.g., United States system). Standardized methods are stipulated in both the United States and the European Union for *E. coli* analysis (Anonymous, 2005).

For bivalve mollusks, these extensive controls do appear to adequately manage the risk from bacterial pathogens associated with fecal pollution (Lees, 2000). However, the historical approach of monitoring for fecal indicators and setting of legislative standards does not appear to adequately control the risk from human enteric viruses, since outbreaks associated with both NoV and HAV continue to occur on a regular basis (Pintó et al., 2009; Shieh et al., 2007; Westrell et al., 2010).

A number of studies have evaluated alternative indicators more physically and behaviorally similar to enteric viruses, such as various bacteriophage species (Dore et al., 2000). However, the main thrust internationally is now towards detection of the virus pathogens directly in the food commodity of concern.

Unfortunately, in the United States there are no equivalent guidelines for the acceptability of waters used for the production and processing of soft fruit. Some producers will evaluate waters periodically for total aerobic bacteria, coliforms, and/or *E. coli*, but there is no legal mandate to do so.

As is the case for shellfish and their harvesting waters, the relationship between the traditional fecal indicators and the presence of human enteric viruses in production and processing waters is tenuous at best. Alternative indicators like the male-specific coliphages may not work, although large-scale studies of their efficacy have not yet been undertaken (Endley et al., 2003). As is the case for shellfish, most international efforts are focusing on direct detection of virus in contaminated products or waters as an alternative to the standard fecal indicators.

VIRUS DETECTION IN BIVALVE MOLLUSKS

The low virus concentrations present in shellfish require the use of laboratory techniques that efficiently recover most of the viruses present from shellfish tissues. The greatest concentration of human enteric viruses being concentrated in the stomach and digestive diverticula, almost all sensitive methods are based on dissected digestive tissues. Two main options are available: (i) methods based on an elution-concentration step of viral particles and (ii) direct elution of virions.

In the elution-concentration methods, the initial steps consist of virus elution from shellfish tissues, recovery of viral particles, and then virus concentration. Some methods use acidic adsorption prior to virus elution, but for most methods viruses are eluted by use of various buffers (e.g., chloroform-butanol or glycine) before concentration by polyethylene glycol (PEG) precipitation or ultracentrifugation (Costafreda et al., 2006; DePaola et al., 2010; Le Guyader et al., 2009; Terio et al., 2010). In the direct elution of virus, particles using proteinase K (Lowther et al., 2010) or even direct lysis of nucleic acid using Trizol or other lysis buffer is used (Nenonen et al., 2009).

Since viruses detected in shellfish are nonculturable, the current detection system is reverse transcription-PCR (RT-PCR). Thus, the second important parameter is nucleic acid extraction and purification. In-house methods (nonstandardized protocols routinely used in laboratories), such as proteinase K capsid lysis followed by purification of nucleic acid using phenol-chloroform and cetyltrimethylammonium bromide (CTAB) precipitation, Trizol treatment, Boom-based methods, or cesium chloride cushion, have been proven efficient to detect viruses in shellfish in various studies including shellfish samples implicated in outbreaks (Atmar et al., 1995; DePaola et al., 2010; Le Guyader et al., 2008; Lowther et al., 2010). Currently, several methods have been adapted for use in commercial kits based on guanidium thiocyanate lysis followed by nucleic acid purification using either a column or paramagnetic silica (Costafreda et al., 2006; Lees, 2000; Terio et al., 2010). The adaptation of these methods to utilize readily available reagents, but with minimum loss of assay sensitivity, has become a priority. The use of a kit, sometimes with automated apparatus and real-time RT-PCR, or quantitative RT-PCR (qRT-PCR), constitutes a major step for standardization. This approach is fundamentally important for shellfish safety and is under evaluation for future European Union regulation.

VIRUS DETECTION IN SOFT FRUITS

There have been several high-profile foodborne disease outbreaks linked, directly or indirectly, to fresh produce, including soft fruits (strawberries, raspberries, and tomatoes) and green onions, over the last decade (Centers for Disease Control and Prevention, 1997; Gaulin et al., 1999; Ponka et al., 1999; Wheeler et al., 2005). The need for reliable methods to detect viruses in foods remains, since current techniques are limited and none are yet commercialized.

Most of the developmental work on the detection of viral contamination of foods has been done using the molluscan shellfish matrix. While these reports have provided insights, soft fruits are different from oysters or clams. In particular, viral contamination in soft fruits is distributed on the surface of the product rather than localized within the fruit. Also, soft fruits lack fat, are high in both simple and complex carbohydrates, and have rough or irregular surfaces. Early studies targeting virus detection in soft fruits (Bidawid et al., 2000) focused on the detection of HAV in strawberries.

Croci and coworkers (2008) authored a comprehensive review of the methods used to detect enteric viral contamination of fresh produce. These authors described the sequential steps of such detection as (i) virus elution/clarification, (ii) virus concentration, (iii) nucleic acid extraction, and (iv) detection. Each of these is discussed below, and a summary is given in Table 1.

The purpose of the elution/clarification step is to facilitate the removal of virus from the surface of the product (elution). This is usually achieved by using eluant solutions of elevated pH and/or solutions that are supplemented with amino acids and/or organic matter, such as those present in beef extract or fetal bovine serum. The principle behind elution is that pH increase reduces the ability of viruses to adhere to matrix components, resulting in release of the virus into a suspension that is relatively free of organic matter. Clarification, or the process of removing

TABLE 1 Summary of the principles of virus concentration, purification, and RNA extraction from soft fruits

Method	Notes
Sample size	Varies from a low of 2–3-cm^2 slices or pieces to a high of 100 g; 25–50 g is a reasonable sample size
Candidate elution buffers	Phosphate-buffered saline 1 M sodium bicarbonate 0.05 M glycine–0.14 N saline, pH range, 7.5–9.5 Beef extract (1–3%) Elution agents frequently combined, such as supplementing glycine-saline with beef extract or fetal bovine serum Additional pectinase treatment
Candidate clarification methods	Filtration: remove residual food matrix, recover filtrate • Cheesecloth, gauze, plastic net • Filter stomacher bags Centrifugation: sediment (with or without coprecipitant) food matrix; recover supernatant • No coprecipitate: 2,500–30,000 × *g*; 5,000–10,000 × *g* benchmark • With coprecipitate: usually CatFlot T or TL, can reduce centrifugation speed to 2,500–3,000 × *g*
Candidate virus concn approaches	Ultracentrifugation (100,000–300,000 × *g* for 0.5 to 2 h) Ultrafiltration PEG precipitation (6–12%) followed by centrifugation Filtration with electronegative membrane filters followed by elution
RNA extraction approaches	Most commonly used approaches are based on guanidinium isothiocyanate with/without secondary silica purification. Commercial kits are available. Proteinase K followed by phenol-chloroform-isoamyl alcohol and ethanol precipitation. Commercial kits are available.

the residual matrix components, can be accomplished by physical separation methods such as filtration or centrifugation. In these cases, the virus-containing filtrate or supernatant, respectively, is recovered for further processing. The decision to use filtration or centrifugation for the clarification process depends on a number of factors, including the specific food matrix, the availability of relevant equipment (for centrifugation), and the experience of the testing facility.

The clarified eluent is usually of relatively high volume (>25 ml, frequently much more), while in naturally contaminated products, the level of virus contamination is anticipated to be quite low. Therefore, one or more additional concentration steps are needed. Both chemical and physical concentration approaches have been used. Chemically based methods rely on the capacity of viruses to behave as proteins in solution. For example, reducing the pH of the solution to below the isoelectric point of the virus favors precipitation (so-called acid precipitation), as does the addition of the protein precipitant PEG.

These are relatively nonspecific methods, which also favor the precipitation of residual matrix components not removed by clarification. The resulting precipitant is usually collected by relatively low speed centrifugation. Ultracentrifugation has been used for virus concentration, but due to the high cost of investment, this technique is not a practical option for most food microbiology testing laboratories. Ultrafiltration has also been used, but filters have a tendency to clog quickly unless ultrafiltration is preceded by enzyme treatments intended to degrade large matrix-associated molecules. Nevertheless, investigators have had success pretreating clarified eluants from soft fruits with pectinase prior to the application of ultrafiltration (Butot et al., 2007b).

Sample volume reductions after a concentration step usually range from 10- to 1,000-fold, resulting in concentrates of 100 μl to 10 ml. Overall, the smaller the volume, the better, as most commercial RNA extraction methods are intended for samples <1 ml in volume. Although heat release of viral RNA (99°C/5 min) was used in the early days of methodological development, this has largely been replaced with nucleic acid extraction, as extraction provides additional sample concentration and also removes residual matrix-associated inhibitors. Increasingly, RNA extraction methods are being supplemented with secondary purification steps, such as silica binding followed by nucleic acid elution (Table 2). Many of the extraction methods have been commercialized over the last decade. A summary of current methods and their detection limits is provided in Table 2. This is by no means complete, but it does provide an indication of the wide variety of methods being used in the research setting. The method of Butot et al. (2007b) as applied to soft fruits has received attention in Europe, while there has been interest in the recirculating ligand capture methods in the United States (Papafragkou et al., 2008). Few have done comprehensive comparisons of these methods. All methods have their own advantages and disadvantages. For example, immunocapture methods tend to be simple, requiring fewer sample manipulations with lower risk of virus loss, although they may be hindered by high specificity since broadly reactive antisera are not available for the human NoV. Presently, there is no universal extraction method that can be applied to all foods, particularly when contamination levels are low.

VIRUS DETECTION IN WATER

The basic steps in virological analysis of water are sampling, concentration, decontamination/removal of inhibitors, and specific virus detection. Sample concentration is a particularly critical step, since the viruses may be present in such low numbers that concentration of the water samples is indispensable to reduce the volume to be assayed to a few milliliters or even microliters.

In drinking water, the virus levels are likely to be so low that optimally hundreds, or even thousands, of liters should be sampled to increase the probability of virus detection. Methods for virus concentration from water samples are depicted in Table 3 and reviewed elsewhere (Wyn-Jones, 2007). A good concentration method should fulfill several requirements: it should be technically simple, be rapid, provide high virus recoveries, be adequate for a wide range of enteric viruses, provide a small volume of concentrate, and be inexpensive. No single method meets all these requirements.

Basically, all available procedures have been evaluated by using samples spiked with known viruses. It is known that the recovery efficiency recorded with experimentally contaminated water dramatically decreases when the method is applied in actual life situations. Additionally, none of the existing concentration procedures has been tested with all of the medically important virus groups; normally, a few specific enteric viruses have been employed to conduct the evaluation trials.

Most of the procedures for concentrating and extracting viruses make use of the properties of the viral proteinaceous macromolecules. Certain protein structures confer on viruses in an aquatic environment the properties of a hydrophilic colloid of an amphoteric nature, whose electric charge varies according to the pH and the ionic force of the environment. Viruses can therefore be adsorbed onto and then detach themselves from different substrates that are positively or negatively charged depending on their pH. Methods based on the adsorption of viruses from the sampled water onto a suitable solid surface, from which they may subsequently be eluted into a much smaller volume, are preferred for use with large-volume samples.

TABLE 2 Summary of candidate methodological approaches to the detection of virus contamination in soft fruits[a]

Commodity and virus(es)	Sample size (g)	General preanalytical processing approach	Detection limits	Reference
Fresh and frozen strawberries; HAV	25	Elution (with pectinase); filtration/ centrifugation; PEG precipitation (with subsequent chloroform-butanol extraction)	2.0 log PFU/25 g	Dubois et al., 2007
Berries; HuNoV, HAV, rotavirus	15	Elution/filtration with pectinase; centrifugation; ultrafiltration	1 $TCID_{50}$ (HAV); 54 qRT-PCR U (HuNoV); 0.02 $TCID_{50}$ (rotavirus) per 15 g	Butot et al., 2007b
Strawberries and raspberries; HAV and HuNoV	50	Elution; filtration using cationically charged filters (nanoalumina); elution	2–20 Viral particles/g	Morales-Rayas et al., 2010
Fruit salad and raspberries; HuNoV	50	Multiple methods, but most effective consisted of elution (with pectinase), centrifugation, chloroform-butanol extraction, and PEG precipitation	10^2–10^3 RT-PCR U per 50 g	Baert et al., 2008a
Blended strawberries and tomato sauce; HAV	30	Adsorption (acid precipitation); elution and Vertrel extraction; PEG precipitation with Vertrel extraction	14 PFU/g (tomato sauce); 33 PFU/g (blended strawberries)	Love et al., 2008
Strawberries; HAV	25	Elution/filtration; recirculating magnetic capture with cationic beads	10^{-1}–10^2 PFU/25 g	Papafragkou et al., 2008

[a] Abbreviations: HuNoV, human norovirus; $TCID_{50}$, 50% tissue culture infective dose; PCR U, PCR units.

Criteria based on the experience and expertise of the user on a given method should be employed to select the most appropriate system. Methods based on positively charged filters (Sobsey and Glass, 1980) and glass wool (Kiulia et al., 2010; Vilagines et al., 1993) are still among the best possibilities. Sampling large volumes requires a two-step concentration procedure, with PEG (Lewis and Metcalf, 1988) and ultrafiltration (Rutjes et al., 2006) as preferred procedures for reconcentration of the primary eluates. Additionally, PEG (Lewis and Metcalf, 1988) as well as lyophilization (Villena et al., 2003) may be used for direct

TABLE 3 Procedures for the concentration of viruses from water samples

Method	Principle	Pros	Cons
Adsorption-elution methods	Ionic charge		
Negatively charged filters		Good recoveries	Requires sample preconditioning
Positively charged filters		Good recoveries	Costly
Glass powder		Inexpensive; good recoveries	Fragile apparatus
Glass wool		Good recoveries	Differences depending on manufacturers
Precipitation methods	Chemical precipitation		
Organic flocculation		Efficient for dirty samples or as secondary concn	Beef extract is inhibitory to RT-PCR enzymes
Ammonium sulfate		Efficient for dirty samples or as secondary concn	High cytotoxicity; inhibitory to RT-PCR enzymes
PEG		Efficient for dirty samples or as secondary concn	Intra-assay variability
Ultracentrifugation	Physical sedimentation	Efficient as secondary concn	Costly
Lyophilization	Freeze-drying	Efficient for dirty samples or as secondary concn; may remove RT-PCR enzyme inhibitors	Costly and time-consuming
Ultrafiltration	Particle size separation	Good recoveries for clean samples	Costly and time-consuming
Magnetic beads	Immunoaffinity	Good recoveries from small volumes	Requires specific assay for each virus; costly; few data available

virus concentration in heavily polluted medium size samples, e.g., sewage, the latter method having the added advantage of removing substances inhibitory to RT-PCR (Gajardo et al., 1995).

RAPID ENUMERATION AND CHARACTERIZATION OF VIRUSES

When virus detection procedures are mentioned, the recurrent issue of detecting infectious or physical particles comes into discussion. Whenever possible, infectious assays coupled with identification methods are preferred for direct assessment of human health risk. The detection of infectious enteroviruses and even astroviruses or rotaviruses may be achieved by cell culture techniques with the appropriate cell lines. However, despite recent reports of cell lines permissive to the propagation of wild-type HAV (Konduru and Kaplan, 2006) and human NoV (Straub et al., 2007), issues related to assay complexity, cost-effectiveness, and even feasibility make their routine use difficult.

Nucleic acid amplification techniques are currently the most widely used methods for detection of viruses in food and water and also enable investigators to gather information on the virus genotypes occurring in the environment and in food products, thus providing the most relevant epidemiological information, particularly important for the implementation and follow-up of vaccination programs in the human population.

Although nucleic acid sequence-based amplification and loop-mediated isothermal amplification techniques have been reported as highly sensitive and specific, respectively (Fukuda et al., 2006; Jean et al., 2004), PCR and RT-PCR remain the current gold standard for virus detection. A further improvement comes from qRT-PCR, which enables not only qualitative determination but also, and particularly, quantitative diagnostic assays (Costafreda et al., 2006; Kageyama et al., 2003). Furthermore, recent evidence indicates that these are more sensitive than the conventional gel-based methods that were developed in the early 1990s. The detection sensitivity and the diagnostic specificity of molecular amplification are highly dependent on the efficiency of the upstream processing methods and on the purity and yield of the RNA.

An additional factor impacting the performance of qRT-PCR is the choice of primers, which is, in fact, very dependent upon the virus. A general rule of thumb is to use broadly reactive primers with high annealing temperature and relatively low degeneracy (to prevent nonspecific amplification). This is easy for some viruses and less so for others. This task, while attainable for HAV, is easier said than done for NoV. Certain RNA regions containing complex multidomain structures, mostly involved in translation and/or replication functions, must be highly conserved. For instance, HAV translation is cap independent and the recruitment of the translation machinery takes place at the internal ribosome entry site contained in the 5′ noncoding region (NCR); hence, the latter region is particularly suitable as a target for amplification in diagnostic assays (Costa-Mattioli et al., 2002a; Costafreda et al., 2006; Jothikumar et al., 2005a).

However, the genetic diversity of human NoV makes universal primer design more challenging, and the absence of a 5′ NCR and, in particular, of the internal ribosome entry site in the calicivirus genomic and subgenomic RNAs makes the approach described for HAV unfeasible. Four regions of the NoV genome have been used for primer design (designated regions A, B, C, and D), but the open reading frame 1 (ORF1)-ORF2 junction (just downstream of region B) seems to be the most conserved and is frequently used for genogroup-specific detection (da Silva et al., 2007; Jothikumar et al., 2005b; Kageyama et al., 2003; Loisy et al., 2005; Svraka et al., 2007).

Yet, the significance of a genome copy still remains controversial, since the virus-specific infectivity or infectious/physical particle ratio is highly variable in food and water samples. Several studies show a lack of correlation between the number of genome copies and infectivity in food and water safety studies (Baert et al., 2008b; El Senousy et al., 2007). On the other hand, a study on seafood reported a correlation between self-reported gastroenteric illness in restaurant consumers and the presence of NoV in the batch of oysters consumed (Lowther et al., 2010). As stated above, no alternative to molecular detection analysis exists for highly health-significant waterborne viruses such as human NoV and HAV.

The aforementioned approaches for the development of broadly reactive detection methods do not provide the ability to fulfill another ultimate goal in diagnostics, which is the possibility to type the isolates, as might be appropriate in outbreak investigations. For HAV, the most widely used region is the amino-terminal 2A region, which is included in a larger fragment corresponding to the VP1-2A junction region (Robertson et al., 1992). However, recent data point to the analysis of the entire VP1 region for improved type discrimination (Costa-Mattioli et al., 2002b; Tallo et al., 2003). For NoV strain comparison, primers corresponding to the capsid region can be used, although these sometimes have a higher degree of degeneracy. It is recommended that multiple primer pairs be used when screening food samples for NoV contamination, as one pair may perform better than another, depending on virus load and the matrix from which the sample was derived (Boxman et al., 2007; Vinje et al., 2003).

Although qRT-PCR provides both amplification and confirmation in a single test, definitive epidemiological association in outbreak investigations relies on direct sequencing of amplicons obtained from clinical and environmental/food samples. Microarray technologies may eventually speed this process. For example, Ayodeji et al. (2009) recently reported a microarray-based approach for the identification of common foodborne viruses, finding that the hybridization profiles generated for HAV are conducive to the identification of closely related strains. Similar studies on the use of microarrays for the simultaneous detection and genotyping of NoV were described by Pagotto et al. (2008).

Multiplexing of amplifications has been applied on occasion (Morales-Rayas et al., 2010) but may reduce detection sensitivity. Seminested amplifications are also commonly applied (Baert et al., 2008a), and these can improve detection limits but with an increased risk for cross-contamination leading to false-positive results. In general, nested amplification is not well suited to real-time amplification approaches. Interestingly, Hu and Arsov (2009) recently developed a method that combines conventional PCR, nested PCR, and real-time PCR in a multistep detection procedure. Specifically, the method involves two consecutive amplifications: the first is a conventional RT-PCR targeting the 5′ NCR of HAV, the second being a real-time PCR using a nested primer pair specific for the first PCR product and a TaqMan probe. This method facilitated the detection of as little as 0.2 PFU of HAV, which can be more sensitive than other RT-PCR techniques. An approach such as this can provide a useful method for detecting HAV at low levels, as long as the risk of cross-contamination is adequately controlled.

STANDARDIZATION OF VIRUS DETECTION METHODS

A major factor limiting the uptake of virus testing into regulatory food controls worldwide is the current absence of any standardized and validated methods. In 2004, the European Committee for Standardization initiated the development of a standard method for detection of NoV and HAV in foodstuffs based on PCR (Lees, 2010). The draft standard developed by a working group of expert European Union laboratories is now well advanced and due for publication in 2012.

As mentioned above, food samples present a challenging matrix and the standard method needs to be capable of extracting low levels of contaminating virus and presenting them in a noninhibitory extract to a sensitive PCR assay. Key aspects of the method were tested by interlaboratory evaluations to ensure robust performance. The method targets the at-risk food matrices of molluscan shellfish, soft fruits, salad vegetables, bottled water, and food surfaces (both the surfaces of foods and food preparation surfaces).

For bivalve mollusks, dissected digestive diverticulum (the digestive gland) is used as the starting material with further enzymatic digestion using proteinase K (Jothikumar et al., 2005b). For food surfaces, swabbing is employed followed by elution into sample buffer (Scherer et al., 2009). For both soft fruit and salad vegetables, viruses are eluted with agitation followed by recovery using PEG/NaCl precipitation (Dubois et al., 2007). For bottled water, viruses are adsorbed to a positively charged membrane, eluted, and then concentrated by ultrafiltration (Butot et al., 2007a).

Following initial sample treatment, all food matrices are then further processed by a common nucleic acid purification and PCR platform. Nucleic acid purification utilizes guanidine thiocyanate to denature viral coat proteins in combination with magnetic silica particles to bind released nucleic acid, which is then purified through successive washing stages before final elution in a small volume.

RT-PCR utilizes a one-step approach using specific primers in order to simplify the procedure as much as possible. However, commercial one-step kits must utilize enzymes specifically engineered for use with low-abundance targets. TaqMan PCR real-time chemistries are stipulated for the amplification because (i) the closed-tube format is less susceptible to contamination, (ii) it is logistically efficient, (iii) it incorporates a probe-based confirmation step, (iv) it is quantitative, and (v) it is more amenable to standardization than conventional PCR.

To maximize the detection limit, real-time PCR assays are done separately for NoV genogroup I, NoV genogroup II, and HAV. Cross-reactive real-time PCR primers and probes are directed in the ORF1-ORF2 junction region for NoV (da Silva et al., 2007; Jothikumar et al., 2005b; Loisy et al., 2005) and in the highly conserved 5′ NCR for HAV (Costafreda et al., 2006). The method is highly sensitive in order to detect the low levels of virus found in environmentally contaminated samples and hence also vulnerable to both cross-contamination (false positives) and potential matrix interferences (false negatives).

Thus, a comprehensive suite of controls was also developed to cover positive and negative process controls, negative RNA extraction control, positive RT-PCR and RT-PCR inhibition controls, and negative and positive PCR controls. The positive process control measures the recovery of virus during the whole extraction and test procedure by using a heterologous nonenveloped positive-sense single-stranded RNA virus spiked into the test sample and assayed in parallel with the target viruses. During the development of the method, interlaboratory studies by the working group successfully utilized the MC_0 strain of Mengo virus (Costafreda et al., 2006; Pintó et al., 2009) as a process control.

The negative process control is a known negative sample that is taken through the entire extraction procedure and analyzed. The RT-PCR inhibition control checks for potential matrix suppression by comparison of amplification of an external RNA template added to the test material and a control well. Taken together, the controls generate data on all aspects of the assay and are utilized to determine the acceptability of test performance against established quality control criteria.

The draft standard incorporates two parts covering both quantitative and qualitative detection. The differences principally relate to the necessary suite of controls and the calibration curves required for determining virus template concentrations. Quantification is based on a plasmid DNA calibration curve for each assay (NoV genogroup I, NoV genogroup II, and HAV) with plasmid DNA concentration measured by spectrometry at 260 nm. Results are reported in the standardized form of detectable virus genome copies per gram of material tested. Qualitative assays will report presence or absence with reference to their limit of detection. Formal validation studies are planned to characterize the method according to international requirements.

CONCLUSIONS AND FUTURE TRENDS

Overall, detection of viruses in foods is done infrequently and usually only in response to known or suspected foodborne disease outbreaks in which there is a strong epidemiological association with one or two candidate foods.

Despite the progress in the adoption of procedures for the detection of viruses in food and water, a decision to test a food product for virus contamination must be approached cautiously due to the following reasons. Although validation work is currently being pursued in Europe, including those focused on bivalve mollusks, soft fruits, and water, at the time of this writing, there are no internationally agreed-upon standardized and validated methods available for the detection of viruses in foods. Other limitations to the routine application of virus detection protocols are problems with demonstrating the association (or lack thereof) between a positive detection signal and virus infectivity; less than optimal recoveries during virus extraction, which may raise detection limits to unacceptably high levels; and finally, the expense and the need for highly trained personnel for implementing virus detection protocols.

Even when foods are tested for potential viral contamination, results should be interpreted with caution, i.e., a negative test result does not rule out product contamination, while a positive test result only suggests that contamination occurred at some point in the farm-to-fork chain but does not confirm the presence of infectious virus in the sample that was screened (D'Souza et al., 2010).

The risk analysis framework laid down by the Codex Alimentarius during the past 10 years has made it increasingly possible to link food safety to public health via risk assessment. However, it is still unclear how these new concepts will be used in future risk analysis. Quantitative risk assessments require quantitative data on viruses in food, and validated standardized procedures should be employed in selected commodities or in outbreak situations, which is particularly relevant in the present situation of global food trade.

However, some issues deserve attention: (i) the representativeness of the analyzed sample, considering that only a tiny amount of material, usually from a very large batch sample, is analyzed; (ii) the type of sample to analyze (e.g., outbreak related, imports, etc.), since it is unrealistic to routinely test all types of food for viruses; and (iii) the need to address a key question: who will cover the costs of the virological assays (e.g., producers, importers, distributors, or consumers)? Despite these pending questions, standardized virological assays will enable the formulation of guidelines to ensure the virological safety of selected commodities in specific scenarios, thus contributing to reduce the incidence of foodborne infections.

REFERENCES

Anonymous. 2005. *Microbiology of Food and Animal Feeding Stuffs—Horizontal Method for the Enumeration of Beta-Glucuronidase-Positive* Escherichia coli. *Part 3. Most Probable Number Technique Using 5-Bromo-4-Chloro-3-Indolyl-Beta-D-Glucuronide. ISO/TS16649-16653.* International Organization for Standardization, Geneva, Switzerland.

Atmar, R. L., F. H. Neill, J. L. Romalde, F. Le Guyader, C. M. Woodley, T. G. Metcalf, and M. K. Estes. 1995. Detection of Norwalk virus and hepatitis A virus in shellfish tissues with the PCR. *Appl. Environ. Microbiol.* **61:**3014–3018.

Ayodeji, M., M. Kulka, S. A. Jackson, I. Patel, M. Mammel, T. A. Cebula, and B. B. Goswami. 2009. A microarray based approach for the identification of common foodborne viruses. *Open Virol. J.* **3:**7–20.

Baert, L., M. Uyttendaele, and J. Debevere. 2008a. Evaluation of viral extraction methods on a broad range of Ready-To-Eat foods with conventional and real-time RT-PCR for Norovirus GII detection. *Int. J. Food Microbiol.* **123:**101–108.

Baert, L., C. E. Wobus, E. Van Coillie, L. B. Thackray, J. Debevere, and M. Uyttendaele. 2008b. Detection of murine norovirus 1 by using plaque assay, transfection assay, and real-time reverse transcription-PCR before and after heat exposure. *Appl. Environ. Microbiol.* **74:**543–546.

Bidawid, S., J. M. Farber, and S. A. Sattar. 2000. Rapid concentration of hepatitis A virus (HAV) from lettuce and strawberries. *J. Virol. Methods* **88:** 175–185.

Boxman, I. L., J. J. Tilburg, N. A. te Loeke, H. Vennema, E. de Boer, and M. Koopmans. 2007. An efficient and rapid method for recovery of norovirus from food associated with outbreaks of gastroenteritis. *J. Food Prot.* **70:**504–508.

Butot, S., T. Putallaz, C. Croquet, G. Lamothe, R. Meyer, H. Joosten, and G. Sanchez. 2007a. Attachment of enteric viruses to bottles. *Appl. Environ. Microbiol.* **73:**5104–5110.

Butot, S., T. Putallaz, and G. Sanchez. 2007b. Procedure for rapid concentration and detection of enteric viruses from berries and vegetables. *Appl. Environ. Microbiol.* **73:**186–192.

Centers for Disease Control and Prevention. 1997. Hepatitis A associated with consumption of frozen strawberries—Michigan, March 1997. *MMWR Morb. Mortal. Wkly. Rep.* **46:**288–295.

Costafreda, M. I., A. Bosch, and R. M. Pinto. 2006. Development, evaluation, and standardization of a real-time TaqMan reverse transcription-PCR assay for quantification of hepatitis A virus in clinical and shellfish samples. *Appl. Environ. Microbiol.* **72:**3846–3855.

Costa-Mattioli, M., S. Monpoeho, E. Nicand, M. H. Aleman, S. Billaudel, and V. Ferre. 2002a. Quantification and duration of viraemia during hepatitis A infection as determined by real-time RT-PCR. *J. Viral Hepat.* **9:**101–106.

Costa-Mattioli, M., J. Cristina, H. Romero, R. Perez-Bercof, D. Casane, R. Colina, L. Garcia, I. Vega, G. Glikman, V. Romanowsky, A. Castello, E. Nicand, M. Gassin, S. Billaudel, and V. Ferre. 2002b. Molecular evolution of hepatitis A virus: a new classification based on the complete VP1 protein. *J. Virol.* **76:**9516–9525.

Croci, L., E. Dubois, N. Cook, D. de Medici, A. Schultz, B. China, S. Rutjes, J. Hoorfar, and W. Van der Poel. 2008. Current methods for extraction and concentration of enteric viruses from fresh fruit and vegetables: towards international standards. *Food Anal. Methods* **1:**73–84.

da Silva, A. K., J. C. Le Saux, S. Parnaudeau, M. Pommepuy, M. Elimelech, and F. S. Le Guyader. 2007. Evaluation of removal of noroviruses during wastewater treatment, using real-time reverse transcription-PCR: different behaviors of genogroups I and II. *Appl. Environ. Microbiol.* **73:** 7891–7897.

DePaola, A., J. L. Jones, J. Woods, W. Burkhardt III, K. R. Calci, J. A. Krantz, J. C. Bowers, K. Kasturi, R. H. Byars, E. Jacobs, D. Williams-Hill, and K. Nabe. 2010. Bacterial and viral pathogens in live oysters: 2007 United States market survey. *Appl. Environ. Microbiol.* **76:**2754–2768.

Dore, W. J., K. Henshilwood, and D. N. Lees. 2000. Evaluation of F-specific RNA bacteriophage as a candidate human enteric virus indicator for bivalve molluscan shellfish. *Appl. Environ. Microbiol.* **66:**1280–1285.

D'Souza, D. H., J. Jean, and L. Jaykus. 2010. Methods for the detection of viral and parasitic protozoan pathogens in food, p. 1–23. *In* Y. H. Hui (ed.), *Handbook of Food Technology and Engineering,* vol. 4. CRC-Taylor and Francis, Boca Raton, FL.

Dubois, E., C. Hennechart, G. Merle, C. Burger, N. Hmila, S. Ruelle, S. Perelle, and V. Ferre. 2007. Detection and quantification by real-time RT-PCR of hepatitis A virus from inoculated tap waters, salad vegetables, and soft fruits: characterization of the method performances. *Int. J. Food Microbiol.* **117:**141–149.

El Senousy, W. M., S. Guix, I. Abid, R. M. Pinto, and A. Bosch. 2007. Removal of astrovirus from water and sewage treatment plants, evaluated by a competitive reverse transcription-PCR. *Appl. Environ. Microbiol.* **73:**164–167.

Endley, S., L. Lu, E. Vega, M. E. Hume, and S. D. Pillai. 2003. Male-specific coliphages as an additional fecal contamination indicator for screening fresh carrots. *J. Food Prot.* **66:**88–93.

Fukuda, S., S. Takao, M. Kuwayama, Y. Shimazu, and K. Miyazaki. 2006. Rapid detection of norovirus from fecal specimens by real-time reverse transcription-loop-mediated isothermal amplification assay. *J. Clin. Microbiol.* **44:**1376–1381.

Gajardo, R., R. M. Pinto, and A. Bosch. 1995. Polymerase chain-reaction amplification and typing of rotavirus in environmental samples. *Water Sci. Technol.* **31:**371–374.

Gaulin, C. D., D. Ramsay, P. Cardinal, and M. A. D'Halevyn. 1999. Epidemic of gastroenteritis of viral origin associated with eating imported raspberries. *Can. J. Public Health* **90:**37–40.

Hu, Y., and I. Arsov. 2009. Nested real-time PCR for hepatitis A detection. *Lett. Appl. Microbiol.* **49:** 615–619.

Jean, J., D. H. D'Souza, and L. A. Jaykus. 2004. Multiplex nucleic acid sequence-based amplification for simultaneous detection of several enteric viruses in model ready-to-eat foods. *Appl. Environ. Microbiol.* **70:**6603–6610.

Jothikumar, N., T. L. Cromeans, M. D. Sobsey, and B. H. Robertson. 2005a. Development and evaluation of a broadly reactive TaqMan assay for rapid detection of hepatitis A virus. *Appl. Environ. Microbiol.* **71:**3359–3363.

Jothikumar, N., J. A. Lowther, K. Henshilwood, D. N. Lees, V. R. Hill, and J. Vinje. 2005b. Rapid and sensitive detection of noroviruses by using TaqMan-based one-step reverse transcription-PCR assays and application to naturally contaminated shellfish samples. *Appl. Environ. Microbiol.* **71:**1870–1875.

Kageyama, T., S. Kojima, M. Shinohara, K. Uchida, S. Fukushi, F. B. Hoshino, N. Takeda, and K. Katayama. 2003. Broadly reactive and highly sensitive assay for Norwalk-like viruses based on real-time quantitative reverse transcription-PCR. *J. Clin. Microbiol.* **41:**1548–1557.

Kiulia, N. M., R. Netshikweta, N. A. Page, W. B. van Zyl, M. M. Kiraithe, A. Nyachieo, J. M. Mwenda, and M. B. Taylor. 2010. The detection of enteric viruses in selected urban and rural river water and sewage in Kenya, with special reference to rotaviruses. *J. Appl. Microbiol.* **109:**818–828.

Konduru, K., and G. Kaplan. 2006. Stable growth of wild-type hepatitis A virus in cell culture. *J. Virol.* **80:**1352–1360.

Lees, D. 2000. Viruses and bivalve shellfish. *Int. J. Food Microbiol.* **59:**81–116.

Lees, D. N. 2010. International standardisation of a method for detection of human pathogenic viruses in molluscan shellfish. *Food Environ. Virol.* **2:**146–155.

Le Guyader, F. S., J. C. Le Saux, K. Ambert-Balay, J. Krol, O. Serais, S. Parnaudeau, H. Giraudon, G. Delmas, M. Pommepuy, P. Pothier, and R. L. Atmar. 2008. Aichi virus, norovirus, astrovirus, enterovirus, and rotavirus involved in clinical cases from a French oyster-related gastroenteritis outbreak. *J. Clin. Microbiol.* **46:**4011–4017.

Le Guyader, F. S., S. Parnaudeau, J. Schaeffer, A. Bosch, F. Loisy, M. Pommepuy, and R. L. Atmar. 2009. Detection and quantification of noroviruses in shellfish. *Appl. Environ. Microbiol.* **75:** 618–624.

Lewis, G. D., and T. G. Metcalf. 1988. Polyethylene glycol precipitation for recovery of pathogenic viruses, including hepatitis A virus and human rotavirus, from oyster, water, and sediment samples. *Appl. Environ. Microbiol.* **54:**1983–1988.

Loisy, F., R. L. Atmar, P. Guillon, P. Le Cann, M. Pommepuy, and F. S. Le Guyader. 2005. Real-time RT-PCR for norovirus screening in shellfish. *J. Virol. Methods* **123:**1–7.

Lopman, B., H. Vennema, E. Kohli, P. Pothier, A. Sanchez, A. Negredo, J. Buesa, E. Schreier, M. Reacher, D. Brown, J. Gray, M. Iturriza, C. Gallimore, B. Bottiger, K. O. Hedlund, M. Torven, C. H. von Bonsdorff, L. Maunula, M. Poljsak-Prijatelj, J. Zimsek, G. Reuter, G. Szucs, B. Melegh, L. Svennson, Y. van Duijnhoven, and M. Koopmans. 2004. Increase in viral gastroenteritis outbreaks in Europe and epidemic spread of new norovirus variant. *Lancet* **363:**682–688.

Love, D. C., M. J. Casteel, J. S. Meschke, and M. D. Sobsey. 2008. Methods for recovery of hepatitis A virus (HAV) and other viruses from processed foods and detection of HAV by nested RT-PCR and TaqMan RT-PCR. *Int. J. Food Microbiol.* **126:**221–226.

Lowther, J. A., J. M. Avant, K. Gizynski, R. E. Rangdale, and D. N. Lees. 2010. Comparison between quantitative real-time reverse transcription PCR results for norovirus in oysters and self-reported gastroenteric illness in restaurant customers. *J. Food Prot.* **73:**305–311.

Lynch, M., J. Painter, R. Woodruff, and C. Braden. 2006. Surveillance for foodborne-disease outbreaks—United States, 1998–2002. *MMWR Surveill. Summ.* **55:**1–42.

Morales-Rayas, R., P. F. Wolffs, and M. W. Griffiths. 2010. Simultaneous separation and detection of hepatitis A virus and norovirus in produce. *Int. J. Food Microbiol.* **139:**48–55.

Nenonen, N. P., C. Hannoun, M. B. Olsson, and T. Bergstrom. 2009. Molecular analysis of an oyster-related norovirus outbreak. *J. Clin. Virol.* **45:**105–108.

Pagotto, F., N. Corneau, K. Mattison, and S. Bidawid. 2008. Development of a DNA microarray for the simultaneous detection and genotyping of noroviruses. *J. Food Prot.* **71:**1434–1441.

Papafragkou, E., M. Plante, K. Mattison, S. Bidawid, K. Karthikeyan, J. M. Farber, and L. A. Jaykus. 2008. Rapid and sensitive detection of hepatitis A virus in representative food matrices. *J. Virol. Methods* **147:**177–187.

Pintó, R. M., M. I. Costafreda, and A. Bosch. 2009. Risk assessment in shellfish-borne outbreaks of hepatitis A. *Appl. Environ. Microbiol.* **75:**7350–7355.

Ponka, A., L. Maunula, C. H. von Bonsdorff, and O. Lyytikainen. 1999. An outbreak of calicivirus associated with consumption of frozen raspberries. *Epidemiol. Infect.* **123:**469–474.

Robertson, B. H., R. W. Jansen, B. Khanna, A. Totsuka, O. V. Nainan, G. Siegl, A. Widell, H. S. Margolis, S. Isomura, K. Ito, T. Ishizu, Y. Moritsugu, and S. M. Lemon. 1992. Genetic relatedness of hepatitis A virus strains recovered from different geographical regions. *J. Gen. Virol.* **73:**1365–1377.

Rutjes, S. A., H. H. J. L. van den Berg, W. J. Lodder, and A. M. de Roda Husman. 2006. Real-time detection of noroviruses in surface water by use of a broadly reactive nucleic acid sequence-based amplification assay. *Appl. Environ. Microbiol.* **72:**5349–5358.

Scherer, K., D. Mäde, L. Ellerbroek, J. Schulenburg, R. Johne, and G. Klein. 2009. Application of a swab sampling method for the detection of norovirus and rotavirus to artificially contaminated food and environmental surfaces. *Food Environ. Virol.* **1:**42–49.

Shieh, Y. C., Y. E. Khudyakov, G. Xia, L. M. Ganova-Raeva, F. M. Khambaty, J. W. Woods, J. E. Veazey, M. L. Motes, M. B. Glatzer, S. R. Bialek, and A. E. Fiore. 2007. Molecular confirmation of oysters as the vector for hepatitis A in a 2005 multistate outbreak. *J. Food Prot.* **70:**145–150.

Sobsey, M. D., and J. S. Glass. 1980. Poliovirus concentration from tap water with electropositive adsorbent filters. *Appl. Environ. Microbiol.* **40:**201–210.

Straub, T. M., K. Höner zu Bentrup, P. O. Orosz-Coghlan, A. Dohnalkova, B. K. Mayer, R. A. Bartholomew, C. O. Valdez, C. J. Bruckner, C. P. Gerba, M. Abbaszadegan, and C. A. Nickerson. 2007. In vitro cell culture infectivity assay for human noroviruses. *Emerg. Infect. Dis.* **13:**396–403.

Svraka, S., E. Duizer, H. Vennema, E. de Bruin, B. van der Veer, B. Dorresteijn, and M. Koopmans. 2007. Etiological role of viruses in outbreaks of acute gastroenteritis in The Netherlands from 1994 through 2005. *J. Clin. Microbiol.* **45:**1389–1394.

Tallo, T., H. Norder, V. Tefanova, K. Ott, V. Ustina, T. Prukk, O. Solomonova, J. Schmidt, K. Zilmer, L. Priimägi, T. Krispin, and L. O. Magnius. 2003. Sequential changes in hepatitis A virus genotype distribution in Estonia during 1994 to 2001. *J. Med. Virol.* **70:**187–193.

Terio, V., V. Martella, P. Moschidou, P. Di Pinto, G. Tantillo, and C. Buonavoglia. 2010. Norovirus in retail shellfish. *Food Microbiol.* **27:**29–32.

Teunis, P. F., C. L. Moe, P. Liu, S. E. Miller, L. Lindesmith, R. S. Baric, J. Le Pendu, and R. L. Calderon. 2008. Norwalk virus: how infectious is it? *J. Med. Virol.* **80:**1468–1476.

Verhoef, L., H. Vennema, W. van Pelt, D. Lees, H. Boshuizen, K. Henshilwood, and M. Koopmans. 2010. Use of norovirus genotype profiles to differentiate origins of foodborne outbreaks. *Emerg. Infect. Dis.* **16:**617–624.

Vilagines, P., B. Sarrette, G. Husson, and R. Vilagines. 1993. Glass wool for virus concentration at ambient water pH level. *Water Sci. Technol.* **27:** 3–4.

Villena, C., W. M. El-Senousy, F. X. Abad, R. M. Pinto, and A. Bosch. 2003. Group A rotavirus in sewage samples from Barcelona and Cairo: emergence of unusual genotypes. *Appl. Environ. Microbiol.* **69:**3919–3923.

Vinje, J., H. Vennema, L. Maunula, C. H. von Bonsdorff, M. Hoehne, E. Schreier, A. Richards, J. Green, D. Brown, S. S. Beard, S. S. Monroe, E. de Bruin, L. Svensson, and M. P. Koopmans. 2003. International collaborative study to compare reverse transcriptase PCR assays for detection and genotyping of noroviruses. *J. Clin. Microbiol.* **41:**1423–1433.

Westrell, T., V. Dusch, S. Ethelberg, J. Harris, M. Hjertqvist, S. N. Jourdan-da, A. Koller, A. Lenglet, M. Lisby, and L. Vold. 2010. Norovirus outbreaks linked to oyster consumption in the United Kingdom, Norway, France, Sweden and Denmark, 2010. *Euro Surveill.* **15**(12)**:**19524.

Wheeler, C., T. M. Vogt, G. L. Armstrong, G. Vaughan, A. Weltman, O. V. Nainan, V. Dato, G. L. Xia, K. Waller, J. Amon, T. M. Lee, A. Highbaugh-Battle, C. Hembree, S. Evenson, M. A. Ruta, I. T. Williams, A. E. Fiore, and B. P. Bell. 2005. An outbreak of hepatitis A associated with green onions. *N. Engl. J. Med.* **353:**890–897.

World Health Organization. 2004. *Guidelines for Drinking-Water Quality*, 3rd ed., vol. 1. *Recommendations.* World Health Organization, Geneva, Switzerland.

Wyn-Jones, P. 2007. The detection of waterborne viruses, p. 177–204. *In* A. Bosch (ed.), *Human Viruses in Water*, vol. 17. Elsevier, Amsterdam, The Netherlands.

PROTOZOAN PARASITES: *CRYPTOSPORIDIUM*, *GIARDIA*, *CYCLOSPORA*, AND *TOXOPLASMA*

B. R. Dixon, R. Fayer, M. Santín, D. E. Hill, and J. P. Dubey

24

The foodborne protozoan parasites of particular concern with respect to public health include *Cryptosporidium* spp., *Giardia duodenalis, Cyclospora cayetanensis*, and *Toxoplasma gondii* (Nichols and Smith, 2002; Dixon, 2003; Dawson, 2005; Dubey, 2009). These organisms commonly infect humans and animals worldwide and are responsible for enteric illness or, in the case of *T. gondii*, flu-like symptoms or neurological problems. The severity and duration of symptoms are largely dependent upon the immunological status of the host. While person-to-person transmission and the waterborne route are much more common for *Cryptosporidium* spp. and *G. duodenalis*, foodborne transmission is reported more frequently and likely accounts for more cases and outbreaks of infection with *C. cayetanensis* and *T. gondii* than the waterborne route, at least in developed countries. In fact, numerous foodborne cases and outbreaks of illness due to infections with all of these parasites have been reported in recent years. While some prepared foods have been implicated, most of these foodborne outbreaks have been associated with cyst- or oocyst-contaminated fresh fruits and vegetables, with the exception of *T. gondii*, which is generally associated with the consumption of raw or poorly cooked meats or organs containing tissue cysts. The total number of reported foodborne illnesses resulting from infection with these organisms is considerably less than the number of illnesses resulting from infection with bacteria and viruses. For example, the combined number of foodborne cases of *Cryptosporidium* spp., *G. duodenalis, C. cayetanensis*, and *T. gondii* was estimated at over 350,000 per year in the United States, which amounted to only 2.6% of the total foodborne illnesses reported (Mead et al., 1999), and this trend continues (Scallan et al., 2011). However, foodborne transmission of protozoan parasites is of growing importance in light of factors such as the global nature of the food trade, international travel, the increased number of immunocompromised and other susceptible individuals, and changes in consumer habits, particularly the consumption of more raw and undercooked foods.

Making an association between the consumption of a suspected food and human illness associated with a parasitic infection is often difficult due to the relatively long in-

B. R. Dixon, Bureau of Microbial Hazards, Food Directorate, Health Canada, Ottawa, ON, K1A 0K9, Canada. *R. Fayer, M. Santín, D. E. Hill, and J. P. Dubey*, Agricultural Research Service, U.S. Department of Agriculture, Beltsville, MD 20705.

Rapid Detection, Identification, and Quantification of Foodborne Pathogens, ed. by J. Hoorfar

cubation period of these pathogens and the problems in diagnosing and reporting human cases. The relatively short shelf life of some foods, particularly produce, adds to the problem. In many cases, the implicated food(s) would have been consumed or disposed of long before any testing could be initiated, unless portions had been frozen. As a result, foods are generally implicated based on epidemiological data rather than actual detection of parasites. When portions of a suspected food are still available, or when surveys are being performed to determine the prevalence of parasites in specific foods, rapid and sensitive methods for the detection of protozoan parasites are required. However, with the exception of *T. gondii* in meats and organs, there are presently no standardized methods for the detection of protozoan parasites in foods, and published techniques are generally inadequate, with low and variable recovery efficiencies (Dixon, 2003). The number of parasites present in or on foods is generally much lower than that found in fecal material, and since protozoans will only multiply in the host or in vitro, enrichment steps such as those used in the detection of bacteria in foods are needed but are not readily available. The recovery of cysts and oocysts is further limited by the fact that generally only a small proportion of a crop or batch can be sampled. Finally, cysts and oocysts are often very difficult to wash off produce and, as a result, their elution and concentration from most types of foods present major challenges. As a result, the incidence of foodborne illness due to infection with protozoan parasites is likely underestimated, and it is often very difficult to determine the level of contamination of raw foods in surveys or to confirm the food source involved in the event of an outbreak.

This chapter will outline the general biology, mechanisms of transmission, and prevalence of infection of *Cryptosporidium* spp., *G. duodenalis, C. cayetanensis*, and *T. gondii* and will focus on their prevalence in foods, reported foodborne outbreaks, and the methods available for their detection, characterization, and control.

CRYPTOSPORIDIUM SPP.

Prevalence

There are 21 valid species of *Cryptosporidium* and >40 other cryptosporidia identified as genotypes (Fayer, 2010). These genetically unique genotypes have been detected in animal feces by molecular methods but lack sufficient additional data to qualify them as unique species at this time. Some species or genotypes of *Cryptosporidium* are found primarily in closely related hosts, such as those found in rodents or those that infect ruminants, while others infect only one known host species. A few other species of *Cryptosporidium*, such as *C. parvum*, have a wide range of hosts. Cryptosporidiosis has been reported in humans in 106 countries (Fayer, 2008). Of the 12 species and genotypes reported to infect humans, the two most prevalent infections result from *Cryptosporidium hominis* (anthroponotic transmission) and *C. parvum* (zoonotic transmission). Cryptosporidiosis is an acute self-limiting disease marked by enteritis (Chalmers and Davies, 2010).

Endogenous stages of the life cycle in gut epithelium begin with asexual multiplication followed by sexual stages leading to fertilization and formation of oocysts that become exogenous and transmit the parasite through direct contact with feces or feces-contaminated water or food.

Surveillance in Foods

Relatively few surveillance studies have been conducted for *Cryptosporidium* associated with foods. In Ontario, Canada, a baseline study of unpasteurized apple cider found *C. parvum* in 2.2% of the samples (Garcia-Vela et al., 2003). In a follow-up study involving the same five producers, *C. parvum* was found in samples of the water from producers 1 and 2 (producer 2 used municipal water), on the washed apples from producer 4, in the fresh cider from producers 1, 2, and 3, and in the finished cider from producer 1, but none of the unwashed apples tested positive for *C. parvum* (Garcia et al., 2006). The fresh cider positive for *C. parvum* was provided by producer 3, who used the cider only for his own consumption. The apples were grounders from an orchard where cows grazed in the fall to eat grounders.

Oocysts have been detected on fresh vegetables in several countries. In Costa Rica, *Cryptosporidium* oocysts were found in 5.2% (4/80) of cilantro leaves, 8.7% (7/80) of cilantro roots, and 2.5% of lettuce samples, and a 1.2% (1/80) incidence was found in other vegetables including carrots, cucumbers, radishes, and tomatoes (Monge and Arias, 1996; Monge et al., 1996). Also in Costa Rica, 25 samples each of lettuce, parsley, cilantro, strawberries, and blackberries were taken in the dry season and 25 in the rainy season from five local agricultural markets and examined by microscopy (Calvo et al., 2004). Oocysts were present in all produce except strawberries. In Peru, lettuce, mint, cilantro, and parsley were found to be contaminated with oocysts (Ortega et al., 1997). In Norway, among 475 samples of fresh vegetables, 5 of 13 lettuce samples from various countries and 14 of 16 mung bean sprout samples grown in Norway from imported seeds were positive for *Cryptosporidium* oocysts (Robertson and Gjerde, 2001a). A larger study of alfalfa, mung beans, radishes, and mixed sprouts, all grown in Norway by the same producer, found 14 of 149 mung bean sprout samples, 6 of 8 100-g batches of mung bean seeds, and one sample of spent irrigation water positive for *Cryptosporidium* (Robertson et al., 2002). In Poland, among fresh vegetables and soft fruits purchased at farmers' markets, 3.6% of leeks, celery, and red and Peking cabbages had oocysts eluted from their surfaces (Rzezutka et al., 2010). Empty oocysts were detected in the majority of samples, with only celery and Peking cabbage harboring intact oocysts. Of 35 fruit samples, none were positive. The farms that grew the oocyst-contaminated samples were located in districts with the highest numbers of cattle, whereas no contaminated produce came from farms in districts that had lower numbers of cattle-owning homesteads.

Shellfish, including oysters, clams, cockles, and mussels, feed on suspended plankton by filtering several liters of water a day through their gills. Oocysts have been detected in the gills and/or stomach of edible shellfish in the United States, the United Kingdom, Ireland, Spain, Portugal, Italy, The Netherlands, Brazil, Japan, and Thailand, although there have been no reports of human infection linked to eating uncooked shellfish (Fayer et al., 2003, 2004; Izumi et al., 2006; Schets et al., 2007; Guiguet Leal et al., 2008; Srisuphanunt et al., 2009).

Foodborne Outbreaks

Three outbreaks have been associated with drinking unpasteurized apple cider. At a school agricultural fair in Maine, 154 persons became infected after drinking freshly pressed cider (Millard et al., 1994). Oocysts were detected in the cider, the cider press, and a calf on the farm that provided the apples. A second outbreak involving 20 confirmed and 11 suspected cases of persons with diarrhea, cramping, vomiting, fever, and bloody diarrhea again implicated cider based on epidemiologic evidence, this time identifying a mill in New York State as the source (Anonymous, 1997). In a third cider-related outbreak, there were 23 laboratory-confirmed cases and 120 probable cases of persons who drank ozonated cider produced at a site in Ohio (Blackburn et al., 2006). Of 12 fecal samples examined by molecular methods, 11 contained *C. parvum* and 1 contained the *Cryptosporidium* cervine genotype (now *C. ubiquitum*). Cider provided by one patient was positive for the *C. parvum* subtype IIaA17G2R1, the same subtype that infected this patient.

Unpasteurized milk has been implicated in five outbreaks. The first outbreak involved 19 persons in semirural Wales and is the most tentative. The authors state, "Despite the absence of statistical proof ... or of confirmed positive findings in the examination of milk, raw milk would appear to provide a possible vehicle for transmission of cryptosporidia" (Casemore et al., 1986). In the United Kingdom, 50 children became ill after drinking milk their school obtained from a producer whose on-farm pasteurization equipment had malfunctioned (Gelletlie et al., 1997). The third outbreak involved 22 Canadians who drank unpasteurized milk while traveling in Mexico (Elsser et al., 1986). A fourth report involved a mother and child who drank unpasteurized goat's milk in Australia (Anonymous, 1984). A fifth report involved eight children in Queensland, Australia, who drank unpasteurized cow's milk, four of whom required hospitalization (Harper et al., 2002).

Outbreaks have been associated with social gatherings. Infection of 50 persons at a gathering in Minnesota was linked to chicken salad prepared by a food handler who also operated a day care facility and had changed a child's diaper earlier that day (Anonymous, 1996). In Spokane, WA, 54 of 62 banquet attendees became ill 3 to 9 days later; 2 were hospitalized and 6 others sought health care (Anonymous, 1998). Fifty-one case-patients recalled eating foods that contained uncooked green onions. The onions had not been washed before delivery, and food workers did not consistently wash them before preparing food. Of two food workers that tested positive for *Cryptosporidium*, one was symptomatic at the same time as the banquet attendees and the other was asymptomatic. A third reported working while experiencing diarrhea but could not remember the dates of the illness. Although the strongest association between illness and a menu item was for items containing uncooked green onions, the data cannot conclusively implicate them because other items could have been contaminated before arriving at the restaurant, or by a food worker, or by cross-contamination during preparation. In Sweden, after a wedding reception at a hotel, 15 of 23 guests and 6 of 8 restaurant staff members developed gastroenteritis due to cryptosporidiosis, and 3 persons were hospitalized with dehydration (Insulander et al., 2008). Because none of the staff reported symptoms around the day the dinner was served, it appeared all persons became infected at the same time. The most probable source of the outbreak appeared to be béarnaise sauce containing chopped fresh parsley.

Outbreaks have also been associated with cafeteria-style settings. In Denmark, illness in 99 company employees, 13 positive for *C. hominis*, was most probably associated with eating peeled whole carrots served in a bowl of water, grated carrots, or red peppers from the company's canteen salad bar possibly contaminated by a person excreting the parasite (Ethelberg et al., 2009). In Washington, DC, gastroenteritis affected 88 students and 4 workers at a university cafeteria (Quiroz et al., 2000). Oocysts were identified in feces of the sick cafeteria workers. A worker with the earliest signs of illness (watery diarrhea) prepared fruits and vegetables. Of 25 specimens submitted for DNA analysis, all were positive for *C. hominis*.

The only suggestion of an association between eating meat (sausage) and acquiring cryptosporidiosis also implicated unpasteurized milk as a possible source of an outbreak (Casemore et al., 1986). This suggestion was based solely on epidemiological data.

Detection Methods

Unless oocysts are extremely abundant in the matrix in which they are collected, they must be concentrated and purified from surrounding debris for detection and identification. Recovery of oocysts from the matrix of irrigation or wash water, soft fruit, salad vegetables, prepared foods, soil, or feces can be problematic, and many oocysts can be lost in the process. Each matrix can contain debris or unknown factors that trap oocysts and inhibit PCRs. Water can contain silt, humic acid, organic material, and free-living microbes. Feces can contain partially digested foodstuffs, mucus, lipids, bacteria, and fungi. Soils such as clays can bind oocysts and significantly decrease recovery. Methods for recovery and concentration of oocysts by sieving, sedimentation, flocculation, formol ethyl acetate, cartridge and membrane filtration, and centrifugation have been reviewed by Zarlenga and Trout (2004). Standardized filtration methods have been described for recovery of oocysts from water (U.S. Environmental Protection Agency, 2001a, 2001b). Removal of oocysts from unwanted matrix debris can employ density gradient centrifugation on high specific gravity solutions to float parasites and pellet heavier debris, or densities may be altered to float lighter debris and pellet the oocysts. Immunomagnetic bead separation can be used alone or following such cleanup procedures (Zarlenga and Trout, 2004).

Identification methods can be grouped into general morphology by microscopy, chemical and immunofluorescent staining methods aiding microscopy, and biochemical and molecular tests. Traditionally, the detection of *Cryptosporidium* oocysts has relied on examination by microscopy. Microscopic observations can be made using bright-field (with or without special stains), phase-contrast, differential interference contrast, or fluorescence microscopy. However, morphology cannot be used to identify species and genotypes of *Cryptosporidium* because their oocyst size ranges overlap, nor can it be used for source tracking. Because not all *Cryptosporidium* species and genotypes that contaminate foods are infectious to humans, molecular methods are required to determine species, genotypes, and subtypes. The preferred gene target for genotyping *Cryptosporidium* is the small-subunit rRNA gene and for subtyping *C. hominis* and *C. parvum* is the GP60 gene (Xiao, 2010). Genotyping and subtyping tools for *Cryptosporidium* are useful in epidemiological investigations of zoonotic transmission and in tracking sources of infection and contamination. As an example of the valuable information obtained by subtyping in investigating an outbreak, in the 1998 Washington, DC, *Cryptosporidium* foodborne outbreak, the same *C. hominis* subtype (IbA10G2) was identified in the food handler and in the outbreak patients (Quiroz et al., 2000).

There is no standardized method for detection of oocysts on foods. Cook et al. (2006) reported detecting oocysts on lettuce and raspberries by concentration and separation using centrifugation followed by immunomagnetic separation and oocyst staining. The optimized method to detect oocysts on artificially contaminated lettuce and raspberries recovered 47 to 71% and 28 to 54% of oocysts ($n = 30$), respectively. Interlaboratory validation of this method yielded correct identification of all artificially contaminated lettuce and raspberry samples at 89.6 and 95%, respectively (Cook et al., 2007).

For removal of oocysts from the surface of artificially contaminated apples, Macarisin et al. (2010b) compared sonication and orbital shaking using three different dissociating buffers, one containing glycine (Cook et al., 2006), another consisting of phosphate-buffered saline with 0.01% Tween 20 (Bohaychuk et al., 2009), and the third consisting of Tris HCl with 0.1% sodium dodecyl sulfate. Apples also were rigorously manually washed with warm water and Dial antibacterial liquid soap and then surface wiped with paper towels. The most efficient removal of oocysts (37.5%) was achieved by rigorous manual washing and also by agitation in an orbital shaker with Tris-HCl with 0.1% sodium dodecyl sulfate buffer, based on detection by nested PCR.

Control Measures

Oocysts can survive under cool moist conditions for several months. At 5 to 10°C, some have remained infectious for over 6 months, but they are quickly rendered inactive by freezing and high temperatures (Fayer, 2008).

Since cattle are a major source of *C. parvum* oocysts, and sheep, goats, and deer also are sources of zoonotic oocysts, dissemination of oocysts in the farming environment can be minimized by restricting these ruminants from fields and orchards where produce is grown. Livestock manure should be thoroughly composted before application to fields where food crops are grown. Livestock should be restricted from waterways and farm ponds. Irrigation waters are likely sources of contamination for fresh produce, considering that 36% of waters used to irrigate crops traditionally eaten raw in the United States and Central America tested positive for *C. parvum* oocysts (Thurston-Enriquez et al., 2002), and irrigation waters examined in Mexico and Norway also contained oocysts (Chaidez et al., 2005; Robertson et al., 2005).

Fruits and vegetables can be contaminated with oocysts during postharvest washing (Garcia et al., 2006), as indicated by the finding of oocysts in wash water tanks in 16% of vegetable packing houses in Mexico (Chaidez et al., 2005). Disinfecting these tanks is difficult because oocysts can be protected within bacterial biofilms in water reservoirs and distribution systems (Helmi et al., 2008) and released later, recontaminating the water (Angles et al., 2007). Under experimental conditions, *C. parvum* oocysts strongly adhere to spinach plants after contact with contaminated water, and some oocysts infiltrated leaves through the stomatal openings and persisted internally even after washing (Macarisin et al., 2010a), again indicating the need for both clean irrigation and wash water. Chemicals used to disinfect drinking and industrial water, such as chlorine and chloramine, are ineffective for disinfecting *Cryptosporidium* at concentrations typically applied (Korich et al., 1990). Some oocysts remained infective after exposure to concentrated laundry bleach for 2 hours (Fayer, 1995). Oocysts in milk and water were rendered inactive by pasteurization (Harp et al., 1996). Food handlers should thoroughly wash their hands, wear clean disposable gloves, and adhere to good hygienic practices, and those who are ill should not handle food. Potential transmission can be reduced or averted by including *Cryptosporidium* in hazard analysis and critical control point plans of industries or providers of fresh produce, and operations in which processing water or ingredient water could contaminate the food product.

GIARDIA DUODENALIS

Prevalence

Giardia duodenalis (synonyms, *G. intestinalis* and *G. lamblia*) is identified by the World Health Organization (WHO) as an organism causing a neglected disease and is one of the most prevalent protozoan diseases worldwide. Prevalence rates for giardiasis range from 2 to 7% in developed countries and 20 to 30% in most developing countries. The CDC estimates there are upwards of 2.5 million cases of giardiasis in the United States annually. Giardiasis is more prevalent in children than in adults, possibly because many individuals seem to have a lasting immunity after infection. *G. duodenalis* is implicated in 25% of all cases of gastrointestinal disease but can also be present asymptomatically. This disease afflicts many homosexual men, both HIV-positive and HIV-negative individuals, presumably due to sexual transmission. The disease is also common in child day care centers, especially in those where diapering is done.

From 2003 to 2005, the total number of cases of giardiasis reported in the United States remained relatively stable, i.e., 20,084, 20,962, and 20,075 for the years 2003, 2004, and 2005, respectively. Compared with other age groups, most cases involved children aged 1 to 9 years and adults aged 30 to 39 years, with peak illness during early summer through early fall (Yoder and Beach, 2010). From 1995 to 2004, *G. duodenalis* was the causal agent of 5 recreational water-associated gastroenteritis outbreaks and 14 drinking water-associated outbreaks. In addition, foodborne outbreaks of giardiasis were linked to infected food handlers and to uninfected food handlers who had diapered infected children prior to handling food (Quick et al., 1992).

Giardia duodenalis is considered a species complex containing several major morphologically similar but genetically distinct assemblages/genotypes identified by a variety of genetic loci including the small-subunit rRNA, β-giardin, glutamate dehydrogenase, triose phosphate isomerase, and other genes.

Eight assemblages/genotypes have been described and are designated by the letters A through H. Genotypes A and B are the most prevalent, found in humans and other mammals. Genotypes C and D are commonly found in canids, including dogs and coyotes; genotype E appears to be specific for artiodactyls; genotype F infects cats; genotype G infects rodents; and genotype H infects seals (Feng and Xiao, 2011).

The trophozoite stage is a flagellated, binucleate protozoan parasite that multiplies in the lumen of the upper digestive tract and causes the illness. It forms a cyst stage that passes from the body and is transmitted to nonimmune hosts primarily through contaminated water, but also via contaminated food or by direct contact with feces containing cysts.

Surveillance in Foods

In Philippine supermarkets and a public market, 1% of the vegetables tested were contaminated with cysts (de Leon et al., 1992). Five percent of vegetables in Costa Rica were contaminated with *Giardia* cysts (Monge and Arias, 1996). In Norway, the contamination rate with *Giardia* was found to be 2% (Robertson and Gjerde, 2000), and *Giardia* cysts were detected in dill, lettuce, mung bean sprouts, radish sprouts, and strawberries (Robertson and Gjerde, 2001a). Several studies on the recovery of parasites from vegetables have been conducted in Iran, where the prevalence of parasites, including *Giardia*, was high in all examined vegetables (Daryani et al., 2008). This could be due to cultivation in gardens fertilized with nontreated night soil. In Ardabil, Iran, 7% of imported vegetables and 8.9% of vegetables cultivated locally were contaminated with *Giardia* (Daryani et al., 2008).

Foodborne Outbreaks

The Centers for Disease Control and Prevention have estimated that 60% of giardiasis cases are associated with contaminated water. A few outbreaks have been traced to food contamination by infected food handlers, and the possibility of infections from contaminated vegetables that are eaten raw cannot be excluded. As with cryptosporidiosis, about 10% of cases of giardiasis are thought to be foodborne (Mead et al., 1999).

In Minnesota, 29 persons at a school gathering became ill after eating home-canned salmon prepared by a person caring for a 1-year-old child who was excreting cysts (Osterholm et al., 1981). In Connecticut, 13 persons became ill after attending a picnic where noodle salad was prepared by a person who had asymptomatic children who were excreting cysts (Petersen et al., 1988). In New Jersey, following a family party, nine persons who had eaten fruit salad became ill, whereas only one person who had not eaten the salad became ill (Porter et al., 1990). The fruit salad preparer had a diapered child and a pet rabbit at home, who were both positive for *Giardia*. In Minnesota, 88 residents at a nursing home who had eaten sandwiches prepared by persons excreting cysts developed giardiasis (White et al., 1989). In New Mexico, 21 persons who attended a church youth group dinner acquired giardiasis (Anonymous, 1989). In Washington, 27 staff of a job training center became ill after attending a day-long meeting in a restaurant that had multiple sanitary violations (Quick et al., 1992). No individual food item could be conclusively associated with illness, but one asymptomatic employee serving ice was excreting cysts, and another had a child excreting cysts. Twenty-seven cases were diagnosed by passive surveillance indicating that sliced raw vegetables served in the employee cafeteria were prepared by a food handler infected with *Giardia* (Mintz et al., 1993). In California, between 2001 and 2008, two foodborne outbreaks of giardiasis involving a total of 80 cases were reported, but neither had a vehicle identified (Cummings and Karon, 2009). Smith et al. (2007) also mention foodborne infection from eating tripe soup made from sheep and Christmas pudding contaminated with rat feces.

Detection Methods

Giardiasis is diagnosed microscopically by demonstration of the organism in stained fecal smears. This technique enables the viewer to observe primarily cysts or, less frequently, the trophozoite stage. Methods to enhance the likelihood of detection include concentration of cysts from stool or environmental specimens followed by staining with colored dyes or fluorescing dyes linked to antibodies. These methods have been developed for detecting waterborne cysts in much the same way as for detecting *Cryptosporidium* oocysts. The use of molecular methods to detect *Giardia* DNA has improved the sensitivity and specificity of detection methods. Separating cysts from complex food matrices that hinder detection has been problematic, and studies have been few, generally following the methods used for *Cryptosporidium* and *Cyclospora*. The recovery efficiency of cysts from contaminated fruits and vegetables was considered poor (Rose and Slifko, 1999) before the studies by Robertson and Gjerde (2000, 2001b). Washing methods, sonication, and immunomagnetic bead separation methods were compared for lettuce, Chinese leaves, and strawberries, with recovery efficiencies reaching 67%. Recovery from bean sprouts was especially difficult.

Control Measures

Survival of *Giardia* cysts in the environment decreases as the temperature increases. At 21°C, some cysts can survive for almost a month. Many cysts can survive for 2 to 3 months in water temperatures of less than 10°C, and a few can survive one freeze-thaw cycle. However, at 54°C cysts are killed in 10 minutes, and boiling kills cysts immediately.

Since most *G. duodenalis* infections are spread by contaminated recreational and drinking water, efforts to reduce or eliminate contamination need to focus on preventing persons with diarrhea, or known *G. duodenalis* infections, from entering recreational waters and stopping backpackers or campers from drinking water directly from rivers, streams, or ponds unless the water is filtered or boiled. Because most foodborne infections with *G. duodenalis* have been traced to food handlers with giardiasis or diarrhea or in association with children with giardiasis, these individuals should not be handling food served to the public or their family members. Since this action would be difficult or impossible to enforce, all food handlers should wear clean disposable gloves when handling foods. A general approach for prevention of foodborne disease is to wash, peel, or cook raw fruits and vegetables before eating.

CYCLOSPORA CAYETANENSIS

Prevalence

The organism responsible for cyclosporiasis was first recognized as a cause of human illness in New Guinea in the late 1970s. This unidentified human pathogen was later described as occurring in travelers to Haiti and Mexico and in a number of subsequent reports as occurring in travelers and inhabitants of regions of endemicity around the world in the 1980s and early 1990s (Ortega and Sanchez, 2010). The causative agent was finally identified as a coccidian protozoan parasite, which was named *Cyclospora cayetanensis* in 1993 (Ortega et al., 1993). While humans appear to be the only host for *C. cayetanensis*, other *Cyclospora* species exist in a variety of mammals, reptiles, and myriapods (Ortega and Sanchez, 2010).

The life cycle of *C. cayetanensis* is similar to that of other coccidians and has been described elsewhere (Mansfield and Gajadhar, 2004; Ortega and Sanchez, 2010). The transmission stage of *C. cayetanensis*, known as an oocyst, is spherical and 8 to 10 μm in diameter. Unsporulated (immature) oocysts are shed into the environment with the feces of infected individuals. While the oocysts do not multiply outside the host, they do undergo sporulation, depending upon conditions such as temperature and humidity, becoming mature and viable within 7 to 15 days (Ortega and Sanchez, 2010).

C. cayetanensis has been identified as a cause of diarrheal illness in natives and travelers to

North, Central, and South America, the Caribbean, Southeast Asia, eastern Europe, the United Kingdom, India, and Africa (Karanja et al., 2007; Ortega and Sanchez, 2010). Most of the prevalence data, however, come from studies done in the regions of endemicity, i.e., Nepal, Haiti, Peru, and Guatemala (Cama, 2006; Ortega and Sanchez, 2010). While most people in developed countries are susceptible to infection, those at greatest risk of infection with *C. cayetanensis* include travelers to regions of endemicity and immunocompromised individuals. In countries of endemicity, susceptibility is restricted to the very young and the very old, suggesting that acquired immunity may play a role (Ortega and Sanchez, 2010). There is a marked seasonality associated with human infections, which is likely associated with temperature, rainfall, humidity, and other factors (Eberhard and Arrowood, 2002; Tram et al., 2008).

Many of the earlier outbreaks were thought to have resulted from the consumption of fecally contaminated drinking water. While there are very few confirmed reports of *C. cayetanensis* transmission through water, this is thought to be a very important mode of transmission. In fact, the first outbreak of cyclosporiasis known to originate in North America was waterborne and occurred at a hospital in Chicago, IL, in 1990. Person-to-person transmission (fecal-oral route) is considered to be unlikely due to the relatively long sporulation period required before oocysts become infective. Zoonotic transmission has also been suggested, and *Cyclospora*-like organisms and/or positive PCR results have been observed for chickens, ducks, dogs, monkeys, chimpanzees, and baboons (Lopez et al., 1999; Chu et al., 2004; Ortega and Sanchez, 2010). However, to date there is little more than anecdotal evidence for this method of transmission.

Before 1995, most infections with *C. cayetanensis* in North America occurred in overseas travelers, and only three small outbreaks had occurred in the United States (Herwaldt et al., 1997). Since then, numerous foodborne outbreaks and cases have been reported in recent years in Canada and the United States (Herwaldt, 2000; Dixon, 2003; Ortega and Sanchez, 2010). It has been estimated that 90% of cyclosporiasis cases in the United States are foodborne, amounting to a total of 14,638 cases each year, including 15 hospitalizations (Mead et al., 1999).

Surveillance in Foods

The contamination of produce and other foods with *C. cayetanensis* oocysts may occur through the use of human feces to fertilize crops, irrigation with contaminated water, use of contaminated water to mix pesticides or wash produce, or contact with contaminated hands or equipment during harvest, packaging, transport, or food preparation (Tram et al., 2008). A number of surveillance studies have reported the presence of this parasite in a variety of fruits and vegetables. For example, *Cyclospora*-like organisms have been found on green leafy vegetables in Nepal (Sherchand et al., 1999). In a survey of vegetables collected from markets in Peru, Ortega et al. (1997) found that of the total vegetables examined, 1.8% and 1.6% were contaminated with *C. cayetanensis* oocysts in the second and third collections, respectively, while none were found to be contaminated in the first collection. The vegetables found to be contaminated included lettuce and two types of herbs. *Cyclospora* spp. were also reported in lettuce from local markets in Egypt (Abou el Naga, 1999) and Costa Rica (Calvo et al., 2004). Anh et al. (2007) reported *Cyclospora* spp. in 8% of water spinach samples collected in Cambodia. Recently, Tram et al. (2008) found *Cyclospora* spp. oocysts in all of seven types of herbs, including lettuce, collected from farms and markets in Vietnam, at rates of between 3.75% and 11.6%. These authors also detected oocysts in irrigation water on farms and in sprinkling water used to keep produce fresh in the markets.

Asian freshwater clams have been found to have the capability of recovering *C. cayetanensis* oocysts from artificially contaminated water and could be used as indicators of water contamination (Graczyk et al., 1998). These clams could also potentially play a role in foodborne transmission.

Foodborne Outbreaks

There has been a distinct seasonality in foodborne outbreaks of cyclosporiasis, which is only partially explained by environmental factors such as temperature and rainfall (Herwaldt, 2000). In North America, most foodborne outbreaks have been reported in the spring to early summer and have been associated with the consumption of raw imported fruits and vegetables. A list of foodborne outbreaks of cyclosporiasis in North America is shown in Table 1. Large multicluster outbreaks occurred in numerous states and in Ontario, Canada, in 1996 and 1997. By means of epidemiological investigations and trace back studies, it was concluded that these large outbreaks were largely associated with the consumption of Guatemalan raspberries, although Guatemalan blackberries may have also been involved in Canadian clusters (Herwaldt, 2000). In May 1998, another outbreak occurred in Ontario (Toronto Public Health et al., 1998), and epidemiological evidence again implicated raspberries from Guatemala. Possibly due to an import ban on Guatemalan raspberries, there were no well-defined clusters of illness due to *C. cayetanensis* infection in the United States during 1998. In May of 1999, outbreaks were reported in Ontario and in Florida. The Ontario outbreak was epidemiologically associated with Guatemalan blackberries, which were accepted without restriction in Canada and the United States. Fresh raspberries from Guatemala were not accepted for importation into Canada during that time, while the United States permitted importation only from Guatemalan farms meeting stricter standards. The Florida outbreak was associated with consumption of fruit, probably berries, from an undefined source. In July 1999, a basil-associated outbreak occurred in Missouri (Lopez et al., 2001). Possible sources of this contaminated basil included farms in the United States and Mexico. For the first time, *C. cayetanensis* was detected in the implicated food when DNA from the organism was detected by PCR on samples of a chicken pasta salad containing basil (Lopez et al., 2001). These PCR results were subsequently confirmed by microscopy. In an outbreak in Philadelphia, PA, in June 2000, the presence of *C. cayetanensis* was again demonstrated by PCR in cake filling containing imported raspberries (Ho et al., 2002). The first well-documented foodborne outbreak of cyclosporiasis in Europe occurred in December 2000 in Germany and was associated with the consumption of salad (Doller et al., 2002). Trace back studies showed that the salad originated from growers within Europe.

Detection Methods

As with other foodborne protozoan parasites, there are no standard methods available for the detection of *C. cayetanensis* in foods. Many of the same methods that are used in clinical diagnoses, however, are now being evaluated and adapted to foods. Detection is very much dependent on the methods used for elution and concentration of the oocysts. Most published methods involve agitation of the fruits or vegetables in water, often in combination with detergents, for eluting oocysts (Ortega et al., 1997; Sherchand et al., 1999; Calvo et al., 2004; Anh et al., 2007; Tram et al., 2008). The subsequent concentration of oocysts generally involves centrifugation and/or flotation. Since it has not yet been possible to produce antibodies against *Cyclospora* spp., immunomagnetic separation unfortunately cannot be used to concentrate the oocysts from food debris as it has for *Cryptosporidium* spp. and *Giardia* spp. Robertson et al. (2000), however, used lectin-coated paramagnetic beads to isolate *C. cayetanensis* oocysts from spiked fruits and vegetables and to minimize debris in the final volume.

A number of surveillance studies have involved the use of microscopy (e.g., brightfield, differential interference contrast, epifluorescence, and acid-fast stain microscopy) for the detection of *Cyclospora* spp. in

TABLE 1 Foodborne outbreaks of cyclosporiasis in North America

Date	Location	No. of cases	Implicated food	Reference
May, 1995	Florida	38	Raspberries (Guatemala?)	Koumans et al., 1998
May–June 1995	New York	32	Fruit?	Herwaldt, 2000
May 1996	Boston, MA	57	Berries	Fleming et al., 1998
May 1996	Charleston, SC	38	Raspberries (Guatemala)	Cáceres et al., 1998
May–June 1996	United States/ Ontario, Canada	1,465	Raspberries (Guatemala)	Herwaldt et al., 1997; Manuel et al., 1999, 2000
March–April 1997	Florida	220	Mesclun (Peru)	Herwaldt, 2000
April–May 1997	United States/ Ontario, Canada	1,012	Raspberries (Guatemala)	DeGraw et al., 1997; Herwaldt et al., 1999
June–July, 1997	Washington, DC	260	Basil	Pritchett et al., 1997
September 1997	Virginia	21	Fruit plate	Herwaldt, 2000
December 1997	Florida	12	Mesclun (Peru)	Herwaldt, 2000
May 1998	Ontario, Canada	192	Raspberries (Guatemala)	Toronto Public Health et al., 1998
May 1999	Ontario, Canada	104	Blackberries?	Herwaldt, 2000
May 1999	Florida	94	Berries	Herwaldt, 2000
July 1999	Missouri	62	Basil (Mexico or United States)	Lopez et al., 2001
July 1999	Vancouver, BC, Canada	15	Unknown	British Columbia Centre for Disease Control, unpublished data
May 2000	Atlanta, GA	21	Raspberries (Guatemala)	Jack Guzewich, FDA, personal communication
June 2000	Philadelphia, PA	54	Raspberries (Guatemala)	Ho et al., 2002
May 2001	Vancouver, BC, Canada	17	Thai basil (United States)	Hoang et al., 2005
January 2002	Vermont	22	Raspberries (Chile?)	Jack Guzewich, FDA, personal communication
June–July 2003	Vancouver, BC, Canada	10	Cilantro?	British Columbia Centre for Disease Control, unpublished data
February 2004	Texas/Illinois	95	Unknown	Ortega and Sanchez, 2010
June–July 2004	Pennsylvania	96	Snow peas (Guatemala)	Centers for Disease Control and Prevention, 2004
April 2005	Florida	592	Basil	Ortega and Sanchez, 2010
April 2005	Ontario, Canada	40	Basil (Peru)	Toronto Public Health, unpublished data
June 2005	Quebec, Canada	229	Basil (Mexico)	Quebec Public Health, unpublished data

foods (Ortega et al., 1997; Sherchand et al., 1999; Calvo et al., 2004; Anh et al., 2007; Tram et al., 2008). Robertson et al. (2000) also used microscopy to detect *C. cayetanensis* oocysts in spiked produce and reported a recovery efficiency of about 12% from mushrooms, lettuce, and raspberries and of 4% from bean sprouts.

PCR methods developed for clinical diagnosis have also been useful in detecting the presence of *C. cayetanensis* in foods. For example, Jinneman et al. (1998) used a nested-PCR assay to amplify a portion of the 18S rRNA gene (Relman et al., 1996), along with restriction fragment length polymorphism, for the detection and identification of *Cyclospora* spp. and *Eimeria* spp. from spiked raspberries, and were able to detect as few as 19 *C. cayetanensis* oocysts per PCR. These authors optimized a DNA template preparation method using a freeze-thaw protocol and the addition of nonfat dried milk to minimize the effect of PCR inhibitors. Orlandi and Lampel (2000) used an extraction-free, filter-based DNA extraction method followed by a modification of Relman's PCR protocol to detect as few as 10 to 30 *C. cayetanensis* oocysts in 100 g of spiked fresh raspberries. Similarly, Steele et al. (2003) were able to detect 40 or fewer *C. cayetanensis* oocysts per 100 g of spiked raspberries or basil by using glass beads to extract DNA, followed by Relman's PCR protocol. More recently, Lalonde and Gajadhar (2008) developed a PCR assay for the detection of *C. cayetanensis* in foods based on the amplification of a portion of the internal transcribed spacer 2 region. This method detected *C. cayetanensis* in basil wash spiked with as little as one oocyst.

The viability of oocysts is also an important consideration in the risk management of foodborne cyclosporiasis. In the absence of animal models or in vitro cultivation for cyclosporiasis, it is very difficult to determine the true infectivity of oocysts; however, their viability may be determined by using a sporulation assay followed by excystation of the oocysts or by electrorotation (Ortega and Sanchez, 2010).

Control Measures

Control measures to reduce the risk of foodborne cyclosporiasis include farm level controls, postharvest disinfection, and consumer level controls (Dixon, 2003). Worker hygiene and water quality on farms, particularly in regions of endemicity, are important factors in the risk of contamination of produce with *C. cayetanensis* oocysts. Following the large foodborne outbreaks of cyclosporiasis in North America in the late 1990s, the United States and Canada worked together with Guatemala to investigate production, packaging, and exporting practices in that country with respect to fresh raspberries, in an effort to identify the source(s) of contamination. A so-called "Model Plan of Excellence" was developed in an attempt to minimize the likelihood of contamination of raspberries with *C. cayetanensis* oocysts (Dixon, 2003). In this hazard analysis and critical control point-based approach, stringent control measures were enforced on raspberry farms and in packing plants exporting fresh berries to North America. Despite these control measures, further outbreaks associated with Guatemalan berries were reported, suggesting that there may have been compliance problems or that the control measures were ineffective or not directed against the actual sources of contamination (Herwaldt, 2000).

There is little information available regarding environmental resistance of *C. cayetanensis* oocysts, although they are considered to be more susceptible to adverse environmental conditions than *Cryptosporidium* spp. For example, sporulation may be retarded in oocysts stored at either 4°C or 37°C for 14 days (Smith et al., 1997). These findings could have important implications regarding risk control during production, storage, and transportation.

While most of the work done on chemical and physical disinfection of protozoans in foods has involved *Cryptosporidium* spp., studies have been done on the effectiveness of commercial freezing, irradiation, UV light, and high hydrostatic pressure against *C. cayetanensis*, or *Eimeria acervulina* as a surrogate organism (Kniel et al., 2007; Ortega and Sanchez, 2010). As with other protozoan parasites, *C. cayetanensis* oocysts are thought to be very resistant to the chemical disinfectants commonly used in the food industry as well as in water treatment.

There are also some important control points at the consumer level. As with all enteric pathogens, good personal hygiene, particularly hand washing, is of considerable importance in reducing the risk of transmission. However, as discussed, person-to-person transmission of cyclosporiasis is unlikely, and oocysts contaminating foods are not immediately infective. While rinsing of fresh fruits and vegetables with water is recommended for reducing the risk of transmission of pathogens, it is unlikely that this practice will remove all *C. cayetanensis* oocysts from the foods. Furthermore, delicate fruits such as raspberries will not tolerate anything more than a gentle rinse. Other treatments such as cooking and freezing may be used as final barriers against transmission, although most of the produce items implicated in outbreaks of cyclosporiasis are typically consumed raw. Further, there are few data available regarding the susceptibility of *C. cayetanensis* to cooking and freezing temperatures. Recent studies, however, suggest that *C. cayetanensis* oocysts are relatively resistant to household freezing and to very short exposures to high temperature but will not sporulate following exposure to extreme temperatures including −70°C, 70°C, and 100°C (Ortega and Sanchez, 2010).

TOXOPLASMA GONDII

Prevalence

Perhaps the most widespread protozoan parasite affecting humans is *Toxoplasma gondii*. *T. gondii*, which is the only species of the genus, infects virtually all warm-blooded animals, including humans, livestock, birds, and marine mammals (Dubey and Beattie, 1988; Hill et al., 2005; Dubey, 2009). Infection in humans occurs worldwide; however, prevalence varies widely from place to place. In the United States, the rate of infection with *T. gondii* appears to be declining. In the most recent serological survey (National Health and Nutrition Examination Survey, 1999–2004), involving over 15,000 people, 10.8% were positive for *T. gondii* antibodies, indicating infection with the parasite (Jones et al., 2007). Prevalence was lower in persons 6 to 11 years of age (3.6%) and increased with age to 15.7% in persons 40 to 49 years of age. The prevalence in women of childbearing age was 11%. In a cross-sectional study of adults, a group known to avoid eating meat (Seventh Day Adventists) had a significantly lower prevalence (18%) than that of the non-Seventh Day Adventists in the study (40%); however, cat ownership and association with oocyst-contaminated environments were not investigated (Roghmann et al., 1999). While healthy adults infected with *T. gondii* are generally asymptomatic or have mild "flu-like" illness, immunosuppressed individuals and congenitally infected infants may develop severe neurological symptoms.

T. gondii oocysts, the environmentally resistant stage of the parasite, are formed in the intestine and shed in the feces of the only definitive host, the cat. After the ingestion of tissue cysts in meat (intracellular bradyzoites surrounded by a cyst wall) by cats, the tissue cyst wall is dissolved, and the released bradyzoites penetrate the epithelial cells of the feline small intestine and initiate development. In the feline intestine, gametes are formed, followed by oocyst wall formation around the fertilized gamete. Mature oocysts are discharged into the intestinal lumen by the rupture of intestinal epithelial cells and released into the environment in feces. An infected cat can shed millions of oocysts each day for up to 1 week, which become infectious outside the cat within 1 to 5 days depending upon aeration and temperature. In all hosts, ingestion of tissue cysts in raw meats or organs or ingestion of oocysts in food, water, or fomites leads to extraintestinal infection and replication by the tachyzoite stage. The tachyzoite multiplies asexually within the host cell by repeated divisions until the host cell ruptures and releases new tachyzoites, which infect new cells. After a few such divisions, *T. gondii* forms tissue cysts, which remain intracellular. The tissue cyst may enclose thousands of the crescent-shaped, slender bradyzoites.

Surveillance in Foods

Aspinall et al. (2002) surveyed 71 meat samples (57 pork, 9 lamb, 4 beef, and 1 pork/beef mixture) obtained from retail stores in the United Kingdom and found 38% (27/71) contained DNA of *T. gondii*. The relative risk to U.S. consumers of acquiring *T. gondii* infection from undercooked meat was recently determined in a nationwide survey of retail chicken, beef, and pork. The survey of 698 retail outlets in 28 metropolitan statistical areas covered 80% of the U.S. population. Only pork was found to harbor viable *T. gondii* tissue cysts, which were isolated from 0.38% of pork samples by cat bioassay, and 0.57% of pork samples were suspected to be infected based on positive enzyme-linked immunosorbent assay (ELISA) results. No beef samples were positive by bioassay or by ELISA, while 1.4% of chickens were positive by ELISA only. The northeastern United States had a higher number of positive pork samples than other regions of the country, reflecting the higher risk of pig infection due to regional management practices (Dubey et al., 2005).

Prevalence in Meat Animals

Most species of livestock, including sheep, goats, and pigs, are susceptible to infection with *T. gondii* (Dubey, 2009). Sheep and goats are important hosts of *T. gondii* in some countries and pose a major risk for human exposure but are minor food commodities in the United States. In a recent study, 25% of lamb hearts collected at a commercial abattoir in the United States contained viable *T. gondii* tissue cysts (Dubey et al., 2008). Similarly, free-range chickens are known to be infected, often at high rates, but have not been indicated as a source of human infection in the United States. Of the major meat animal species investigated thus far in the United States, pigs are the only species shown to frequently harbor the parasite. The prevalence of *T. gondii* in pigs varies, but generally exceeds 10 to 20% in most countries. Infection rates are higher in breeding populations than in market pigs, reflecting that length of exposure is a factor in acquiring *T. gondii* infection. A number of serologic surveys for *T. gondii* have been conducted in the United States (Dubey, 1990; NAHMS Swine 1995, 2000, as reported in Hill et al., 2010) and have demonstrated a steep decline in seroprevalence in hogs from 23% in 1984 to 2.6% in 2006. The reason for this decline in prevalence likely reflects the increased use of confinement housing systems in the U.S. pork industry. A study of pigs raised under outdoor conditions in the northeastern United States found an infection rate of nearly 50%. Virtually all edible portions of an animal can harbor viable *T. gondii* tissue cysts, and tissue cysts can survive in food animals for years.

Transmission of *T. gondii* to pigs on the farm occurs in a manner similar to that of human infection: ingestion of oocysts from the environment; consumption of infected animals such as mice, birds, and other wildlife; consumption of undercooked meat scraps; and through in utero transmission. In several studies of management factors, outside housing of swine, access of cats to swine, infection in local cats and mice, and small herd size were positively correlated with *T. gondii* infection. Prevention of infection in swine is accomplished through on-farm adherence to good production practices, which includes adopting an effective rodent control program, biosecure housing and feed, elimination of feral cats, prompt carcass removal, and stringent boot hygiene in barns.

Foodborne Outbreaks

Consumption of raw or undercooked venison has been associated with outbreaks of clinical toxoplasmosis and retinitis in previously healthy deer hunters (Sacks et al., 1983; Ross et al., 2001). Uncooked pork was identified as the source of two outbreaks in healthy adults in Korea resulting in chorioretinitis (three patients) and lymphadenopathy (five patients) (Choi et al., 1997). Raw goat milk has been implicated in several outbreaks of human toxoplasmosis (Riemann et al., 1975; Sacks et al., 1982; Chiari and Neves, 1984; Skinner et al., 1990). In a case-control study of 148 recently (<6 months) infected individuals, Jones et al. (2009) identified elevated risk of infection associated with eating raw ground beef, rare lamb, and locally produced cured, dried, or smoked meat; eating raw oysters, clams, or mussels; working with meat; and drinking unpasteurized goat's milk. Hundreds of outbreaks have resulted from contamination of unfiltered, untreated drinking water contaminated with cat feces (Dubey and Jones, 2008; Jones and Dubey, 2010).

Detection Methods

Environmental contamination with *T. gondii* oocysts is widespread as a result of fecal contamination of soil and ground water by domestic and feral cats in the United States. It is likely, therefore, that some human infections result from contamination of fresh fruit and produce with soil or water containing infectious oocysts; however, there are no cases in the literature in which infection from this source has been definitively determined. There are currently no described methods for detecting oocysts on produce.

It is not possible to detect the microscopic *T. gondii* tissue cysts by visual inspection. The most definitive methods for detecting tissue cysts in pigs and other livestock are bioassays of tissues in mice or cats. Cat bioassay is considered the gold standard of detection methods, since cats become infected with very low numbers of parasites. Bioassay requires either (i) inoculation of digested samples of tissue into mice, or (ii) feeding of tissue to cats and monitoring oocyst output in feces for 2 to 3 weeks; both bioassay methods require several weeks to determine the presence of parasites and are therefore not suitable for testing slaughtered animals. We compared the sensitivities of six indirect methods for detection of *T. gondii* in naturally and experimentally infected pigs; the presence of anti-*Toxoplasma* antibody or parasite DNA was presumed to indicate infection and therefore the presence of tissue cysts (Hill et al., 2006). A total of 69 pork samples were tested by cat bioassay, direct PCR, seminested PCR, real-time PCR, ELISA (35 sera and 69 tissue fluid samples), and the modified agglutination test (35 sera and three tissue fluid samples). Thirty-nine *T. gondii*-positive samples were detected in the cat bioassay out of 69 samples tested. Using the cat bioassay as the gold standard (i.e., assuming the cat bioassay detected 100% of positive samples), the detection methods ranked as follows in descending order of sensitivity (indicated in parentheses): serum ELISA (100%) > serum modified agglutination test (80.64%) > tissue fluid ELISA (76.9%) > real-time PCR (20.51%) > seminested PCR (12.82%) > direct PCR (0%).

Control Measures

Currently, there are no programs for the slaughter inspection of pigs for toxoplasmosis. In addition, no regulations requiring that pork be inspected for *T. gondii* or undergo further processing are mandated to inactivate the parasite. However, many of the methods that are in place for processing pork for inactivation of *Trichinella spiralis* (trichinae) (U.S. Code of Federal Regulations, 9CFR318.10 [U.S. Department of Agriculture, Food Safety and Inspection Service, 2003]) are also effective for the inactivation of *T. gondii*. While these regulations remain in effect, processed pork products should be safe for human consumption without further treatment. The following discussion summarizes the inactivation of *T. gondii* in fresh meats by processing methods.

Cooking

T. gondii is killed in 336 seconds at 49°C, in 44 seconds at 55°C, and in 6 seconds at 61°C. These times and temperatures apply only when the product reaches and maintains temperatures evenly distributed throughout the meat (Dubey et al., 1990). The U.S. Department of Agriculture advises consumers to cook pork to an internal temperature of 71°C; thus, guidance to consumers adds a substantial margin of error. The use of microwaves is not effective in killing *T. gondii*, probably because of uneven heating throughout the meat. People handling meat should wash their hands thoroughly with soap and water before they go to other tasks (Dubey and Beattie, 1988; Lopez et al., 2000). Tasting meat while cooking or while seasoning should be avoided. All cutting boards, sink tops, knives, and other materials coming in contact with uncooked meat should be washed with soap and water. Washing is effective because the stages of *T. gondii* in meat are killed by contact with soap and water.

Freezing

Although tissue cysts remained viable in pork at temperatures slightly below freezing (11.2 days at −6.7°C and 22.4 days at −3.9 and −1.0°C), parasites were inactivated almost instantaneously at temperatures of −9.4°C and lower (Kotula et al., 1991). There is no evidence that there are strains of *T. gondii* with different freezing susceptibilities.

Curing

Knowledge of the effects of various curing processes on *T. gondii* is limited, and additional studies are needed to determine the effectiveness of curing for the destruction of *T. gondii* in pork and pork products. Pumping of pork products with salt solutions containing 2% NaCl or ≥1.4% potassium or sodium lactate has been shown to inactivate *T. gondii* tissue cysts in pork (Hill et al., 2004).

Other Treatments

T. gondii tissue cysts were rendered noninfectious by treatment with 40 to 50 krad of cesium-137, indicating that irradiation is a suitable method for eliminating the risk of this parasite in pork products. High-pressure processing at 300 MPa or greater renders tissue cysts in meat noninfectious.

Humans can also be exposed to oocysts in the environment through eating unwashed fruits and vegetables and drinking water contaminated with cat feces. Sporulated *T. gondii* oocysts can survive in the environment for 18 months and for 4.5 years at 4°C. To prevent infection by inadvertent consumption of oocysts, hands should always be thoroughly washed after handling any material potentially contaminated with soil or cat feces. Unfiltered surface water should not be consumed due to the risk of contamination with cat feces containing oocysts. High-pressure processing at 550 MPa or greater of fresh fruits and vegetables renders oocysts noninfectious, as does irradiation at 0.5 kGy. Freezing of sporulated oocysts at −20°C for 24 hours or heating to 60°C for 1 minute kills oocysts.

CONCLUSIONS AND FUTURE TRENDS

Cryptosporidium spp., *Giardia duodenalis, Cyclospora cayetanensis*, and *Toxoplasma gondii* are protozoan parasites commonly reported in humans and animals worldwide. Numerous foodborne outbreaks of illness associated with a variety of foods contaminated with these pathogens have been reported in recent years. Foods may become contaminated at any one of a number of points from farm level to consumer level.

There are no standard methods for the elution, concentration, and detection of protozoan parasites in foods, and the methods that are available are often limited in their sensitivity due to the difficulty in eluting and concentrating cysts and oocysts from the food item or due to the presence of PCR inhibitors. The most commonly employed methods for detection of cysts and oocysts in foods include a variety of microscopical techniques, such as bright-field microscopy with or without staining, differential interference contrast microscopy, and immunofluorescence microscopy, or PCR-based protocols. *T. gondii* tissue cysts are generally detected in foods of animal origin through the inoculation or feeding of samples to mice or cats, followed by serological or histological examination, although numerous PCRs and ELISAs have also been developed.

Although the control of illness due to foodborne protozoan infections will be much more efficient through the identification and management of sources of contamination than through routine testing of foods, further studies are needed to improve methods for recovery, detection, molecular characterization, and viability testing. In addition to food surveillance studies, these methods are required for outbreak investigations, source tracking, and providing data for risk assessments, in support of risk management decisions such as product recalls, trace backs, and import policies.

REFERENCES

Abou el Naga, I. F. 1999. Studies on a newly emerging protozoal pathogen: *Cyclospora cayetanensis*. *J. Egypt. Soc. Parasitol.* **29:**575–586.

Angles, M. L., J. P. Chandy, P. T. Cox, I. H. Fisher, and M. R. Warnecke. 2007. Implications of biofilm-associated waterborne *Cryptosporidium* oocysts for the water industry. *Trends Parasitol.* **23:** 352–356.

Anh, V. T., N. T. Tram, L. T. Klank, P. D. Cam, and A. Dalsgaard. 2007. Faecal and protozoan parasite contamination of water spinach (*Ipomoea aquatica*) cultivated in urban wastewater in Phnom Penh, Cambodia. *Trop. Med. Int. Health* **12**(Suppl. 2):73–81.

Anonymous. 1984. Cryptosporidiosis surveillance. *Wkly. Epidemiol. Rec.* **59:**72–73.

Anonymous. 1989. Epidemiologic notes and reports common-source outbreak of giardiasis—New Mexico. *MMWR Morb. Mortal. Wkly. Rep.* **38:** 405–407.

Anonymous. 1996. Foodborne outbreak of diarrheal illness associated with *Cryptosporidium parvum*—Minnesota, 1995. *MMWR Morb. Mortal. Wkly. Rep.* **45:**783–784.

Anonymous. 1997. Outbreaks of *Escherichia coli* O197:H7 infection and cryptosporidiosis associated with drinking unpasteurized apple cider—Connecticut and New York, October 1996. *MMWR Morb. Mortal. Wkly. Rep.* **46:**4–8.

Anonymous. 1998. Foodborne outbreak of cryptosporidiosis—Spokane, Washington, 1997. *MMWR Morb. Mortal. Wkly. Rep.* **47:**565–567.

Aspinall, T. V., D. Marlee, J. E. Hyde, and P. F. Sims. 2002. Prevalence of *Toxoplasma gondii* in commercial meat products as monitored by polymerase chain reaction—food for thought? *Int. J. Parasitol.* **32:**1193–1199.

Blackburn, B. G., J. M. Mazurek, M. Hlavsa, J. Park, M. Tillapaw, M. Parrish, E. Salehi, W. Franks, E. Koch, F. Smith, L. Xiao, M. Arrowood, V. Hill, A. da Silva, S. Johnston, and J. L. Jones. 2006. Cryptosporidiosis associated with ozonated apple cider. *Emerg. Infect. Dis.* **12:** 684–686.

Bohaychuk, V. M., G. E. Gensler, R. K. King, K. I. Manninen, O. Sorensen, and J. T. Wu. 2009. A microbiological survey of selected Alberta-grown fresh produce from farmers' markets in Alberta, Canada. *J. Food Prot.* **72:**415–420.

Cáceres, V. M., R. T. Ball, S. A. Somerfeldt, R. L. Mackey, S. E. Nichols, W. R. Mackenzie, and B. L. Herwaldt. 1998. A foodborne outbreak of cyclosporiasis caused by imported raspberries. *J. Fam. Pract.* **47:**231–234.

Calvo, M., M. Carazo, M. L. Arias, C. Chaves, R. Monge, and M. Chinchilla. 2004. Prevalence of *Cyclospora* sp., *Cryptosporidium* sp., microsporidia and fecal coliform determination in fresh fruit and vegetables consumed in Costa Rica. *Arch. Latinoam. Nutr.* **54:**428–432.

Cama, V. 2006. Coccidian parasites, p. 33–55. *In* Y. R. Ortega (ed.), *Foodborne Parasites*. Springer, New York, NY.

Casemore, D. P., E. G. Jessop, D. Douce, and F. B. Jackson. 1986. *Cryptosporidium* plus *Campylobacter*: an outbreak in a semi-rural population. *J. Hyg.* (London) **96:**95–105.

Centers for Disease Control and Prevention. 2004. Outbreak of cyclosporiasis associated with snow peas—Pennsylvania, 2004. *MMWR Morb. Mortal. Wkly. Rep.* **53:**876–878.

Chaidez, C., M. Soto, P. Gortares, and K. Mena. 2005. Occurrence of *Cryptosporidium* and *Giardia* in irrigation water and its impact on the fresh produce industry. *Int. J. Environ. Health Res.* **15:**339–345.

Chalmers, R. M., and A. P. Davies. 2010. Minireview: clinical cryptosporidiosis. *Exp. Parasitol.* **124:**138–146.

Chiari, C., and D. P. Neves. 1984. Human toxoplasmosis acquired by ingestion of goat's milk. *Mem. Inst. Oswaldo Cruz* **79:**337–340.

Choi, W.-Y., H. W. Nam, N. H. Kwak, W. Huh, Y. R. Kim, M. W. Kang, S. Y. Cho, and J. P. Dubey. 1997. Foodborne outbreaks of human toxoplasmosis. *J. Infect. Dis.* **175:**1280–1282.

Chu, D.-M. T., J. B. Sherchand, J. H. Cross, and P. A. Orlandi. 2004. Detection of *Cyclospora cayetanensis* in animal fecal isolates from Nepal using an FTA filter-base polymerase chain reaction method. *Am. J. Trop. Med.* **71:**373–379.

Cook, N., C. A. Paton, N. Wilkinson, R. A. Nichols, K. Barker, and H. V. Smith. 2006. Towards standard methods for the detection of

Cryptosporidium parvum on lettuce and raspberries. Part 1. Development and optimization of methods. *Int. J. Food Microbiol.* **109:**215–221.

Cook, N., C. A. Paton, N. Wilkinson, R. A. Nichols, K. Barker, and H. V. Smith. 2007. Towards standard methods for the detection of *Cryptosporidium parvum* on lettuce and raspberries. Part 2. Validation. *Int. J. Food Microbiol.* **109:**222–228.

Cummings, K., and A. Karon. 2009. Epidemiologic summary of giardiasis in California, 2001–2008. California Department of Public Health, Sacramento. http://www.cdph.ca.gov/data/statistics/Documents/giardiasis-episummary.pdf.

Daryani, A., G. H. Ettehad, M. Sharif, L. Ghorbani, and H. Ziaei. 2008. Prevalence of intestinal parasites in vegetables consumed in Ardabil, Iran. *Food Control* **19:**790–794.

Dawson, D. 2005. Foodborne protozoan parasites. *Int. J. Food Microbiol.* **103:**207–227.

DeGraw, E., S. Heber, and A. Rowan. 1997. Update: outbreaks of cyclosporiasis—United States and Canada, 1997. *MMWR Morb. Mortal. Wkly. Rep.* **46:**521–523.

de Leon, W. U., R. B Monzon, A. A. Aganon, R. E. Arceo, E. J. Ignacio, and G. Santos. 1992. Parasitic contamination of selected vegetables sold in metropolitan Manila, Philippines. *Southeast Asian J. Trop. Med. Public Health* **23:**162–164.

Dixon, B. R. 2003. The prevalence and control of foodborne protozoan parasites, p. 31–76. *In* B. W. Blais (ed.), *Current Challenges in Food Microbiology*. Research Signpost, Kerala, India.

Doller, P. C., K. Dietrich, N. Filipp, S. Brockmann, C. Dreweck, R. Vonthein, C. Wagner-Wiening, and A. Wiedenmann. 2002. Cyclosporiasis outbreak in Germany associated with the consumption of salad. *Emerg. Infect. Dis.* **8:**992–994.

Dubey, J. P. 1990. Status of toxoplasmosis in pigs in the United States. *J. Am. Vet. Med. Assoc.* **196:**270–274.

Dubey, J. P. 2009. *Toxoplasmosis of Animals and Humans*, 2nd ed. CRC Press, Boca Raton, FL.

Dubey, J. P., and C. P. Beattie. 1988. *Toxoplasmosis of Animals and Man*. CRC Press, Boca Raton, FL.

Dubey, J. P., D. E. Hill, J. L. Jones, A. W. Hightower, E. Kirkland, J. M. Roberts, T. Lehmann, C. Sreekumar, M. C. B. Vianna, O. C. H. Kwok, S. K. Shen, and H. R. Gamble. 2005. Prevalence of viable *Toxoplasma gondii* in beef, chicken, and pork from retail meat stores in the United States. *J. Parasitol.* **91:**1082–1093.

Dubey, J. P., and J. L. Jones. 2008. *Toxoplasma gondii* infections in humans and animals in the United States. *Int. J. Parasitol.* **38:**1257–1278.

Dubey, J. P., A. W. Kotula, A. Sharar, C. D. Andrews, and D. S. Lindsay. 1990. Effect of high temperature on infectivity of *Toxoplasma gondii* tissue cysts in pork. *J. Parasitol.* **76:**201–204.

Dubey, J. P., N. Sundar, D. Hill, G. V. Velmurugan, L. A. Bandini, O. C. Kwok, D. Majumdar, and C. Su. 2008. High prevalence and abundant atypical genotypes of *Toxoplasma gondii* isolated from lambs destined for human consumption in the USA. *Int. J. Parasitol.* **38:**999–1006.

Eberhard, M. L., and M. J. Arrowood. 2002. *Cyclospora* spp. *Curr. Opin. Infect. Dis.* **15:**519–522.

Elsser, K. A., M. Moricz, and E. M. Proctor. 1986. *Cryptosporidium* infections: a laboratory survey. *Can. Med. Assoc. J.* **135:**211–213.

Ethelberg, S., M. Lisby, L. S. Vestergaard, H. L. Enemark, K. E. Olsen, C. R. Stensvold, H. V. Nielsen, and K. Molbak. 2009. A foodborne outbreak of *Cryptosporidium hominis* infection. *Epidemiol. Infect.* **137:**348–356.

Fayer, R. 1995. Effect of sodium hypochlorite exposure on infectivity of *Cryptosporidium parvum* oocysts for neonatal BALB/c mice. *Appl. Environ. Microbiol.* **61:**844–846.

Fayer, R. 2008. Biology, p. 1–42. *In* R. Fayer and L. Xiao (ed.), *Cryptosporidium and Cryptosporidiosis*, 2nd ed. CRC Press and IWA Publishing, Boca Raton, FL.

Fayer, R. 2010. Taxonomy and species delimitation in *Cryptosporidium*. *Exp. Parasitol.* **124:**90–97.

Fayer, R., J. P. Dubey, and D. S. Lindsay. 2004. Zoonotic protozoa: from land to sea. *Trends Parasitol.* **20:**531–536.

Fayer, R., J. M. Trout, E. J. Lewis, M. Santin, L. Zhou, A. A. Lal, and L. Xiao. 2003. Contamination of Atlantic coast commercial shellfish with *Cryptosporidium*. *Parasitol. Res.* **89:**141–145.

Feng, Y., and L. Xiao. 2011. Zoonotic potential and molecular epidemiology of *Giardia* species and giardiasis. *Clin. Microbiol. Rev.* **24:**110–140.

Fleming, C. A., D. Caron, J. E. Gunn, and M. A. Barry. 1998. A foodborne outbreak of *Cyclospora cayetanensis* at a wedding. *Arch. Intern. Med.* **158:**1121–1125.

Garcia, L., J. Henderson, M. Fabri, and M. Oke. 2006. Potential sources of microbial contamination in unpasteurized apple cider. *J. Food Prot.* **69:**137–144.

Garcia-Vela, L. A., B. Forrest, A. Mahdi, M. Wynne, and T. Baker. 2003. *Baseline Study of Ontario Unpasteurized Apple Cider*. Food Inspection Branch, Ontario Ministry of Agriculture and Food, Ontario, Canada.

Gelletlie, R., J. Stuart, N. Soltanpoor, R. Armstrong, and G. Nichols. 1997. Cryptosporidiosis associated with school milk. *Lancet* **350:**1005–1006.

Graczyk, T. K., Y. R. Ortega, and D. B. Conn. 1998. Recovery of waterborne oocysts of *Cyclospora cayetanensis* by Asian freshwater clams (*Corbicula fluminea*). *Am. J. Trop. Med. Hyg.* **59:**928–932.

Guiguet Leal, D. A., M. A. Pereira, R. M. Bueno Franco, N. Branco, and R. C. Neto. 2008. First report of *Cryptosporidium* spp. oocysts in oysters (*Crassostrea rhizophorae*) and cockles (*Tivela mactroides*) in Brazil. *J. Water Health* **6:**527–532.

Harp, J. A., R. Fayer, B. A. Pesch, and G. J. Jackson. 1996. Effect of pasteurization on infectivity of *Cryptosporidium parvum* oocysts in water and milk. *Appl. Environ. Microbiol.* **62:**2866–2868.

Harper, C. M., N. A. Cowell, B. C. Adams, A. J. Langley, and T. D. Wohlsen. 2002. Outbreak of *Cryptosporidium* linked to drinking unpasteurised milk. *Commun. Dis. Intell.* **26:**449–450.

Helmi, K., S. Skraber, C. Gantzer, R. Willame, L. Hoffmann, and H. Cauchie. 2008. Interactions of *Cryptosporidium parvum, Giardia lamblia,* vaccinal poliovirus type 1, and bacteriophages phiX174 and MS2 with a drinking water biofilm and a wastewater biofilm. *Appl. Environ. Microbiol.* **74:**2079–2088.

Herwaldt, B. L. 2000. *Cyclospora cayetanensis*: a review, focusing on the outbreaks of cyclosporiasis in the 1990s. *Clin. Infect. Dis.* **31:**1040–1057.

Herwaldt, B. L., M.-L. Ackers, and the *Cyclospora* Working Group. 1997. An outbreak in 1996 of cyclosporiasis associated with imported raspberries. *N. Engl. J. Med.* **336:**1548–1556.

Herwaldt, B. L., M. J. Beach, and the *Cyclospora* Working Group. 1999. The return of *Cyclospora* in 1997: another outbreak of cyclosporiasis in North America associated with imported raspberries. *Ann. Intern. Med.* **130:**210–220.

Hill, D. E., S. Chirukandoth, and J. P. Dubey. 2005. Biology and epidemiology of *Toxoplasma gondii* in man and animals. *Anim. Health Res. Rev.* **6:** 41–61.

Hill, D. E., S. Chirukandoth, J. P. Dubey, J. K. Lunney, and H. R. Gamble. 2006. Comparison of detection methods for *Toxoplasma gondii* in naturally and experimentally infected swine. *Vet. Parasitol.* **141:**9–17.

Hill, D. E., C. Haley, B. Wagner, H. R. Gamble, and J. P. Dubey. 2010. Seroprevalence of and risk factors for *Toxoplasma gondii* in the U.S. swine herd using sera collected during the National Animal Health Monitoring Survey (Swine 2006). *Zoonoses Public Health* **57:**53–59.

Hill, D. E., C. Sreekumar, H. R. Gamble, and J. P. Dubey. 2004. Effect of commonly used enhancement solutions on the viability of *Toxoplasma gondii* tissue cysts in pork loin. *J. Food Prot.* **67:** 2230–2233.

Ho, A. Y., A. S. Lopez, M. G. Eberhart, R. Levenson, B. S. Finkel, A. J. da Silva, J. M. Roberts, P. A. Orlandi, C. C. Johnson, and B. L. Herwaldt. 2002. An outbreak of cyclosporiasis associated with imported raspberries in Pennsylvania in 2000. *Emerg. Infect. Dis.* **8:**783–788.

Hoang, L. M. N., M. Fyfe, C. Ong, J. Harb, S. Champagne, B. Dixon, and J. Isaac-Renton. 2005. Outbreak of cyclosporiasis in British Columbia associated with imported Thai basil. *Epidemiol. Infect.* **133:**23–27.

Insulander, M., B. de Jong, and B. Svenungsson. 2008. A food-borne outbreak of cryptosporidiosis among guests and staff at a hotel restaurant in Stockholm County, Sweden, September 2008. *Euro Surveill.* **13:**1–2.

Izumi, T., K. Yagita, T. Endo, and T. Ohyama. 2006. Detection system of *Cryptosporidium parvum* oocysts by brackish water benthic shellfish (*Corbicula japonica*) as a biological indicator in river water. *Arch. Environ. Contam. Toxicol.* **51:**559–566.

Jinneman, K. C., J. H. Wetherington, W. E. Hill, A. M. Adams, J. M. Johnson, B. J. Tenge, N.-L. Dang, R. L. Manger, and M. M. Wekell. 1998. Template preparation for PCR and RFLP of amplification products for the detection and identification of *Cyclospora* sp. and *Eimeria* spp. oocysts directly from raspberries. *J. Food Prot.* **61:** 1497–1503.

Jones, J. L., V. Dargelas, J. Roberts, C. Press, J. S. Remington, and J. G. Montoya. 2009. Risk factors for *Toxoplasma gondii* infection in the United States. *Clin. Infect. Dis.* **49:**878–884.

Jones, J. L., and J. P. Dubey. 2010. Waterborne toxoplasmosis—recent developments. *Exp. Parasitol.* **124:**10–25.

Jones, J. L., D. Kruszon-Moran, K. Sanders-Lewis, and M. Wilson. 2007. *Toxoplasma gondii* infection in the United States, decline from the prior decade. *Am. J. Trop. Med. Hyg.* **77:**405–410.

Karanja, R. M., W. Gatei, and N. Wamae. 2007. Cyclosporiasis: an emerging public health concern around the world and in Africa. *African Health Sci.* **7:**62–67.

Kniel, K. E., A. E. H. Shearer, J. L. Cascarino, G. C. Wilkins, and M. C. Jenkins. 2007. High hydrostatic pressure and UV light treatment of produce contaminated with *Eimeria acervulina* as a *Cyclospora cayetanensis* surrogate. *J. Food Prot.* **70:**2837–2842.

Korich, D. G., J. R. Mead, M. S. Madore, N. A. Sinclair, and C. R. Sterling. 1990. Effects of ozone, chlorine dioxide, chlorine, and monochloramine on *Cryptosporidium parvum* oocyst viability. *Appl. Environ. Microbiol.* **56:**1423–1428.

Kotula, A. W., J. P. Dubey, A. K. Sharar, C. D. Andrews, S. K. Shen, and D. S. Lindsay. 1991.

Effect of freezing on infectivity of *Toxoplasma gondii* tissue cysts in pork. *J. Food Prot.* **54:**687–690.

Koumans, E. H. A., D. J. Katz, J. M. Malecki, S. Kumar, S. P. Wahlquist, M. J. Arrowood, A. W. Hightower, and B. L. Herwaldt. 1998. An outbreak of cyclosporiasis in Florida in 1995: a harbinger of multistate outbreaks in 1996 and 1997. *Am. J. Trop. Med. Hyg.* **59:**235–242.

Lalonde, L. F., and A. A. Gajadhar. 2008. Highly sensitive and specific PCR assay for reliable detection of *Cyclospora cayetanensis* oocysts. *Appl. Environ. Microbiol.* **74:**4354–4358.

Lopez, A., V. J. Dietz, M. Wilson, and J. L. Jones. 2000. Preventing congenital toxoplasmosis. *MMWR Morb. Mortal. Wkly. Rep.* **49:**59–75.

Lopez, A. S., D. R. Dodson, M. J. Arrowood, P. A. Orlandi, A. J. da Silva, J. W. Bier, S. D. Hanauer, R. L. Kuster, S. Oltman, M. S. Baldwin, K. Y. Won, E. M. Nace, M. L. Eberhard, and B. L. Herwaldt. 2001. Outbreak of cyclosporiasis associated with basil in Missouri in 1999. *Clin. Infect. Dis.* **32:**1010–1017.

Lopez, F. A., J. Manglicmot, T. M. Schmidt, C. Yeh, H. V. Smith, and D. A. Relman. 1999. Molecular characterization of *Cyclospora*-like organisms from baboons. *J. Infect. Dis.* **179:**670–676.

Macarisin, D., G. Bauchan, and R. Fayer. 2010a. *Spinacia oleracea* L. leaf stomata harboring *Cryptosporidium parvum* oocysts: a potential threat to food safety. *Appl. Environ. Microbiol.* **76:**555–559.

Macarisin, D., M. Santín, G. Bauchan, and R. Fayer. 2010b. Infectivity of *Cryptosporidium parvum* oocysts after storage of experimentally contaminated apples. *J. Food Prot.* **73:**1824–1829.

Mansfield, L. S., and A. A. Gajadhar. 2004. *Cyclospora cayetanensis*, a food- and waterborne coccidian parasite. *Vet. Parasitol.* **126:**73–90.

Manuel, D. G., S. Neamatullah, R. Shahin, D. Reymond, J. Keystone, J. Carlson, C. Le Ber, B. L. Herwaldt, and D. H. Werker. 2000. An outbreak of cyclosporiasis in 1996 associated with consumption of fresh berries—Ontario. *Can. J. Infect. Dis.* **11:**86–92.

Manuel, D. G., R. Shahin, W. Lee, and M. Grmusa. 1999. The first reported cluster of foodborne cyclosporiasis in Canada. *Can. J. Public Health* **90:**399–402.

Mead, P. S., L. Slutsker, V. Dietz, L. F. McCaig, J. S. Bresee, C. Shapiro, P. M. Griffin, and R. V. Tauxe. 1999. Food-related illness and death in the United States. *Emerg. Infect. Dis.* **5:**607–625.

Millard, P. S., K. F. Gensheimer, D. G. Addiss, D. M. Sosin, G. A. Beckett, A. Houck-Jankoski, and A. Hudson. 1994. An outbreak of cryptosporidiosis from fresh-pressed apple cider. *JAMA* **272:**1592–1596.

Mintz, E. D., M. Husan-Wragg, P. Mshar, M. L. Cartter, and J. L. Hadler. 1993. Foodborne giardiasis in a corporate office setting. *J. Infect. Dis.* **167:** 250–253.

Monge, R., and M. L. Arias. 1996. Presence of various pathogenic microorganisms in fresh vegetables in Costa Rica. *Arch. Latinoam. Nutr.* **46:**292–294.

Monge, R., M. Chinchilla, and L. Reyes. 1996. Seasonality of parasites and intestinal bacteria in vegetables that are consumed raw in Costa Rica. *Rev. Biol. Trop.* **44:**369–375.

Nichols, R., and H. Smith. 2002. Parasites: *Cryptosporidium, Giardia* and *Cyclospora* as foodborne pathogens, p. 453–478. *In* C. de W. Blackburn and P. J. McClure (ed.), *Foodborne Pathogens: Hazards, Risk Analysis and Control.* CRC Press, Washington, DC.

Orlandi, P. A., and K. A. Lampel. 2000. Extraction-free, filter-based template preparation for rapid and sensitive PCR detection of pathogenic parasitic protozoa. *J. Clin. Microbiol.* **38:** 2271–2277.

Ortega, Y. R., C. R. Roxas, R. H. Gilman, N. J. Miller, L. Cabrera, C. Taquiri, and C. R. Sterling. 1997. Isolation of *Cryptosporidium parvum* and *Cyclospora cayetanensis* from vegetables collected in markets of an endemic region in Peru. *Am. J. Trop. Med. Hyg.* **57:**683–686.

Ortega, Y. R., and R. Sanchez. 2010. Update on *Cyclospora cayetanensis*, a food-borne and waterborne parasite. *Clin. Microbiol. Rev.* **23:**218–234.

Ortega, Y. R., C. R. Sterling, R. H. Gilman, V. A. Cama, and F. Diaz. 1993. *Cyclospora* species—a new protozoan pathogen of humans. *N. Engl. J. Med.* **328:**1308–1312.

Osterholm, M. T., J. C. Forfang, R. L. Ristinen, A. G. Dean, J. W. Washburn, J. R. Godes, R. A. Rude, and J. G. McCullough. 1981. An outbreak of foodborne giardiasis. *N. Engl. J. Med.* **304:**24–28.

Petersen, L. R., M. L. Cartier, and J. L. Hadler. 1988. A foodborne outbreak of *Giardia lamblia. J. Infect. Dis.* **157:**846–848.

Porter, J. D. H., C. Garney, D. Heymann, and W. Parkin. 1990. Foodborne outbreak of *Giardia lamblia. Am. J. Public Health* **60:**1259–1260.

Pritchett, R., C. Gossman, V. Radke, et al. 1997. Outbreak of cyclosporiasis—northern Virginia-Washington, DC-Baltimore, Maryland, metropolitan area, 1997. *JAMA* **278:**538–539.

Quick, R., K. Paugh, D. Addiss, J. Kobayashi, and R. Baron. 1992. Restaurant-associated outbreak of giardiasis. *J. Infect. Dis.* **166:**673–676.

Quiroz, E. S., C. Bern, J. R. MacArthur, L. Xiao, M. Fletcher, M. J. Arrowood, D. K. Shay, M. E. Levy, R. Glass, and A. Lal. 2000.

An outbreak of cryptosporidiosis linked to a foodhandler. *J. Infect. Dis.* **181:**695–700.

Relman, D. A., T. M. Schmidt, A. Gajadhar, M. Sogin, J. Cross, K. Yoder, O. Sethabutr, and P. Echeverria. 1996. Molecular phylogenetic analysis of *Cyclospora*, the human intestinal pathogen, suggest that it is closely related to *Eimeria* species. *J. Infect. Dis.* **173:**440–445.

Riemann, H. P., M. E. Meyer, J. H. Theis, G. Kelso, and D. E. Behymer. 1975. Toxoplasmosis in an infant fed unpasteurized goat milk. *J. Pediatr.* **87:**573–576.

Robertson, L. C., and B. Gjerde. 2000. Isolation and enumeration of *Giardia* cysts, *Cryptosporidium* oocysts, and *Ascaris* eggs from fruits and vegetables. *J. Food Prot.* **63:**775–778.

Robertson, L. C., and B. Gjerde. 2001a. Occurrence of parasites on fruits and vegetables in Norway. *J. Food Prot.* **64:**1793–1798.

Robertson, L. C., and B. Gjerde. 2001b. Factors affecting recovery efficiency in isolation of *Cryptosporidium* oocysts and *Giardia* cysts from vegetables for standard method development. *J. Food Prot.* **64:** 1799–1805.

Robertson, L. C., G. S. Johannessen, B. K. Gjerde, and S. Loncarevic. 2002. Microbiological analysis of seed sprouts in Norway. *Int. J. Food Microbiol.* **75:**119–126.

Robertson, L. J., B. Gjerde, and A. T. Campbell. 2000. Isolation of *Cyclospora* oocysts from fruits and vegetables using lectin-coated paramagnetic beads. *J. Food Prot.* **63:**1410–1414.

Robertson, L. J., J. D. Greig, B. Gjerde, and A. Fazil. 2005. The potential for acquiring cryptosporidiosis or giardiosis from consumption of mung bean sprouts in Norway: a preliminary step-wise risk assessment. *Int. J. Food Microbiol.* **98:**291–300.

Roghmann, M.-C., C. T. Faulkner, A. Lefkowitz, S. Patton, J. Zimmerman, and J. G. Morris. 1999. Decreased seroprevalence for *Toxoplasma gondii* in Seventh Day Adventists in Maryland. *Am. J. Trop. Med. Hyg.* **60:**790–792.

Rose, J. B., and T. R. Slifko. 1999. *Giardia, Cryptosporidium,* and *Cyclospora* and their impact on foods: a review. *J. Food Prot.* **62:**1059–1070.

Ross, R. D., L. A. Stec, J. C. Werner, M. S. Blumenkranz, L. Glazer, and G. A. Williams. 2001. Presumed acquired ocular toxoplasmosis in deer hunters. *Retina* **21:**226–229.

Rzezutka, A., R. A. Nichols, L. Connelly, A. Kaupke, I. Kozyra, N. Cook, S. Birrell, and H. V. Smith. 2010. *Cryptosporidium* oocysts on fresh produce from areas of high livestock production in Poland. *Int. J. Food Microbiol.* **139:**96–101.

Sacks, J. J., D. G. Delgado, H. O. Lobel, and R. L. Parker. 1983. Toxoplasmosis infection associated with eating undercooked venison. *Am. J. Epidemiol.* **118:**832–838.

Sacks, J. J., R. R. Roberto, and N. F. Brooks. 1982. Toxoplasmosis infection associated with raw goat's milk. *JAMA* **248:**1728–1732.

Scallan, E., R. M. Hoekstra, F. J. Angulo, R. V. Tauxe, M.-A. Widdowson, S. L. Roy, J. L. Jones, and P. M. Griffin. 2011. Foodborne illness acquired in the United States—major pathogens. *Emerg. Infect. Dis.* **17:**7–15.

Schets, F. M., H. H. van den Berg, G. B. Engels, W. J. Lodder, and A. M. de Roda Husman. 2007. *Cryptosporidium* and *Giardia* in commercial and non-commercial oysters (*Crassostrea gigas*) and water from the Oosterschelde, The Netherlands. *Int. J. Food Microbiol.* **113:**189–194.

Sherchand, J. B., J. Cross, M. Jimba, S. Sherchand, and M. P. Shrestha. 1999. Study of *Cyclospora cayetanensis* in health care facilities, sewage water and green leafy vegetables in Nepal. *Southeast Asian J. Trop. Med. Public Health* **30:**58–63.

Skinner, L. J., A. C. Timperley, D. Wightman, J. M. Chatterton, and D. Ho-Yen. 1990. Simultaneous diagnosis of toxoplasmosis in goats and goatowner's family. *Scand. J. Infect. Dis.* **22:**359–361.

Smith, H. V., S. M. Caccio, N. Cook, R. A. B. Nichols, and A. Tait. 2007. *Cryptosporidium* and *Giardia* as foodborne zoonoses. *Vet. Parasitol.* **149:** 29–40.

Smith, H. V., C. A. Paton, M. M. A. Mtambo, and R. W. A. Girdwood. 1997. Sporulation of *Cyclospora* sp. oocysts. *Appl. Environ. Microbiol.* **63:** 1631–1632.

Srisuphanunt, M., V. Wiwanitkit, W. Saksirisampant, and P. Karanis. 2009. Detection of *Cryptosporidium* oocysts in green mussels (*Perna viridis*) from shell-fish markets of Thailand. *Parasite* **16:** 235–239.

Steele, M., S. Unger, and J. Odumeru. 2003. Sensitivity of PCR detection of *Cyclospora cayetanensis* in raspberries, basil, and mesclun lettuce. *J. Microbiol. Methods* **54:**277–280.

Thurston-Enriquez, J. A., P. Watt, S. E. Dowd, R. Enriquez, I. L. Pepper, and C. P. Gerba. 2002. Detection of protozoan parasites and microsporidia in irrigation waters used for crop production. *J. Food Prot.* **65:**378–382.

Toronto Public Health, et al. 1998. Outbreak of cyclosporiasis—Ontario, Canada, May 1998. *MMWR Morb. Mortal. Wkly. Rep.* **47:**806–809.

Tram, N. T., L. M. N. Hoang, P. D. Cam, P. T. Chung, M. W. Fyfe, J. L. Isaac-Renton, and C. S. L. Ong. 2008. *Cyclospora* spp. in herbs and water samples collected from markets and farms in Hanoi, Vietnam. *Trop. Med. Int. Health* **13:**1415–1420.

U.S. Department of Agriculture, Food Safety and Inspection Service. 2003. *Prescribed Treatment of Pork and Products Containing Pork To Destroy Trichinae. U.S. Code of Federal Regulations, 9CFR318.10.* U.S. Government Printing Office, Washington, DC. http://edocket.access.gpo.gov/cfr_2003/9CFR318.10.htm.

U.S. Environmental Protection Agency. 2001a. *Method 1622*: Cryptosporidium *in Water by Filtration/IMS/FA/ United States. EPA 821-R01-026.* Office of Water, U.S. Environmental Protection Agency, Washington, DC.

U.S. Environmental Protection Agency. 2001b. *Method 1623*: Cryptosporidium *and* Giardia *in Water by Filtration/IMS/FA/ United States. EPA 821-R01-025.* Office of Water, U.S. Environmental Protection Agency, Washington, DC.

White, K. E., C. W. Hedberg, L. M. Edmonson, D. B. W. Jones, M. T. Osterholm, and K. L. MacDonald. 1989. An outbreak of giardiasis in a nursing home with evidence for multiple modes of transmission. *J. Infect. Dis.* **15:**298–303.

Xiao, L. 2010. Molecular epidemiology of cryptosporidiosis: an update. *Exp. Parasitol.* **124:**80–89.

Yoder, J. S., and M. J. Beach. 2010. *Cryptosporidium* surveillance and risk factors in the United States. *Exp. Parasitol.* **124:**31–39.

Zarlenga, D. S., and J. M. Trout. 2004. Concentrating, purifying and detecting waterborne parasites. *Vet. Parasitol.* **126:**195–217.

FOOD SERVICE AND CATERING

VI

PRACTICAL SAMPLING PLANS, INDICATOR MICROORGANISMS, AND INTERPRETATION OF TEST RESULTS FROM TROUBLESHOOTING

J. L. Kornacki

25

Food service can be defined as those entities responsible for meals prepared outside the home. Included in this definition are restaurants, schools, catering companies, hospital cafeterias, and commercial or institutional kitchens operating within other venues. Food service distributors typically supply foods in bulk to these entities. Consequently, these institutions have a need to understand and implement appropriate microbiological specifications, such as sampling plans, rapid testing, and interpretation of data.

SAMPLING CONSIDERATIONS

In food service operations as well as the rest of the food industry, the question, "How many samples should I require be taken and tested?" is an important question that is frequently asked and is particularly important in the context of establishing supplier microbiological criteria. Related to that question are deeper considerations and questions including the following: (i) Is the target organism likely to be evenly distributed in my food sample? and (ii) if the organism is evenly distributed, does that mean one sample is enough?

Another critical concern relates to interpretation of results, particularly as they relate to retesting, and will address the questions, "Is retesting an appropriate means to verify the accuracy of my test data? And, if so, when is it appropriate?" This chapter will address these issues.

Number of Samples

How many samples should be taken to provide confidence in the acceptability of lot testing? No amount of finished product testing other than testing of an entire lot with a perfect method can guarantee that it is free of the target organism. Given that no method is perfect and complete destructive testing of entire lots is unacceptable, there is a need to introduce other systems in addition to mere lot testing to ensure food safety and quality. Such systems may include third-party good manufacturing practices (GMPs)-based inspectional audits at the bulk food manufacturer, in-plant microbiological risk assessments, and hazard analysis and critical control points (HACCP)-based verification audits for environmental and other controls (Kornacki, 2009a, 2009b, 2010a). However, these approaches are beyond the scope of this chapter. Approaches to in-plant microbiological risk assessment can be

J. L. Kornacki, Kornacki Microbiology Solutions, Inc., P.O. Box 163, McFarland, WI 53558.

Rapid Detection, Identification, and Quantification of Foodborne Pathogens, ed. by J. Hoorfar

found elsewhere (Kornacki, 2009a, 2009b, 2010a).

Foods are rarely contaminated uniformly unless they are nonviscous fluids and well mixed. However, even in those instances, selected microorganisms (e.g., *Clostridium perfringens*) can be prolific at forming clumps and microscopic colonies. If the number of bacterial cells or clumps in a given well-mixed, nonviscous fluid product is low, e.g., in 1% of samples, then there is in theory a 1-in-100 chance that the sample taken will test positive for the microorganism present (assuming a perfect method). In solid foods, this situation is more complicated. In most instances, the distribution of microorganisms in a solid food is apt to be sporadic, resulting from the dispersion of microbial growth niches at indeterminate times into the food stream depending on a number of variables, which often include lack of mixing to evenly distribute the microbial population. Factors that contribute to the likelihood of contamination of foods and food ingredients from microbial growth niches include the number of niches in the processing environment, the microbial population in each niche, the proximity of the niches to the product stream, the spatial relationship of the niches to the product stream, the degree of niche disruption, and the exposure of the product stream to the processing environment (Gabis and Faust, 1988). All of these factors play into the safety and quality of the ingredients and foods received by the food service distributor and hence the foodservice operation. The net effect is that the statistics associated with recovery of microbes from foods are often worse than what is depicted in tables that assume homogeneous distribution of contaminants and detection with perfect methods. Consequently, a single 25-g sample tested from a given lot of solid food is unlikely to provide a meaningful result.

MICROBIOLOGICAL CRITERIA

There are three types of microbiological criteria. These include (i) standards, (ii) industry guidelines, and (iii) specifications (National Research Council, 1985). Standards are legal and mandatory and typically relate to pathogens. However, guidelines are advisory. Regulators or industry uses these to determine if hygienic provisions that may be related to GMPs or critical control points have been met. They may include organisms of no public health significance. Microbiological specifications are purchase requirements set by industry.

Establishing Appropriate Microbiological Criteria

A number of entities have published approaches to establishment of microbiological criteria and include, among others, the International Commission on the Microbiological Specifications for Foods (ICMSF), FAO/WHO, the U.S. National Advisory Committee on Microbiological Criteria for Foods, the U.S. Food and Drug Administration (FDA), the U.S. Department of Agriculture, the European Commission, and other governments.

Every specification should include at least the following:

1. Identity of the food
2. Specification of the contaminant of concern
3. The method that must be used to detect the target organism(s)
4. Definition of the sampling plan
5. Establishment of the microbiological limits appropriate to the food and consistent with the sampling plan

The scheme recommended by ICMSF (International Commission on the Microbiological Specifications for Foods, 2002) is one of the most comprehensive of such sampling plans and is useful for the establishment of specifications by the food service provider or distributor. However, before reliance upon the ICMSF scheme to establish a specification, the food service entity establishing the specification should ensure that it is not in conflict with the requirements of its customer or the testing requirements of the country to which it is distributing. In addition, a number of factors should be considered including the following (items in italics are addressed at least in part by the ICMSF scheme):

- Evidence for potential hazard. These can be derived from epidemiologic data and an appropriate hazard analysis
- Knowledge of natural or acquired microflora
- *Ability of the food to support growth*
- *Effect of processing on the microflora of the food*
- *Potential for contamination and/or growth during processing, handling, storage, or distribution of the food*
- *Health status of the consumer*
- State in which the food is distributed (e.g., solid or liquid)
- *Potential for abuse by the consumer*
- *Spoilage potential, utility, and GMPs*
- *Manner of preparation for ultimate consumption*
- Reliability of assays for detection of microbes
- Costs and benefits associated with the application of the microbiological criterion

In this scheme, foods are categorized into 15 cases based on three conditions (i.e., decreased risk, no change, or increased risk) associated with both the intrinsic properties of the food (e.g., pH, water activity, etc.) and extrinsic factors associated with the food (e.g., cooking, irradiation, or subsequent handling after a lethal step) and also with five levels related to the severity of the target agent (International Commission on the Microbiological Specifications for Foods, 2002). A discussion of the factors that influence microbial survival and growth can be found elsewhere (Kornacki, 2010b). These 15 cases are depicted in Fig. 1.

In Fig. 1, the term "n" refers to the number of samples to be taken and tested per lot, and the term "c" refers to the subset of those samples that can fall into a marginally acceptable range. The definition of the term "lot" is extremely important. It is "that quantity of product produced, handled, and stored within a limited time period under uniform conditions" (International Commission on the Microbiological Specifications for Foods, 2002).

The marginal range is described by two parameters, *m* and *M* (not shown in Fig. 1): *M* is any population of microorganisms deemed unacceptable in any number of samples, and "*m*" is the population associated with a threshold for marginal acceptability. Hence, marginal quality is defined as the allowable

Increasing Risk from Product or Environment →

Increasing Microbial or Toxic Risk ↓

Agent	**Conditions that Decrease Risk**	**Conditions that Do Not Change Risk**	**Conditions that Increase Risk**
Spoilage n=5	Case 1 c=3	Case 2 c=2	Case 3 c=1
Indicator n=5	Case 4 c=3	Case 5 c=2	Case 6 c=1
Moderate	Case 7 n=5, c=2	Case 8 n=5, c=1	Case 9 n=10, c=1
Serious c=0	Case 10 n=5	Case 11 n=10	Case 12 n=20
Severe c=0	Case 13 n=15	Case 14 n=30	Case 15 n=60

n= number of samples per lot, c= maximum number of samples allowed between m and M.

FIGURE 1. ICMSF Attributes Sampling Plans. (Amended from Kornacki, 2010e.)

number of samples (c) with populations of bacteria between *m* and *M*. For example, if one has a product wherein *M* is 100,000 per gram and *m* is 10,000 per gram, then c is the maximum number of samples that are allowed by the plan to fall into this marginal range between 10,000 and 100,000 per gram. This type of approach addresses the issue of nonhomogeneity of contamination in a given lot, which is a situation very likely to occur with industrially processed foods and ingredients (Kornacki, 2010e). However, this approach is only applicable to cases 1 through 9, wherein the microbes or toxic agents are of lower risk than in cases 10 through 15 (Fig. 1). Such plans are called three-class-attributes sampling plans, because food lots are defined in three ways: acceptable, marginal, and unacceptable. The value for c is always set to zero for "serious" and "severe" pathogens or agents (cases 10 through 15). Hence, cases 10 through 15 describe a two-class plan, as there is no acceptable marginal lot if it contains any sample with a highly invasive pathogen or toxin that at any level can result in illness.

ORGANISMS, TARGET PATHOGENS, AND INDICATORS

The ICMSF (2002) provided a list of those organisms and agents that fit into the moderate, severe, and serious pathogen categories represented in cases 7 through 15 (Tables 1 through 5). However, it is left to the industry to determine which indicator or spoilage organisms should be included in cases 1 through 6.

Organisms of Concern in Food Service Products

Food service organizations want assurance that low-dose (usually invasive) pathogens are not present in processed ingredients and ready-to-eat products they receive and later prepare. Hence, appropriate specifications for organisms like *Salmonella* (especially in dry products, e.g., dried milk powders or dried foods of animal origin) and *Listeria monocytogenes* (more significant in high water activity ready-to-eat foods, e.g., frankfurters) should be established. ICMSF and other statistically based sampling plans may be cost-prohibitive at 15, 30, or 60 samples tested per lot, but there exist validated compositing schemes (pooled samples) that can reduce the number of tests done on these samples (U.S. Food and Drug Administration, 2003; McMahon et al., 2005; Curiale, 2000). However, appropriate cooking and handling will provide an additional barrier to the potential for foodborne illness with these organisms. Consequently, organisms or their toxins that present the greater potential for concern would be those that can survive a thermal (e.g., cooking) process.

TABLE 1 ICMSF severe hazards for the general population (cases 13 through 15)[a]

Botulinum neurotoxin (*Clostridium botulinum, C. butyricum, C. barati*)
Brucella abortus, B. suis, B. melitensis
Enterohemorrhagic *E. coli*
Salmonella enterica serovar Typhi, Paratyphi A, B, and C
Shigella dysenteriae I
Burkholderia cocovenenans
Vibrio cholerae (O1 and O139)
Mycobacterium bovis
Aflatoxins (from *Aspergillus flavus* and *A. parasiticus*)
Variant Creutzfeldt-Jakob disease and bovine spongiform encephalopathy
Taenia solium

[a] Adapted from International Commission on the Microbiological Specifications for Foods, 2002.

Microorganisms of Principal Concern

Pathogens

Given the above, the microorganisms and toxins of greatest concern will be heat-resistant spores such as *Clostridium perfringens* and *Bacillus cereus* or heat-resistant toxins produced by certain strains of staphylococci (e.g., staphylococcal enterotoxins). The above pathogens are covered in other chapters of the present book. The following is focused on indicator microorganisms.

TABLE 2 ICMSF severe hazards for restricted populations (cases 13 through 15)[a]

Hazard	Restricted population
Campylobacter jejuni serovar O:19 and associated serotypes	Guillain-Barré syndrome
Enteropathogenic and enterotoxigenic (not enterohemorrhagic) *E. coli*	Infants
C. perfringens type C (enteritis necrotans)	Protein-deficient persons
Clostridium botulinum (types A and B)	Infants
Enterobacter sakazakii	Infants
Listeria monocytogenes	Immunocompromised
Vibrio vulnificus	Patients with liver disease
Hepatitis A virus	Patients with liver disease
Cryptosporidium parvum	Immunocompromised

[a] Adapted from International Commission on the Microbiological Specifications for Foods, 2002.

Indicator Microorganisms

The term "indicator" has been used in various ways and can include (i) an indicator of unsanitary conditions of manufacture, (ii) an indicator of the microbiological quality of the food, (iii) an index of the likely presence of a pathogen, and (iv) a surrogate organism (Kornacki, 2010d). It is well established that there are no good index microorganisms when testing a food on a lot-by-lot basis (Kornacki and Johnson, 2001). Surrogate microorganisms may be used to validate the efficacy of critical control points and are nonpathogens with survival or growth properties that can be used to model the behavior of pathogens; these fall outside the scope of this chapter (Kornacki, 2010d). In this chapter, we shall use the term "indicator" in the first and second definitions above. Any number of indicator microorganisms can be considered; however, a few selected indicators are listed below.

TABLE 3 ICMSF serious microbiological and viral hazards (cases 10 through 12)[a]

Salmonella enteritidis serovar Typhimurium and other non-typhoidal serovars (salmonellosis)
Yersinia enterocolitica, Y. pseudotuberculosis
Shigella flexneri, S. boydii, S. sonnei (nondysentery shigellosis)
Listeria monocytogenes (general population)
Hepatitis A virus (if no liver disease involved)
Arcobacter butzleri, A. cryaerophilus
Cryptosporidium parvum
Cyclospora cayetanensis

[a] Adapted from International Commission on the Microbiological Specifications for Foods, 2002.

Escherichia coli. Conventional methods involve most-probable-number and direct-count methods, which rely on the ability of the organism to grow in the presence of bile salts (e.g., lauryl sulfate tryptose broth or brilliant green lactose bile broth) and produce acid and gas followed by the ability to grow in EC broth (FDA Bacteriological Medium M49 [U.S. Food and Drug Administration, 2003]) at elevated temperature (either 44.5°C for shellfish or 45.5 ± 0.2°C for most foods,

TABLE 4 ICMSF serious hazards (cases 10 through 12) associated with toxins[a]

Trichothecene toxins (produced by *Fusarium graminarium* and related spp.)
Zearalenone (produced by *F. graminarium* and related spp.)
Fumonisins (produced by *Fusarium moniliforme* and related spp.)
Ochratoxin A (produced by *Penicillium verrucosum, Aspergillus ochraceus* and related spp., and *A. carbonarius* and related spp.)

[a] Adapted from International Commission on the Microbiological Specifications for Foods, 2002.

TABLE 5 ICMSF moderate hazards (cases 7 through 9)[a]

Bacillus cereus
Clostridium perfringens (type A)
Escherichia coli (enteropathogenic and enterotoxigenic *E. coli*) (not enterohemorrhagic strains)
Staphylococcal enterotoxins (*Staphylococcus aureus*)
Vibrio cholerae non-O1 and non-O139
Vibrio parahaemolyticus
Small round structured virus (Norwalk virus)
Biogenic amines (e.g., histamine)

[a] Adapted from International Commission on the Microbiological Specifications for Foods, 2002.

depending upon the commodity (Kornacki and Johnson, 2001). Microorganisms that produce gas in brilliant green lactose bile broth at 35°C within 48 h of incubation are considered coliforms via the FDA method, whereas those coliforms that can grow in EC broth at 44.5 or 45.5°C are considered fecal coliforms. Traditional confirmation relies upon the correct IMViC reactions of these organisms. IMViC is an acronym for indole, methyl-red, Voges-Proskauer, and citrate reactions, described below.

The term "fecal coliform" is a misnomer in this author's opinion, as most "fecal" coliforms can also be found in the natural environment. However, the natural reservoir of *E. coli*, a fecal coliform, is believed to be the intestinal tract of man and animals (Kornacki and Johnson, 2001). Nevertheless, secondary nonfecal transmission of *E. coli* in the industrial food processing environment has been observed by this author (e.g., *E. coli*-contaminated ceiling-mounted overhead refrigerant coils contaminated, via aspiration, ripening cheese from pasteurized milk in exposed vessels). Thus, *E. coli* is the closest to a "fecal coliform" that one can get. Sampling for *E. coli* would fall within the ICMSF categories 4 to 6; hence, five samples per lot are recommended. Foods that have been previously cooked should have undetectable *E. coli* when tested by the most-probable-number (e.g., <3 CFU per gram) or direct-count (<10 per gram) methods.

E. coli can be grouped into biotypes I and II. Biotype I organisms yield results of positive, positive, negative, and negative for the four reactions of the IMViC biochemical series, respectively, and hence can convert indole to tryptophan, ferment glucose to stable acid intermediates detected by the methyl-red acid/base indicator, do not produce 2,3 butanediol and/or acetoin from glucose (negative Voges-Proskauer reaction), and do not utilize citrate as their sole carbon and energy source. Biotype II organisms show an IMViC reaction pattern of negative, positive, negative, and negative, respectively.

E. coli may fit into classical enteropathogenic serotypes, and serotyping approaches to characterize the somatic (O) and flagellar (H) antigen have been developed.

Interpretation of Results. Serotyping of "generic" *E. coli* is not used commonly, unless large populations (>1 million CFU per gram) of these are recovered from foods. These large populations have been associated with foodborne illness (Kornacki and Marth, 1982). One exception is the finding of *E. coli* O157:H7 (or other recognized enterohemorrhagic strains), which, at any level, would have public health significance. However, *E. coli* O157:H7 has not been shown to grow at 44.5°C, nor do they react with the MUG reagent (4-methylumbelliferyl-beta-D-glucuronide). Hence, a finding of generic *E. coli* is unlikely to be confirmed as *E. coli* O157:H7, and the effort and time involved are unproductive unless a specific test for this organism is done.

Other Indicator Organism Assays

Other assays to consider for use are the aerobic plate count, the coliform count, *Enterococcus* count, and the anaerobic plate count (typically for ICMSF cases 1 through 3, except if used as an indicator of potential spoilage).

SPOILAGE DEFECTS/ORGANISMS

A variety of spoilage defects may exist, depending upon the product, and could include excess acidity, often from *Lactobacillus* spp. or other acidophiles, and gassiness (e.g., from heterofermentative *Lactobacillus* spp., *Clostridium* spp., coliforms, yeasts, and *Bacillus* spp.), as well as color (e.g., from pigmented yeasts or molds or modified oxidation reduction potential in some foods from certain lactic acid bacteria) and flavor defects. A review of spoilage organisms can be found elsewhere (Clavero, 2010). Sampling for these organisms can be done according to the ICMSF cases 1 through 3.

Retesting

Retesting for qualitatively detected organisms (e.g., those recovered by enrichment techniques) is unreliable and should not be done except when there is clear evidence for an "equivocal" result (Kornacki, 2010c, 2010e). This is because of the likelihood of sporadic, nonuniform contamination of foods. Retesting of quantitative assays is essentially obviated by using the ICMSF sampling schemes, as no fewer than five subsamples per lot are recommended in any of the cases.

REFERENCES

Clavero, R. 2010. Solving microbial spoilage problems in processed foods, chapt. 3, p. 63–78. *In* J. L. Kornacki (ed.), *Principles of Microbiological Troubleshooting in the Industrial Food Processing Environment.* Springer, New York, NY.

Curiale, M. 2000. Validation of the use of composite sampling for *Listeria monocytogenes* in ready-to-eat meat and poultry products. http://www.amif.org/ht/d/sp/i/26883/pid/26883#. Accessed 22 June 2010.

Gabis, D. A., and R. E. Faust. 1988. Controlling microbial growth in the food-processing environment. *Food Technol.* **December:**81–82, 89.

International Commission on the Microbiological Specifications for Foods. 2002. *Microorganisms in Foods*, vol. 7. Springer, New York, NY.

Kornacki, J. L., and E. H. Marth. 1982. Foodborne illness caused by *Escherichia coli*: a review. *J. Food Prot.* **45:**1051–1067.

Kornacki, J. L., and J. Johnson. 2001. Enterobacteriaceae, coliforms, and *Escherichia coli* as quality and safety indicators, chapt. 8. *In* F. P. Downes and K. Ito (ed.), *Compendium of Methods for the Microbiological Examination of Foods.* American Public Health Association, Washington, DC.

Kornacki, J. L. 2009a. The missing element in microbiological food safety inspection approaches, part 1. *Food Safety Magazine* **2009:**(February–March).

Kornacki, J. L. 2009b. The missing element in microbiological food safety inspection approaches, part 2. *Food Safety Magazine* **2009:**(April–May).

Kornacki, J. L. 2010a. Where are these contaminants found?, chapt. 4, p. 79–102. *In* J. L. Kornacki (ed.), *Principles of Microbiological Troubleshooting in the Industrial Food Processing Environment.* Springer, New York, NY.

Kornacki, J. L. 2010b. What factors are required for microbes to grow, survive and die?, chapt. 5, p. 103–116. *In* J. L. Kornacki (ed.), *Principles of Microbiological Troubleshooting in the Industrial Food Processing Environment.* Springer, New York, NY.

Kornacki, J. L. 2010c. Where do I start?, chapt. 6, p. 117–123. *In* J. L. Kornacki (ed.), *Principles of Microbiological Troubleshooting in the Industrial Food Processing Environment.* Springer, New York, NY.

Kornacki, J. L. 2010d. How do I sample the environment and equipment?, chapt. 7, p. 125–136. *In* J. L. Kornacki (ed.), *Principles of Microbiological Troubleshooting in the Industrial Food Processing Environment.* Springer, New York, NY.

Kornacki, J. L. 2010e. How many samples do I take?, chapt. 8, p. 137–149. *In* J. L. Kornacki (ed.), *Principles of Microbiological Troubleshooting in the Industrial Food Processing Environment.* Springer, New York, NY.

McMahon, W. A., A. M. McNamara, A. M. Schultz, P. Hall, L. Ledenbach, and M. Clark. 2005. *Validation of the Use of Composite Sampling for the Detection of* Listeria monocytogenes *in a Variety of Food Products.* Int. Assoc. Food Prot. Annu. Meet., 17 August 2005, Baltimore, MD.

National Research Council. 1985. *An Evaluation of the Role of Microbiological Criteria for Foods and Food Ingredients.* National Academic Press, Washington, DC.

U.S. Food and Drug Administration. 2003. Food sampling preparation of sample homogenate, chapt. 1. *In Bacteriological Analytical Manual.* http://www.fda.gov/Food/ScienceResearch/LaboratoryMethods/BacteriologicalAnalyticalManualBAM/UCM063335. Accessed 23 June 2010.

CLOSTRIDIUM PERFRINGENS IN FOOD SERVICE

R. G. Labbé and K. Grant

26

CHARACTERISTICS OF THE ORGANISM

Clostridium perfringens toxins are responsible for a variety of human and veterinary diseases. The public health importance of the organism is reflected in the fact that to date the complete gene sequences of three pathogenic type A strains have been published (Myers et al., 2006; Shimizu et al., 2002). *C. perfringens* is a prolific toxin producer with more than a dozen known to date (Labbe, 2000; Petit et al., 1999) including the more recently identified beta 2 toxin (Gilbert et al., 1997). *C. perfringens* isolates are commonly classified into one of five types, A to E, based on their ability to produce four so-called lethal toxins, alpha, beta, epsilon, and iota. Certain isolates, mostly type A, produce an enterotoxin, CPE, responsible for foodborne illness. Strains of toxin types B through E are always disease associated. On the other hand, type A strains are found in the normal flora of the intestinal tract of humans and animals. While enterotoxigenic type A strains are best known for their association with foodborne illness, this toxigenic type also causes nonfoodborne human gastrointestinal disease, including antibiotic-associated diarrhea and sporadic diarrhea (Carman, 1997; Collie et al., 1998; McClane, 2007).

In animals, type A strains are responsible for a number of diseases such as necrotic enteritis in chickens (Cooper and Songer, 2009), enteritis in piglets (van Damme-Johnsten et al., 1990), and other disease states (Sawires and Songer, 2006).

Early evidence for the association of *C. perfringens* and foodborne illness appeared in the 1950s. Conclusive proof accumulated in the 1970s demonstrating the role of an enterotoxin and its physiological effects. Among the attributes associated with its ability to cause foodborne illness are (i) its wide distribution in the natural environment, (ii) its ability to form heat-resistant spores allowing its selective survival during food processing procedures, (iii) a rapid growth rate at relatively high temperatures, and (iv) its ability to produce an enterotoxin responsible for foodborne illnesses.

In addition to the enterotoxigenic nature of certain isolates, the two defining characteristics of *C. perfringens*'s role in foodborne illness are its potential growth rate and its ability to produce heat-resistant spores. Spore formation by

R. G. Labbé, Department of Food Science, University of Massachusetts, Amherst, MA 01003. *K. Grant*, Health Protection Agency, Centre for Infections, London NW9 5HT, United Kingdom.

Rapid Detection, Identification, and Quantification of Foodborne Pathogens, ed. by J. Hoorfar

C. perfringens is an important characteristic because (i) spores are able to survive cooking procedures while eliminating competitors, (ii) it is an essential classification tool, and (iii) enterotoxin formation is associated with sporulation. The generally accepted temperature range for growth of *C. perfringens* is 15 to 51°C, with an optimum of 43 to 46°C (Juneja et al.,. 1999). This organism has the shortest generation time of any bacterial pathogen. For example, generation times of 7 to 8 minutes have been reported in autoclaved ground beef (Labbe and Huang, 1995; Willardsen et al., 1979).

CHARACTERISTICS OF *C. PERFRINGENS* FOODBORNE ILLNESS

Foodborne illness due to *C. perfringens* is neither an intoxication nor an infection. It is considered a toxicoinfection, in which ingestion of vegetative cells is required, followed by production of an enterotoxin in vivo. Illness typically results from the ingestion of large numbers ($>10^8$) of enterotoxin-positive vegetative cells in temperature-abused foods. A sufficient number survive stomach passage and subsequently sporulate in the small intestine. The enterotoxin (CPE) is produced during the sporulation process and is released together with the mature spore. Thus, an elevated number of fecal spores is another criterion for confirmation of its role in outbreaks. CPE is the only example of a sporulation-associated toxin affecting humans.

As mentioned above, the clinical symptoms associated with foodborne illness are due to an intracellular enterotoxin produced in the small intestine during sporulation. In laboratory media, CPE and spores are released from the mother cell. The kinetics of these events corresponds to the time required for onset of symptoms in clinical cases. While CPE can represent a large (up to ~20%) proportion of total intracellular protein (Labbe, 1981; Garcia-Alvarado et al., 1992; Czeczulin et al., 1993), its function remains elusive.

Foodborne illness associated with this organism consists of an incubation period of 8 to 24 h with diarrhea and abdominal pain as typical symptoms; fever is rarely present. Symptoms subside within 24 hours as the toxin is passed from the small intestine. Residual cramping may persist for an additional 1 to 2 days. The removal of unbound toxin and normal turnover of intestinal cells to which the toxin has bound are presumably responsible for the brief nature of the illness.

EPIDEMIOLOGY

As with other agents of foodborne illness, the incidence of *C. perfringens* as the etiologic agent is widely underreported. In countries in which it is a notifiable disease, such as England and Wales, the incidence is severalfold greater than in the United States, where there is no such requirement.

Food service establishments such as restaurants, hospitals, prisons, schools, and caterers are the most likely sources, without any seasonal prevalence. Food prepared in such institutions is often done well in advance and held for later serving, allowing for the role of improper holding. This explains the large case-to-outbreak ratio, for example, 52, for the period 1998 to 2002, significantly higher than for other bacterial foodborne disease agents (Centers for Disease Control and Prevention, 2005).

C. perfringens lacks the genetic ability to produce 13 of the 20 essential amino acids and is therefore typically associated with high-protein foods such as meat and poultry (Brynestad and Granum, 2002). Responsible vehicles are known to contain cells or spores of this organism. These can survive a cooking procedure, germinate, and resume growth during slow cooling.

Surveys on the incidence of *C. perfringens* in foods were initially conducted in the 1960s and 1970s. At the time, it was not known that not all isolates were enterotoxin positive. In hindsight, this limits the usefulness of such early surveillance studies, except to demonstrate the widespread distribution of the organism in general. More-recent surveys indicate that CPE-positive strains are uncommon in retail foods (Miki et al., 2008; Wen and McClane, 2004).

The criteria for establishing an outbreak of *C. perfringens* type A food poisoning include (i) $>10^6$ spores (per g feces from ill individuals), (ii) $>10^5$ cells/g in incriminated food, or (iii) presence of the CPE in feces of ill individuals (Labbe and Juneja, 2006). Earlier criteria included the presence of a common serotype in feces and suspect foods. However, the antigenic heterogeneity of strains of *C. perfringens* has limited the availability of antisera, so that this is not a commonly employed epidemiological tool.

ISOLATION AND IDENTIFICATION

Cultural Procedures

Classical Methods

The currently recommended procedure for enumerating *C. perfringens* in suspected outbreak foods involves the use of a sulfite-cycloserine agar under anaerobic conditions. Detailed procedures for isolation and confirmation of *C. perfringens* are available (Food and Drug Administration, 1998; U.S. Department of Agriculture, 1998; International Organization for Standardization, 2004). The effectiveness of this method has been supported by comparative studies (de Jong et al., 2003). In the case of an outbreak, elevated spore counts in stool samples can be easily determined by heating samples before plating. Enterotoxigenic isolates with the *cpe* gene on plasmids are more heat sensitive than isolates possessing the chromosome-borne *cpe* gene (see below). This suggests the use of alcohol may be a better selective method than heat before plating. Black colonies on selective media are confirmed by using the reactions of nitrate reduction, gelatin hydrolysis, liquefaction of gelatin, and lack of motility. Other confirmatory reactions proposed are the acid phosphatase reaction and the reverse CAMP (an acronym composed of the initials of the original authors) tests (Eisgruber et al., 2003; Labbe, 1989). Both are relatively simple but not used frequently.

For routine enumeration of low amounts of generic *C. perfringens* in foods, an iron-milk medium (IMM) is a rapid and economical alternative when used in conjunction with a most-probable-number protocol (Lin and Labbe, 2003). A unique stormy fermentation within 18 h at 45°C is said to be adequate for confirmation (Abeyta et al., 1985). Stormy fermentation is the production of an acid curd with subsequent disruption of the curd by gas. In the case of shellfish, the IMM method has been adopted as first action by AOAC International (Abeyta and Wetherington, 1994).

A commercial rapid miniaturized test is the RAPID ID 32A identification system. It is based on evaluation of the action of bacterial enzymes by use of chromogenic substrates and allows the identification of anaerobes within 4 h without the need for anaerobic identification. In one assessment, it was able to correctly identify all 49 *C. perfringens* strains tested. Overall, 95% of *Clostridium* species were correctly identified by this test (Sperner et al., 1999a). Neither IMM nor RAPID ID 32A specifically detects *cpe*-positive *C. perfringens*.

Fluorogenic Culture Media

A number of selective media that incorporate chromogenic or fluorogenic substrates into selective media for foodborne pathogens are now available (Orenga et al., 2009). A membrane filtration method using a chromogenic agar is available for enumeration of *C. perfringens* in aqueous environments (Mueller-Spitz et al., 2009). For enumeration of *C. perfringens* from foods, a supplement to tryptose sulfite cycloserine (TSC) agar is available (Merck), which consists of cycloserine and 4-methylumbelliferyl-phosphate (MUP). In the presence of an acid or alkaline phosphatase, the phosphate group of the MUP is separated and the fluorescent methylumbelliferone is detected by UV light. However, some investigators (Eisgruber et al., 2003; Sartory et al., 2006) noted that a number of other *Clostridium* species showed a positive reaction when MUP was incorporated in a selective *C. perfringens* medium. On the other hand, these workers demonstrated the effectiveness of an acid phosphatase test using naphthyl phosphate and *o*-dianisidine applied directly onto colonies leading to the formation of a visible azo dye (but requiring surface plating). This avoids the neutral conditions of selective agars, which would allow expression of both acid and alkaline phosphatases. At least two studies (Eisgruber et al., 2003; Sartory et al., 2006) have shown the superiority of acid phosphatase over motility-nitrate-lactose-gelatin medium for confirmation. Of 114 environmental isolates, 94.7% were confirmed by this method compared to 91.2% by motility-nitrate-lactose-gelatin medium tests (Sartory et al., 2006).

In Situ Methods for *cpe*-Positive *C. perfringens*

The impetus for this work was the use of the membrane filtration method for enumeration of *C. perfringens* in water. The standard TSC cultural method does not distinguish *cpe*-positive colonies, requiring the confirmatory steps mentioned above. In addition, the lengthy procedures involved in confirming ent^+ *C. perfringens*, i.e., selective isolation, biochemical tests, and confirmation of enterotoxigenicity of isolates, have prompted the development of methods for direct testing of food samples for *C. perfringens*.

An early development described a nonisotopic colony hybridization assay said to detect as few as 10 CFU of *C. perfringens* per g (Baez and Juneja, 1995). *C. perfringens* colonies appearing on cellulose nitrate filters on TSC agar were identified by using anti-digoxigenin-alkaline phosphatase-conjugated antibody. The procedure, which employed a 364-bp gene probe, required a total assay time of 2 days. A nonspecific bacterial flora of at least 10^6 CFU/g in hamburger did not affect the specificity of the assay.

A related technique, hydrophobic grid membrane filter-colony hybridization, allowed enumeration of *cpe* gene-carrying *C. perfringens* spores and may be useful in examining outbreak stools in which large numbers of competing microorganisms are expected to be present (Heikinheimo et al., 2004). Using a hydrophobic grid membrane instead of a cellulose nitrate filter, up to 1,600 separate colonies could be screened for the presence of the *cpe* gene, which, as in the colony hybridization method, was detected by using a *cpe*-specific digoxigenin-labeled DNA probe.

In this procedure, heated fecal extracts are filtered on a hydrophobic grid, which is placed on TSC agar plates and held anaerobically for 36 to 42 h. By not preheating the samples, the technique could, in principle, be used to enumerate *cpe*-positive *C. perfringens* in outbreak foods. However, the authors did notice some obstruction of the filtration apparatus by the fecal extract, and similar limitations could be encountered in diluted food homogenates. A related in situ procedure used a fluorescently labeled rRNA-targeted oligonucleotide probe to detect microcolonies of *C. perfringens* on hydrophilic membrane filters, which were then placed on TSC agar (Shimizu et al., 2009). As little as 9 h was required for the test. No significant difference was noted between viable counts detected by this method and by the plate count method on TSC agar. However, the probe did not specifically detect *cpe*-positive colonies.

Detection of Enterotoxin

One criterion for confirming *C. perfringens* as the etiological agent of an outbreak is the detection of the enterotoxin in outbreak stools. The assays can also be employed to confirm the ability of isolates to express CPE, provided sporulation in a laboratory medium is achieved. These assays are based on latex agglutination or enzyme-linked immunosorbent assay formats. Stools of healthy individuals contain undetectable levels of CPE. However, if the objective is to detect CPE in suspect isolates, the instructions included with commercial detection kits may be misleading by implying that a single sporulation medium will induce sporulation in all isolates. Indeed, sporulation is highly strain dependent.

A number of sporulation media have been proposed over several decades. Most but not all (Meyer and Tholozan, 1999) are variations of Duncan and Strong's original peptone/ starch formulation with or without adjuncts (caffeine or theophylline) (de Jong et al., 2002; Le Marc et al., 2008; for a summary, see Hsieh and Labbé, 2007). Fortunately, low levels of spores, sufficient for culture carriage, are usually produced in commercially available cooked-meat medium.

Molecular Methods

PCR Detection of Toxin Genes

The development and introduction of molecular methods for identifying and typing *C. perfringens* have not only improved the ability to identify and investigate outbreaks of *C. perfringens* diarrheal disease but also enabled investigators to reassess disease transmission. One of the most significant outcomes of the development and use of molecular methods has been the ability to replace the use of animal assays for the detection of *C. perfringens* lethal toxins as a method of toxin typing *C. perfringens*. While detection of *C. perfringens* lethal toxin genes does not detect toxin production itself, it indicates the potential of a strain to produce toxin and, as such, provides a rapid, cost-effective, and ethically acceptable means of toxin typing strains of *C. perfringens*. There are a number of PCR-based assays, both conventional and real time, that have been described in the literature for the detection of the *C. perfringens* lethal toxin genes alpha, beta, epsilon, and iota. Frequently, these assays are multiplexed to detect other *C. perfringens* toxin genes such as *cpe* and β_2 (Garmory et al., 2000; Baums et al., 2004; Albini et al., 2008; Gurjar et al., 2007; van Asten et al., 2008).

As previously stated, *C. perfringens* diarrheal disease is an underreported form of gastroenteritis, mainly due to its mild clinical symptoms, which are generally of a short and self-limiting nature. The diagnosis of *C. perfringens* diarrheal illness and the identification and investigation of outbreaks can be complex due to a variety of factors including delay in patients reporting illness, with consequent delays in investigation of the incident; difficulties in obtaining food and fecal samples for testing; and the involvement of multiple agencies, sometimes from different geographical regions, which often leads to a lack of coordination.

Symptoms of *C. perfringens* diarrheal disease are the direct result of the action of CPE on the human gastrointestinal tract, and disease diagnosis can be performed by demonstrating the presence of CPE in fecal specimens from patients with classical disease symptoms. However, specimens must be taken within 2 days of onset of symptoms to maximize the likelihood of toxin being present, and obtaining specimens within this time frame may not (for the reasons outlined above) always be possible. Demonstrating high numbers of *C. perfringens* vegetative cells in food consumed by patients can also assist in diagnosis, as can isolation of the organism from feces of patients. However, isolation of *C. perfringens* on its own is insufficient evidence for confirmation of food poisoning because not all strains of *C. perfringens* produce enterotoxin, and only those capable of doing so have the potential to cause disease.

Also, nontoxigenic *C. perfringens* can be part of the normal human gastrointestinal flora in some individuals and may be at high levels in elderly persons. It is important, therefore, to demonstrate that food and fecal isolates of *C. perfringens* are able to, or have the potential to, produce enterotoxin. *C. perfringens* only produces enterotoxin when vegetative cells sporulate, and in *C. perfringens* food poisoning sporulation and enterotoxin production occur in the human small intestine. Getting *C. perfringens* to sporulate and produce enterotoxin under the artificial conditions in the laboratory is not always easy, there being no single method guaranteed to induce sporulation in all strains.

The introduction of molecular methods has enabled the rapid detection of the CPE gene by PCR, which indicates the potential of strains to produce enterotoxin and thus their potential to cause diarrheal disease. Prior to the use of PCR techniques for enterotoxin gene detection, the ability of isolates to produce enterotoxin was not routinely looked for in outbreak situations, and this undoubtedly made such investigations more complex and inconclusive.

Generally, the detection of enterotoxin genes by PCR is coupled with the detection of the *C. perfringens* alpha-toxin gene. This gene encodes the *C. perfringens* alpha-toxin, a phospholipase or lecithin that is detected in the Nagler reaction, the classic microbiological test for identifying strains as *C. perfringens*. There are a number of duplex-PCR assays described in the literature for detecting the enterotoxin and alpha-toxin genes of *C. perfringens*, including both simple conventional block-based and real-time PCR formats (Fach and Popoff, 1997; Grant et al., 2008).

Such assays can be used to screen out isolates that do not possess the enterotoxin gene as not playing a role in the disease process and allowing only those isolates in which the enterotoxin gene is detected from cases and foods to be further investigated by subtyping.

Subtyping

The subtyping of *C. perfringens* is important for the confirmation of foodborne disease, as finding the same subtype from suspect foods and fecal specimens provides supportive evidence of a common source. Traditionally, serotyping has been used to distinguish different *C. perfringens* strains, but this was usually only performed by specialist reference laboratories. Unfortunately, there were problems with cross-reaction and autoagglutination of strains, which meant that approximately 20% of all strains were nontypeable. Today, serotyping of *C. perfringens* to demonstrate that food isolates are the same as those from human cases and that isolates from outbreaks are of the same type has been largely superseded by the use of molecular fingerprinting techniques. A variety of methods are available including pulsed-field gel electrophoresis (PFGE) (Maslanka et al., 1999; Sperner et al., 1999b), ribotyping (Schalch et al., 2003), amplified fragment length polymorphism (Keto-Timonen et al., 2006), randomly amplified polymorphic DNA (LeFlon-Guibout et al., 1999), multilocus sequence typing (MLST) (Chalmers et al., 2008), and multiple-locus variable-number tandem repeat analysis (MLVA) (Sawires and Songer, 2005). Such techniques have the advantage of generally being applicable to all strains, are reproducible, and do not rely on the use of animals for production of reagents, thus minimizing the variation that this can incur.

While PFGE is considered the gold standard by certain governmental agencies for subtyping many bacterial pathogens in epidemiological investigations, it does have its shortcomings for use with *Clostridium* species including *C. perfringens* (Schalch et al., 2003). This is mainly due to DNA degradation problems from endogenous nucleases, common among clostridial isolates. PFGE is also a lengthy procedure, requiring skill to perform. Other more rapid molecular typing methods are now available to discriminate strains effectively for real-time epidemiological investigations.

Amplified fragment length polymorphism is reported to be an effective, rapid, and reproducible alternative to PFGE; it is a whole-genome fingerprinting method based on selective amplification of restriction fragments (McLauchlin et al., 2000; Vos et al., 1995). It is advantageous for extensive fingerprinting databases; with automation, high throughput is achieved, and the procedure overcomes problems of extracellular DNases (Keto-Timonen et al., 2006). Using purified DNA, analyses can be completed within two working days.

Ribotyping, a method based on restriction fragment length polymorphisms in the rRNA genes, has been found to be as effective as PFGE for typing isolates of *C. perfringens*. Automated ribotyping equipment using commercially available equipment has been used to subtype *C. perfringens* in poultry (Craven et al., 2003; De Cesare et al., 2009).

Randomly amplified polymorphic DNA is a rapid PCR technique of moderate cost and time requirements that may be a suitable tool in epidemiological investigations. It avoids the need for knowledge of DNA sequences of the target organism and is based on amplification of DNA sequences using a single primer of arbitrary sequence. Using this technique, LeFlon-Guibout et al. (1999) were able to identify 25 unrelated *C. perfringens* strains with a discriminatory index of 0.97 compared to 0.99 using zymotyping, a technique based on detection of esterases following their separation by electrophoresis. However, zymotyping is not as easy to perform, is more time-consuming, and is not used routinely for subtyping *C. perfringens* isolates.

MLVA is a newer subtyping method with increasingly widespread application to subtyping of gram-positive and -negative pathogens (Lindstedt, 2005). Here, the presence of short, highly conserved, repetitive DNA sequences at specific regions throughout the bacterial genome is exploited for strain differentiation. The specific regions containing the so-called "variable number repeats" are amplified by PCR, with amplicon size relating to the number of repeats present. The number of repeats at each region or locus is characteristic of different strains. MLVA has been found to be useful for phylogenetic and epidemiological investigations of *C. perfringens* isolates from veterinary sources and may provide a rapid, high-throughput method for strain typing in *C. perfringens* diarrheal outbreaks (Chalmers et al., 2008; Sawires and Songer, 2005).

MLST, a DNA sequence-based subtyping procedure, is based on nucleotide differences present in a set of housekeeping genes. While the technique involves DNA sequencing, the data generated are unambiguous, and its interpretation is less subjective than the DNA fragment analysis methods described above. MLST has been applied to a range of foodborne pathogens and has been found to be particularly useful for phylogenetic analyses. With *C. perfringens*, MLST has been used to investigate the clonality of veterinary isolates (Chalmers et al., 2008; Jost et al., 2006) and more recently to characterize enterotoxigenic human strains (Deguchi et al., 2009). However, its usefulness as an effective, rapid typing method for human *C. perfringens* diarrheal outbreaks needs to be fully evaluated.

LOCATION OF THE ENTEROTOXIN GENE IN *C. PERFRINGENS* ISOLATES

In strains of *C. perfringens* that possess the *cpe* gene, it is usually located either on the chromosome or on a large plasmid. Until relatively recently, isolates from outbreaks of food poisoning were usually found to have the *cpe* gene located on the chromosome, while isolates from sporadic cases and antibiotic-associated diarrhea were found to carry the *cpe* gene on a plasmid (Tanaka et al., 2007; Lahti et al., 2008). Vegetative cells and spores of Type A chromosome-borne *cpe* isolates from food poisoning outbreaks survive heating and low temperatures better than do cells or spores of plasmid-borne *cpe* isolates (Anderson et al., 2004; Li and McClane, 2006; Sarker et al., 2000).

For example $D_{100°C}$ values for spores of chromosomal *cpe* isolates were nearly 60-fold higher than those of plasmid *cpe* isolates. This appears to provide a logical explanation for the association predominantly of chromosomal *cpe* strains with food poisoning outbreaks. However, recent molecular investigations of *C. perfringens* isolates from Japan and Europe have shown that approximately 25 to 30% of food poisoning incidents are caused by plasmid *cpe* strains (Nakamura et al., 2004; Lahti et al., 2008; Grant et al., 2008). As strains with plasmid *cpe* genes have been shown to be less heat resistant, this has raised the question of whether such strains are transmitted in the same way as chromosomal *cpe* strains, i.e., are present in food prior to cooking.

As plasmid *cpe*$^+$ strains are less likely to survive cooking temperatures, does contamination of food occur after cooking? A survey of food handlers in Finland revealed a prevalence of *cpe*-positive *C. perfringens* of 18%, using PCR detection of enrichment cultures. In this study, *cpe*$^+$ isolates were obtained from approximately one-half of the PCR positives, with an average number of *cpe*$^+$ isolates of 2.5×10^2/g with an approximately 100-fold majority by the dominant *cpe*-negative population (Heikinheimo et al., 2006).

For the majority of *cpe*$^+$ strains, the gene was found to be plasmid-borne, suggesting that healthy humans could be a source of plasmid *cpe*$^+$ strains in food poisoning incidents. Another study found that one-half of healthy donors possessed on average approximately 10^6 cells of *C. perfringens*/g with a broad range of carriage (Carman et al., 2008). A similar number of donors (7%) carried *cpe*$^+$ isolates, and these were all plasmid *cpe* strains. Japanese investigators have also found predominantly plasmid *cpe*$^+$ isolates in feces from healthy individuals (Tanaka et al., 2007), and further investigations are required to establish if such carriers are reservoirs of plasmid *cpe*$^+$ isolates for food poisoning transmission.

CONCLUSIONS

The application of molecular techniques has had a tremendous impact on the diagnosis and investigation of *C. perfringens* food poisoning and nonfoodborne outbreaks. Reference laboratories are now able to determine rapidly and easily if isolates have the potential to cause diarrheal disease and through the use of discriminatory molecular typing methods provide evidence that *cpe*-positive strains from humans and foods have a common origin. The ability to investigate the genetic diversity within different *C. perfringens* strains has enabled us to question and investigate the phylogeny of *C. perfringens* strains as well as diarrheal disease transmission, and this will ultimately lead to a greater understanding of the way in which this important pathogen causes diarrheal disease.

REFERENCES

Abeyta, C., Jr., A. Michalovski, and M. Wekell. 1985. Differentiation of *Clostridium perfringens* from related clostridia in iron milk medium. *J. Food Prot.* **48:**130–134.

Abeyta, C., Jr., and J. Wetherington. 1994. Iron milk medium method for recovering *Clostridium perfringens* from shellfish: collaborative study. *J. AOAC Int.* **77:**351–356.

Albini, S., I. Brodard, A. Jaussi, N. Wollschlaeger, J. Frey, R. Miserez, and C. Abril. 2008. Real-time multiplex PCR assays for reliable detection of *Clostridium perfringens* toxin genes in animal isolates. *Vet. Microbiol.* **127:**179–185.

Anderson, K., T. Hansen, and S. Knøchel. 2004. Growth of heat-treated enterotoxin-positive *Clostridium perfringens* and the implication for safe cooling rates. *J. Food Prot.* **67:**83–89.

Baez, L., and V. Juneja. 1995. Nonradioactive colony hybridization assay for detection and enumeration of enterotoxigenic *Clostridium perfringens* in raw beef. *Appl. Environ. Microbiol.* **61:**807–810.

Baums, C., U. Schotte, G. Amtsberg, and R. Goethe. 2004. Diagnostic multiplex PCR for toxin genotyping of *Clostridium perfringens* isolates. *Vet. Microbiol.* **100:**11–16.

Brynestad, S., and P. E. Granum. 2002. *Clostridium perfringens* and foodborne infections. *Int. J. Food Microbiol.* **74:**195–202.

Carman, R. 1997. *Clostridium perfringens* in spontaneous and antibiotic-associated diarrhea of man and other animals. *Rev. Med. Microbiol.* **8**(Suppl. 1)**:**S43–S45.

Carman, R. J., S. Sayeed, J. Li, C. Genheimer, M. Hiltonsmith, T. Wilkins, and B. McClane. 2008. *Clostridium perfringens* toxin types in the feces of healthy North Americans. *Anaerobe* **14:**102–108.

Centers for Disease Control and Prevention. 2005. Surveillance for foodborne-disease outbreaks—United States, 1998–2002. *MMWR Surveill. Summ.* **55**(SS-10)**:**1–34.

Chalmers, G., H. Bruce, D. Hunter, V. Parreira, R. Kulkari, Y.-F. Jiang, J. Prescott, and P. Boerlin. 2008. Multilocus sequence typing analysis of *Clostridium perfringens* isolates from necrotic enteritis outbreaks in broiler chicken populations. *J. Clin. Microbiol.* **46:**3957–3964.

Collie, R., J. Kokai-Kun, and B. McClane. 1998. Phenotypic characterization of enterotoxigenic *Clostridium perfringens* isolates from non-foodborne human gastrointestinal diseases. *Anaerobe* **4:**69–79.

Cooper, K. K., and J. G. Songer. 2009. Necrotic enteritis in chickens: a paradigm of enteric infection by *Clostridium perfringens* type A. *Anaerobe* **15:**55–60.

Craven, S., N. Cox, J. Bailey, and D. Cosby. 2003. Incidence and tracking of *Clostridium perfrin-*

gens through an integrated broiler chicken operation. *Avian Dis.* **47:**707–711.

Czeczulin, J., P. Hanna, and B. McClane. 1993. Cloning, nucleotide sequencing and expression of the *Clostridium perfringens* enterotoxin gene in *Escherichia coli. Infect. Immun.* **61:**3429–3439.

De Cesare, A., G. Borilova, I. Svobodova, V. Bodioli, and G. Manfreda. 2009. *Clostridium perfringens* occurrence and ribotypes in healthy broilers reared in different European countries. *Poultry Sci.* **88:**1850–1857.

Deguchi, A., K. Miyamoto, T. Kuwahara, Y. Miki, I. Kaneko, J. Li, B. McClane, and S. Akimoto. 2009. Genetic characterization of type A enterotoxigenic *Clostridium perfringens* strains. *PLoS ONE* **4**(5):e5598. doi:10.1371/journal.pone.0005598.

De Jong, A. E. I., R. Beumer, and F. Rombouts. 2002. Optimizing sporulation of *Clostridium perfringens. J. Food Prot.* **65:**1457–1462.

De Jong, A. E. I., G. Eijhusen, E. J. F. Brouwer-Post, M. Grand, T. Johansson, T. Kärkkäinen, J. Marugg, P. H. in't Veld, F. H. M. Warmerdam, G. Würner, A. Zicavo, F. M. Rombouts, and R. R. Beumer. 2003. Comparison of media for enumeration of *Clostridium perfringens* from foods. *J. Microbiol. Methods* **54:**359–366.

Eisgruber, H., P. Geppert, B. Sperner, and A. Stolle. 2003. Evaluation of different methods for the detection of *Clostridium perfringens* phosphatases. *Int. J. Food Microbiol.* **82:**81–86.

Fach, P., and M. Popoff. 1997. Detection of enterotoxigenic *Clostridium perfringens* in food and fecal samples with a duplex PCR and the slide agglutination test. *Appl. Environ. Microbiol.* **63:** 4232–4236.

Food and Drug Administration. 1998. *Bacteriological Analytical Manual*, 8th ed. Food and Drug Administration, Gaithersburg, MD. www.fda.gov/Food / ScienceResearch / LaboratoryMethods / BacteriologicalAnalyticalManualBAM/default.htm.

Garcia-Alvarado, J., R. Labbe, and M. Rodriguez. 1992. Sporulation and enterotoxin production by *Clostridium perfringens* type A at 37 and 43°C. *Appl. Environ. Microbiol.* **58:**1411–1414.

Garmory, H., N. Chanter, N. French, D. Beuschel, J. Songer, and R. Titball. 2000. Occurrence of *Clostridium perfringens* beta 2-toxin amongst animals, determined using genotyping and subtyping PCR assays. *Epidemiol. Infect.* **124:**61–67.

Gilbert, M., C. Jolivet-Reynaud, and M. Popoff. 1997. Beta 2 toxin, a novel toxin produced by *Clostridium perfringens.* **203:**56–73.

Grant, K., S. Kenyon, I. Nwafor, J. Plowman, C. Ohai, R. Halford, M. Peck, and J. McLauchin. 2008. The identification and characterization of *Clostridium perfringens* by real-time PCR, location of enterotoxin gene and heat resistance. *Foodborne Pathog. Dis.* **5:**629–639.

Gurjar, A., N. Yennawar, H. Yennawar, K. Rajashankar, N. Hegde, and B. Jayarao. 2007. Expression, crystallization and preliminary X-ray diffraction studies of recombinant *Clostridium perfringens* ss 2-toxin. *Acta Crystallogr. F* **63:**484–487.

Heikinheimo, A., M. Lindstrom, and H. Korkeala. 2004. Enumeration and isolation of *cpe*-positive *Clostridium perfringens* spores from feces. *Clin. Microbiol.* **42:**3992–3997.

Heikinheimo, A., M. Lindstrom, P. E. Granum, and H. Korkeala. 2006. Humans as reservoir for enterotoxin gene-carrying *Clostridium perfringens* type A. *Emerg. Infect. Dis.* **12:**1724–1729.

Hsieh, P., and R. Labbé. 2007. Influence of peptone source on sporulation of *Clostridium perfringens. J. Food Prot.* **70:**1730–1734.

International Organization for Standardization. 2004. *Microbiology of Food and Animal Feeding Stuffs—Horizontal Method for the Enumeration of* Clostridium perfringens—*Colony Count Technique. ISO 7937:2004.* International Organization for Standardization, Geneva, Switzerland.

Jost, B. H., H. T. Trinh, and J. G. Songer. 2006. Clonal relationships among *Clostridium perfringens* of porcine origin as determined by multilocus sequence typing. *Vet. Microbiol.* **116:**158–165.

Juneja, V., R. Whiting, H. Marks, and O. Snyder. 1999. Predictive model for growth of *Clostridium perfringens* at temperature applicable to cooling of cooked meat. *Food Microbiol.* **16:**335–349.

Keto-Timonen, R., A. Heikinheimo, E. Eerola, and H. Korkeala. 2006. Identification of *Clostridium* species and DNA fingerprinting of *Clostridium perfringens* by amplified fragment length polymorphism analysis. *J. Clin. Microbiol.* **44:**4057–4065.

Labbe, R. 1981. Enterotoxin formation by *Clostridium perfringens* type A in a defined medium. *Appl. Environ. Microbiol.* **41:**315–317.

Labbe, R. 1989. *Clostridium perfringens*, p. 191–234. *In* M. Doyle (ed.), *Foodborne Bacterial Pathogens.* Marcel Dekker, New York, NY.

Labbe, R. 2000. *Clostridium perfringens*, p. 1110–1135. *In* B. Lund, T. Baird-Parker, and G. Gould (ed.), *The Microbiological Safety of Food.* Aspen, Gaithersburg, MD.

Labbe, R., and T. Huang. 1995. Generation times and modeling of enterotoxin-positive and enterotoxin-negative strains of *Clostridium perfringens* in laboratory media and ground beef. *J. Food Prot.* **58:**1303–1306.

Labbe, R., and V. Juneja. 2006. *Clostridium perfringens* gastroenteritis, p. 137–184. *In* H. Rieman and D. Cliver (ed.), *Foodborne Infections and Intoxications*, 3rd ed. Academic Press, New York, NY.

Lahti, P., A. Heikinheimo, T. Johansson, and H. Korkeala. 2008. *Clostridium perfringens* type A strains carrying plasmid-borne enterotoxin gene (genotype IS *1151-cpe* or IS *1470*-like-*cpe*) as a common cause of food poisoning. *J. Clin. Microbiol.* **46:**371–373.

LeFlon-Guibout, V., J.-L. Pons, B. Heym, and M.-H. Nicolas-Chanoine. 1999. Typing of *Clostridium perfringens* strains by use of random amplified polymorphic DNA (RAPD) system in comparison with zymotyping. *Anaerobe* **3:**245–250.

Le Marc, Y., J. Plowman, C. Aldus, M. Munoz-Cuevas, J. Baranyi, and M. Peck. 2008. Modelling the growth of *Clostridium perfringens* during the cooling of bulk meat. *Int. J. Food Microbiol.* **128:** 41–50.

Li, J., and B. McClane. 2006. Further comparisons of temperature effects on growth and survival of *Clostridium perfringens* type A isolates carrying a chromosomal or plasmid-borne enterotoxin gene. *Appl. Environ. Microbiol.* **72:**4561–4568.

Lin, Y. T., and R. Labbe. 2003. Enterotoxigenicity and genetic relatedness of *Clostridium perfringens* isolates from retail foods in the United States. *Appl. Environ. Microbiol.* **69:**1642–1646.

Lindstedt, B.-A. 2005. Multiple-locus variable number tandem repeats analysis for fingerprinting of pathogenic bacteria. *Electrophoresis* **26:**2567–2582.

Maslanka, S., J. Kerr, G. Williams, J. Barbavee, L. Carson, J. M. Miller, and B. Swaminathan. 1999. Molecular subtyping of *Clostridium perfringens* by pulsed-field gel electrophoresis to facilitate foodborne disease outbreak investigations. *J. Clin. Microbiol.* **37:**2209–2214.

McClane, B. 2007. *Clostridium perfringens*, p. 423–444. *In* M. Doyle and L. Beuchat (ed.), *Food Microbiology: Fundamentals and Frontiers*, 3rd ed. ASM Press, Washington, DC.

McLauchlin, J., G. Ripabelli, M. Brett, and E. Threlfall. 2000. Amplified fragment length polymorphism (AFLP) analysis of *Clostridium perfringens* for epidemiological typing. *Int. J. Food Microbiol.* **56:**21–28.

Meyer, M., and J. Tholozan. 1999. A new growth and sporulation medium for *Clostridium perfringens*. *Lett. Appl. Microbiol.* **28:**98–102.

Miki, Y., K. Miyamoto, I. Eko-Hirano, K. Fujiuchi, and S. Akimoto. 2008. Prevalence and characterization of enterotoxin gene-carrying *Clostridium perfringens* isolates from retail meat products in Japan. *Appl. Environ. Microbiol.* **74:**5366–5372.

Mueller-Spitz, S., L. Stewart, and S. McLellan. 2009. Reliability of mCP method for identification of *Clostridium perfringens* from fecal polluted aquatic environments. *J. Appl. Microbiol.* **108:**1994–2002.

Myers, G., D. Rasko, J. Cheung, J. Ravel, R. Seshadri, R. DeBoy, Q. Ren, J. Varga, M. Awad, L. Brinkac, S. Daugherty, D. Haft, R. Dodson, R. Madupu, W. Nelson, M. Rosovitz, S. Sullivan, H. Khouri, G. Dimitrov, K. Watkins, S. Mulligan, J. Benton, D. Radune, D. Fisher, H. Atkins, T. Hiscox, B. H. Jost, S. Billington, J. G. Songer, B. McClane, R. Titball, J. Rood, S. Melville, and T. Paulsen. 2006. Skewed genomic variability in strains of the toxigenic bacterial pathogen *Clostridium perfringens*. *Genome Res.* **16:**1031–1040.

Nakamura, M., A. Kato, D. Tanaka, Y. Gyobu, S. Higaki, T. Karasawa, and T. Yamgishi. 2004. PCR identification of the plasmid-borne enterotoxin gene (*cpe*) in *Clostridium perfringens* strains isolated from food poisoning outbreaks. *Int. J. Med. Microbiol.* **294:**261–265.

Orenga, S., A. James, M. Manafi, J. Perry, and D. Pincus. 2009. Enzymatic substrates in microbiology. *J. Microbiol. Methods* **79:**139–155.

Petit, L., M. Gilbert, and M. Popoff. 1999. *Clostridium perfringens*: toxinotype and genotype. *Trends Microbiol.* **7:**104–110.

Sarker, M., R. Shivers, S. Sparks, V. Juneja, and B. McClane. 2000. Comparative experiments to examine the effects of heating on vegetative and cells and spores of *Clostridium perfringens* isolates carrying plasmid versus chromosomal enterotoxin genes. *Appl. Environ. Microbiol.* **66:**3234–3240.

Sartory, D., R. Waldock, C. Davies, and A. Field. 2006. Evaluation of acid phosphatase as a confirmatory test for *Clostridium perfringens* isolated from water. *Lett. Appl. Microbiol.* **42:**418–424.

Sawires, Y., and J. G. Songer. 2005. Multiple-locus variable-number tandem repeat analysis for strain typing of *Clostridium perfringens*. *Anaerobe* **11:** 262–272.

Sawires, Y., and J. G. Songer. 2006. *Clostridium perfringens*: insight into virulence evolution and population structure. *Anaerobe* **12:**23–43.

Schalch, B., L. Bader, H.-P. Schau, R. Bergmann, A. Rometsch, G. Maydl, and S. Keβler. 2003. Molecular typing of *Clostridium perfringens* from a foodborne disease outbreak in a nursing home: ribotyping versus pulsed-field gel electrophoresis. *J. Clin. Microbiol.* **41:**892–895.

Shimizu, S., M. Ootsubo, Y. Kubosau, I. Fuchizawa, Y. Kawai, and K. Yamazaki. 2009. Fluorescent in situ hybridization in combination with filter cultivation (FISHFC) method for specific detection and enumeration of viable *Clostridium perfringens*. *Food Microbiol.* **26:**425–431.

Shimizu, T., K. Ohtani, H. Hirakawa, K. Ohshima, A. Yamashita, T. Shiba, N. Ogasawara, M. Hattori, S. Kuhara, and H. Hayashi. 2002.

Complete genome sequence of *Clostridium perfringens*, an anaerobic flesh-eater. *Proc. Natl. Acad. Sci. USA* **99:**996–1001.

Sperner, B., H. Eisgruber, and A. Stolle. 1999a. Use of the RAPID 32A system for rapid identification of *Clostridium* species in food hygiene. *Int. J. Food Microbiol.* **52:**169–180.

Sperner, B., B. Schalch, H. Eisgruber, and A. Stolle. 1999b. Short protocols for pulsed-field gel electrophoresis of a variety of *Clostridium* species. *FEMS Immunol. Med. Microbiol.* **24:**287–292.

Tanaka, D., K. Kimata, M. Shimizu, J. Isobe, J. Watahiki, T. Karaswa, T. Yamagishi, S. Kuramoto, T. Serikawa, M. Yamada, K. Yamaoka, M. Tokoro, T. Fukao, M. Matsumoot, R. Hiramatsu, C. Monma, and Y. Nagai. 2007. Genotyping of *Clostridium perfringens* isolates collected from food poisoning outbreaks and healthy individuals in Japan based on the *cpe* locus. *Jpn. J. Infect. Dis.* **60:**68–610.

U.S. Department of Agriculture. 1998. *Microbiology Laboratory Guidebook*, 3rd ed. U.S. Department of Agriculture, Washington, DC. www.fsis.usda.gov./Science/Microbiological_Lab_Guidebook/Index.asp.

van Asten, A., J. Allaart, A. Grone, P. Gloudemans, D. Houwers, and A. Gröne. 2008. A new PCR followed by MboI digestion for the detection of all variants of the *Clostridium perfringens* cpb2 gene. *Vet. Microbiol.* **127**:412–416.

van Damme-Johnsten, M., J. Haagsma, and S. Notermans. 1990. Testing strains of *Clostridium perfringens* type A isolated from diarrheic piglets for the presence of the enterotoxin gene. *Vet. Res.* **126:** 191–192.

Vos, P., R. Hogers, M. Bleeker, M. Reijians, et al. 1995. AFLP: a new technique for DNA fingerprinting. *Nucleic Acids Res.* **23:**4407–4414.

Wen, Q., and B. McClane. 2004. Detection of enterotoxigenic *Clostridium perfringens* type A isolates in American retail foods. *Appl. Environ. Microbiol.* **70:**2685–2691.

Willardsen, R., F. Busta, and C. Allen. 1979. Growth of *Clostridium perfringens* in three different beef media and fluid thioglycollate medium at static and constantly rising temperature. *J. Food Prot.* **42:** 144–148.

HEPATITIS A VIRUS IN READY-TO-EAT FOODS

D. H. D'Souza, K. E. Kniel, and L. Jaykus

27

Hepatitis A virus (HAV), a member of the *Picornaviridae* family, is a small (27 to 32 nm in diameter) nonenveloped virus enclosed in a capsid with an icosahedral symmetry. The virion genome consists of a single, positive-sense linear strand of RNA (7.5 kb) that is divided into three regions, the middle of which encodes the functional proteins of the virus. Human isolates of HAV constitute a single serotype, and while the genome sequence between strains is relatively conserved, there are regions that provide some degree of sequence heterogeneity. In particular, the RNA sequences corresponding to the junction between the genes encoding the VP1 structural protein and the 2A nonstructural protein have been used for HAV genotyping. Based on this technique, isolates can be differentiated into seven unique genotypes, which differ from one another in sequence homology by no more than 15%. Of these seven genotypes (designated I to VII), only genotypes I, II, III, and VII are associated with human disease, with genogroups I and III predominating (Arauz-Ruiz et al., 2001).

HAV causes a different disease than do most other foodborne pathogens. Of particular note is the fact that the mean incubation period for clinical disease is 28 days, with a range of 2 to 7 weeks. This means that even in instances where a food vehicle is epidemiologically implicated, that product is seldom available for testing. Symptomatic hepatitis A infection begins with generalized malaise, which is often accompanied by gastrointestinal symptoms that last from 5 to 7 days. This is followed by the icteric phase, which presents as darkening of the urine, pale-colored stools, jaundice, and right upper quadrant pain with hepatomegaly. Icteric disease occurs in fewer than 10% of children younger than 6 years of age but in up to 50% of older children and in 70 to 80% of adults. In most cases, infection is mild and self-limiting, with long-term complications or death relatively rare. However, disease severity does increase with age (Hollinger and Emerson, 2001).

Immunity after HAV infection is complete and considered to be lifelong in duration. Prior to the introduction of vaccines in the late 1990s, postexposure prophylactic admin-

D. H. D'Souza, Department of Food Science and Technology, University of Tennessee-Knoxville, 2605 River Dr., Room 102 FSPB, Knoxville, TN 37996-4591. *K. E. Kniel*, Department of Animal and Food Sciences, University of Delaware, Newark, DE 19716. *L. Jaykus*, Department of Food, Bioprocessing and Nutrition Sciences, North Carolina State University, 339-A Schaub Hall, Raleigh, NC 27695-7624.

Rapid Detection, Identification, and Quantification of Foodborne Pathogens, ed. by J. Hoorfar

istration of immunoglobulins was recommended for patrons and food handlers thought to have been exposed to HAV. More recently, vaccines are being used for protection from infection, which is believed to last for about 20 years. Routine vaccination of all food handlers and restaurant employees is still considered cost-prohibitive, but in the United States, HAV vaccination was recently recommended for children up to 2 years of age (FitzSimons et al., 2010).

EPIDEMIOLOGY

Traceback is exceedingly challenging in HAV outbreak investigations due to the long incubation period of an average of 30 days (range, 15 to 50 days), as described above, as well as the expense involved with HAV outbreak investigations (World Health Organization, 2000). In recent years, a number of high-profile HAV outbreaks have occurred in association with the consumption of contaminated produce items, where environmental contamination was likely and food handler contamination less likely (Wheeler et al., 2005; Hutin et al., 1999). Nonetheless, the contamination of ready-to-eat (RTE) and otherwise prepared foods with HAV most frequently comes from poor hand-washing practices of infected food handlers after toilet use (Papafragkou et al., 2006). Many HAV outbreaks occur as the result of a single infected food handler at a single food establishment, and many of these outbreaks may be unrecognized and unreported. However, HAV outbreaks linked to poor personal hygiene of infected food handlers have been reported in association with highly handled foods such as salads, sandwiches, and bakery products (Sair et al., 2002a). Table 1 lists some of the outbreaks associated with HAV and RTE foods.

An important period of consideration of HAV transmission is the time when an infected food handler sheds virus in the feces. HAV, with an incubation period of 15 to 50 days, appears to be more easily transmitted during the later stages of incubation. This means that food workers in retail grocery and restaurant environments with acute HAV infection can readily contaminate RTE products if they do not practice adequate personal hygiene (D'Souza et al., 2007). The survival and transmission of HAV in the environment also depend on factors that affect the stability of the organisms, which include relative humidity, temperature, degree of inoculum drying, type of suspending medium (fecal or other matrices, such as food), virus type, and the type of surface contaminated (Abad et al., 1997; D'Souza et al., 2007). HAV was found to be stable for months to years when held at freezing temperatures in laboratory studies and for up to 2 years in frozen strawberries associated with a multistate outbreak (Niu et al., 1992). HAV was also found to be resistant to drying (Grove et al., 2006). HAV was found to remain infectious after being held at pH 1 for 5 h and in acidic marinade (pH, ~3.75) at 4°C for 4 weeks (D'Souza et al., 2007; Hewitt and Greening, 2004).

Hands play a significant role in transmission of HAV in RTE foods. Mbithi et al. (1992) studied the survival of HAV on finger pads and were able to detect 16 to 30% of the initially recoverable viruses after 4 h. Virus transfer is also found to be dependent on pressure and friction. An average 9% of dried HAV inoculum was transferred from artificially contaminated finger pads to lettuce by using pressures of 0.3 kg/cm^2 (2,700 g/9 cm^2) (Bidawid et al., 2000a). The resilient nature of HAV showing survival at various temperatures, pHs, and pressures and their ease of transfer all play significant roles in the contamination of foods and in foodborne outbreaks.

CONTROL STRATEGIES

Control of HAV in RTE foods is an important factor in the prevention of HAV-related foodborne outbreaks. Washing of produce is one of the most common decontamination methods (Papafragkou et al., 2006). However, the wash waters should be devoid of fecal contamination. Similarly, ice used for food transport and/or storage should be produced with clean water and under hygienic conditions.

TABLE 1 HAV outbreaks associated with produce and RTE foods

Food type	Sample	Detection method	Source	Reference
Bread	Clinical	Serology	Food handler	Warburton et al., 1991
Sandwiches	Clinical	Serology	Food handler	Centers for Disease Control and Prevention, 1993
Lettuce	Clinical	Serology	Food handler	Centers for Disease Control and Prevention, 1993
Bakery items: sugar-glazed baked goods	Clinical	Serology	Food handler	Weltman et al., 1996; Schenkel et al., 2006
Strawberries	Clinical	Serology, RT-PCR	At harvest and picking	Hutin et al., 1999
Green onions	Clinical	Serology	Pre- or postharvest or distribution system	Wheeler et al., 2005
Blueberries	Clinical and food	RT-PCR, sequencing	Food handlers or polluted groundwater	Calder et al., 2003
Raspberries	Clinical	Serology	Pre- or postharvest	Fiore et al., 2004
Orange juice	Clinical	Serology (genotype 1b)	Food handler, sewage contamination of machinery or fruit; 351 cases of tourists from nine European countries	Frank et al., 2005, 2007
Delicatessen products	Clinical	Serology	Food handler	Schmid et al., 2009
Liver paté	Clinical	Serology and RT-PCR	Food handler; unhygienic conditions	Schmid et al., 2009; Schwarz et al., 2008
Raw beef	Clinical	Serology, RT-PCR, sequencing	Food handler	Robesyn et al., 2009

Water used for washing is often treated with sanitizers, mostly chlorine, chlorine dioxide, organic acids, and ozone; and simple home use remedies such as vinegar (acetic acid) can be somewhat effective (Butot et al., 2008). Washing fresh produce can reduce but may not completely eliminate foodborne viruses, especially those that have penetrated through produce skin or that might have entered tissues through cuts and abrasions (D'Souza et al., 2007).

The morphology and physiology of food surfaces can make cleaning or disinfection difficult. Decontamination of leafy greens can be challenging because of their rough and ridged leaves, and soft fruits like raspberries and blackberries have many porous and uneven surfaces that can entrap viruses (Richards, 2001; D'Souza et al., 2007). HAV was found to be more readily inactivated (ca. 96%) when strawberries were washed in tap water containing 2 ppm chlorine dioxide (ClO_2) for 30 min compared to 30 s (ca. 67% inactivation) (Mariam and Cliver, 2000). Current decontamination strategies are not very efficacious, and some proposed agents (such as hydrogen peroxide) cause unacceptable sensory alterations to the product. Additional research is needed to determine efficacious chemical disinfection methods for HAV on foods.

Employee and public education is important to prevent HAV contamination by RTE foods, and strict hygienic practices need to be followed. Another factor in hand hygiene is fingernail washing, as virus-containing feces can adhere to nails and transmit the spread of HAV (D'Souza et al., 2007). Hand disinfection requires the proper use of effective disinfectants along with regular compliance (Sattar et al., 2002). Hand-washing agents differ in their inactivation effectiveness against foodborne viruses, and new formulations are being researched to determine inactivation efficiency. Mbithi et al. (1993) found that a medicated liquid soap containing 0.3% triclosan was the most effective hand-washing agent and reduced HAV from finger pads by 92%. Reports for inactivation of viruses by alcohol remain quite variable, and most often water and soap alone are reported to work best (Bidawid et al., 2000b). Recently, a formulation containing 70% ethanol (vol/vol) with a synergistic blend of polyquaternium-37 and citric acid, as well as gelling and skin-conditioning ingredients, was shown to reduce HAV titers by 1.75 $\log_{10}$ 50% tissue culture infective dose ($TCID_{50}$)/ml after a mere 30-s exposure (Macinga et al., 2008). A combination of hand washing and hand sanitizing is favored, as together they can better reduce viruses than either practice alone. Residual moisture on hands after hand washing can promote virus transfer; hence, hand drying after washing can be critical (D'Souza et al., 2007).

To prevent cross-contamination involving virus-contaminated surfaces, retailers and consumers use contact surface disinfection, but the efficacy of a chemical disinfection treatment varies with virus type, surface type, and study design (Mbithi et al., 1990, 1991; Sattar, 2004). The presence of fecal material also probably protects the viruses from the action of disinfectants and thus increases virus survival (D'Souza et al., 2007). Commonly used chemical disinfectants in both institutional and domestic environments do not appear to effectively inactivate HAV (details on decontamination are reviewed by Papafragkou et al. [2006]).

NOVEL PROCESSING APPROACHES TO PREVENT HAV TRANSMISSION VIA FOODS

The HAV capsid protects the viral genome from degradation by extreme conditions of the environment and the gastrointestinal tract (Grove et al., 2006). Most processing conditions target denaturation or disruption of the viral capsid (such that host cell attachment is prevented) or RNA destruction to prevent viral replication. Inactivation by exposure to UV, hypochlorite (1.2 to 1.25 ppm), or heating at 72°C results from conformational changes of capsid proteins that affect antigenic sites and receptor attachment sites (Grove et al., 2006). HAV (>6 log) is inactivated by heating at 70°C for 4 min in phosphate-buffered saline (PBS) (Parry and Mortimer, 1984). Heat inactivation may be affected by properties of the food matrix; however, fat content in milk did not affect thermal inactivation of HAV (Bidawid et al., 2000a). These traditional processing methods have several disadvantages including adverse sensory changes of the treated foods. Novel processing approaches that require low energy and maintain the texture, color, and appearance of foods while retaining nutritive value are increasingly researched for viral inactivation and include high hydrostatic pressure (HHP), high-pressure homogenization, high-intensity ultrasound, irradiation, pulsed electric fields, and high-intensity pulsed light (Table 2).

Kingsley et al. (2002) found that HAV in cell culture medium could be inactivated to nondetectable levels at 450 MPa of HHP for 5 min at ambient temperature from initial titers of 10^7 infectious units ($TCID_{50}$) and that increased salinity protected HAV from high pressure. HAV could be also be inactivated by HHP in contaminated strawberry puree and sliced green onions (Kingsley et al., 2005). A dose of 1.0 J/cm^2 of high-intensity pulsed light could inactivate HAV by at least 4.8 log (Roberts and Hope, 2003). Fino and Kniel (2008) found that UV light inactivation of HAV on strawberries, green onions, and lettuce was most affected by the food matrix. Greater inactivation occurred for HAV on green onions and lettuce than on strawberries (Nuanualsuwan et al., 2002; Fino and Kniel, 2008). Generally, the UV dose to inactivate HAV on produce was less than that necessary for the liquid systems. This may be due to the increased release of phenols and antimicrobial compounds by the produce upon exposure to UV light (Fino and Kniel, 2008). Cost reduction associated with these novel processing technologies along with the advantages of maintaining quality and nutritional value appear to be attractive for their eventual commercialization.

Detection in Foods

Typically, detection of viruses in foods requires sequential steps of (i) sampling, (ii) virus concentration, (iii) detection, and (iv) confirmation.

Sampling

For effective sampling, a large and representative sample size is needed. A sample size of 25 to 100 g is usually recommended, which is large enough to provide adequate representation yet small enough to work with in the

TABLE 2 Processing technologies for inactivation of HAV

Method	Food type	HAV reduction	Reference
Vacuum drying and steam blanching	Berries	1–2.4 log $TCID_{50}$/ml	Butot et al., 2009
Heat (60°C for 10 h)	Culture	3–5 log FFU	Shimasaki et al., 2009
Heat (85°C for <0.5 min)	Skim and homogenized milk, cream	5 log PFU	Bidawid et al., 2000a
Hydrostatic pressure (420 MPa)		3–5 log FFU (strain dependent-strains used: KRM238, KRM003 (subgenotype IIIB), KRM031 (IA), and TKM005 (IB)	Shimasaki et al., 2009
Hydrostatic pressure (400 MPa for 1 min at 50°C)		4.7 log PFU/ml	Kingsley et al., 2006
Radiation			
UV light; PBS, 42 mWs/cm^2		1.0 log $TCID_{50}$/ml	Nuanualsuwan et al., 2002
UV light, cell culture media:			Fino and Kniel, 2008
10 mWs/cm^2		1.0 log $TCID_{50}$/ml	
75 mWs/cm^2		4.0 log $TCID_{50}$/ml	
40 mWs/cm^2	Lettuce	4–5 log $TCID_{50}$/ml	
40 mWs/cm^2	Strawberries	1–1.5 log $TCID_{50}$/ml	
40 mWs/cm^2	Green onions	4.0 log $TCID_{50}$/ml	

laboratory (D'Souza et al., 2006). To further increase the chances for detection, the analyst may choose to obtain multiple samples of the suspected or implicated product. For complex food products such as sandwiches, it may be easier to divide the product into its component parts, processing each part separately for virus concentration and detection. Although the viruses may be relatively stable in the food matrix, food samples should nonetheless be refrigerated upon collection and processed immediately for virus recovery upon receipt by the testing facility. Freezing can be used for extended storage, but it is recommended that the sample be kept at −80°C, as very low temperatures are necessary for long-term maintenance of capsid structure and hence virus infectivity (D'Souza et al., 2006).

Virus Concentration

Concentration of viruses from foods is carried out for several reasons including (i) to decrease the sample volume; (ii) to concentrate and purify the viral particles from the food, especially since their levels in contaminated food can be quite low and their infectious dose can be as low as 10 to 100 infectious particles (Nainan et al., 2006); and (iii) to remove inhibitory or cytotoxic compounds that can compromise detection downstream. Cesium chloride gradients have typically been used to concentrate and purify viruses for molecular and microscopic analyses from clinical samples that use ultracentrifugation. Ultracentrifugation methods for HAV concentration from water samples have been proposed but are time-consuming and still remain expensive (Di Pasquale et al., 2010). HAV was concentrated from 1.5 liters of bottled water by using a Convective Interaction Media quaternary amine-strong anion exchanger in combination with ultracentrifugation to recover up to 37.5% of the inoculated virus (Kovac et al., 2009). For food samples, typical schemes include (i) extraction-concentration or (ii) adsorption-elution-concentration. Generally, sample volume reductions of 10 to 1,000-fold can be achieved. Various buffers with varying pHs and concentrations have also been studied to determine HAV recovery from food matrices followed by concentration using centrifugation, filtration, ultrafiltration, adsorption, elution, solvent extraction, precipitation, and/or organic flocculation (Butot et al., 2007; D'Souza et al., 2007). Details about concentration schemes can be found in D'Souza and Jaykus (2006), D'Souza et al. (2007), and in Table 3 as adapted from Sanchez et al. (2007).

Ultrafiltration helps to reduce sample volumes and also purifies the sample and can be applied as a final concentration step. Polyethylene glycol (PEG) is a commonly used precipitant, with acid precipitation being another viable alternative. Precipitation makes use of the property that viruses behave as proteins and "fall out" of solution with the removal of water (PEG) or upon exposure to pH near the isoelectric point of the virus (acid precipitation) (D'Souza et al., 2007).

Antibody capture of intact viruses or concentration precipitation agents such as ProCipitate and Viraffinity to eliminate polysaccharides are also used (Sair et al., 2002a). The cationic detergent cetyltrimethylammonium bromide is also used to eliminate polysaccharides, while Sephadex, cellulose, or Chelex can remove salts and proteins (D'Souza and Jaykus, 2006). Lipid components of foods can be removed by organic solvents such as chloroform, trichlorotrifluoroethane (Freon), and (perhaps environmentally friendly) solvents such as Vertrel.

Flocculating agents commonly utilized in water treatment are used in the organic flocculation method and by interacting with organic matter in food matrices can cause viruses to form a gelatinous "floc," which is separated by centrifugation. However, all these reported concentration schemes can be time-consuming and labor-intensive and have low viral recovery rates, which decrease the detection sensitivity of the assay being used. Methods that are rapid and efficient in the concentration of viral particles from a variety of food matrices will benefit rapid detection

TABLE 3 Viral concentration methods for RTE foods[a]

Sample	Wt (g)	Method	Detection limit	Reference
Ham, turkey, roast beef	20–40	Guanidinium-phenol	10 PCR U/wash	Schwab et al., 2000
Lettuce, strawberries	Unknown	PBS, IMS, and positively charged filters	10 PFU	Bidawid et al., 2000c
Lettuce, hamburger	6	Alkaline elution and PEG precipitation	10 PFU	Sair et al., 2002b
Cilantro	5	Homogenization and phenol extraction	6 $TCID_{50}$	Goswami et al. 2002
Vegetables, fruits	100	Alkaline elution, PEG precipitation	50 $TCID_{50}$	Dubois et al., 2002
Green onion, strawberry rinse	25	PBS elution and IMS	0.5 PFU	Shan et al., 2005
Green onions	25	Alkaline elution and PEG precipitation	1 $TCID_{50}$	Guévremont et al. 2006
Raspberries, strawberries	60–90	Alkaline elution and ultracentrifugation	10^3–10^4 PCR U; 5.6 × 10^2–10^3 $TCID_{50}$	Rzezutka et al. 2006
Vegetables	25	PBS elution; filtration	50 PFU	Dubois et al. 2006
Vegetables and berries	15	Alkaline elution and ultrafiltration	1.2 $TCID_{50}$ or 190 RNA copies	Butot et al. 2007
Deli turkey, cake, milk, salsa	25	Cationic beads	10^2 PFU	Papafragkou et al., 2008; Hirneisen et al., 2009
Lettuce, strawberries, raspberries, green onions	50	Alkaline elution and cationically charged filters	2–20 particles/g	Morales-Rayas, 2010

[a] Adapted from Sanchez et al., 2007.

of contaminated products. As an example, a prototype virus concentration method applied to hamburgers is shown in Fig. 1 (Sair et al., 2002b).

Bacteria are often isolated by immunomagnetic concentration methods. Similar techniques have been tested for the recovery of HAV by using a recirculating magnetic device (Papafragkou et al., 2008; Hirneisen et al., 2009). Papafragkou et al. (2008) described this application in a combined effort between two independent laboratories (one in the United

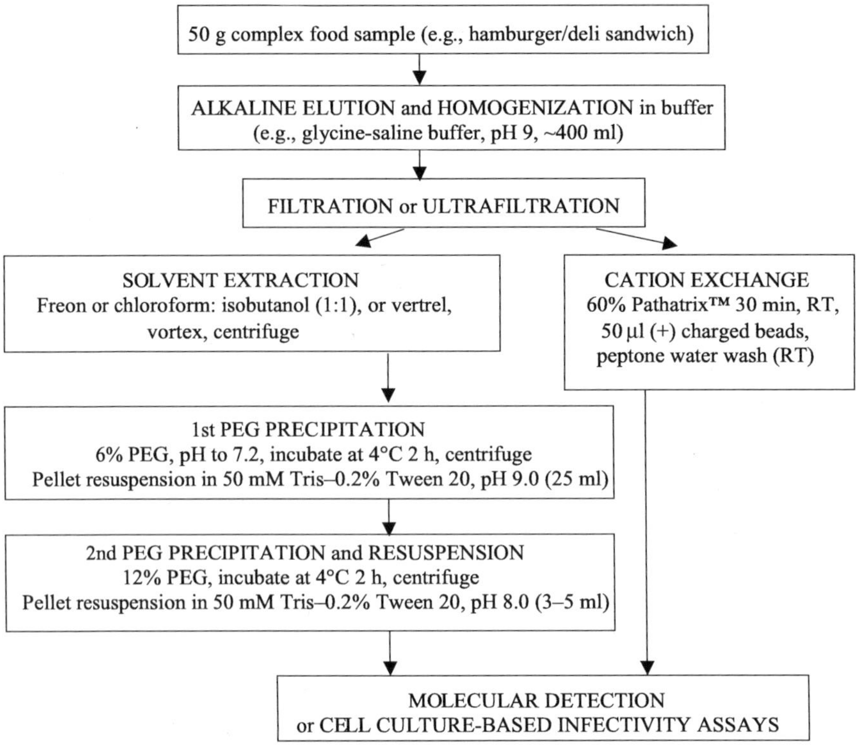

FIGURE 1 Scheme for HAV concentration from complex foods (Sair et al., 2002b).

States and the other in Canada) for HAV concentration from 25-g samples of artificially contaminated lettuce, strawberries, green onions, deli turkey, oysters, and cake with frosting. Hirneisen et al. (2009) utilized the same system to recover virus from salsa and milk. In both studies, it appeared that HAV and another picornavirus were readily amenable to this viral concentration method. Internal process controls are recommended to monitor the efficiency of virus recovery and the presence of residual matrix-associated inhibitors (Sanchez et al., 2007). Extensive research is needed to further improve viral recoveries from various food matrices and to decrease overall sampling costs to make routine foodborne viral diagnostics possible.

Detection

Following concentration and purification of HAV from food matrices, detection of the virus can be carried out using the following assays: (i) mammalian cell culture infectivity assays (these, however, are used only as a research tool and not for wild-type HAV); (ii) immunological methods; or (iii) nucleic acid-based molecular methods.

Mammalian Cell Culture-Based Infectivity Assays. The lab-adapted strain of HAV, HM-175, is propagated in the laboratory by using FRhK-4 (fetal rhesus monkey kidney-derived) cells, which is ineffective for the detection of wild-type HAV (D'Souza et al., 2007). The main drawbacks, besides the lack of susceptible host cells for wild-type HAV, include time (>2 weeks to obtain confirmed results), labor, skilled technical staff, and cost. Other compounding factors include the low levels of viral contamination in food that causes infection, emphasizing the need for sensitive detection assays/methods. In addition, food matrix inhibitors can be toxic to the cell lines and interfere with assay sensitivity and specificity.

Immunological Methods. When cell culture hosts were unavailable, immunological methods were used in the past, but they are currently not widely used. These methods are specific to the viral capsid and include radioimmunoassay, enzyme immunoassay, enzyme-linked immunosorbent assay, immunoelectron microscopy, and radioimmunofocus assay. Schwab et al. (2001) developed immunological methods for HAV detection in shellfish. The disadvantages include low sensitivity and general reagent unavailability.

Nucleic Acid-Based Molecular Assays. Nucleic acid-based molecular assays are now increasingly common for viral detection from foods. An additional step during nucleic acid-based molecular assays can include purification of the viral RNA from the food matrix after concentration. When immunocapture is used, direct heat release of RNA may be carried out for the reverse transcriptase (RT) PCR (antibody capture RT-PCR) or RNA-based assays without the need for RNA purification. However, use of antibodies can increase the assay cost and also remains dependent on the availability of antibodies and immunological reagents. Methods to purify and extract RNA include TRIzol (guanidine isothiocyanate)-based assays and commercial kits such as Qiagen RNeasy kits, etc. The quality and quantity of the extracted RNA are crucial in determining assay sensitivity. Sair et al. (2002b) compared guanidinium isothiocyanate, microspin column, QIAshredder Homogenizer, and TRIzol for the extraction of HAV from spiked lettuce and hamburger samples. They found that TRIzol with the QIAshredder Homogenizer gave the best RNA yield with detection limits of <10 and <100 PFU of HAV per 6 g of hamburger and lettuce, respectively. Recently, three extraction methods were compared for HAV detection, based on chemical agent with organic phase (TRIzol method), adsorbent paper (Flinders Technology Associates card), and a silica-based membrane column (Qiagen Extraction kit), with the Qiagen kit showing improved detection by 1 to 2 log (Brassard et al., 2009). It is recommended that all RT-PCR assays or molecular methods include an internal amplification control in their reaction in order to prevent false negatives that may arise due to food matrix inhibitors, enzyme or reaction failure, or instrument malfunction.

RT-PCR. Amplification of viral RNA currently remains the most sensitive and widely used approach for foodborne viral detection (Morales-Rayas et al., 2010), requiring two enzymes: RT that converts mRNA to cDNA and DNA polymerase to amplify DNA. Routine RT-PCR analysis requires the removal of inhibitory compounds that may be present in the food matrix such as carbohydrates, lipids, chelators, or adverse pH as described above. Traditional RT-PCR approaches using various primers are described in Table 4. This method requires agarose gel electrophoresis to detect the specific size amplicon followed by confirmation, which increases assay time and cost. Hence, traditional RT-PCR methods are now replaced by real-time RT-PCR approaches.

Confirmation Methods

Confirmation methods are needed during the testing of food and environmental samples in order to eliminate false positives, because nonspecific products can be amplified from the food matrix components themselves. Confirmation methods include (i) direct sequencing of amplified PCR product, (ii) restriction digestion of the amplified product, (iii) Southern hybridizations, and (iv) nested PCR. These are briefly described below since they are now rarely used. As mentioned above, real-time approaches are now being increasingly used instead.

Direct Sequencing. Direct sequencing of amplicons from samples is usually carried out with clinical samples to confirm the pres-

TABLE 4 Traditional RT-PCR primers for the detection of HAV in RTE foods

Primer	Sequence	Location	Size (bp)	Detection limits/ application	Reference
H1	GGAAATGTCTCAGGTACTTTCTTG (−)	2389–2413	247	Fruits/vegetables	Le Guyader et al., 2000
H2	GTTTTGCTCCTCTTTATCATGCTATG (+)	2167–2192			Dubois et al., 2002
HAV-R	CTCCAGAATCATCTCAAC (−)	2208–2226	192	Lettuce, hamburger	Leggitt and Jaykus, 2000; Sair et al., 2002a, 2002b
HAV-L	CAGCACATCAGAAAGGTGAG (+)	2035–2054			
BG7	CCGAAACTGGTTTCAGCTGAGG (−)	7125–7104	276	Produce (nested PCR)	Goswami et al., 2002
BG8	CCTCTGGGTCTCCTTGTACAGC (+)	6850–6871			
BG7a	CTGGTTTCAGCTGAGGYA (−)	7120–7102	264		
BG8a	GGTCTCCTTGTACAGCTT (+)	6856–6873			
P1	CAGGGGCATTTAGGTTT (−)	669–685	415	Produce (nested PCR)	Croci et al., 2002
P2	CATATGTATGGTATCTCAACAA (+)	1063–1084			
P3	TGATAGGACTGCAGTGACT (−)	807–825	211		
P4	CCAATTTGCAACTTCATG (+)	1000–1018			

ence of the target as well as for epidemiological purposes. However, sequencing is also carried out from food samples implicated in outbreaks to confirm the presence of the target in the food matrix. Sequencing can be time-consuming and expensive, but with the advances in sequencing technology, the cost is no longer as prohibitive as it used to be. Sequencing requires purification of the amplicon and a DNA sequencer (such as an ABI Prism that uses fluorescence). Viral metagenomics and novel technologies such as pyrosequencing, whereby DNA polymerization is detected and quantified through the light emitted by the release of attached pyrophosphates, are also being researched but have not yet been applied to foods.

Restriction Digestion. Restriction digestion is another confirmation method that uses restriction enzymes that cut specific sequences within the target amplified PCR product to confirm the target nucleic acid.

Hybridization. Hybridization involves binding of specific labeled oligoprobes that are internal to the amplicon sequence. The amplicons obtained from the RT-PCR assay are bound onto solid membranes such as nylon or nitrocellulose and then annealed with the specific probe. Northern hybridizations (as well as dot blot formats) have been reported by Jean et al. (2004) for detection of HAV in cell culture and from spiked deli meats and lettuce samples.

Nested PCR. Nested PCR is a method that uses a second pair of primers internal to the first RT-PCR amplified product. Nested PCR confirms the presence of the amplicon and also simultaneously improves detection sensitivity as the specific products are amplified in the second PCR. The drawback is the potential for false positives from cross-contamination due to opening of tubes, and therefore, it is not recommended by researchers. Caution and extreme care must therefore be used along with appropriate controls to prevent cross-contamination of PCR products. This method has been applied to detect HAV in produce (Table 4) (Croci et al., 2002; Goswami et al., 2002).

Real-Time RT-PCR Approaches

Table 5 lists the real-time RT-PCR approaches used for HAV detection.

SYBR Green I Assays

A reported SYBR Green I real-time RT-PCR assay was able to detect HAV RNA up to 5 $TCID_{50}$/g in all seeded oyster samples (Casas et al., 2007) by using primers based on the 5′ untranslated region (UTR) of HAV to yield an ~90-bp product. HAV was also detected in all eight tested water samples ranging from 90 to 3,523 copies of HAV/liter of ocean water (Brooks et al., 2005).

TaqMan Assays

TaqMan assays using RT-PCR with TaqMan probes are summarized in Table 5. By combining conventional PCR, nested PCR, and real-time PCR techniques, a specific nested real-time PCR assay for the detection of HAV was developed by Hu and Arsov (2009). The first PCR step involved a conventional RT-PCR using primers specific for HAV 5′ noncoding region (NCR). The second PCR involved a nested primer pair specific for the first PCR product and a TaqMan probe, capable of detecting as low as 0.02 PFU of HAV culture and 63 PFU from green onions (Hu and Arsov, 2009).

Molecular Beacons

A specific molecular beacon real-time RT-PCR was developed for HAV targeting a 125-bp, highly conserved 5′ noncoding region of HAV with a detection limit of 1 PFU (Abd El Galil et al., 2004). When combined with immunomagnetic separation (IMS), this RT-PCR assay detected as few as 20 PFU in seeded groundwater samples and can potentially be applied for the rapid detection of HAV in contaminated foods or water.

NASBA Assay

Nucleic acid sequence-based amplification (NASBA) assay is an isothermal RNA amplification method requiring three enzymes, i.e., RT, T7 RNA polymerase, and RNase H, and two target sequence-specific oligonucleotide primers (one containing the bacteriophage T7 promoter sequence at its 5′ end) acting in concert to amplify target sequences $>10^8$-fold, requiring only a simple water bath (Compton, 1991). NASBA has been successfully used to detect as little as 0.4 ng HAV RNA/ml, in comparison to 4 ng/ml, by RT-PCR using primers targeting the VP1 and VP2 capsid genes (Jean et al., 2001). A detection limit of 2 PFU was obtained with NASBA on crude HAV viral lysates, and was also detected from spiked samples of wastewater, lettuce, and blueberries by these researchers. Jean et al. (2002) also used a multiplex NASBA system to simultaneously detect rotavirus and HAV to obtain 268- and 474-base RNA products, respectively, where 400 PFU/ml of HAV was detected with decreased time and cost compared to monoplex systems. Furthermore, three previously reported primer sets for target RNA amplification of 474, 371, and 165 nucleotides were used to simultaneously detect HAV and genogroup I and genogroup II noroviruses, respectively (Jean et al., 2004). Spiked RTE foods (deli sliced turkey and lettuce) also showed simultaneous detection of all three viruses with detection limits for HAV at 10^3 PFU/9 cm^2 in both food commodities. A specific molecular beacon NASBA assay was reported for the real-time detection and quantification of HAV based on a 202-bp, highly conserved 5′ noncoding region by Abd El Galil et al. (2005) with a detection limit of 1 PFU. Combined with IMS, as little as 10 PFU from seeded lake water samples was detected, showing promise for detection from food or water.

RT-LAMP

Reverse transcription loop-mediated isothermal amplification (RT-LAMP) is an amplification method, similar to NASBA in requiring

TABLE 5 Representative real-time RT-PCR approaches for HAV detection[a]

Region	Sequence	Method, detection limit	Reference
5′ UTR	GCGGCGGATATTGGTGAG (+)	TaqMan	Di Pasquale et al., 2010
(M86379)	CAATGCATCCACTGGATGAGA (−)	5 $TCID_{50}$/ml	
		FCV as process control	
	Probe: FAM-TTAAGACAAAAACCATTCAACGCCGGAG-TAMRA		
5′ NCR	AGGCTACGGGTGAAACCTCTTAG (+)	Nested TaqMan, 0.2 PFU	Hu and Arsov, 2009
	CGCCGCTGTTACCCTATCC (−)		
	FAM-AATACTTCTATGAAGGATGCC		
VP1	(Not specified) 90-bp product	FRET probes, 2 $TCID_{50}$/ml or 500 RNA copies/ml, Roche Kit, LC RT-PCR	Sánchez et al., 2006
5′ NCR	GGAGAGCCCTGGAAGAAAG (−)	TaqMan	Costafreda et al., 2006
	TCACCGCCGTTTGCCTAG (+)	10 copies ssRNA/rn	
	Probe: TTAATTCCTGCAGGTTCAGG	0.05 $TCID_{50}$	
VP3-VP1	TAC AGA GCA GAA TGT TCC TGA TCC (+)	SYBR Green 76 bp, T_m = 81°C	Brooks et al., 2005
	TCC CCT ATT GGC TTT CCC TT (−)		
5′ NCR	ATCTTCCACAAGGGGTAG (+)	Molecular beacon, 1 PFU/rn	Abd El Galil et al., 2005
	CGGCGTTGAATGGTTTTT (−)		
	MB FAM-CTTGCGGGATAGGGTAACAGCGGCGGCGCAAG-DABCYL		
5′ NCR	GGTAGGCTACGGGTGAAAC (+)	TaqMan, 40 copies/rn	Jothikumar et al., 2005
	AACAACTCACCAATATCCGC (−)		
	ProbeFAM-CTTAGGCTAATACTTCTATGAAGAGATGC-BHQ (+)	0.05 PFU/rn	
5′ NCR	TTTCCGGAGCCCCTCTTG (+)	TaqMan, 5 copies/rn	Costa-Mattioli et al., 2002
(M14707)	AAAGGGAAATTTAGCCTATAGCC (−)		
	AAAGGGAAAATTTAGCCTATAGCC (−)		
	Probe: FAM-ACTTGATACCTCACCGCCGTTTGCCT-TAMRA		

[a] Abbreviations: FAM, 6-carboxyfluorescein; FCV, feline calicivirus; FRET, fluorescent resonance energy transfer; rn, reaction; ssRNA, single-stranded RNA; TAMRA, 6-carboxytetramethylrhodamine; T_m, melting temperature.

a one-temperature, one-step amplification system. It differs from NASBA in that it requires only two enzymes, i.e., RT and *Bst* DNA polymerase, and visualization is by turbidity or fluorescence. The amplification relies on an autocycling strand displacement DNA synthesis by the *Bst* DNA polymerase large fragment and approximately six specific primers, including two inner primers, two outer primers, and two loop primers that recognize six distinct regions on the target gene (Notomi et al., 2000). The visual detection is based on the formation of insoluble magnesium pyrophosphate, which can be observed visually or by using a simple turbidimeter. A one-step, single-tube, real-time RT-LAMP assay was developed for detecting specific sequences of the HAV UTR in 50 min under isothermal conditions at 62.5°C, using a set of seven primers (Yoneyama et al., 2007). The RNAs of three cell-adapted HAV strains from different subgenotypes (IA, IB, and IIIB) were equally amplified with detection limits of 0.4 to 0.8 focus-forming units (FFU)/reaction. The results of the calibration using the WHO international standard indicated that this RT-LAMP assay had sensitivity similar to that of conventional RT-PCR and LightCycler PCR assays using fecal clinical samples. This assay was very suitable for detection and quantitation of most prevalent HAV genotypes (Yoneyama et al., 2007). However, the application of this assay to foods needs further research. The cross-reactivity with the food matrices and reaction sensitivity in the presence of food matrix inhibitors need to be determined to eliminate false positives and false negatives, respectively.

Biosensors

Biosensors are detection devices that combine biological components with physicochemical detectors. A biosensor instrument has been used for HAV antibody detection in human serum samples, using linear and branched synthetic peptides related to the VP3 capsid protein. Linear and dimeric VP3(110-121) multiple antigen peptides were found to be the most sensitive for serological studies of HAV-infected patient serum (Gómara et al., 2000). Recently, an integrated automatic electrochemical immunosensor array was reported for the simultaneous detection of 5-type hepatitis virus antigens (i.e., hepatitis A, hepatitis B, hepatitis C, hepatitis D, and hepatitis E) with a detection limit of ≤1.0 ng/ml within 5 min (Tang et al., 2010). This electrochemical sensor array contained immobilized hepatitis virus antibodies using nanogold particles and protein A as matrices to capture specific viruses. The change in electric potential before and after binding was monitored by a two-electrode system for detection within 5 min. The chip-based immunosensor array can provide high-throughput multianalyte immunoassays in point-of-care testing of mini-lab-on-a-chip devices (Tang et al., 2010). However, its application to foods needs much research.

Microarrays

Microarrays are increasingly being investigated for use in identification of target genes. The technique uses specific oligonucleotide probes immobilized on a solid surface and hybridization of target-labeled amplified viral cDNA to immobilized probes followed by detection. The presence of a change in the nucleotide sequence in the target is detected by the absence or the reduction of hybridization to the wild-type probes around the change or by the ratio of the signals generated by a mutant against a reference strain (Ayodeji et al., 2009). High-density microarrays can contain up to hundreds of thousands of oligonucleotide probes, instead of a few dozen, in a single array, thereby expanding the power of identification. A reported oligonucleotide array (microarray) incorporating 13,000 spots representing selected strains of HAV, human coxsackieviruses A and B, genogroups I and II of norovirus, and human rotavirus gene segments 3, 4, 10, and 11 was able to identify closely related strains (Ayodeji et al., 2009) using 29-base oligonucleotides. PCR-amplified products from multiple HAV strains were labeled with biotin, hybridized to the array, and detected using Cy3-labeled streptavidin, followed by laser scanning. The hybridization data were plotted as fluorescence intensity against each oligonucleotide in the array, to identify the strain most likely represented in the amplified cDNA target. This microarray-based virus identification has certain advantages over PCR and sequencing. However, much research is still needed before microarrays can be applied for HAV detection in foods as they have the same challenges as PCR in sample preparation, inhibitor removal, low volumes, etc. Once the contaminated food is identified and confirmed for the presence of HAV, proper control measures must be taken.

FEASIBILITY OF TESTING AND INTERPRETATION OF TEST RESULTS

A decision to test a food product for virus contamination must be approached cautiously. Although some validation work is currently being pursued in Europe, the virus detection methods under consideration for standardization are restricted to certain commodities (soft fruits, bottled water, and molluscan shellfish). At the time of this writing, there were no fully vetted standard methods for the detection of HAV in RTE foods, and because the RTE category consists of many different types of foods, each of which may require slight modifications of virus extraction protocols, this is unlikely to change in the near future. Overall, detection of HAV in RTE foods is done infrequently and usually only in response to known or suspected foodborne disease outbreaks in which there is a strong epidemiological association with one or two candidate foods. Other limitations to the routine application of virus detection protocols are problems with the association (or lack thereof) between a positive detection signal and virus infectivity; suboptimal recoveries during virus extraction that may raise detection limits to unacceptably high levels; the expense and need for highly trained personnel for implementing virus detection protocols; and the frequent absence of potentially contaminated product because of the extended incubation period for HAV infection.

Even when foods are tested for potential HAV contamination, results should be interpreted with caution, i.e., a negative test result does not ensure the absence of contamination in the product, and a positive test result suggests that contamination occurred at some point in the farm-to-fork chain but does not prove the presence of infectious virus in the sample that was screened. Progress continues to be made towards improving detection methods for viruses in foods, but they still lag behind the clinical realm (D'Souza and Jaykus, 2006).

CONCLUSIONS

It can be concluded that efficient concentration of virus from foods remains product dependent. However, advances can be made towards using universal concentration schemes with validation among various laboratories and by using round-robin testing approaches. Continuing collaborative efforts will help facilitate transfer of technology between academia, government, and industry. This will enable the application of universal schemes for improved and sensitive detection from food matrices. The eventual goal will be to decrease the cost associated with the rapid and sensitive assays to make routine testing of foods for viral contamination a reality. Thus, rapid and sensitive detection of contaminated foods will prevent their release into the market, preventing foodborne viral outbreaks and product recalls. Also, the use of effective control strategies and stringent hygienic practices along with public education will continue to remain key factors in the control of HAV transmission through RTE foods.

REFERENCES

Abad, F. X., R. M. Pinto, and A. Bosch. 1997. Disinfection of human enteric viruses on fomites. *FEMS Microbiol. Lett.* **156:**107–111.

Abd El Galil, K. H., M. A. El Sokkary, S. M. Kheira, A. M. Salazar, M. V. Yates, W. Chen,

and A. Mulchandani. 2004. Combined immunomagnetic separation-molecular beacon-reverse transcription-PCR assay for detection of hepatitis A virus from environmental samples. *Appl. Environ. Microbiol.* **70:**4371–4374.

Abd El Galil, K. H., M. A. El Sokkary, S. M. Kheira, A. M. Salazar, M. V. Yates, W. Chen, and A. Mulchandani. 2005. Real-time nucleic acid sequence-based amplification assay for detection of hepatitis A virus. *Appl. Environ. Microbiol.* **71:**7113–7116.

Arauz-Ruiz, P., L. Sundqvist, Z. Garcia, L. Taylor, K. Visona, H. Norder, and L. O. Magnius. 2001. Presumed common source outbreaks of hepatitis A in an endemic area confirmed by limited sequencing within the VP1 region. *J. Med. Virol.* **65:**449–456.

Ayodeji, M., M. Kulka, S. A. Jackson, I. Patel, M. Mammel, T. A. Cebula, and B. B. Goswami. 2009. A microarray based approach for the identification of common foodborne viruses. *Open Virol. J.* **3:**7–20.

Bidawid, S., J. M. Farber, S. A. Sattar, and S. Hayward. 2000a. Heat inactivation of hepatitis A virus in dairy foods. *J. Food Prot.* **63:**522–528.

Bidawid, S., J. M. Farber, and S. A. Sattar. 2000b. Contamination of foods by food handlers: experiments on hepatitis A virus transfer to food and its interruption. *Appl. Environ. Microbiol.* **66:** 2759–2763.

Bidawid, S., J. M. Farber, and S. A. Sattar. 2000c. Rapid concentration and detection of hepatitis A virus from lettuce and strawberries. *J. Virol. Methods* **88:**175–185.

Brassard, J., L. Lamoureux, M. J. Gagné, E. Poitras, Y. L. Trottier, and A. Houde. 2009. Comparison of commercial viral genomic extraction kits for the molecular detection of foodborne viruses. *Can. J. Microbiol.* **55:**1016–1019.

Brooks, H. A., R. M. Gersberg, and A. K. Dhar. 2005. Detection and quantification of hepatitis A virus in seawater via real-time RT-PCR. *J. Virol. Methods* **127:**109–118.

Butot, S., T. Putallaz, and G. Sánchez. 2007. Procedure for rapid concentration and detection of enteric viruses from berries and vegetables. *Appl. Environ. Microbiol.* **73:**186–192.

Butot, S., T. Putallaz, and G. Sánchez. 2008. Effects of sanitation, freezing and frozen storage on enteric viruses in berries and herbs. *Int. J. Food Microbiol.* **126:**30–35.

Butot, S., T. Putallaz, R. Amoroso, and G. Sánchez. 2009. Inactivation of enteric viruses in minimally processed berries and herbs. *Appl. Environ. Microbiol.* **75:**4155–4161.

Calder, L., G. Simmons, C. Thornley, P. Taylor, K. Pritchard, G. Greening, and J. Bishop. 2003. An outbreak of hepatitis A associated with consumption of raw blueberries. *Epidemiol. Infect.* **131:**745–751.

Casas, N., F. Amarita, and I. M. de Marañón. 2007. Evaluation of an extracting method for the detection of hepatitis A virus in shellfish by SYBR-Green real-time RT-PCR. *Int. J. Food Microbiol.* **120:**179–185.

Centers for Disease Control and Prevention. 1993. Foodborne hepatitis A—Missouri, Wisconsin, and Alaska, 1990–1992. *MMWR Morb. Mortal. Wkly. Rep.* **42:**526–534.

Compton, J. 1991. Nucleic acid sequence-based amplification. *Nature* **350:**91–92.

Costafreda, M. I., A. Bosch, and R. M. Pintó. 2006. Development, evaluation, and standardization of a real-time TaqMan reverse transcription-PCR assay for quantification of hepatitis A virus in clinical and shellfish samples. *Appl. Environ. Microbiol.* **72:**3846–3855.

Costa-Mattioli, M., S. Monpoeho, E. Nicand, M. H. Aleman, S. Billaudel, and V. Ferré. 2002. Quantification and duration of viraemia during hepatitis A infection as determined by real-time RT-PCR. *J. Viral Hepat.* **9:**101–106.

Croci, L., D. De Medici, G. Morace, A. Fiore, C. Scalfaro, F. Beneduce, and L. Toti. 2002. The survival of hepatitis A in fresh produce. *Int. J. Food Microbiol.* **73:**29–34.

Di Pasquale, S., M. Paniconi, B. Auricchio, L. Orefice, A. C. Schultz, and D. De Medici. 2010. Comparison of different concentration methods for the detection of hepatitis A virus and calicivirus from bottled natural mineral waters. *J. Virol. Methods* **165:**57–63.

D'Souza, D. H., C. L. Moe, and L. A. Jaykus. 2007. Foodborne viral pathogens, p. 581–607. *In* M. P. Doyle and L. R. Beuchat (ed.), *Food Microbiology: Fundamentals and Frontiers*, 3rd ed. ASM Press, Washington, DC.

D'Souza, D. H., J. Jean, and L. Jaykus. 2006. Methods for the detection of viral and parasitic protozoan pathogens in foods, p. 188-1–188-23. *In* Y. H. Hui (ed.), *Handbook of Food Technology and Engineering*, vol. 4. CRC-Taylor and Francis, Boca Raton, FL.

D'Souza, D. H., and L. Jaykus. 2006. Molecular approaches for the detection of foodborne viral pathogens. *In* J. Maurer (ed), *PCR Methods in Foods*. Springer, New York, NY.

Dubois, E., C. Agier, O. Traoré, C. Hennechart, G. Merle, C. Crucière, and H. Laveran. 2002. Modified concentration method for the detection of enteric viruses on fruits and vegetables by reverse transcriptase-polymerase chain reaction or cell culture. *J. Food Prot.* **65:**1962–1969.

Dubois, E., C. Hennechart, N. Deboosère, G. Merle, O. Legeay, C. Burger, M. Le Calvé, B. Lombard, V. Ferré, and O. Traoré. 2006. Intra-laboratory validation of a concentration method adapted for the enumeration of infectious F-specific RNA coliphage, enterovirus, and hepatitis A virus from inoculated leaves of salad vegetables. *Int. J. Food Microbiol.* **108:**164–171.

Fino, V. R., and K. E. Kniel. 2008. UV light inactivation of hepatitis A virus, Aichi virus, and feline calicivirus on strawberries, green onions, and lettuce. *J. Food Prot.* **71:**908–913.

Fiore, A. E. 2004. Hepatitis A transmitted by food. *Clin. Infect. Dis.* **38:**705–715.

FitzSimons, D., G. Hendrickx, A. Vorsters, and P. Van Damme. 2010. Hepatitis A and E: update on prevention and epidemiology. *Vaccine* **28:**583–588.

Frank, C., J. Walter, M. Muehlen, A. Jansen, U. van Treeck, A. M. Hauri, I. Zoellner, E. Schreider, O. Hamouda, and K. Stark. 2005. Large outbreak of hepatitis A in tourists staying at a hotel in Hurghada, Egypt, 2004—orange juice implicated. *Euro Surveill.* **10**(6)**:**E050609.2.

Frank, C., J. Walter, M. Muehlen, A. Jansen, U. van Treeck, A. M. Hauri, I. Zoellner, M. Rakha, M. Hoehne, O. Hamouda, E. Schreier, and K. Stark. 2007. Major outbreak of hepatitis A associated with orange juice among tourists, Egypt, 2004. *Emerg. Infect. Dis.* **13:**156–158.

Gómara, M. J., G. Ercilla, M. A. Alsina, and I. Haro. 2000. Assessment of synthetic peptides for hepatitis A diagnosis using biosensor technology. *J. Immunol. Methods* **246:**13–24.

Goswami, B. B., M. Kulka, D. Ngo, P. Istafanos, and T. A. Cebula. 2002. A polymerase chain reaction-based method for the detection of hepatitis A virus in produce and shellfish. *J. Food Prot.* **65:**393–402.

Grove, S. F., A. Lee, T. Lewis, C. M. Stewart, H. Chen, and D. G. Hoover. 2006. Inactivation of foodborne viruses of significance by high pressure and other processes. *J. Food Prot.* **69:**957–968.

Guévremont, E., J. Brassard, A. Houde, C. Simard, and Y. L. Trottier. 2006. Development of an extraction and concentration procedure and comparison of RT-PCR primer systems for the detection of hepatitis A virus and norovirus GII in green onions. *J. Virol. Methods* **134:**130–135.

Hewitt, J., and G. E. Greening. 2004. Survival and persistence of norovirus, hepatitis A virus, and feline calicivirus in marinated mussels. *J. Food Prot.* **67:**1743–1750.

Hirneisen, K. A., D. G. Hoover, and K. E. Kniel. 2009. Isolation and infectivity of potential foodborne viral pathogens by immunomagnetic capture. *Food Prot. Trends* **29:**564–570.

Hollinger, F. B., and S. U. Emerson. 2001. Hepatitis A virus, p. 799–840. *In* D. M. Knipe, P. M. Howley, D. E. Griffin, R. A. Lamb, M. A. Martin, B. Roizman, and S. E. Straus (ed.), *Fields Virology*. Lippincott Williams & Wilkins, Philadelphia, PA.

Hu, Y., and I. Arsov. 2009. Nested real-time PCR for hepatitis A detection. *Lett. Appl. Microbiol.* **49:**615–619.

Hutin, Y. J., V. Pool, E. H. Cramer, O. V. Nainan, J. Weth, I. T. Williams, S. T. Goldstein, K. F. Gensheimer, B. P. Bell, C. N. Shapiro, M. J. Alter, and H. S. Margolis. 1999. A multistate foodborne outbreak of hepatitis A. *N. Engl. J. Med.* **340:**595–601.

Jean, J., B. Blais, A. Darveau, and I. Fliss. 2001. Detection of hepatitis A virus by the nucleic acid sequence-based amplification technique and comparison with reverse transcription-PCR. *Appl. Environ. Microbiol.* **67:**5593–5600.

Jean, J., B. Blais, A. Darveau, and I. Fliss. 2002. Simultaneous detection and identification of hepatitis A virus and rotavirus by multiplex nucleic acid sequence-based amplification (NASBA) and microtiter plate hybridization system. *J. Virol. Methods* **105:**123–132.

Jean, J., D. H. D'Souza, and L. Jaykus. 2004. Multiplex nucleic acid sequence-based amplification (NASBA) for the simultaneous detection of enteric viruses in ready-to-eat food. *Appl. Environ. Microbiol.* **70:**6603–6610.

Jothikumar, N., T. L. Cromeans, M. D. Sobsey, and B. H. Robertson. 2005. Development and evaluation of a broadly reactive TaqMan assay for rapid detection of hepatitis A virus. *Appl. Environ. Microbiol.* **71:**3359–3363.

Kingsley, D. H., D. G. Hoover, E. Papafragkou, and G. P. Richards. 2002. Inactivation of hepatitis A virus and a calicivirus by high hydrostatic pressure. *J. Food Prot.* **65:**1605–1609.

Kingsley, D. H., D. Guan, and D. G. Hoover. 2005. Pressure inactivation of hepatitis A virus in strawberry puree and sliced green onions. *J. Food Prot.* **68:**1748–1751.

Kingsley, D. H., D. Guan, D. G. Hoover, and H. Chen. 2006. Inactivation of hepatitis A virus by high-pressure processing: the role of temperature and pressure oscillation. *J. Food Prot.* **69:**2454–2459.

Kovac, K., I. Gutiérrez-Aguirre, M. Banjac, M. Peterka, M. Poljsak-Prijatelj, M. Ravnikar, J. Z. Mijovski, A. C. Schultz, and P. Raspor. 2009. A novel method for concentrating hepatitis A virus and caliciviruses from bottled water. *J. Virol. Methods* **162:**272–275.

Leggitt, P. R., and L. A. Jaykus. 2000. Detection methods for human enteric viruses in representative foods. *J. Food Prot.* **63:**1738–1744.

Le Guyader, F., L. Haugarreau, L. Miossec, E. Dubois, and M. Pommepuy. 2000. Three-year study to assess human enteric viruses in shellfish. *Appl. Environ. Microbiol.* **66:**3241–3248.

Macinga, D. R., S. A. Sattar, L. A. Jaykus, and J. W. Arbogast. 2008. Improved inactivation of nonenveloped enteric viruses and their surrogates by a novel alcohol-based hand sanitizer. *Appl. Environ. Microbiol.* **74:**5047–5052.

Mariam, T. W., and D. O. Cliver. 2000. Hepatitis A virus control in strawberry products. *Dairy Food Environ. Sanit.* **20:**612–616.

Mbithi, J. N., V. S. Springthorpe, and S. A. Sattar. 1990. Chemical disinfection of hepatitis A virus on environmental surfaces. *Appl. Environ. Microbiol.* **56:**3601–3604.

Mbithi, J. N., V. S. Springthorpe, and S. A. Sattar. 1991. Effect of relative humidity and air temperature on survival of hepatitis A virus on environmental surfaces. *Appl. Environ. Microbiol.* **57:** 1394–1399.

Mbithi, J. N., V. S. Springthorpe, and S. A. Sattar. 1993. Comparative in vivo efficiencies of hand-washing agents against hepatitis A virus (HM-175) and poliovirus type 1 (Sabin). *Appl. Environ. Microbiol.* **59:**3463–3469.

Mbithi, J. N., V. S. Springthorpe, J. R. Boulet, and S. A. Sattar. 1992. Survival of hepatitis A virus on human hands and its transfer on contact with animate and inanimate surfaces. *J. Clin. Microbiol.* **30:**757–763.

Morales-Rayas, R., P. F. Wolffs, and M. W. Griffiths. 2010. Simultaneous separation and detection of hepatitis A virus and norovirus in produce. *Int. J. Food Microbiol.* **139:**48–55.

Nainan, V. O., G. Xia, G. Vaughon, and H. S. Margolis. 2006. Diagnosis of hepatitis A virus infection: a molecular approach. *Clin. Microbiol. Rev.* **19:**63–79.

Niu, M. T., L. B. Polish, B. H. Robertson, B. K. Khanna, B. A. Woodruff, C. N. Shapiro, M. A. Miller, J. D. Smith, J. K. Gedrose, M. J. Alterand, and H. S. Margolis. 1992. Multistate outbreak of hepatitis A associated with frozen strawberries. *J. Infect. Dis.* **166:**518–524.

Notomi, T., H. Okayama, H. Masubuchi, T. Yonekawa, K. Watanabe, N. Amino, and T. Hase. 2000. Loop-mediated isothermal amplification of DNA. *Nucleic Acids Res.* **28**(12)**:**E63.

Nuanualsuwan, S., T. Mariam, S. Himathongkham, and D. O. Cliver. 2002. Ultraviolet inactivation of feline calicivirus, human enteric viruses and coliphages. *Photochem. Photobiol.* **76:** 406–410.

Papafragkou, E., M. Plante, K. Mattison, S. Bidawid, K. Karthikeyan, J. M. Farber, and L. A. Jaykus. 2008. Rapid and sensitive detection of hepatitis A virus in representative food matrices. *J. Virol. Methods* **147:**177–187.

Papafragkou, E., D. H. D'Souza, and L. Jaykus. 2006. Foodborne viruses: prevention and control. *In* S. M. Goyal (ed.), *Food Virology*. Kluwer Academic/Plenum Publishers, New York, NY.

Parry, J. V., and P. P. Mortimer. 1984. The heat sensitivity of hepatitis A virus determined by a simple tissue culture method. *J. Med. Virol.* **14:**277–283.

Richards, G. P. 2001. Food-borne pathogens: enteric virus contamination of foods through industrial practices: a primer on intervention strategies. *J. Ind. Microbiol. Biotechnol.* **27:**117–125.

Roberts, P., and A. Hope. 2003. Virus inactivation by high intensity broad spectrum pulsed light. *J. Virol. Methods* **1:**61–65.

Robesyn, E., K. De Schrijver, E. Wollants, G. Top, J. Verbeeck, and M. Van Ranst. 2009. An outbreak of hepatitis A associated with the consumption of raw beef. *J. Clin. Virol.* **44:**207–210.

Rzezutka, A., M. D'Agostino, and N. Cook. 2006. An ultracentrifugation-based approach to the detection of hepatitis A virus in soft fruits. *Int. J. Food Microbiol.* **108:**315–320.

Sair, A. I., D. H. D'Souza, and L. Jaykus. 2002a. Human enteric viruses as causes of foodborne disease. *Comp. Rev. Food Sci. Saf.* **1:**73–89.

Sair, A. I., D. D'Souza, C. L. Moe, and L. A. Jaykus. 2002b. Improved detection of human enteric viruses in foods by RT-PCR. *J. Virol. Methods* **100:**57–69.

Sánchez, G., S. Populaire, S. Butot, T. Putallaz, and H. Joosten. 2006. Detection and differentiation of human hepatitis A strains by commercial quantitative real-time RT-PCR tests. *J. Virol. Methods* **132:**160–165.

Sanchez, G., A. Bosch, and R. M. Pinto. 2007. Hepatitis A virus detection in food: current and future prospects. *Lett. Appl. Microbiol.* **45:**1–5.

Sattar, S. A. 2004. Microbicides and the environmental control of nosocomial viral infections. *J. Hosp. Infect.* **56:**564–569.

Sattar, S. A., V. S. Springthorpe, J. Tetro, R. Vashon, and B. Keswick. 2002. Hygienic hand antiseptics: should they not have activity and label claims against viruses? *Am. J. Infect. Control* **30:**355–372.

Schenkel, K., V. Bremer, C. Grabe, U. Van Treeck, E. Schreier, M. Höhne, A. Ammon, and K. Alpers. 2006. Outbreak of hepatitis A in two federal states of Germany: bakery products as vehicle of infection. *Epidemiol. Infect.* **134:**1292–1298.

Schmid, D., R. Fretz, G. Buchner, C. König, H. Perner, R. Sollak, A. Tratter, M. Hell, M. Maass, M. Strasser, and F. Allerberger. 2009. Foodborne outbreak of hepatitis A, November 2007–January 2008, Austria. *Eur. J. Clin. Microbiol. Infect. Dis.* **28:**385–391.

Schwab, K. J., F. H. Neill, F. Le Guyader, M. K. Estes, and R. L. Atmar. 2001. Development of a reverse transcription-PCR-DNA enzyme immunoassay for detection of "Norwalk-like" viruses and hepatitis A virus in stool and shellfish. *Appl. Environ. Microbiol.* **67:**742–749.

Schwab, K. J., F. H. Neill, R. L. Fankhauser, N. A. Daniels, S. S. Monroe, D. A. Bergmire-Sweat, M. K. Estes, and R. L. Atmar. 2000. Development of methods to detect "Norwalk-like viruses" (NLVs) and hepatitis A virus in delicatessen foods: application to a food-borne NLV outbreak. *Appl. Environ. Microbiol.* **66:**213–218.

Schwarz, N. G., M. Revillion, A. M. Roque-Afonso, E. Dussaix, M. Giraud, C. Liberpre, E. Couturier, and E. Delarocque Astagneau. 2008. A food-borne outbreak of hepatitis A virus HAV) infection in a secondary school in Upper Normandy, France, in November 2006. *Euro Surveill.* **13**(22)**:**pii=18885.

Shan, X. C., P. Wolffs, and M. W. Griffiths. 2005. Rapid and quantitative detection of hepatitis A virus from green onion and strawberry rinses by use of real-time reverse transcription-PCR. *Appl. Environ. Microbiol.* **71:**5624–5626.

Shimasaki, N., T. Kiyohara, A. Totsuka, K. Nojima, Y. Okada, K. Yamaguchi, J. Kajioka, T. Wakita, and T. Yoneyama. 2009. Inactivation of hepatitis A virus by heat and high hydrostatic pressure: variation among laboratory strains. *Vox Sang.* **96:**14–19.

Tang, D., J. Tang, B. Su, J. Ren, and G. Chen. 2010. Simultaneous determination of five-type hepatitis virus antigens in 5 min using an integrated automatic electrochemical immunosensor array. *Biosens. Bioelectron.* **25:**1658–1662.

Warburton, A. R., T. G. Wreghitt, A. Rampling, R. Buttery, K. N. Ward, K. R. Perry, and J. V. Parry. 1991. Hepatitis A outbreak involving bread. *Epidemiol. Infect.* **106:**199–202.

Weltman, A. C., N. M. Bennett, D. A. Ackman, J. H. Misage, J. J. Campana, L. S. Fine, A. S. Doniger, G. J. Balzano, and G. S. Birkhead. 1996. An outbreak of hepatitis A associated with a bakery, New York, 1994: the 1968 "West Branch, Michigan" outbreak repeated. *Epidemiol. Infect.* **117:**333–341.

Wheeler, C., T. M. Vogt, G. L. Armstrong, G. Vaughan, A. Weltman, O. V. Nainan, V. Dato, G. Xia, K. Waller, J. Amon, T. M. Lee, A. Highbaugh-Battle, C. Hembree, S. Evenson, M. A. Ruta, I. T. Williams, A. E. Fiore, and B. P. Bell. 2005. An outbreak of hepatitis A associated with green onions. *N. Engl. J. Med.* **353:** 890–897.

World Health Organization. 2000. Hepatitis A vaccines: WHO position paper. *Wkly. Epidemiol. Rec.* **75:**38–42.

Yoneyama, T., T. Kiyohara, N. Shimasaki, G. Kobayashi, Y. Ota, T. Notomi, A. Totsuka, and T. Wakita. 2007. Rapid and real-time detection of hepatitis A virus by reverse transcription loop-mediated isothermal amplification assay. *J. Virol. Methods* **145:**162–168.

CONCLUSIONS

VII

FUTURE TRENDS IN RAPID METHODS: WHERE IS THE FIELD MOVING, AND WHAT SHOULD WE FOCUS ON?

J. Hoorfar, B. B. Christensen, F. Pagotto, K. Rudi, A. Bhunia, and M. Griffiths

28

BEGINNING WITH THE SAMPLES

Proper sample preparation is a precondition for advanced laboratory diagnostics. The newer methods that claim less time to reach a result have a major limitation with how much volume they can take prior to the rapid analysis for the target molecule. PCRs, which require tens to hundreds of microliters, represent a good example of the problems facing the modern food microbiologist. For example, how can we get an accurate representation of 25 grams of cheese into the 100-μl sample that will go into a real-time, reverse transcriptase reaction? As reviewed earlier in this work, this issue should be the center of focus: here, using some of the technologies described may be useful, although nucleic acid-based methods currently require 10^3 to 10^4 organisms in the enrichment broth (which includes food particles, etc.).

Irrespective of whether the target is DNA, RNA, or proteins, it is not currently possible to overcome the problem posed by sample size without some sort of enrichment step, which may not be replaceable in the foreseeable future. This is especially important as the scientific community is now entering the so-called "nanotechnology" phase. Pathogens that are known to enter the so-called viable but nonculturable state or that are stressed pose further challenges when the enrichment period normally required for bacterial pathogens is shortened or eliminated to reach levels amenable to detection by various platforms.

Clearly, foodborne viruses and parasites (cysts and oocysts) pose perhaps the greatest challenges, as they cannot be enriched at all. For these organisms, sample preparation is the key step in being able to apply any downstream detection strategy. Unless the sample preparation method can isolate the 10 oocysts or 100 norovirus particles required, for example, no amount of advanced downstream technology will help with the analyses required. This detection limit may be reached by combining simple filtration using nanoalumina filters with real-time PCR (Morales-Rayas et al., 2009).

J. Hoorfar and B. B. Christensen, National Food Institute, Technical University of Denmark (DTU), Mørkhøj Bygade 19, DK-2860 Søborg, Denmark. *F. Pagotto*, Bureau of Microbial Hazards Health Products and Food Branch, P/L 2204E Room E412 Health Canada, 251 Sir Frederick Banting Driveway, Ottawa, Ontario K1A 0K9, Canada. *K. Rudi*, Nofima Norwegian Food Research Institute, Osloveien 1, 1430 Ås, Norway. *A. Bhunia*, Department of Food Science, Purdue University, 745 Agriculture Mall Dr., West Lafayette, IN 47907. *M. Griffiths*, Canadian Research Institute for Food Safety, University of Guelph, 43 McGilvray Street, Guelph, Ontario N1G 2W1, Canada.

Rapid Detection, Identification, and Quantification of Foodborne Pathogens, ed. by J. Hoorfar

Although enrichment is a limitation in terms of assay speed, it is necessary for several reasons beyond increasing bacterial numbers, including diluting the effects of inhibitors, allowing the differentiation of viable from nonviable cells and allowing for resuscitation of injured cells.

However, alternatives to enrichment may exist. Tolba et al. (2010) and Cademartiri et al. (2010) have described methods to immobilize bacteriophages on paramagnetic beads and surfaces so that they retain infectivity. These immobilized bacteriophages can be used to capture host bacterial cells, and following propagation, bacteriophage nucleic acid can be detected by real-time PCR. This provides two cycles of amplification of the target: the first due to replication of the phage in the host and the second by PCR amplification of unique phage nucleic acid sequences. Using this technique, it is possible to detect one cell of the host bacterium within 4 hours. Because the bacteriophage requires metabolically active cells to propagate, the assay will only detect live cells.

REAL-TIME TESTING

In order to be truly useful, microbiological testing must provide results in "real time." While this is difficult to achieve in food systems where the prevalence and incidence of the target are invariably low and it is not homogeneously distributed throughout the food, there are approaches, particularly for bacteria, that may allow detection without the need for enrichment or crude sample pretreatments.

Among these approaches, the use of signal amplification techniques has received attention. If it is possible to generate results in close to real time, then applications of microbiological testing could be extended to online monitoring and on-farm monitoring as well as to devices that consumers could use to verify the safety and quality of foods.

An increasing number of monitoring devices are becoming more compact to facilitate on-site detection of pathogens in more-real time. In general, diagnostic technologies are becoming mobile, digital, virtual (software-based), and personal. This development means, among others, that the use of these devices can be easier at the source of infection (field, farm, animal, slaughterhouse, etc.). In clinical and veterinary settings, this is point-of-care diagnostics, but in food testing this development will result in on-site, online, or in-line monitoring systems that obviate the need to ship samples to testing laboratories. An example of this system is the continuous monitoring of *Campylobacter* in the air of chicken houses (Olsen et al., 2009).

In the ideal situation, a noninvasive, handheld scanning device (similar to barcode scanners at local grocery stores) would be used to quickly and accurately assess all aspects of a food commodity. This would include the presence or absence of pathogens and their associated toxins down to a single-cell level, even when they are not uniformly distributed in a food matrix. This may not be as far out of reach as one might think. Devices have been developed based on molecular fluorescence spectroscopy that can analyze microbial signature in wounds (DaCosta et al., 2007).

Advances in genome sequencing and bioinformatics, for example, single nucleotide polymorphisms and the various assays that have been designed to address them, can allow for an assay to detect pathogenicity features unique to a foodborne pathogen.

Alternatively, assays can be designed to detect pathogenicity islands, virulence factors, etc., that might preclude the need to isolate the intact organism. In this light, the key feature will be the ability to generate a molecular fingerprint within the detection signal so that it may be possible to help epidemiological investigations. This was tried for norovirus, using a DNA microarray platform assay (Pagotto et al., 2008).

Most rapid methods can be completed within a few minutes to a few hours, so they are more rapid than traditional methods. However, such methods still lack sufficient diagnostic sensitivity and specificity for direct testing, and foods still need to be culture enriched prior to analysis, ultimately increasing the total detection time.

AUTOMATION AND MULTIPATHOGEN DETECTION

Another key issue is automation, where the key drivers are miniaturization and multiple testing (which means one instrument is flexible enough to test for many pathogens, but also many pathogens can be detected with one test). The downstream application, whether detection of nucleic acid, protein, or specific analyte, has seen an increase in methods used normally in other fields such as chemistry, biochemistry, and physics. These include Fourier transform infrared spectroscopy, aptamers, biosensors, etc.

Almost all existing rapid methods are designed to detect a single target, which makes them ideal for use in quality control programs to quickly screen large numbers of food samples for the presence of a particular pathogen or toxin. A positive result by a rapid method, however, is only regarded as presumptive. Still, the emphasis in food microbiology is shifting from methods that detect a single target to methods that detect the presence of multiple pathogens simultaneously.

Strategies to detect multiple pathogens or to analyze microbial diversity in food or a clinical specimen with a single assay platform are highly desirable for food safety and food biosecurity applications. It would not only cut costs for pathogen testing but also aid in providing total microbiological food safety.

Such screening capabilities would also allow researchers to study microbial diversity in a sample, food processors to release products in a timely manner, and regulatory agencies to make prompt decisions about whether the food is safe for consumption. In the event of an outbreak due to known or unknown agents, it is necessary to first provide a general assessment of the nature of the pathogens/toxins present in a product to initiate corrective or preventive actions such as treatment regimen, further processing, disposal of product, etc. At a later time, more-rigorous laboratory-based assay procedures can be initiated to determine the identity of the agent by using molecular or phenotypic array-based methods.

As such, it is likely that molecular amplification assays including PCR, loop-mediated isothermal amplification, and nucleic acid sequence-based amplification will dominate future diagnostic developments when multiplex detection is a requirement. However, multipathogen detection requires that different classes of microorganisms (bacteria, viruses, and parasites) be detected at the same time. Such requirements will pose challenges to the current rapid methods, due to issues associated with the fact that some microorganisms (bacteria) can be enriched, while others (viruses and parasites) are difficult, if not impossible, to culture. For example, researchers have developed a detection array containing 388,000 probes that together can screen for about 2,000 viruses and 900 bacteria (Jiang et al., 2008).

UNTAPPED POTENTIAL OF PCR FOR THE INDUSTRY

Quantification by real-time PCR is an established technique. Despite this, PCR is still mostly used only to declare food products free from pathogens. That is one of the main strengths of PCR testing: screening a large number of samples for negative results to rapidly dispatch food products to the market. For the food producers, the more advanced use of rapid tests would be to directly link patients that have eaten such foods to the pathogen, but such studies are difficult to conduct.

One of the unexploited potentials of real-time PCR for the food industry is to gain information on the number of pathogens present in the food or food production environment. This could help food producers to assess the impact of hygiene measures or the implementation of a new quality assurance program. When personnel involved in quality assurance observed a reduction of the contamination they have been fighting, they would be encouraged to continue the efforts. Fortunately, the new generation of PCR technologies takes us closer to achieving this aim but does not achieve it completely.

Until now, it was believed that the enrichment step, which is necessary to increase the number of the target pathogens, prevented us from enumerating the pathogens present in the initial sample *before* it was enriched. Recent works involving *Campylobacter* and *Salmonella* indicate that this obstacle can be overcome by a short enrichment step, with the analysis being performed while the target pathogens are still in the log phase of growth (Josefsen et al., 2004; Krämer et al., 2010; see also chapters 13 and 15).

Another new application of PCR is its use by routine analysis laboratories to determine the source of contamination, whether raw material, personnel, equipment, or the food processing environment. By simultaneous detection and subtyping of pathogens, it will be possible for food producers to accumulate in-house data on persistent strains. Adding DNA-cutting enzymes to post-PCR tubes could be easily done, but at the present time, this requires highly trained personnel. There is no question that this will be made much easier in the future, when different versions of the same PCR kit will be commercially available.

Recent developments in real-time PCR have made it possible to carry out high-throughput source tracing for routine purposes in the food industry. The technique called high-resolution melting curve analysis takes advantage of differences in the melting curves of multiple DNA products, all in the same PCR tube. The software included in the PCR instrument produces several distinct graphs, each representing one subtype of target pathogen. This is all done without any additional laboratory work (Rudi et al., 2007a; Tajiri-Utagawa et al., 2009).

CULTURE-INDEPENDENT DETECTION AND TYPING

Methods for detecting and tracing foodborne pathogens are increasingly being influenced by the "omics" technologies. In particular, the genomics technologies have reached a stage where they are used on a routine basis in many laboratories. These technologies are expected to move towards culture-independent detection and characterization techniques based on the purification of total DNA from diagnostic samples and subsequent metagenomic analysis.

Nevertheless, the limitations of the information obtained when using the science of "omics" must be recognized. For example, DNA-based assays detect genetic targets (such as virulence genes), which only indicate that bacteria with those gene sequences are present. However, they do not indicate whether the gene is actually expressed or whether the bacterium is viable.

The need imposed by existing legislation for isolation of the target bacteria in situations including food recalls also means that cultural methods may need to be used in addition to DNA-based methods during outbreak investigations. Whereas nucleic acid-based methods cannot detect proteins such as enterotoxins, which can cause foodborne intoxications, proteomic techniques, including protein arrays, will be increasingly used by the food industry. As these various technologies become accepted, obtaining an isolate may not be a regulatory requirement in the future.

METAGENOMICS

The emphasis of future testing developments should be on the use of metagenomics (gene-based) rather than the phenotypic methods used today. One challenge will be to identify genotypes that are associated with infection. To accomplish that, a significant investment in bioinformatics will be required in order to translate large amounts of metagenomics data into a form that provides useful diagnostic, surveillance, and epidemiologic information.

Infectious diseases are caused by various pathogens, including yet-unidentified microorganisms. Because procedures for detecting and identifying pathogens vary according to the target microorganism, diagnostic methods require a variety of media, reagents, and skills. As an alternative approach, DNA sequencers can determine more than 100 megabases of DNA sequences per run, and new sequencing technologies eliminate the bacterial cloning step used in traditional Sanger sequencing. Instead, they amplify single isolated DNA molecules and analyze them with computers capable of massive parallel processing.

As a result, an increasing number of so-called metagenomic methods tap into the potential of novel sequences to promptly detect and identify various infectious pathogens (Nakamura et al., 2008). However, these sequencing-based methods are not without drawbacks, as it has been pointed out that care should be taken to avoid bias caused by different sample preparation protocols (Morgan et al., 2010).

FUTURE RISK ASSESSMENTS BASED ON METAGENOMICS

The food safety risk analysis tool as first described by the Food and Agriculture Organization/World Health Organization (1995) states that it should be the role of the official bodies to use risk analysis to determine realistic and achievable risk levels of foodborne hazards. In recent years, this has been followed up by a discussion on setting risk-based metrics, i.e., food safety objectives, performance objectives, and microbiological criteria (van Schothorst et al., 2009).

However, by moving towards online testing, it may also be appropriate to change the risk assessment concept towards online product risk assessments based on the microbiological testing. This requires the development of mathematical risk assessment tools, which are part of the testing scheme, so that the microbiological test result is immediately modeled into an estimation of the public health outcome.

The new "omics" technologies, which will shift "isolate-based diagnostics" towards "gene diagnostics," will further challenge the currently applied risk assessment methodologies, because bacterial detection and characterization will no longer be based on isolation of a specific bacterial strain but rather on detection of genes or gene pools from a sample. These genes need not necessarily come from the same bacterial strain but may represent several strains and even different bacterial species in the same sample.

Therefore, the development of omics-based diagnostic tools should be linked to mathematical models that relate the information obtained from gene-based diagnostic assays to the risks posed to public health. One of the benefits of this approach would be improved source attribution, i.e., models for quantifying the relative contribution of different food sources to a particular disease burden.

BIOTRACING

A rapid method by itself can be useful, but it will be more beneficial to food producers, regulators, and control authorities if it is part of an integrated source-tracing system.

In outbreak situations, epidemiologists point to a certain food product as being suspected to have caused the outbreak. In most cases, this is indicated through case-control studies, interviewing the affected patients and a number of control cases from the same subpopulation. However, in many cases it is not possible to isolate the microorganism that caused the infection.

In the future, the source identification may be based on software using Bayesian belief networks to point to the source, even though no isolate is obtained. This new concept, referred to as biotracing, can combine laboratory results obtained from testing with information collected during the entire food production process, in order to make sound decisions about possible product recalls or determine the source of possible contamination (Hoorfar et al., 2010).

ONE GLOBAL VALIDATION

Despite the enormous supply of commercial kits and published protocols surrounding rapid methods, surprisingly, many laboratories still use standard culture-based methods. Is this because there is a lack of demand for faster methods? The answer is no. This is partly caused by the history of microbiology and laboratory traditions, but also because of the lack of a universal validation infrastructure; it can be costly to develop a new method, but it also requires significant resources to thoroughly validate a new method against standard culture-based methods.

As explained earlier in this work, new detection methods must perform as well as, or better than, standard culture-based methods through extensive comparative and collaborative trials, and even then it may be difficult to obtain regulatory approval for their use.

Numerous studies have shown that rapid assays perform better in some foods than in others, mainly due to differences in the bacterial microflora and the sample matrix. Thus, all rapid assays should be carefully validated in the exact type of food that is intended for routine testing.

The way ahead is the creation of a single validation body with universally accepted criteria. No organization is more suitable to take over this task than the United Nations. Although the International Organization for Standardization has published Standard 16140, this body does not have the infrastructural capacity to carry out the validation.

One solution would be for the United Nations organizations WHO and FAO to delegate the actual validation work to a series of regional private or public institutions, which could carry out the necessary comparative and collaborative trials. The benefit of this system for test procedures and industry is to have only one validation that is recognized globally.

RETURN ON INVESTMENT

The cost and technical requirements of rapid assays remain challenging obstacles to surmount. Rapid assays need to be cost-effective to facilitate their employment within the food chain. Similarly, although such methods have been developed to facilitate their ease of use, the need for sample manipulation and enrichment protocols prior to detection often complicates their adoption for food testing. This means that the majority of rapid assays are still largely confined to the laboratory, where they are performed by skilled technicians. This again adds to the cost of setup and training.

Partly due to the cost, the industry may be unwilling to adopt advanced detection technologies. Market forces play a crucial role as to who would be willing to invest in education and technology adoption. The value of a commodity increases significantly as it moves through the supply chain from the farm to the retail level. The return on investment into a particular intervention strategy or program at the farm level can be small compared with the return of investment at the retail stage. Thus, the increased value of the commodity at the retail level provides the necessary impetus for ensuring that the commodity meets the food safety expectations of the consumer.

From an industrial point of view, the necessary incentive rests with the retail industry to invest in education and adoption of new technologies to protect their brand name. In contrast, from a regulatory and public health perspective, the best value for the money is to prevent pathogens from entering the production chain from the primary production source.

EMERGING FOOD CONTAMINATIONS

The human gut has through millions of years of evolution been adapted to commensal bacteria in food, water, and the environment (Ley et al., 2008). Over the last century, we have seen the introduction of new food preservation and hygiene measures to combat foodborne pathogens. These have dramatically altered the commensal microbiota we are exposed to, potentially leading to the emergence of diseases related to lack of bacterial exposure, such as allergy and other immunologically related diseases (Blaser and Falkow, 2009).

The incidences of chronic immunological gut-related disease such as inflammatory bowel disease are increasing dramatically in the developed world, at a rate far exceeding that of infectious diseases associated with traditional foodborne pathogens (Binder, 2004).

A major analytical challenge is the wide and undefined range of microorganisms that we are exposed to through food and beverages. A major disadvantage is that we lack the proper analytical tools to describe complete microbial diversity and to differentiate between beneficial and harmful bacteria (Rudi et al., 2007b). A potential future direction, however, could be to define the normal microbiota in different types of food and beverages and to use deviations from normal microbiota as an indicator of improper production (Rudi et al., 2007a). It may then be possible to detect any incidences that are related to increased risk of pathogen contamination.

With the emerging technologies becoming available at a lower price, it will soon be realistic to rapidly screen the complete microbiota (Brugere et al., 2009). This would enable us to preserve and define the beneficial commensal microflora while combating pathogens.

In the future, arguably more important than the direct pathogen detection will be the monitoring of the complete microbiota in order to detect potential changes and to correlate historic changes with emergence of new diseases. For instance, such data would have been very useful in identifying potential underlying microbial causes for the increase of immunologically related disorders seen today, such as type 1 diabetes, multiple sclerosis, food allergies, etc. (Blaser and Falkow, 2009).

CONCLUSIONS

Notwithstanding the limitations of rapid assays, the development of this class of detection method continues to increase. Assays will need to be developed to focus on "overlooked" pathogens including viruses and parasites. Biosensors, at the center of the marriage between biology and electronics, should be designed with the food matrix in mind and should be able to test larger, rather than smaller, volumes of samples. Most importantly, research must commence to find a truly viable alternative to bacterial enrichment. Such an invention will revolutionize the food diagnostic field, leading to true real-time testing during the food production.

ACKNOWLEDGMENTS

This work was supported in part by the European Union-funded Integrated Project BIOTRACER (Contract 036272) under the 6th RTD Framework. We thank Iben Bang-Berthelsen for editorial assistance.

REFERENCES

Binder, V. 2004. Epidemiology of IBD during the twentieth century: an integrated view. *Best. Pract. Res. Clin. Gastroenterol.* **18:**463–479.

Blaser, M. J., and S. Falkow. 2009. What are the consequences of the disappearing human microbiota? *Nat. Rev. Microbiol.* **7:**887–894.

Brugere, J. F., A. Mihajlovski, M. Missaoui, and P. Peyret. 2009. Tools for stools: the challenge of assessing human intestinal microbiota using molecular diagnostics. *Exp. Rev. Mol. Diagn.* **9:**353–365.

Cademartiri, R., H. Anany, I. Gross, R. Bhayani, M. Griffiths, and M. A. Brook. 2010. Immobilization of bacteriophages on modified silica particles. *Biomaterials* **31:**1904–1910.

DaCosta, R. S., B. C. Wilson, and N. E. Marcon. 2007. Fluorescence and spectral imaging. *Sci. World J.* **7:**2046–2071.

Food and Agriculture Organization/World Health Organization. 1995. *Application of Risk Analysis to Food Standard Issues. Report of the Joint FAO/WHO Expert Consultation, 13 to 17 March 1995.* Food and Agriculture Organization/World Health Organization, Geneva, Switzerland.

Hoorfar, J., M. Wagner, K. Jordan, and G. C. Barker. 2010. Biotracing: a novel concept in food safety. *In* S. Brul, P. Fratamico, and T. McMeekin (ed.), *Tracing Pathogens in the Food Chain.* Woodhead Publishing, Cambridge, United Kingdom.

Jiang, C., S. Gardner, K. McLoughlin, N. Mulakken, M. Alegria-Hartman, P. Banda, P. Williams, P. Gu, M. Wagner, C. Manohar, and T. Slezak. 2008. A functional gene array for detection of bacterial virulence elements. *PloS ONE* **3**(5)**:**e2163.

Josefsen, M. H., N. R. Jacobsen, and J. Hoorfar. 2004. Enrichment followed by quantitative PCR for rapid detection and as a tool for quantitative risk assessment of foodborne thermotolerant campylobacters. *Appl. Environ. Microbiol.* **70:**3588–3592.

Krämer, N., C. Löfström, H. Vigre, J. Hoorfar, C. Bunge, and B. Malorny. 2010. A novel strategy to obtain quantitative data for modelling: combined enrichment and real-time PCR for enumeration of salmonellae from pig carcasses. *Int.*

J. Food Microbiol. Epub ahead of print. doi:10.1016/j.ijfoodmicro.2010.08.026.

Ley, R. E., M. Hamady, C. Lozupone, P. J. Turnbaugh, R. R. Ramey, J. S. Bircher, M. L. Schlegel, T. A. Tucker, M. D. Schrenzel, R. Knight, and J. I. Gordon. 2008. Evolution of mammals and their gut microbes. *Science* **320:**1647–1651.

Morales-Rayas, R., P. F. G. Wolffs, and M. W. Griffiths. 2009. Ion-exchange filtration and real-time PCR to detect a norovirus surrogate in food. *J. Food Prot.* **72:**2178–2183.

Morgan, J. L., A. E. Darling, and J. A. Eisen. 2010. Metagenomic sequencing of an in vitro-simulated microbial community. *PLoS ONE* **5**(4): e10209.

Nakamura, S., N. Maeda, I. M. Miron, M. Yoh, K. Izutsu, C. Kataoka, T. Honda, T. Yasunaga, T. Nakaya, J. Kawai, Y. Havashizaki, T. Horii, and T. Iida. 2008. Metagenomic diagnosis of bacterial infections. *Emerg. Infect. Dis.* **14:**1784–1786.

Olsen, K. N., M. Lund, J. Skov, L. S. Christensen, and J. Hoorfar. 2009. Detection of *Campylobacter* bacteria in air samples for continuous, real-time monitoring of *Campylobacter* colonization in broiler flocks. *Appl. Environ. Microbiol.* **75:**2074–2078.

Pagotto, F., N. Corneau, K. Mattison, and S. Bidawid. 2008. Development of a DNA microarray for the simultaneous detection and genotyping of noroviruses. *J. Food Prot.* **71:**1434–1441.

Rudi, K., G. H. Kleiberg, R. Heiberg, and J. T. Rosnes. 2007a. Rapid identification and classification of bacteria by 16S rDNA restriction fragment melting curve analyses (RFMCA). *Food Microbiol.* **24:**474–481.

Rudi, K., M. Zimonja, P. Trosvik, and T. Naes. 2007b. Use of multivariate statistics for 16S rRNA gene analysis of microbial communities. *Int. J. Food Microbiol.* **120:**95–99.

Tajiri-Utagawa, E., M. Hara, K. Takahashi, M. Watanabe, and T. Wakita. 2009. Development of a rapid high-throughput method for high-resolution melting analysis for routine detection and genotyping of noroviruses. *J. Clin. Microbiol.* **47:** 435–440.

Tolba, M., O. Minikh, L. Y. Brovko, S. Evoy, and M. W. Griffiths. 2010. Oriented immobilization of bacteriophages for biosensor applications. *Appl. Environ. Microbiol.* **76:**528–535.

van Schothorst, M. A., M. H. Zwietering, T. Ross, R. L. Buchanan, M. B. Cole, and the International Commission on Microbiological Specifications for Foods. 2009. Relating microbiological criteria to food safety objectives and performance objectives. *Food Control* **20:**967–979.

INDEX